FUNDAMEN

MOLECU

FUNDAMENTALS OF
MOLECULAR EVOLUTION

Second Edition

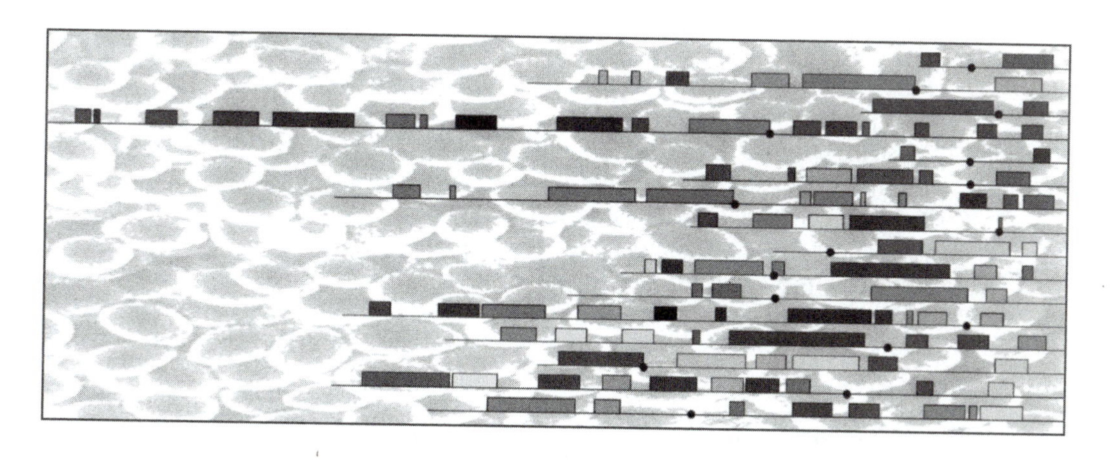

Dan Graur
TEL AVIV UNIVERSITY

Wen-Hsiung Li
UNIVERSITY OF CHICAGO

SINAUER ASSOCIATES, INC., *Publishers*
Sunderland, Massachusetts

FRONT COVER

Dozens of duplicated gene-containing DNA segments within the completely sequenced yeast genome attest to a tumultuous evolutionary history that includes such momentous events as whole-genome duplication, progressive loss of paralogous genes, and at least 100 gene order rearrangements. The two copies of each duplicated region are shown in the same color on a background of yeast (*Saccharomyces cerevisiae*) cells. Data from Wolfe and Shields 1997 (Nature 387: 708–713). Image designed by Dan Graur and Varda Wexler.

FUNDAMENTALS OF MOLECULAR EVOLUTION *Second Edition*

 For information or to order, address:

Sinauer Associates, Inc., PO Box 407
23 Plumtree Road, Sunderland, MA, 01375 U.S.A.
FAX: 413-549-1118
Internet: publish@sinauer.com; www.sinauer.com

Library of Congress Cataloging-in-Publication Data

Graur, Dan, 1953-
Fundamentals of molecular evolution / Dan Graur, Wen-Hsiung Li. -- 2nd ed.
p. cm.
Previous ed. published under: Fundamentals of molecular evolution/ Wen-Hsiung Li. Sunderland, Mass. : Sinauer, 1991.
Includes bibliographical references and index.
ISBN 0-87893-266-6
1. Molecular evolution. I. Li, Wen-Hsiung, 1942- . II. Title
QH325.L65 1999 99-16706
572.8'38--dc21 CIP

Printed in U.S.A.

9 8 7

To our mentors,

James F. Crow, Wendell H. Fleming, Masatoshi Nei, and David Wool

Contents

Chapter 8 *Genome Evolution* 367

Preface

The purpose of this book is to describe the dynamics of evolutionary change at the molecular level, the driving forces behind the evolutionary process, and the effects of the various molecular mechanisms on the long-term evolution of genomes, genes, and their products. In addition, the book is meant to provide basic methodological tools for comparative and phylogenetic analyses of molecular data from an evolutionary perspective.

The study of molecular evolution has its roots in two disparate disciplines: population genetics and molecular biology. Population genetics provides the theoretical foundation for the study of evolutionary processes, whereas molecular biology provides the empirical data. Thus, to understand molecular evolution it is essential to acquire some basic knowledge of both the theory of population genetics and the practice of molecular biology. We emphasize that, although individuals are the entities affected by natural selection and other evolutionary processes, it is populations and genes that change over evolutionary times.

We set out to write a book for "beginners" in molecular evolution. At the same time, we have tried to maintain the standards of the scientific method and to include quantitative treatments of the issues at hand. Therefore, in describing evolutionary phenomena and mechanisms at the molecular level, both mathematical and intuitive explanations are provided. Neither is meant to be at the expense of the other; rather, the two approaches are intended to complement each other and to help the reader achieve a more complete grasp of the issues. We have not attempted to attain encyclopedic completeness, but have provided a large number of examples to support and clarify the many theoretical arguments and methodological discussions.

ACKNOWLEDGMENTS

We are indebted to many colleagues, students, and friends for helping us put this edition together. Their comments, suggestions, corrections, and discussions greatly improved this work. For their help and encouragement, we thank Loren Babcock, Sara Barton, Giorgio Bernardi, T. Cavalier-Smith, Susan Cropp, David De Lorenzo, Joseph Felsenstein, Irit Gat, Takashi Gojobori, Inar Graur, Mina Graur, Or Graur, Sara Greenberg, Inna Gur-El, Einat Hazkani, Dov Herskowits, Varda Herskowits, David Hewett-Emmett, Toshimichi Ikemura, Harvey Itano, Amanda Ko, Tal Kohen, Giddy Landan, Jean Lobry, Ora Manheim, David Mindell, Tom Nagylaki, Norihiro Okada, Ron Ophir, Naomi Paz, Tal Pupko, Marc Robinson, William J. Schull, Noburu Sueoka, Tomohiko Suzuki, Nicholas Tourasse, Varda Wexler, Ken Wolfe, and David Wool. We owe special thanks to Andrew G. Clark, Michael W. Nachman, and J. Bruce Walsh for thoroughly reviewing the entire manuscript and for their valuable suggestions. Support from the National Institutes of Health and the Da'at Consortium have made this endeavor possible. Part of the work was done while W.-H. Li was a faculty member of the University of Texas at Houston, and the support from the Betty Wheless Trotter professorship is gratefully acknowledged.

Dan Graur
TEL AVIV UNIVERSITY

Wen-Hsiung Li
UNIVERSITY OF CHICAGO

Introduction

A curious aspect of the theory of evolution is that everybody thinks he understands it. I mean philosophers, social scientists, and so on. While, in fact, very few people actually understand it as it stands, even as it stood when Darwin expressed it, and even less as we now are able to understand it.

Jacques Monod (1975)

At a meeting of the Linnean Society of London on July 1, 1858, Charles Darwin proposed a theory of evolution by means of natural selection. His monumental treatise, *The Origin of Species*, was published a year later. Darwin's theory revolutionized not only biological thinking, but also politics, sociology, and moral philosophy. The weakest part of Darwin's theory was its inability to account for the transfer of biological information from generation to generation. Although it was clear to him that natural selection could be meaningfully defined only in terms of heritable traits, the nature of heredity was poorly understood at that time.

Seven years after the publication of *The Origin of Species*, Gregor Johann Mendel, a monk at the Augustinian monastery of St. Thomas at Brünn in Moravia, published a paper entitled "Versuche über Pflanzenhybriden" ("Experiments in Plant Hybridization"), in which he established the existence of elementary "characters" of heredity and established the statistical laws governing their transmission from one generation to the next. Mendel, however, did not even speculate on the nature of these "characters." Indeed, he was not even sure whether the units of heredity have a material basis or merely represent "vital interactions."

In 1869, Johann Friedrich Miescher, a Swiss chemistry student working on pus cells in Tübingen castle, discovered a phosphorus-containing substance, which he called "nuclein." For a while Miescher entertained the "wild" notion that nuclein—the term was changed to "nucleic acid" by biochemist Richard Altmann in 1889—might have something to do with heredity, but he soon abandoned this "absurd" idea and instead spent more than 30 years studying protamines, a group of very basic proteins in sperm cells. Little did Darwin, Mendel, and Miescher know how intimately connected their discoveries were destined to become one hundred years later.

As it turned out, DNA molecules are not only the key to heredity, but, as one of the pioneers of molecular evolution, Emile Zuckerkandl, phrased it, they are "documents of evolutionary history." In fact, the DNA of every living organism is an accumulation of historical records not unlike an Egyptian palimpsest. Admittedly, the information contained in these records is in a disorderly multilayered state, scattered, and at times fragmentary. Some of it is hidden or camouflaged beyond recognition, and parts of it are lost without a trace. Charles Darwin's words on the imperfections of the evolutionary record sound as accurate and pertinent in the molecular era as they were in the last century: "A history of the world, imperfectly kept and written in a changing dialect. Of this history we possess the last volume alone. Of this volume, only here and there a short chapter has been preserved; and of each page only here and there a few lines." The purpose of molecular evolution is to unravel these historical records, fill in the missing gaps, put the information in order, and decipher its meaning.

Since each evolutionary process leaves its distinctive marks on the genetic material, it is possible to use molecular data not only to reconstruct the chronology of evolution but also to identify the driving forces behind the evolutionary process. Spectacular achievements in molecular biology, such as gene cloning, DNA sequencing, amplification of DNA by the polymerase chain reaction, and restriction endonuclease fragment analysis have, in a sense, placed scientists in a new and privileged position. We can peer into a previously unseen world where genes evolve by such processes as gene duplication, shuffling of DNA, nucleotide substitution, transposition, and gene conversion. In this world, genomes appear both static and fluid, sometimes experiencing little change over long periods of time, and at other times changing dramatically during the geological equivalent of the blink of an eye.

By studying the genetic material we can also attempt to build a classification of the living world that is based not so much on taxonomic convenience but on phylogenetic fact. As opposed to traditional fields of evolutionary inquiry (such as comparative anatomy, morphology, and paleontology), which out of necessity restrict themselves to the study of relationships among relatively closely related organisms, we can now build gigantic family trees connecting vertebrates, insects, plants, fungi, and bacteria, and can trace their common ancestry to times truly and geologically immemorial.

For many years, biochemists and molecular biologists regarded evolutionary studies as an aggregate of wild speculations, unwarranted assumptions, and undisciplined methodology. Roger Stanier, for instance, claimed that evolution "constitutes a kind of metascience, which has the same intellectual fascination for some biologists that metaphysical speculation possessed for medieval scholastics." In his speech before the members of the Society for General Microbiology in 1970, he admonished that evolutionary studies are but "a relatively harmless habit, like eating peanuts," unless such studies are pursued seriously, in which case they become "a vice." While such assessments have never been accurate, the introduction of molecular methods has undoubtedly put evolution on a much more solid footing, and has turned it into a science in which relevant parameters are measured, counted, or computed from empirical data, and theories are tested against objective reality. Conjectures in evolutionary studies today serve the same purpose as in physics; they are quantitative working hypotheses meant to encourage experimental work, so that the theory can be verified, refined, or refuted. One of the main aims of this book is to show that by strengthening the factual basis of the field, evolutionary studies have achieved what Sir William Herschel (1792–1871) called the true goal of all natural sciences, namely to phrase its propositions "not vaguely and generally, but with all possible precision in place, weight, and measure."

Today we are witnessing what Jim Bull and Holly Wichman called "a revolution in evolution." The revolution has been advanced on several fronts through "changes in technology, expansion of theory, and novel methodological approaches." Moreover, the field has now become socially, economically, and politically relevant. Molecular evolutionary principles are being used to counteract the development of resistance in pests and pathogens, to infer routes of disease infection, to develop new drugs through directed exon shuffling and in vitro selection, and to identify organisms capable of cleaning toxic wastes. Indeed, phylogenetic trees (the epitome of evolutionary iconography) have even been admitted as evidence in criminal courts. Molecular evolution, like nuclear physics in the 1940s, impacts society at all levels, from medicine and industry to economy and warfare.

Molecular evolution encompasses two areas of study: the evolution of macromolecules, and the reconstruction of the evolutionary history of genes and organisms. By "the evolution of macromolecules" we refer to the characterization of the changes in the genetic material (DNA or RNA sequences) and its products (proteins or RNA molecules) during evolutionary time, and to the rates and patterns with which such changes occur. This area of study also attempts to unravel the mechanisms responsible for such changes. The second area, also known as "molecular phylogenetics," deals with the evolutionary history of organisms and macromolecules as inferred from molecular data and the methodology of tree reconstruction.

It might appear that the two areas of study constitute independent fields of inquiry, for the object of the first is to elucidate the causes and effects of evolutionary change in molecules, while the second uses molecules merely as a tool to reconstruct the biological history of organisms and their genetic constituents. In practice, however, the two disciplines are intimately related, and progress in one area facilitates progress in the other. For instance, phylogenetic knowledge is essential for determining the order of changes in the molecular characters under study. Conversely, knowledge of the pattern and rate of change of a given molecule is crucial in attempts to reconstruct the evolutionary history of organisms.

Traditionally, a third area of study, prebiotic evolution, or the "origin of life," is included within the framework of molecular evolution. This subject, however, involves a great deal of speculation and is less amenable to quantitative treatment. Moreover, the rules that govern the process of information transfer in prebiotic systems (i.e., systems devoid of replicable genetic material) are not known at the present time. Therefore, this book will not deal with the origin of life. Interested readers may consult Oparin (1957), Cairns-Smith (1982), Dyson (1985), Loomis (1988), and Eigen and Winkler-Oswatitch (1996).

Molecular evolution is a relatively new scientific discipline, in which much of the basic data is lacking, and where new findings often solve few problems but raise many new ones. Studies on such topics as modular evolution, transposable elements, intron loss and sliding, gene resurrection, and exonization were unthinkable in the 1960s. Alternative splicing and RNA editing were unknown in the 1970s. The division of prokaryotes into Archaea and Bacteria was enunciated only in the 1980s, and the 1990s gave us inteins, chirochores, and complete genomic sequences. Indeed, there are signs of change and upheaval in molecular evolution, chief among them a plurality of buzzwords currently in vogue. We have "hypertexts" and "DNA syntax," "genomics" and "post-genomics," "biocomputing" and "bioinformatics," "molecular data mining," and the perennial and inevitable "paradigm shift." Something is bound to happen, but what?

Judging from the past, scientists are the worst of prophets. For example, decades of study on the structure of genes in bacteria prompted Nobel laureate Jacques Monod to claim that what is true for *Escherichia coli* is true for the elephant too ("*Ce qui est vrai pour* E. coli *est vrai pour l'éléphant.*"). He was wrong. The discovery of split genes in eukaryotes has opened previously unthinkable venues of scientific research. A similar case is illustrated by the example of the much-awaited genome-based phylogenetics. For years, scientists expected that genomic sequences would solve the hard phylogenetic problems, and that the genome era would be a phylogenetic boon. They were wrong. Gene content proved to change too little, and gene order proved to change too much, and both turned out to be of limited phylogenetic usefulness. Who would have thought that, of all the genetic elements in the genome, it would be the fickle transposable elements that would shed light on difficult phylogenetic issues?

There are only two predictions that we are willing to make about the future of molecular evolution. The first concerns old controversies. Issues such as the neutralist–selectionist controversy or the antiquity of introns, will continue to be debated with varying degrees of ferocity, and roars of "The Neutral Theory Is Dead," and "Long Live the Neutral Theory" will continue to reverberate, sometimes in the title of a single article. Our second prediction is that the most interesting findings in molecular evolution will be those which at present belong to the realm of the unimaginable, the unpredictable, and the inarticulable. And, while prudence compels us to exercise caution as far as predictions are concerned, this is the one prediction we are most confident about.

Chapter 1

Genes, Genetic Codes, and Mutation

This chapter provides some basic background in molecular biology that is required for studying evolutionary processes at the DNA level. In particular, we discuss the structure of the hereditary material, the attributes of the different kinds of prokaryotic and eukaryotic genes and their products, the structure of genetic codes, and the various types of mutation.

Nucleotide Sequences

The hereditary information of all living organisms, with the exception of some viruses, is carried by **deoxyribonucleic acid (DNA)** molecules. DNA usually consists of two complementary **strands** twisted around each other to form a right-handed double helix. Each chain is a linear polynucleotide consisting of four kinds of **nucleotides**. There are two **purines**, **adenine** (**A**) and **guanine** (**G**), and two **pyrimidines**, **thymine** (**T**) and **cytosine** (**C**). (By convention, purines are abbreviated as **R** and pyrimidines as **Y**. All one-letter abbreviations are listed in Table 1.1.)

The two chains of DNA are joined throughout their lengths by hydrogen bonds between pairs of nucleotides. A purine always pairs with a pyrimidine: adenine pairs with thymine by means of two hydrogen bonds, also referred to as the **weak bond**, and guanine pairs with cytosine by means of three hydrogen bonds, the **strong bond** (Figure 1.1). The complementary pairing of two nucleotides in the double helix is indicated by a colon (e.g., G:C). G:C and A:T are also called the **canonical base pairs**.

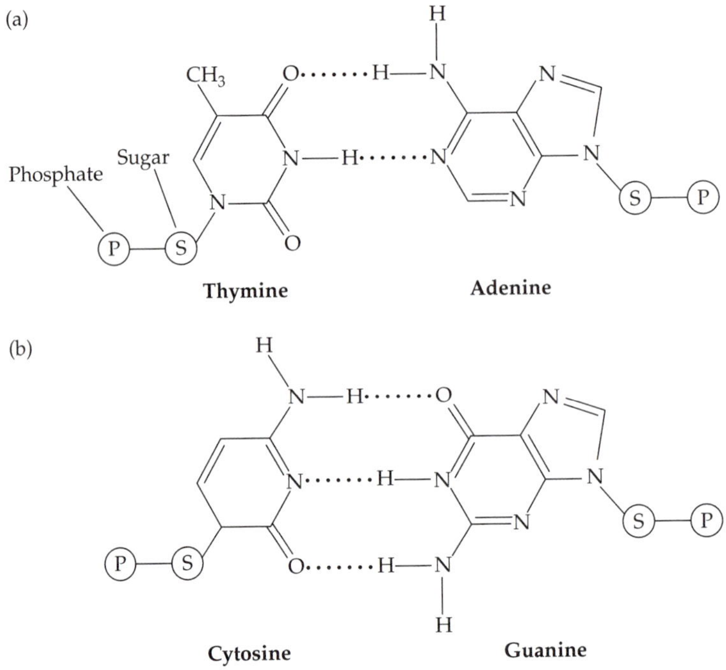

FIGURE 1.1 Complementary base pairing by means of hydrogen bonds (dotted lines) between (a) thymine and adenine (weak bond), and (b) cytosine and guanine (strong bond).

Each nucleotide in a DNA sequence contains a pentose sugar (deoxyribose), a phosphate group, and a purine or pyrimidine base. The backbone of the DNA molecule consists of sugar and phosphate moieties, which are covalently linked in tandem by asymmetrical 5′—3′ phosphodiester bonds. Consequently, the DNA molecule is polarized, one end having a phosphoryl radical (—P) on the 5′ carbon of the terminal nucleotide, the other possessing a free hydroxyl (—OH) on the 3′ carbon of the terminal nucleotide. The direction of the phosphodiester bonds determines the molecule's character; thus, for instance, the sequence 5′—G—C—A—A—T—3′ is different from the sequence 3′—G—C—A—A—T—5′. By convention, the nucleotides in a DNA sequence are written in the order of transcription, i.e., from the 5′ end to the 3′ end. Because of complementarity, only one strand is written. Relative to a specified location or a nucleotide in a DNA sequence, the 5′ and the 3′ directions are also referred to as **upstream** and **downstream**, respectively.

The double helical form of DNA has two strands in antiparallel array, or reverse complementarity (Figure 1.2). The **heavy strand** is the one that contains more than 50% of the heavier nucleotides, the purines A and G. The **light strand** is the one that contains more than 50% of the lighter nucleotides, the pyrimidines C and T.

Ribonucleic acid (RNA) is found as either a single- or double-stranded molecule. RNA differs from DNA by having ribose instead of deoxyribose as

TABLE 1.1 One-letter abbreviations for the DNA alphabet

Symbol	Description
A	Adenine
C	Cytosine
T	Thymine
G	Guanine
W	Weak bonds (A, T)
S	Strong bonds (C, G)
R	Purines (A, G)
Y	Pyrimidines (C, T)
K	Keto (T, G)
M	Amino (A, C)
B	C, G, or T
D	A, G, or T
H	A, C, or T
V	A, C, or G
N	A, C, T, or G
—	No nucleotide (gap symbol)

its backbone sugar moiety, and by using the nucleotide **uracil** (**U**) in place of thymine. Adenine, cytosine, guanine, thymine, and uracil are referred to as the **standard nucleotides**. Some functional RNA molecules, most notably tRNAs, contain nonstandard nucleotides, i.e., nucleotides other than A, C, U, and G, that have been derived by chemical modification of standard nucleotides after the transcription of the RNA. In addition to the canonical complementary base pairs, which in RNA are G:C and A:U, the G:U pair is also stable in RNA. (In DNA, stable pairing does not occur between G and T.)

The length of a single-stranded nucleic acid is measured in number of nucleotides. That of a double-stranded sequence is measured in **base pairs** (**bp**), thousands of base pairs (**kilobases, Kb**), or millions of base pairs (**megabases, Mb**).

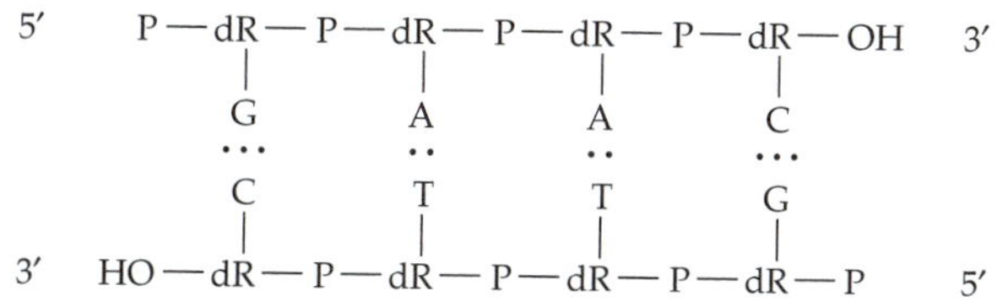

```
5'   P—dR—P—dR—P—dR—P—dR—OH   3'
       |     |     |     |
       G     A     A     C
      ...    ..    ..   ...
       C     T     T     G
       |     |     |     |
3'  HO—dR—P—dR—P—dR—P—dR—P    5'
```

FIGURE 1.2 Schematic representation of the antiparallel structure of double-stranded DNA. P, phosphate; dR, deoxyribose; OH, hydroxyl; —, covalent bond; ••, weak bond; •••, strong bond. The DNA sequence here may be written in compact form as GAAC.

GENOMES AND DNA REPLICATION

The entire complement of genetic material carried by an individual is called the **genome**. The portion of the genome that contains genes is called **genic DNA**. In addition to genes, most genomes also contain various quantities of **nongenic DNA** (Chapter 8). By definition, nongenic DNA does not contain genes. Because of our imperfect knowledge of the genome, however, "nongenic DNA" is sometimes used as an *ad hoc* term that may include genic sequences whose function has not yet been determined. The entire set of transcribed sequences produced by the genome is called the **transcriptome**; the entire set of proteins encoded by the genome is called the **proteome**.

All organisms replicate their DNA before every cell division. DNA replication occurs at elongation rates of about 500 nucleotides per second in bacteria and about 50 nucleotides per second in vertebrates. During DNA replication, each of the two DNA strands serves as a template for the formation of a new strand. Because each of the two daughters of a dividing cell inherits a double helix containing one "old" and one "new" strand, DNA is said to be replicated **semiconservatively**.

Replication starts at a structure called a **replication bubble** or **origin of replication**, a local region where the two strands of the parental DNA helix have been separated from each other. Replication proceeds in both directions as two **replication forks**. Because DNA replication occurs only in the 5′-to-3′ direction, one of the strands, known as the **leading strand**, serves as a template for a continuously synthesized daughter strand (Figure 1.3). The other strand, the **lagging strand**, is replicated in discontinuous small pieces known as **Okazaki fragments**, which are subsequently ligated. Okazaki fragments are longer in bacteria (1,000–2,000 nucleotides) than in vertebrates (100–200 nucleotides).

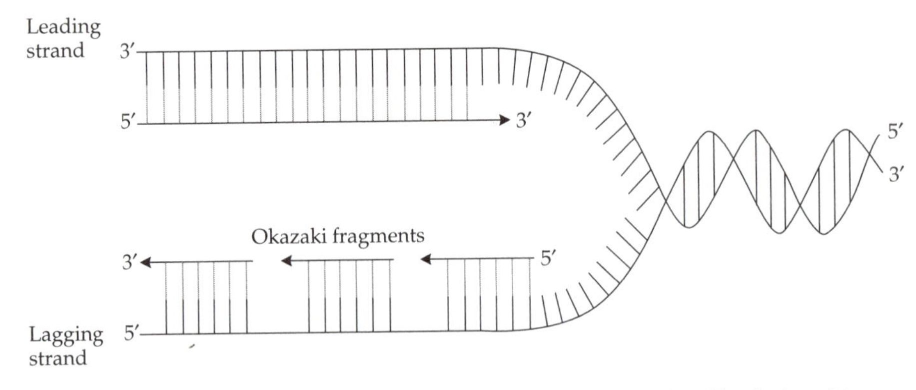

FIGURE 1.3 Double-stranded DNA replication. The synthesis of both daughter strands occurs only in the 5′-to-3′ direction. The leading strand serves as a template for a continuously synthesized daughter strand (long arrow). The lagging strand is replicated into discontinuous Okazaki fragments (short arrows) that are subsequently ligated to one another.

Most bacterial genomes have one origin of replication, and a typical bacterial genome can replicate in less than 40 minutes. In eukaryotic cells, many replication origins exist. They are usually spaced at intervals of 30,000 to 300,000 bp from one another. Replication in eukaryotic cells may take several hours.

GENES AND GENE STRUCTURE

Traditionally, a gene was defined as a segment of DNA that codes for a polypeptide chain or specifies a functional RNA molecule. Recent molecular studies, however, have altered our perception of genes, and we shall adopt a somewhat vaguer (or fuzzier) definition. Accordingly, a **gene** is a sequence of genomic DNA or RNA that is essential for a specific function. Performing the function may not require the gene to be translated or even transcribed.

Three types of genes are recognized: (1) **protein-coding genes**, which are transcribed into RNA and subsequently translated into proteins, (2) **RNA-specifying genes**, which are only transcribed, and (3) **untranscribed genes**. Protein-coding genes and RNA-specifying genes are also referred to as **structural** or **productive genes**. (Note that some authors restrict the definition of "structural genes" to include only protein-coding genes.)

Transcription is the synthesis of a complementary RNA molecule on a DNA template. The transcription of productive genes is carried out by DNA-dependent RNA polymerases. Transcription in eubacteria is carried out by only one type of RNA polymerase. In archaebacteria, several types of DNA-dependent RNA polymerases exist, but in each of the archaebacterial species studied to date only one type is used. (The distinction between eubacteria and archaebacteria is discussed in Chapter 5.) In each eukaryotic species, three types of RNA polymerase are employed in the transcription of nuclear genes. Ribosomal RNA (rRNA) genes are transcribed by RNA polymerase I, protein-coding genes are transcribed by RNA polymerase II, and small cytoplasmic RNA genes, such as genes specifying transfer RNAs (tRNAs), are transcribed by RNA polymerase III. Some small nuclear RNA genes (e.g., U2) are transcribed by polymerase II, others (e.g., U6) by polymerase III. The snRNA gene U3 is transcribed by polymerase II in animals but by polymerase III in plants.

Protein-coding genes

A standard eukaryotic protein-coding gene consists of transcribed and untranscribed parts. The untranscribed parts are designated according to their location relative to the transcribed parts as **5′** and **3′ flanking regions**.

The 5′ flanking region contains several specific sequences called **signals** that determine the initiation, tempo, timing, and tissue specificity of the transcription process. Because these regulatory sequences promote the transcription process, they are also referred to as promoters, and the region in which they reside is called the promoter region. The promoter region consists of the

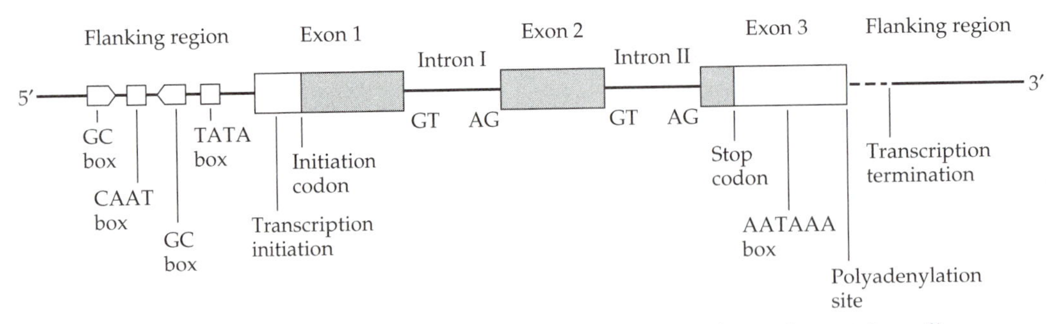

FIGURE 1.4 Schematic representation of a typical eukaryotic protein-coding gene and its constituent parts. By convention, the 5′ end is at the left. Protein-coding regions are shaded. The dashed line indicates that the polyadenylation site may not be identical to the transcription termination site, although in the majority of cases it is. The regions are not drawn to scale.

following signals: the TATA or Hogness-Goldberg box, located 19–27 bp upstream of the startpoint of transcription; the CAAT box further upstream; and one or more copies of the GC box, consisting of the sequence GGGCGG or a closely related variant and surrounding the CAAT box (Figure 1.4). The CAAT and GC boxes, which may function in either orientation, control the initial binding of the RNA polymerase, whereas the TATA box controls the choice of the startpoint of transcription. However, none of the above signals is uniquely essential for promoter function. Some genes do not possess a TATA box and thus do not have a unique startpoint of transcription. Other genes possess neither a CAAT box nor a GC box; their transcription initiation is controlled by other elements in the 5′ flanking region.

The 3′ flanking region contains signals for the termination of the transcription process. Because of our incomplete knowledge of the regulatory elements, which may be found at considerable distances upstream or downstream from the transcribed regions of the gene, it is impossible at present to delineate with precision the points at which a gene begins and ends.

The transcribed RNA is referred to as **precursor messenger RNA** (**pre-mRNA**). Since the different pre-mRNAs produced at any given time in the cell differ greatly from each other in length and molecular weight, the pre-mRNAs in the nucleus are collectively denoted as **heterogeneous nuclear RNA (hnRNA)**. The DNA strand from which the pre-mRNA is transcribed is called the **antisense strand**. The untranscribed complementary strand, which is identical in sequence with the pre-mRNA, is called the **sense strand**. The transcription of protein-coding genes in eukaryotes starts at the **transcription initiation site** (the cap site in the RNA transcript), and ends at the **transcription termination site**, which may or may not be identical with the **polyadenylation** or **poly(A)-addition** site in the mature **messenger RNA** (**mRNA**) molecule. In other words, termination of transcription may occur further downstream from the polyadenylation site.

Pre-mRNA contains exons and introns. **Introns**, or intervening sequences, are those transcribed sequences that are spliced out during the processing of pre-mRNA. All the other sequences, which remain in the mRNA following splicing, are referred to as **exons**. In addition to splicing, the maturation of pre-mRNA into mRNA also consists of capping the 5′ end (i.e., the addition of a methylated guanine), degradation of those nucleotides that may be located downstream of the polyadenylation site, and polyadenylation of the 3′ end (i.e., the addition of 100–200 adenine residues). The last two processes do not occur in all eukaryotic protein-coding genes. Extensive transcriptional and posttranscriptional modifications of mRNA may also occur.

Exons or parts of exons that are translated are referred to as **protein-coding exons** or **coding regions**. The first and last coding regions of a gene are flanked by 5′ and 3′ **untranslated regions**, respectively. Both the 5′ and 3′ untranslated regions are transcribed. The 3′ untranslated region is usually much longer than the 5′ untranslated region. For instance, in the human pre-prourokinase gene, the 3′ untranslated region (>850 bp) is more than 10 times longer than the 5′ untranslated region (79 bp). Untranslated regions may contain regulatory sequences involved in processing the pre-mRNA or regulating the translation process. One such sequence is the AATAAA or Proudfoot-Brownlee box, which is located about 20 bp upstream of the transcription termination site and serves as a polyadenylation signal. Some untranslated parts may also be involved in determining the turnover rate of the mRNA molecule in the cell.

According to the specific mechanism with which the intron is cleaved out of the pre-mRNA, we may classify introns into two main categories. Introns in the nuclear genes that are transcribed by RNA polymerase II are called **non-self-splicing** or **spliceosomal introns**, because they are cleaved out of the pre-mRNA by means of an enzymatic complex called the spliceosome. These introns differ from several types of **self-splicing introns**, found mainly in mitochondrial and chloroplast genomes, which are cleaved out without the help of exogenous gene products.

The **splicing sites** or **junctions** of spliceosomal introns are determined to a large extent by nucleotides at the 5′ and 3′ ends of each intron, known as the **donor** and **acceptor sites**, respectively. With few exceptions, all eukaryotic nuclear introns begin with GT and end in AG (the **GT–AG rule**), and these sequences have been shown to be essential for the correct excision and splicing of introns (Figure 1.4). Exon sequences adjacent to introns may also contribute to the determination of the splicing site. In addition, each intron contains a specific sequence called the TACTAAC box, located approximately 30 bp upstream of the 3′ end of the intron. This sequence is highly conserved among yeast nuclear genes, though it is more variable in the genes of multicellular eukaryotes. Splicing involves the cleavage of the 5′ splice junction and the creation of a lariat through a phosphodiester bond between the G at the 5′ end of the intron and the A in the sixth position of the TACTAAC box. Subsequently, the 3′ splice junction is cleaved and the two exons are ligated.

The number of introns varies greatly from gene to gene. Some genes possess dozens of introns; others (e.g., most histone and olfactory receptor genes) are devoid of introns altogether. The distribution of intron sizes in vertebrate genes is very broad, and the longest introns extend out to hundreds of kilobases. The dis-

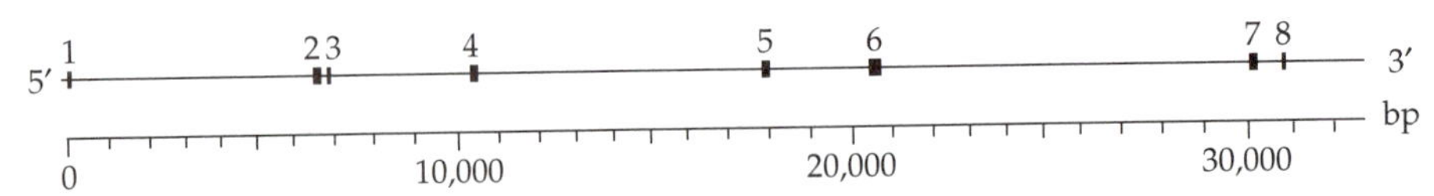

FIGURE 1.5 The localization of the eight exons in the human factor IX gene. The vertical bars represent the exons. Only the transcribed region is shown. The exons and introns are drawn to scale. The total length of the exons is 1,386 nucleotides as opposed to 29,954 nucleotides for the total length of introns. The 5′ untranslated region (30 nucleotides) is much shorter than the 3′ untranslated region (1,389 nucleotides). Only about 4% of the pre-mRNA sequence actually encodes the protein. Note the proximity of several exons (e.g., 2 and 3, and 7 and 8) to each other, as opposed to the remoteness of other neighboring exons (e.g., 6 and 7). Data from Yoshitake et al. (1985).

tribution of exon sizes is much narrower, peaking at around 150 bp. Exons are not distributed evenly over the length of the gene. Some exons are clustered; others are located at great distances from neighboring exons (Figure 1.5).

The vast majority of protein-coding genes in vertebrates consist mostly of introns. One such extreme example is the human dystrophin gene, which has 79 exons spanning over 2.3 Mb; the mRNA is only 12 Kb, leading to an exon-to-intron ratio of about 1:200, or 0.5%. In other eukaryotes (e.g., *Drosophila melanogaster*, *Arabidopsis thaliana*, and *Caenorhabditis elegans*), introns are fewer and smaller. Not all introns interrupt coding regions; some occur in untranslated regions, mainly in the region between the transcription initiation site and the translation initiation codon.

Protein-coding genes in eubacteria are different from those in eukaryotes in several respects. Most importantly, they do not contain introns, i.e., they are co-linear with the protein product (Figure 1.6). Promoters in eubacteria contain a –10 sequence and a –35 sequence, so named because they are located, respectively, 10 bp and 35 bp upstream of the initiation site of transcription. The former, also known as the Pribnow box, consists of the sequence TATAAT or a similar variant, while the latter consists of the sequence TTGACA or its variant. A prokaryotic promoter may also contain other specific sequences further upstream from the –35 sequence.

A set of structural genes in prokaryotes (and more rarely in eukaryotes) may be arranged consecutively to form a unit of genetic expression that is transcribed into one molecule of polycistronic mRNA and subsequently translated into different proteins. Such a unit usually contains genetic elements that control the coordinated expression of the genes belonging to the unit. This arrangement of genes is called an **operon**.

RNA-specifying genes

RNA-specifying genes are transcribed into RNAs but are not translated into proteins. In eukaryotes, these genes may contain introns that must be spliced out before the RNA molecule becomes functional. For example, many nuclear

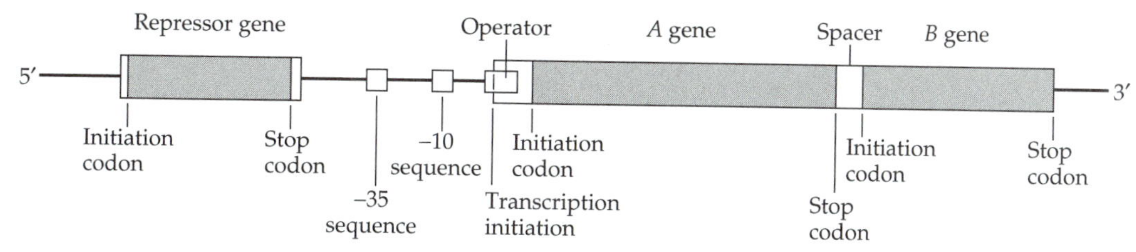

FIGURE 1.6 Schematic structure of an induced prokaryotic operon. Genes *A* and *B* are protein-coding genes and are transcribed into a single messenger RNA. The repressor gene encodes a repressor protein, which binds to the operator and prevents the transcription of the structural genes *A* and *B* by blocking the movement of the RNA polymerase. The operator is a DNA region at least 10 bases long, which may overlap the transcribed region of the genes in the operon. By binding to an inducer (a small molecule), the repressor is converted to a form that cannot bind the operator. The RNA polymerase can then initiate transcription. Protein-coding regions are shaded. The regions are not drawn to scale.

tRNA-specifying genes possess very short introns. Sequence elements involved in the regulation of transcription of some RNA-specifying genes are sometimes included within the sequence specifying the functional end product. In particular, all eukaryotic tRNA-specifying genes contain an internal transcriptional start signal recognized by RNA polymerase III (Chapter 7). Some regulatory elements, e.g., small nucleolar RNA (snoRNA), are located within the introns of rRNA-specifying genes.

Posttranscriptional modifications of RNA

Whether they are derived from protein-coding genes or RNA-specifying genes, many RNA molecules are modified following transcription. Such modifications may include the insertion or deletion of standard nucleotides, the modification of a standard nucleotide into another (e.g., C → U), the modification of a standard nucleotide into a nonstandard one, and the enzymatic addition of terminal sequences of ribonucleotides to either the 5′ or the 3′ end. These processes, which ultimately result in the production of an RNA molecule that is not complementary to the DNA sequence from which it has been transcribed, are collectively known as **RNA editing**. In some cases, RNA editing can be quite extensive, so that the resulting RNA may bear little resemblance to the DNA sequence from which it has been transcribed (Figure 1.7). In such cases, the template gene is called a **cryptogene** (literally, a hidden gene).

Untranscribed genes

Our knowledge of untranscribed genes is less advanced than that of the other types of genes. Several types of untranscribed genes or families of genes have

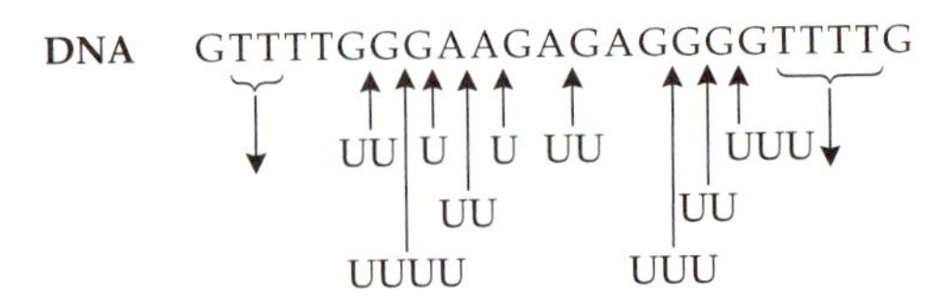

FIGURE 1.7 Comparison of a representative region of the sense strand of the mitochondrial gene encoding subunit III of cytochrome *c* oxidase in *Trypanosoma brucei* with the mRNA after editing. Insertions (U's) and deletions (T's) are indicated by upward and downward arrows, respectively. Data from Feagin et al. (1988).

been tentatively identified. **Replicator genes** specify the sites for initiation and termination of DNA replication. This term may apply to the specific sequences at the origins of replicons, which are the units of DNA replication in eukaryotic genomes. These sequences may serve as binding sites for specific initiator or repressor molecules, or as recognition sites for the enzymes involved in DNA replication. **Recombinator genes** provide specific recognition sites for the recombination enzymes during meiosis. **Telomeric sequences** are short, repetitive sequences at the end of many eukaryotic chromosomes that provide it with a protective cap against such eventualities as terminal exonucleolytic degradation. **Segregator genes** provide specific sites for attachment of the chromosomes to the spindle machinery during meiotic and mitotic segregation. Within this category one may include centromeric sequences, which consist of short, repetitive sequences similar to those of telomeres. **Attachment sites** are sequences that bind structural proteins, enzymes, hormones, metabolites, and other molecules with a high specificity for certain DNA sequences. Within this category, one may include matrix attachment regions that bind the nonhistone proteins that form the nuclear matrix or scaffold, as well as enhancers, which bind specific proteins that activate the transcription of distantly located downstream protein-coding genes. **Constructional sites** are sequence elements involved in the determination of structural features of the chromosome, such as supercoils and fragile sites.

Many more untranscribed genes are likely to exist, some of which could be independent of structural genes in both function and location and could be involved in complex regulatory functions, such as the ontogenetic development of multicellular organisms.

Pseudogenes

A **pseudogene** is a nongenic DNA segment that exhibits a high degree of similarity to a functional gene but which contains defects, such as nonsense and frameshift mutations, that prevent it from being expressed properly. Pseudogenes are marked by the prefix ψ followed by the name of the gene to which they exhibit sequence similarity, e.g., ψβ-globin. In computerized databanks, the suffix P is used instead, e.g., CA5P for the α-carbonic anhydrase pseudogene 5.

Most pseudogenes are not transcribed; a significant minority, however, undergo transcription but not translation, and a handful of pseudogenes may even be translated. Pseudogenes are ubiquitous at all genomic locations and in all organisms, although some organisms tend to harbor more pseudogenes than others. There are two main types of pseudogenes: unprocessed and processed. The two types are created by different molecular evolutionary processes and will therefore be dealt with separately, in Chapters 6 and 7, respectively.

AMINO ACIDS

Amino acids are the elementary structural units of proteins. All proteins in all organisms are constructed from the 20 primary amino acids listed in Table 1.2. Each amino acid has an **—NH_2 (amine) group** and a **—COOH (carboxyl) group** on either side of a central carbon called the **α carbon** (Figure 1.8). Also attached to the α carbon are a hydrogen atom and a **side chain**, also denoted as the **—R group**. Because of the possibility of forming two nonidentical, mirror-image enantiomers around the α carbon, all amino acids (with the exception of glycine) can occur in two optical isomeric forms: levorotatory (L) and dextrorotatory (D). Only L-amino acids are used in the process of translation of mRNAs into proteins. (D-isomers and L-amino acids other than the primary ones are sometimes found in proteins, but they are products of posttranslational modifications.) The side chains of the amino acids vary in size, shape, charge, hydrogen-bonding capacity, composition, and chemical reactivity. It is the side chains that distinguish one amino acid from another. Indeed, the classification of amino acids is made on the basis of their —R groups.

The simplest and smallest amino acid is **glycine**, which has only a hydrogen as its side chain (molecular weight = 75). Next in size is **alanine**, having a methyl as its —R group. **Valine** has a three-carbon-long side chain. A four-carbon-long side chain is found in both **leucine** and **isoleucine**. These large, aliphatic (i.e., containing no rings) side chains are hydrophobic (literally, water-fearing)—that is, they have an aversion to water and tend to cluster in the internal part of a protein molecule, away from the aqueous environment of the cell. **Proline** (an imino acid rather than an amino acid) also has an aliphatic side chain, but its side chain loops back to create a second bond to the nitrogen in its amine group. This ring forces a contorted bend on the polypeptide chain.

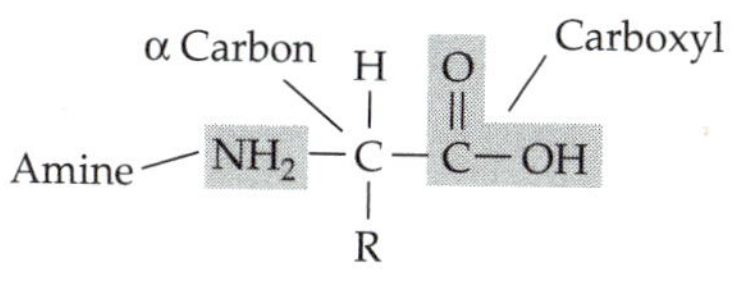

α Carbon
Carboxyl
Amine — NH_2 — C(H)(R) — C(=O) — OH

FIGURE 1.8 Structure of an amino acid. It contains a central α carbon, an amine group, a carboxyl group, a hydrogen, and a side chain (denoted by R).

TABLE 1.2 Primary amino acids, their three- and one-letter abbreviations, their ionized structure at pH 6.0 to 7.0, and their molecular weights

Amino acid	Ionized structure	Three-letter abbreviation	One-letter abbreviation	Molecular weight
Alanine	$^{+}H_3N{-}C(COO^-)(H){-}CH_3$	Ala	A	89.1
Arginine	$^{+}H_3N{-}C(COO^-)(H){-}CH_2{-}CH_2{-}CH_2{-}NH{-}C(=NH_2^+){-}NH_2$	Arg	R	174.2
Asparagine	$^{+}H_3N{-}C(COO^-)(H){-}CH_2{-}C(=O){-}NH_2$	Asn	N	132.1
Aspartic acid	$^{+}H_3N{-}C(COO^-)(H){-}CH_2{-}C(=O){-}O^-$	Asp	D	133.1
Cysteine	$^{+}H_3N{-}C(COO^-)(H){-}CH_2{-}SH$	Cys	C	121.2
Glutamic acid	$^{+}H_3N{-}C(COO^-)(H){-}CH_2{-}CH_2{-}C(=O){-}O^-$	Glu	E	147.1

TABLE 1.2 CONTINUED

Amino acid	Ionized structure	Three-letter abbreviation	One-letter abbreviation	Molecular weight
Glutamine	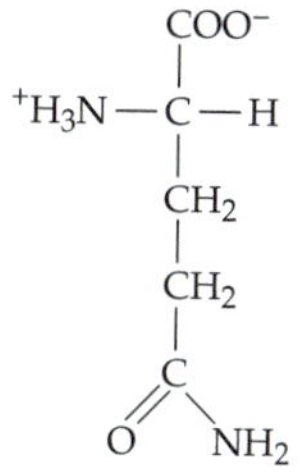 COO^- ; $^+H_3N-C-H$; CH_2 ; CH_2 ; $C(=O)NH_2$	Gln	Q	146.2
Glycine	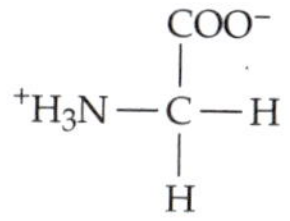 COO^- ; $^+H_3N-C-H$; H	Gly	G	75.1
Histidine	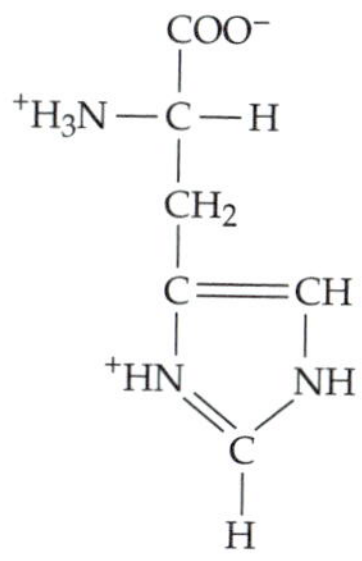 COO^- ; $^+H_3N-C-H$; CH_2 ; $C=CH$; ^+HN, NH ; C ; H	His	H	155.2
Isoleucine	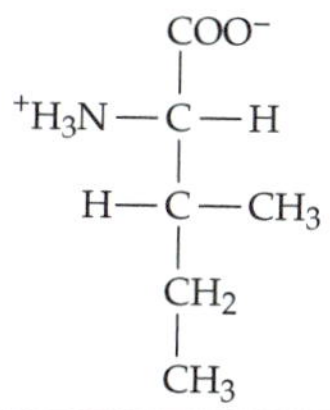 COO^- ; $^+H_3N-C-H$; $H-C-CH_3$; CH_2 ; CH_3	Ile	I	131.2
Leucine	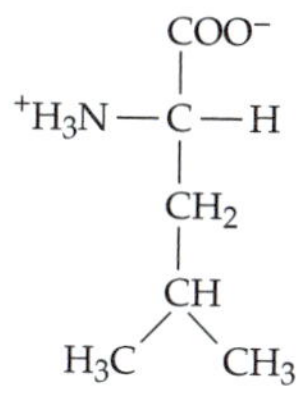 COO^- ; $^+H_3N-C-H$; CH_2 ; CH ; H_3C, CH_3	Leu	L	131.2

TABLE 1.2 CONTINUED

Amino acid	Ionized structure	Three-letter abbreviation	One-letter abbreviation	Molecular weight
Lysine	COO^- / $^+H_3N-C-H$ / CH_2 / CH_2 / CH_2 / CH_2 / NH_3^+	Lys	K	146.2
Methionine	COO^- / $^+H_3N-C-H$ / CH_2 / CH_2 / S / CH_3	Met	M	149.2
Phenylalanine	COO^- / $^+H_3N-C-H$ / CH_2 / phenyl ring	Phe	F	165.2
Proline	COO^- / $^+H_2N-C-H$ / H_2C CH_2 / CH_2 (ring)	Pro	P	115.1
Serine	COO^- / $^+H_3N-C-H$ / $H-C-OH$ / H	Ser	S	105.1
Threonine	COO^- / $^+H_3N-C-H$ / $H-C-OH$ / CH_3	Thr	T	119.1

TABLE 1.2 CONTINUED

Amino acid	Ionized structure	Three-letter abbreviation	One-letter abbreviation	Molecular weight
Tryptophan	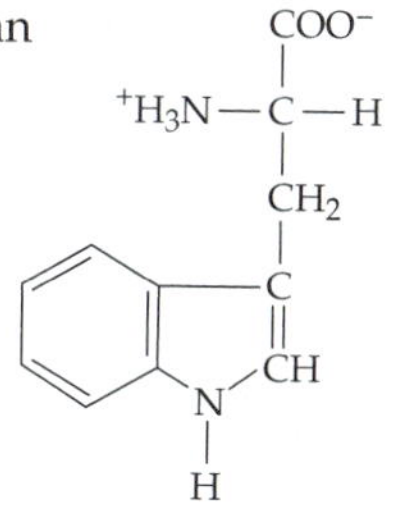 COO^-; $^+H_3N-C-H$; CH_2; C; CH; N; H (indole ring)	Trp	W	204.2
Tyrosine	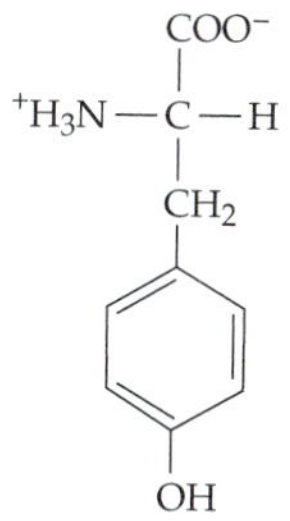 COO^-; $^+H_3N-C-H$; CH_2; (benzene ring); OH	Tyr	Y	181.2
Valine	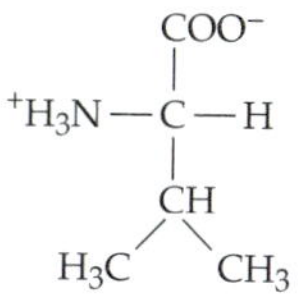 COO^-; $^+H_3N-C-H$; CH; H_3C CH_3	Val	V	117.2

Three amino acids have aromatic (benzene-like) side chains. **Phenylalanine** contains a phenyl ring attached to a methylene group. The aromatic ring of **tyrosine** contains a hydroxyl group, which makes this amino acid less hydrophobic than the other amino acids mentioned so far. **Tryptophan**, the largest amino acid (molecular weight = 204), has an indole ring joined to a methylene group. Phenylalanine and tryptophan are highly hydrophobic.

A sulfur atom is present in the —R groups of **cysteine** and **methionine**. Both these sulfur-containing amino acids are hydrophobic; however, the sulfhydryl in cysteine is reactive and, through oxidation, may form a disulfide bridge (**cystine**) with another cysteine. **Serine** and **threonine** contain aliphatic hydroxyl groups and are therefore much more hydrophilic (water-loving) and reactive than alanine and valine (which are the dehydroxylated versions of serine and threonine, respectively).

Five amino acids have very polar side chains and are therefore highly hydrophilic. **Lysine** and **arginine** are positively charged at neutral pH, and their side chains are among the largest of the 20 amino acids. The imidazole ring of **histidine** can be either uncharged or positively charged, depending on the local environment. At pH 6.0, over 50% of histidine molecules are positively

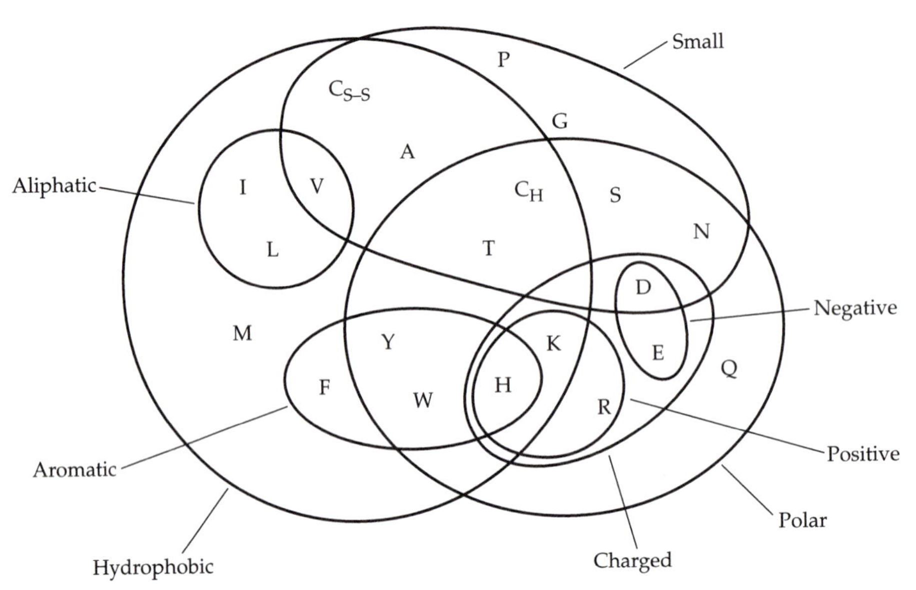

FIGURE 1.9 A Venn diagram showing the division of the 20 primary amino acids into overlapping categories according to size, structure of the side chain, polarity, charge, and hydrophobicity. Note the placement of cysteine in two places, as reduced cysteine (C_H) and as cystine (C_{S-S}). See Table 1.2 for the one-letter abbreviations of the amino acids. Modified from Taylor (1986).

charged; at pH 7.0, less than 10% have a positive charge. Two amino acids contain acidic side chains: **aspartic acid** and **glutamic acid**. They are nearly always negatively charged at physiological pH values. Uncharged derivatives of glutamic acid and aspartic acid are **glutamine** and **asparagine**, respectively, which contain a terminal amide group in place of a carboxylate.

Amino acids may be classified into several overlapping categories by using different criteria. Figure 1.9 is a Venn diagram of the amino acids classified according to size, polarity, charge, and hydrophobicity.

PROTEINS

A protein is a macromolecule that consists of one or more polypeptide chains, each of which has a unique, precisely defined amino acid sequence. In a polypeptide chain, the α-carboxyl group of one amino acid is joined to the α-amino group of another amino acid by a **peptide bond**. For example, Figure 1.10 shows the formation of a dipeptide, glycylalanine, from two amino acids (glycine and alanine), accompanied by the loss of a water molecule. Note that in this dipeptide, glycine acts as the carboxylic acid (its amino end is free), while alanine acts as the amine (its carboxylic end is free). If the roles were reversed, then a different dipeptide, alanylglycine, would have been formed. In other words, polypeptides are polar molecules. The standard presentation of

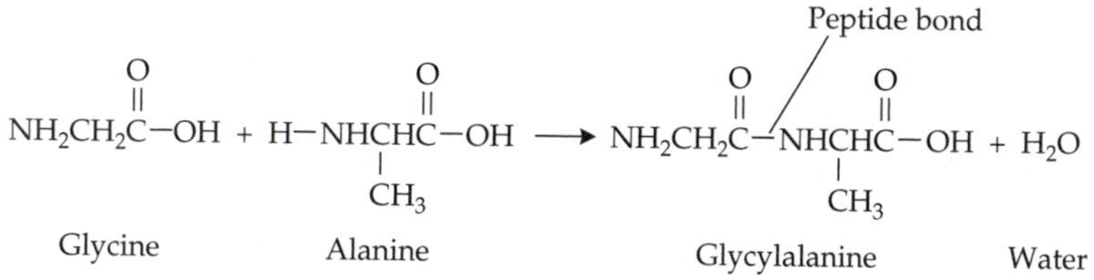

Peptide bond

$$\mathrm{NH_2CH_2\overset{\overset{\displaystyle O}{\|}}{C}{-}OH} + \mathrm{H{-}NH\underset{\underset{\displaystyle CH_3}{|}}{C}H\overset{\overset{\displaystyle O}{\|}}{C}{-}OH} \longrightarrow \mathrm{NH_2CH_2\overset{\overset{\displaystyle O}{\|}}{C}{-}NH\underset{\underset{\displaystyle CH_3}{|}}{C}H\overset{\overset{\displaystyle O}{\|}}{C}{-}OH} + \mathrm{H_2O}$$

Glycine Alanine Glycylalanine Water

FIGURE 1.10 Formation of a dipeptide from two amino acids.

a sequence is from the amino end to the carboxyl end. For instance, glycylalanine may be written GA, and alanylglycine as AG. Each amino acid in a polypeptide is called a **residue**. The "left" end of a polypeptide, containing the free amino group, is termed the **amino** or **N terminus**. The "right" end, with the free carboxyl group, is termed the **carboxyl** or **C terminus**. Polypeptides vary greatly in size, from a few amino acids to such brobdingnagian proteins as the 26,926-amino-acid-long titin, which is found in muscle cells.

Four levels of structural organization are usually mentioned when dealing with proteins. The **primary structure** is simply the linear arrangement of amino acid residues along the polypeptide sequence. The **secondary structure** refers to the spatial arrangement, or **folding**, of amino acid residues that are near to one another in the primary structure. Some of these arrangements are of a regular kind giving rise to periodical structures. One such periodical secondary structure is the **α helix**, a rodlike entity whose main chain coils tightly as a right-handed screw, with all the side chains sticking outward in a helical array. The very tight structure of the α helix is stabilized by same-strand hydrogen bonds between —NH groups and —CO groups spaced at four-amino-acid-residue intervals. Another periodical structural motif is the **β-pleated sheet**, in which parallel or antiparallel loosely coiled β strands are stabilized by hydrogen bonds between —NH and —CO groups from adjacent strands. Some protein regions remain in **random coil**, i.e., possess no regular pattern of secondary structure. Different proteins have varying proportions of α helix, β-pleated sheet, and random coil. For instance, the major protein in hair, keratin, is made almost entirely of α helices. In contrast, silk is a protein made entirely of β-pleated sheets.

The three-dimensional structure of a protein is termed **tertiary structure**, which refers to the spatial arrangement of amino acid residues that are not adjacent to one another in the linear primary sequence. The tertiary structure is formed by the packing of local features, such as α helices and β sheets, by covalent and noncovalent forces, such as hydrogen bonds, hydrophobic interactions, salt bridges between positively and negatively charged residues, as well as disulfide bonds between pairs of cysteines. In the process of three-dimensional folding, amino acid residues with hydrophobic side chains tend to be buried on the inside of the protein, while those with hydrophilic side chains tend to be on the outside. This tendency is particularly strong in globular proteins, which usually function within an aqueous environment.

A protein may consist of more than one polypeptide. Such proteins must be assembled after each individual polypeptide has assumed its tertiary structure. Each polypeptide chain in such a protein is called a **subunit**. For example, hemoglobin, the oxygen carrier in blood, consists of four subunits: two α-globins and two β-globins. Pyruvate dehydrogenase, a mitochondrial protein involved in energy metabolism, consists of 72 subunits. Proteins containing more than one polypeptide chain exhibit an additional level of structural organization. **Quaternary structure** refers to the spatial arrangement of the subunits and the nature of their contacts.

Once a polypeptide chain is formed, the molecule automatically folds itself in response to hydrogen bonds, salt bridges, and hydrophobic and hydrophilic interactions. The primary structure of a protein uniquely determines its secondary and tertiary structures. However, our knowledge of protein folding is not yet sufficient to predict accurately the secondary and tertiary structures of a protein from its primary structure. This is a challenging problem in protein chemistry today.

Some proteins require nonprotein components to carry out their function. Such components are called **prosthetic groups**. The complex of a protein and its prosthetic group is called the **holoprotein**; the protein without the prosthetic group is called the **apoprotein**.

TRANSLATION AND GENETIC CODES

The synthesis of a protein involves a process of decoding, whereby the genetic information carried by an mRNA molecule is translated into primary amino acids through the use of **transfer RNA (tRNA)** mediators. A transfer RNA is a small molecule, usually 70–90 nucleotides in length. Each of the 20 amino acids has at least one type of tRNA assigned to it; most have several tRNAs. Each tRNA is designed to carry only one amino acid. Twenty specific enzymes, called **aminoacyl-tRNA synthetases**, couple each amino acid to the 3′ end of its appropriate tRNA. **Translation** involves the sequential recognition of adjacent nonoverlapping triplets of nucleotides, called **codons**, by a complementary sequence of three nucleotides (the **anticodon**) in the tRNA. A tRNA is demarcated by a superscript denoting the amino acid it carries and a subscript identifying the codon it recognizes. Thus, a tRNA that recognizes the codon UUA and carries the amino acid leucine is designated $\text{tRNA}^{\text{Leu}}_{\text{UUA}}$.

Translation starts at an **initiation codon** and proceeds until a **stop** or **termination codon** is encountered. The phase in which a sequence is translated is determined by the initiation codon and is referred to as the **reading frame**. In the translational machinery at the interface between the ribosome, the charged tRNA, and the mRNA, each codon is translated into a specific amino acid, which is subsequently added to the elongating polypeptide. Stop codons are recognized by cytoplasmic proteins called **release factors**, which terminate the translation process.

Following translation, the creation of a functional protein may involve posttranslational processes, such as modifications of the primary amino acids,

TABLE 1.3 The universal genetic code

Codon	Amino acid	Codon	Amino acid	Codon	Amino acid	Codon	Amino acid
UUU	Phe	UCU	Ser	UAU	Tyr	UGU	Cys
UUC	Phe	UCC	Ser	UAC	Tyr	UGC	Cys
UUA	Leu	UCA	Ser	UAA	Stop	UGA	Stop
UUG	Leu	UCG	Ser	UAG	Stop	UGG	Trp
CUU	Leu	CCU	Pro	CAU	His	CGU	Arg
CUC	Leu	CCC	Pro	CAC	His	CGC	Arg
CUA	Leu	CCA	Pro	CAA	Gln	CGA	Arg
CUG	Leu	CCG	Pro	CAG	Gln	CGG	Arg
AUU	Ile	ACU	Thr	AAU	Asn	AGU	Ser
AUC	Ile	ACC	Thr	AAC	Asn	AGC	Ser
AUA	Ile	ACA	Thr	AAA	Lys	AGA	Arg
AUG	Met	ACG	Thr	AAG	Lys	AGG	Arg
GUU	Val	GCU	Ala	GAU	Asp	GGU	Gly
GUC	Val	GCC	Ala	GAC	Asp	GGC	Gly
GUA	Val	GCA	Ala	GAA	Glu	GGA	Gly
GUG	Val	GCG	Ala	GAG	Glu	GGG	Gly

removal of terminal sequences at both ends, intra- or intercellular transport, the addition of prosthetic groups, and protein splicing.

The correspondence between the codons and the amino acids is determined by a set of rules called the **genetic code**. With few exceptions, the genetic code for nuclear protein-coding genes is **universal**, i.e., the translation of almost all eukaryotic nuclear genes and prokaryotic genes is determined by the same set of rules. (Because exceptions do exist, the genetic code is not truly universal, and some biologists prefer the term "standard genetic code.")

The universal genetic code is given in Table 1.3. Since a codon consists of three nucleotides, and since there are four different types of nucleotides, there are $4^3 = 64$ possible codons. In the universal genetic code, 61 of these code for specific amino acids and are called **sense codons**, while the remaining three are **stop codons**. The stop codons in the universal genetic code are UAA, UAG, and UGA.

With very rare exceptions, the genetic code is **unambiguous**, i.e., one codon can code for only one amino acid. Exceptions usually involve termination codons, which are sometimes translated by charged tRNAs, so that translation continues beyond the original termination codon until the next stop codon is encountered. This process is called translational readthrough, and the tRNAs involved in the readthrough are called termination suppressors. In several organisms, natural termination suppressors are known. For example, in *Escherichia coli*, the UGA codon can direct the incorporation of the unusual amino acid selenocysteine instead of signaling the termination of the translation process.

Since there are 61 sense codons and only 20 primary amino acids in proteins, most amino acids (18 out of 20) are encoded by more than one codon. Such a code is referred to as a **degenerate code**. The different codons specifying the same amino acid are called **synonymous codons**. Synonymous codons that differ from each other at the third position only are referred to as a **codon family**. For example, the four codons for valine (GUU, GUC, GUA, GUG) form a four-codon family. In contrast, the six codons for leucine are divided into a four-codon family (CUU, CUC, CUA, and CUG) and a two-codon family (UUA and UUG).

The first amino acid in most eukaryotic and archaebacterial proteins is a methionine encoded by the initiation codon AUG. This amino acid is usually removed in the mature protein. Most eubacterial genes also use the AUG codon for initiation, but the amino acid initiating the translation process is a methionine derivative called formylmethionine. Alternative initiation codons are known in both prokaryotes and eukaryotes. For example, in eubacteria, the codons GUG, UUG, and AUU, in order of decreasing frequency, are also used. In addition to AUG, the codons AUA, GUG, UUG, AUC, and AAG are used for initiation of translation in the yeast *Saccharomyces cerevisiae*.

The universal genetic code is also used in the independent process of translation employed by the genomes of plastids, such as the chloroplasts of vascular plants. In contrast, most animal mitochondrial genomes use codes that are different from the universal genetic code. One such example, the vertebrate mitochondrial code, is shown in Table 1.4. Note that two of the codons that specify serine in the universal genetic code are used as termination codons, and that tryptophan and methionine are each encoded by two codons rather than one.

A few prokaryotic genomes have been shown to use alternative genetic codes. For example, eubacterial species belonging to the genus *Mycoplasma* use UGA to code tryptophan. Deviations from the universal genetic code have also been observed in the nuclear genomes of a few eukaryotes. For example, ciliates belonging to the genera *Paramecium* and *Tetrahymena* use UAA and UAG to code for glutamine, and in the yeast *Candida cylindracea*, CUG codes for serine. As a rule, there are only minor differences between the universal genetic code and the alternative genetic codes.

In some organisms, some codons may never appear in protein-coding genes. These are called **absent codons**. For some codons, there is no appropriate tRNA that can pair with them. These are called **unassigned codons**. For example, in the genome of the Gram-positive eubacterium *Micrococcus luteus*, the codons AGA and AUA are unassigned (Kano et al. 1993). A similar situation was found for the codon CGG in the genome of *Mycoplasma capricolum* (Oba et al. 1991). Unassigned codons differ from stop codons by not being recognized by release factors. Thus, upon encountering an unassigned codon, translation stalls and the polypeptide remains attached to the ribosome.

TABLE 1.4 The vertebrate mitochondrial genetic code[a]

Codon	Amino acid	Codon	Amino acid	Codon	Amino acid	Codon	Amino acid
UUU	Phe	UCU	Ser	UAU	Tyr	UGU	Cys
UUC	Phe	UCC	Ser	UAC	Tyr	UGC	Cys
UUA	Leu	UCA	Ser	UAA	Stop	**UGA**	**Trp**
UUG	Leu	UCG	Ser	UAG	Stop	UGG	Trp
CUU	Leu	CCU	Pro	CAU	His	CGU	Arg
CUC	Leu	CCC	Pro	CAC	His	CGC	Arg
CUA	Leu	CCA	Pro	CAA	Gln	CGA	Arg
CUG	Leu	CCG	Pro	CAG	Gln	CGG	Arg
AUU	Ile	ACU	Thr	AAU	Asn	AGU	Ser
AUC	Ile	ACC	Thr	AAC	Asn	AGC	Ser
AUA	**Met**	ACA	Thr	AAA	Lys	**AGA**	**Stop**
AUG	Met	ACG	Thr	AAG	Lys	**AGG**	**Stop**
GUU	Val	GCU	Ala	GAU	Asp	GGU	Gly
GUC	Val	GCC	Ala	GAC	Asp	GGC	Gly
GUA	Val	GCA	Ala	GAA	Glu	GGA	Gly
GUG	Val	GCG	Ala	GAG	Glu	GGG	Gly

[a]Differences from the universal genetic code are shown in **boldface**.

MUTATION

DNA sequences are normally copied exactly during the process of replication. Rarely, however, errors in either DNA replication or repair occur giving rise to new sequences. These errors are called **mutations**. Mutations are the ultimate source of variation and novelty in evolution. Four aspects of the mutational process will be dealt with in this section: (1) the molecular mechanisms responsible for the creation of mutations, (2) the effects that each type of mutation has on the hereditary material and its products, (3) the spatial and temporal attributes of mutations, and (4) the randomness of mutations.

Mutations can occur in either somatic or germline cells. Since somatic mutations are not inherited, we can disregard them in an evolutionary context, and throughout this book the term "mutation" will be used exclusively to denote mutations in germline cells. Some organisms (e.g., plants) do not have a sequestered germline, so the distinction between somatic and germline mutations is not absolute.

Mutations may be classified by the length of the DNA sequence affected by the mutational event. For instance, mutations may affect a single nucleotide (**point mutations**) or several adjacent nucleotides (**segmental muta-**

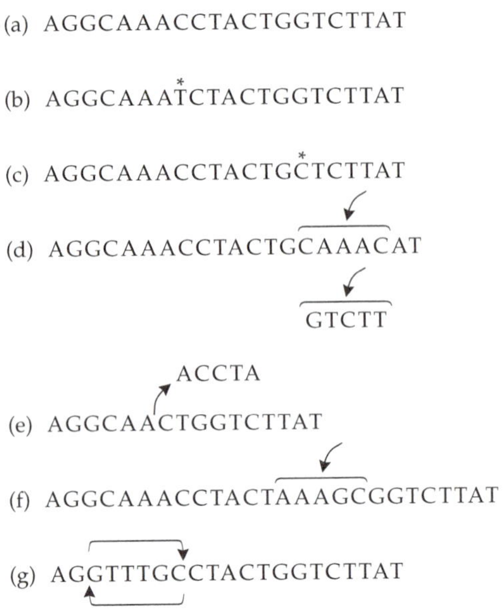

(a) AGGCAAACCTACTGGTCTTAT

(b) AGGCAAATCTACTGGTCTTAT

(c) AGGCAAACCTACTGCTCTTAT

(d) AGGCAAACCTACTGCAAACAT

GTCTT

ACCTA

(e) AGGCAACTGGTCTTAT

(f) AGGCAAACCTACTAAAGCGGTCTTAT

(g) AGGTTTGCCTACTGGTCTTAT

FIGURE 1.11 Types of mutations. (a) Original sequence. (b) Transition from C to T. (c) Transversion from G to C. (d) Recombination, the exchange of the sequence GTCTT by CAAAC. (e) Deletion of the sequence ACCTA. (f) insertion of the sequence AAAGC. (g) Inversion of 5′—GCAAAC—3′ to 5′—GTTTGC—3′.

tions). Mutations may also be classified by the type of change caused by the mutational event into: (1) **substitution mutations**, the replacement of one nucleotide by another; (2) **recombinations**, the exchange of a sequence with another; (3) **deletions**, the removal of one or more nucleotides from the DNA; (4) **insertions**, the addition of one or more nucleotides to the sequence; and (5) **inversions**, the rotation by 180° of a double-stranded DNA segment comprising two or more base pairs (Figure 1.11).

Substitution mutations

Substitution mutations are divided into transitions and transversions. **Transitions** are changes between A and G (purines), or between C and T (pyrimidines). **Transversions** are changes between a purine and a pyrimidine. There are four types of transitions, A → G, G → A, C → T, and T → C, and eight types of transversions, A → C, A → T, C → A, C → G, T → A, T → C, G → C, and G → T.

Substitution mutations occurring in protein-coding regions may be classified according to their effect on the product of translation—the protein. A mutation is said to be **synonymous** if it causes no change in the amino acid specified (Figure 1.12a). Otherwise it is **nonsynonymous**. A change in an amino acid due to a nonsynonymous nucleotide change is called a **replacement**. The

```
(a) Ile   Cys   Ile   Lys   Ala   Leu   Val   Leu   Leu   Thr
    ATA   TGT   ATA   AAG   GCA   CTG   GTC   CTG   TTA   ACA
                                          ↓
    ATA   TGT   ATA   AAG   GCA   CTG   GTA   CTG   TTA   ACA
    Ile   Cys   Ile   Lys   Ala   Leu   Val   Leu   Leu   Thr

(b) Ile   Cys   Ile   Lys   Ala   Asn   Val   Leu   Leu   Thr
    ATA   TGT   ATA   AAG   GCA   AAC   GTC   CTG   TTA   ACA
                                        ↓
    ATA   TGT   ATA   AAG   GCA   AAC   TTC   CTG   TTA   ACA
    Ile   Cys   Ile   Lys   Ala   Asn   Phe   Leu   Leu   Thr

(c) Ile   Cys   Ile   Lys   Ala   Asn   Val   Leu   Leu   Thr
    ATA   TGT   ATA   AAG   GCA   AAC   GTC   CTG   TTA   ACA
                      ↓
    ATA   TGT   ATA   TAG   GCAAACGTCCTGTTAACA
    Ile   Cys   Ile   Stop
```

FIGURE 1.12 Types of substitution mutations in a coding region: (a) synonymous, (b) missense, and (c) nonsense.

terms synonymous and **silent mutation** are often used interchangeably, because in the great majority of cases, synonymous changes do not alter the amino acid sequence of a protein and are therefore not detectable at the amino acid level. However, a synonymous mutation may not always be silent. It may, for instance, create a new splicing site or obliterate an existing one, thus turning an exonic sequence into an intron or vice versa, and causing a different polypeptide to be produced (Chapter 6). For example, a synonymous change from the glycine codon GGT to GGA in codon 25 of the first exon of β-globin has been shown to create a new splice junction, resulting in the production of a frameshifted protein of abnormal length (Goldsmith et al. 1983). Such a change is obviously not "silent." Therefore, it is advisable to distinguish between the two terms. Of course, all silent mutations in a protein-coding gene are synonymous.

Nonsynonymous or **amino acid-altering mutations** are further classified into missense and nonsense mutations. A **missense mutation** changes the affected codon into a codon that specifies a different amino acid from the one previously encoded (Figure 1.12b). A **nonsense mutation** changes a sense codon into a termination codon, thus prematurely ending the translation process and ultimately resulting in the production of a truncated protein (Figure 1.12c). Codons that can mutate to a termination codon by a single nucleotide change, e.g., UGC (Tyr), are called **pretermination codons**. Mutations in stop codons that cause translation to continue, sometimes up to the polyadenylation site, are also known.

TABLE 1.5 Relative frequencies of different types of mutational substitutions in a random protein-coding sequence

Substitution	Number	Percent
Total in all codons	549	100
Synonymous	134	25
Nonsynonymous	415	75
Missense	392	71
Nonsense	23	4
Total in first codons	183	100
Synonymous	8	4
Nonsynonymous	175	96
Missense	166	91
Nonsense	9	5
Total in second codons	183	100
Synonymous	0	0
Nonsynonymous	183	100
Missense	176	96
Nonsense	7	4
Total in third codons	183	100
Synonymous	126	69
Nonsynonymous	57	31
Missense	50	27
Nonsense	7	4

Each of the sense codons can mutate to nine other codons by means of a single nucleotide change. For example, CCU (Pro) can experience six nonsynonymous mutations, to UCU (Ser), ACU (Thr), GCU (Ala), CUU (Leu), CAU (His), or CGU (Arg), and three synonymous mutations, to CCC, CCA, or CCG. Since the universal genetic code consists of 61 sense codons, there are $61 \times 9 = 549$ possible substitution mutations. If we assume they occur with equal frequency, and that all codons are equally frequent in coding regions, we can compute the expected proportion of the different types of substitution mutations from the genetic code. These are shown in Table 1.5.

Because of the structure of the genetic code, synonymous changes occur mainly at the third position of codons. Indeed, almost 70% of all the possible nucleotide changes at the third position are synonymous. In contrast, all the nucleotide changes at the second position of codons are nonsynonymous, and so are most nucleotide changes at the first position (96%).

In the vast majority of cases, the exchange of a codon by a synonym, i.e., one that codes for the same amino acid, requires only one or at most two synonymous nucleotide changes. The only exception to this rule is the exchange of a serine codon belonging to the four-codon family (UCU, UCC, UCA, and UCG) by one belonging to the two-codon family (AGU and AGC). Such an event requires two nonsynonymous nucleotide changes.

Substitution mutations are thought to arise mainly from the mispairing of bases during DNA replication. As part of their original formulation of DNA replication, Watson and Crick (1953) suggested that transitions might be due to the formation of purine–pyrimidine mispairs (e.g., A:C), in which one of the bases assumes an unfavored tautomeric form, i.e., enol instead of keto in the case of guanine and thymine, or imino instead of amino in the case of adenine and cytosine (Figure 1.13). Later, Topal and Fresco (1976) proposed that purine–purine mispairs can also occur, but pyrimidine–pyrimidine mispairs cannot. The purine–pyrimidine mispairs are A*:C, A:C*, G*:T and G:T*, and the purine–purine mispairs are A*:A, A*:G, G*:A, and G*:G, in which the asterisk denotes an unfavored tautomeric form. Thus, there are two pathways via which substitution mutations can arise. In the first, transitions arise from purine–pyrimidine mispairing and can occur on either strand. For example, the transition A:T → G:C can arise from one of four possible mispairs: A*:C, A:C*, G:T*, or G*:T. In the second, transversions arise from purine–purine mispairing, but this can only occur if the purine resides on the template strand. For instance, the transversion A:T ↔ T:A can arise only from A*:A, where the unfavored tautomer A* is on the template strand.

Recombination

There are two types of homologous recombination: **crossing over** (**reciprocal recombination**) and **gene conversion** (**nonreciprocal recombination**). Reciprocal recombination involves the even exchange of homologous sequences between homologous chromosomes, thereby producing new combinations of adjacent sequences while at the same time retaining both variants involved in the recombination event. Nonreciprocal recombination, on the other hand, involves the uneven replacement of one sequence by another, a process resulting in the loss of one of the variant sequences involved in the recombination event. Both types of homologous recombination are thought to involve a molecular intermediate called the **Holliday structure** or **junction** (Figure 1.14). The resolution of the Holliday structure results in the formation of stretches of mismatched double-stranded DNA called **heteroduplexes**. The mismatches in the heteroduplex are recognized and excised by cellular enzymes. Then, using the complementary strand as a template, DNA polymerase fills the resulting gap. The way in which the Holliday structure is resolved (cut) and the manner in which the heteroduplex is repaired determine the outcome of the recombination (Figure 1.15).

Crossing over and gene conversion involve recombination between homologous sequences, and therefore belong to a class of molecular events called **homologous** or **generalized recombination**. **Site-specific recombination**, in contrast, involves the exchange of a sequence (usually a very short

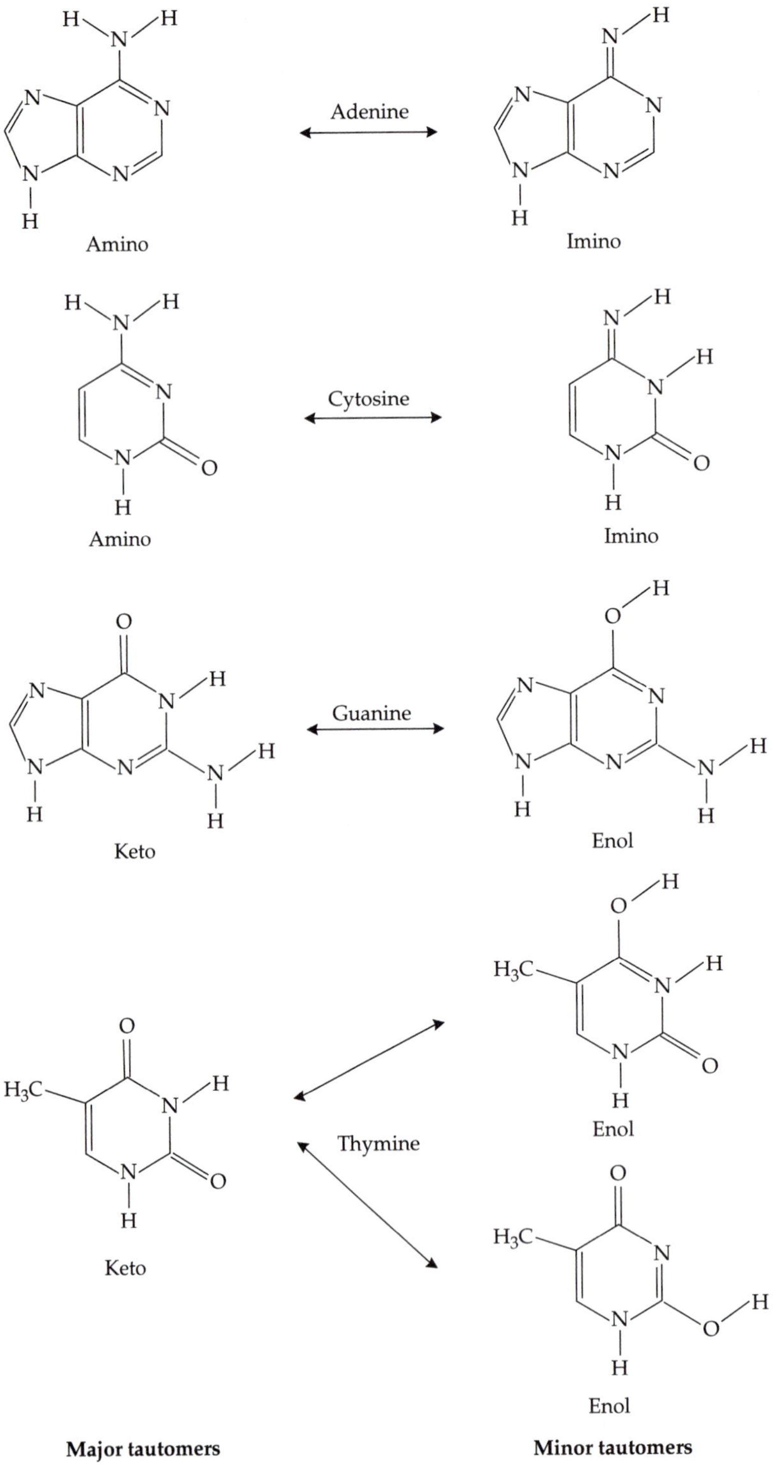

Adenine

Amino

Imino

Cytosine

Amino

Imino

Guanine

Keto

Enol

Thymine

Keto

Enol

Enol

Major tautomers

Minor tautomers

◀ **FIGURE 1.13 Amino ↔ imino and keto ↔ enol tautomers. Adenine and cytosine are usually found in the amino form, but rarely assume the imino configuration. Guanine and thymine are usually found in the keto form, but rarely form the enol configuration. Thymine has two enol tautomers. Each of the minor tautomers can assume two rotational forms (not shown). Modified from Mathews and van Holde (1990).**

one, comprising no more than a few nucleotides) by another (usually a long one that bears no similarity to the original sequence). Site-specific recombination is responsible, for instance, for the integration of phage genomes into bacterial chromosomes. From a mutational point of view, it is therefore proper to treat site-specific recombination as a type of insertion.

Reciprocal recombination in conjunction with substitution mutations is a powerful generator of genetic variability. For example, a recombination between sequences 5′—AACT—3′ and 5′—CACG—3′ may result in two new sequences: 5′—AACG—3′ and 5′—CACT—3′. The more alleles there are, the more alleles will come into being through recombination, and the rate of generating new genetic variants may become quite high. Golding and Strobeck (1983) referred to this phenomenon as "variation begets variation."

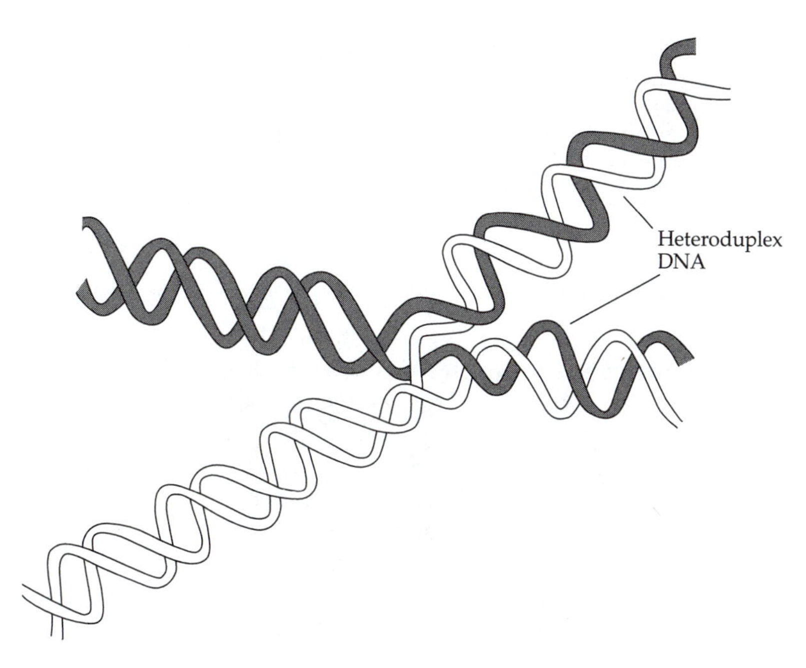

FIGURE 1.14 The Holliday structure. Note the heteroduplex DNA composed of mismatched chromatid strands (gray and white strands).

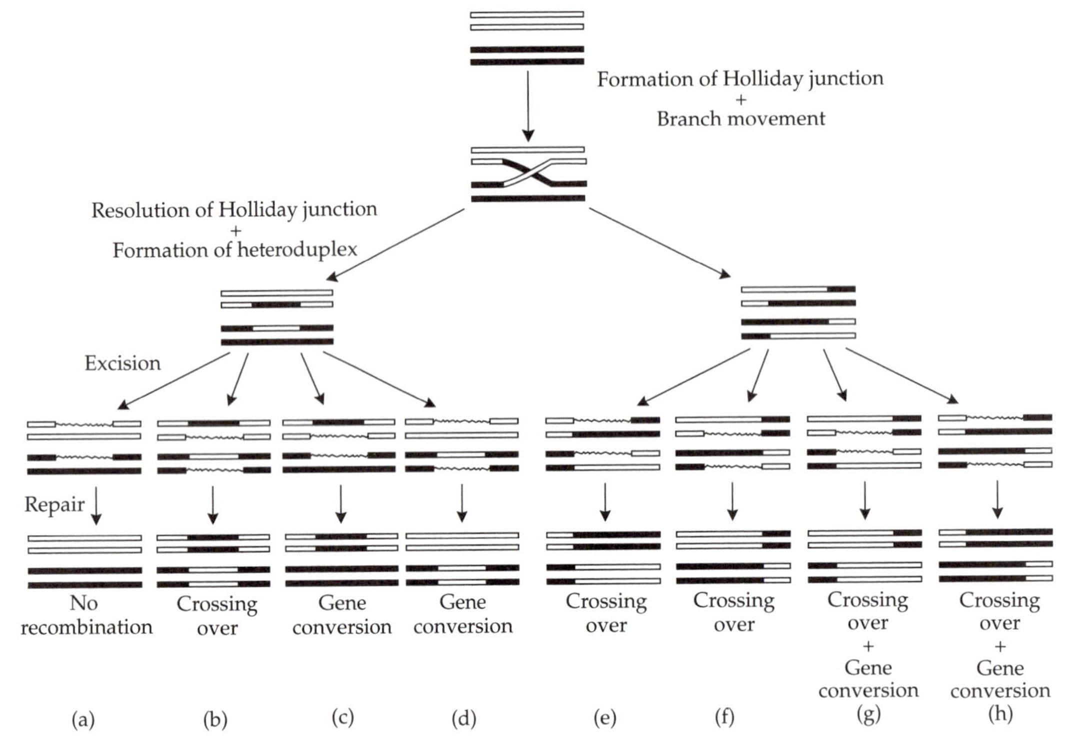

FIGURE 1.15 Possible outcomes of the resolution of a Holliday junction and the subsequent excision and mismatch repair of heteroduplex DNA. Each "ribbon" represents one strand of a double helix. Double-stranded regions of black and white strands denote mismatches. Wavy lines denote the location of the excision and mismatch repair. Note that depending on the type of resolution and the choice of strands for excision and repair, we obtain either no recombination (a), crossing over (b, e, f), gene conversion (c, d), or crossing over plus gene conversion (g, h).

Deletions and insertions

Deletions and insertions can occur by several mechanisms. One mechanism is **unequal crossing over**. Figure 1.16a shows a simple model, in which unequal crossing over between two chromosomes results in the deletion of a DNA segment in one chromosome and a reciprocal addition in the other. The chance of unequal crossing over is greatly increased if a DNA segment is duplicated in tandem (Figure 1.16b) because of the higher probability of misalignment.

Another mechanism is **intrastrand deletion**, a type of site-specific recombination that arises when a repeated sequence pairs with another in the same orientation on the same chromatid, and an intrachromosomal crossing over event occurs as a consequence (Figure 1.17). For example, in *Escherichia coli*, spontaneous deletions in the *lacI* gene often appear to be due to intrastrand recombination involving small regions of similarity. The precise excision of transposable elements frequently involves recombination between direct re-

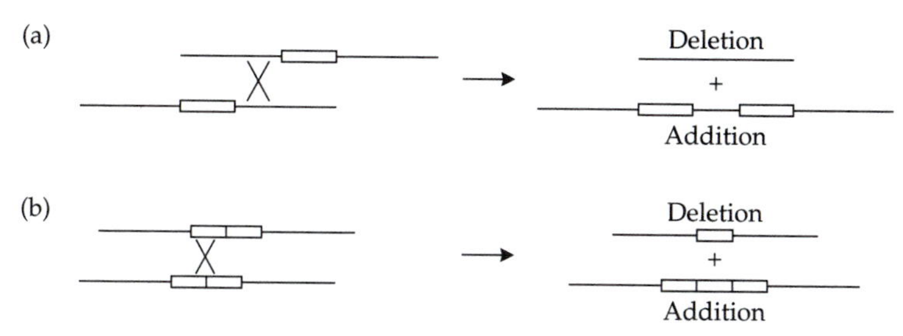

FIGURE 1.16 (a) Unequal crossing over resulting in the deletion of a DNA sequence in one of the daughter strands and the duplication of the same sequence in the other strand. (b) When a DNA segment is duplicated in tandem, the chance of misalignment increases, as does the chance of unequal crossing over. A box denotes a particular stretch of DNA. Modified from Li (1997).

peats, 5–9 base pairs long, flanking the element (Chapter 7). Similarly, intrastrand deletion is responsible for the reduction in the number of repeats in tandem arrays, such as simple repetitive DNA and satellite DNA (Chapter 8).

A third mechanism is **replication slippage**, or **slipped-strand mispairing.** This type of event occurs in DNA regions that contain contiguous short re-

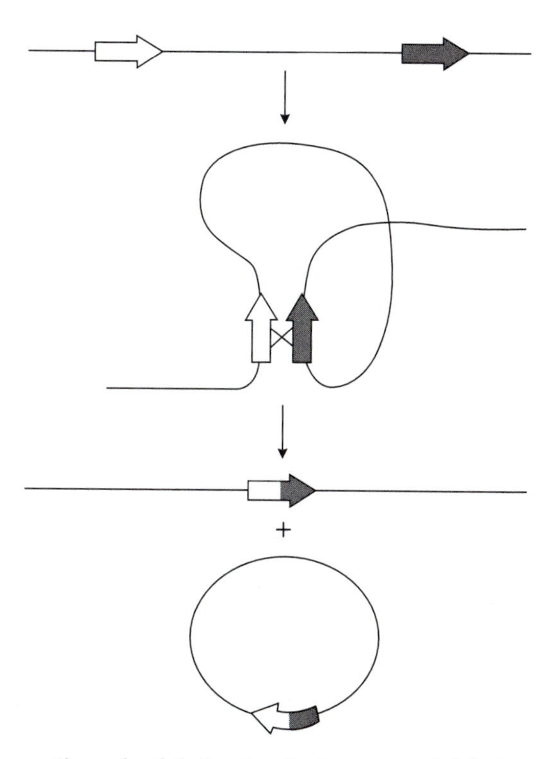

FIGURE 1.17 Generation of a deletion by the intrastrand deletion process. The repeated sequences (arrows) that are oriented in the same direction recombine (×) to produce a genomic deletion (left) and an extrachromosomal element (right).

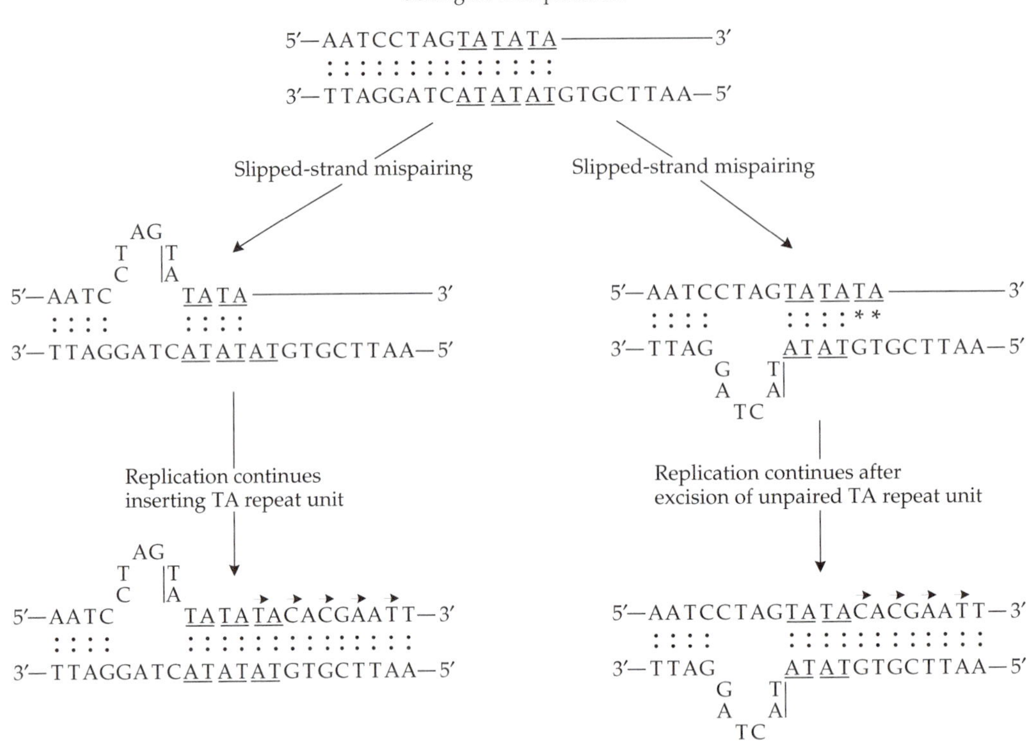

peats. Figure 1.18a shows that, during DNA replication, slippage can occur because of mispairing between neighboring repeats, and that slippage can result in either deletion or duplication of a DNA segment, depending on whether the slippage occurs in the $5' \rightarrow 3'$ direction or in the opposite direction. Figure 1.18b shows that slipped-strand mispairing can also occur in nonreplicating DNA. A fourth mechanism responsible for the insertion or deletion of DNA sequences is DNA transposition (Chapter 7).

Deletions and insertions are collectively referred to as **indels** (short for **in**sertion-or-**del**etion), because when two sequences are compared, it is impossible to tell whether a deletion has occurred in one or an insertion has occurred in the other (Chapter 3). The number of nucleotides in an indel ranges from one or a few nucleotides to contiguous stretches involving thousands of nucleotides. The lengths of the indels essentially exhibit a bimodal type of frequency distribution, with short indels (up to 20–30 nucleotides) most often being caused by errors in the process of DNA replication, such as the slipped-strand mispairing discussed above, and with long insertions or deletions occurring mainly because of unequal crossing over, site-specific recombination, DNA transposition, or horizontal gene transfer (Chapter 7).

In a coding region, an indel that is not a multiple of three nucleotides causes a shift in the reading frame so that the coding sequence downstream of

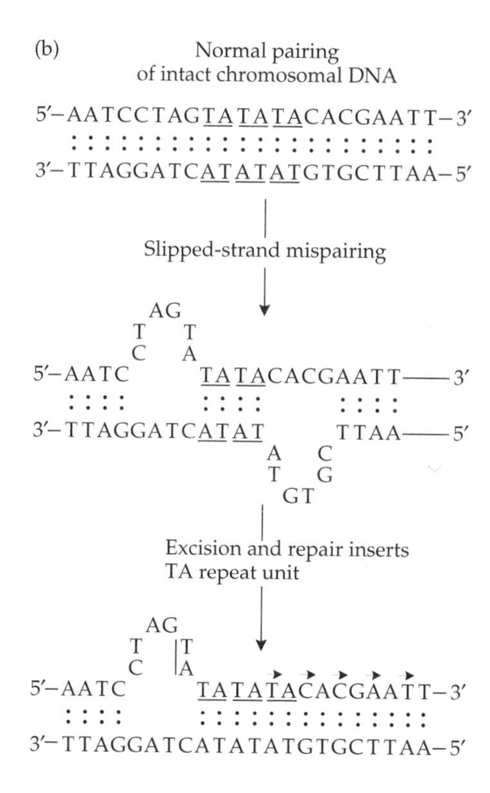

FIGURE 1.18 Generation of duplications or deletions by slipped-strand mispairing between contiguous repeats (underlined). Small arrows indicate direction of DNA synthesis. Dots indicate pairing. (a) A two-base slippage in a TA repeat region during DNA replication. Slippage in the 3′ → 5′ direction results in the insertion of one TA repeat unit (left panel). Slippage in the other direction (5′ → 3′) results in the deletion of one repeat unit (right panel). The deletion shown in the right panel results from excision of the unpaired repeat unit (asterisks) at the 3′ end of the growing strand, presumably by the 3′ → 5′ exonuclease activity of DNA polymerase. (b) A two-base slippage in a TA repeat region in nonreplicating DNA. Mismatched regions form single-stranded loops, which may be targets of excision and mismatch repair. The outcome (a deletion or an insertion) will depend on which strand is excised and repaired and which strand is used as template in the DNA repair process. Modified from Levinson and Gutman (1987).

the gap will be read in the wrong phase. Such a mutation is known as a **frameshift mutation**. Consequently, indels may not only introduce numerous amino acid changes, but may also obliterate the termination codon or bring into phase a new stop codon, resulting in a protein of abnormal length (Figure 1.19).

Inversions

Inversion is a type of DNA rearrangement that can occur following either of two processes: (1) chromosome breakage and rejoining, or (2) intrachromosomal crossing over between two homologous segments that are oriented in opposite directions (Figure 1.20). The vast majority of known inversions involve very long stretches of DNA, usually hundreds or thousands of nucleotides in length.

Mutation rates

Because of the rarity of mutations, the rate of spontaneous mutation is very difficult to determine directly, and at the present time only a few such estimates exist at the DNA sequence level (e.g., Drake et al. 1998). The rate of mutation, however, can be estimated indirectly from the rate of substitution in pseudogenes (Chapter 4). Li et al. (1985a) and Kondrashov and Crow (1993) estimated the average rate of mutation in mammalian nuclear DNA to be 3 to 5×10^{-9} nucleotide substitutions per nucleotide site per year. The mutation rate, however,

(a) Lys Ala Leu Val Leu Leu Thr Ile Cys Ile Stop
AAG GCA CTG GTC CTG TTA ACA ATA TGT ATA **TAA** TACCATCGCAATATGAAAATC
↓
G

AAG GCA CTG TCC TGT **TAA** CAATATGTATATAATACCATCGCAATATGAAAATC
Lys Ala Leu Phe Cys Stop

(b) Lys Ala Leu Val Leu Leu Thr Ile Cys Ile Stop
AAG GCA CTG GTC CTG TTA ACA ATA TGT ATA **TAA** TACCATCGCAATATGAAAATC
↓
A

AAG GCA CTG GTC CTG TTA ACA ATA TGT ATT AAT ACC ATC GCA ATA **TGA** AAA
Lys Ala Leu Val Leu Leu Thr Ile Cys Ile Asn Thr Ile Ala Ile Stop

(c) Lys Ala Axn Val Leu Leu Thr Ile Cys Ile Stop
AAG GCA AAC GTC CTG TTA ACA ATA TGT ATA **TAA** TACCATCGCAATAGGG
↑
G

AAG GCA AAC GGT CCT GTT AAC AAT ATG TAT ATA ATA CCA TCG CAA **TAG** GG
Lys Ala Asn Gly Pro Val Asn Asn Met Tyr Ile Ile Pro Ser Gln Stop

(d) Lys Ala Asn Val Leu Leu Thr Ile Cys Ile Stop
AAG GCA AAC GTC CTG TTA ACA ATA TGT ATA **TAA** TACCATCGCAATAGGG
↑
GA

AAG GCA AAC GAG TCC TGT **TAA** CAATATGTATATAATACCATCGCAATAGGG
Lys Ala Asn Glu Ser Cys Stop

FIGURE 1.19 Examples of frameshifts in reading frames. (a) Deletion of a G causes premature termination. (b) Deletion of an A obliterates a stop codon. (c) Insertion of a G obliterates a stop codon. (d) Insertion of the dinucleotide GA causes premature termination of translation. Stop codons are shown in bold type.

varies enormously with genomic region; in microsatellites, for instance, the rate of insertion and deletion in humans was estimated to be greater than 10^{-3} (Brinkmann et al. 1998).

The rate of mutation in mammalian mitochondrial DNA has been estimated to be at least 10 times higher than the average nuclear rate (Brown et al. 1982). RNA viruses have error-prone polymerases (i.e., lacking proofreading mechanisms), and they lack an efficient postreplication repair mechanism of mutational damage. Thus, their rate of mutation is several orders of magnitude higher than that of organisms with DNA genomes. Gojobori and Yokoyama (1985), for instance, estimated the rates of mutation in the influenza A virus and the Moloney murine sarcoma virus to be on the order of

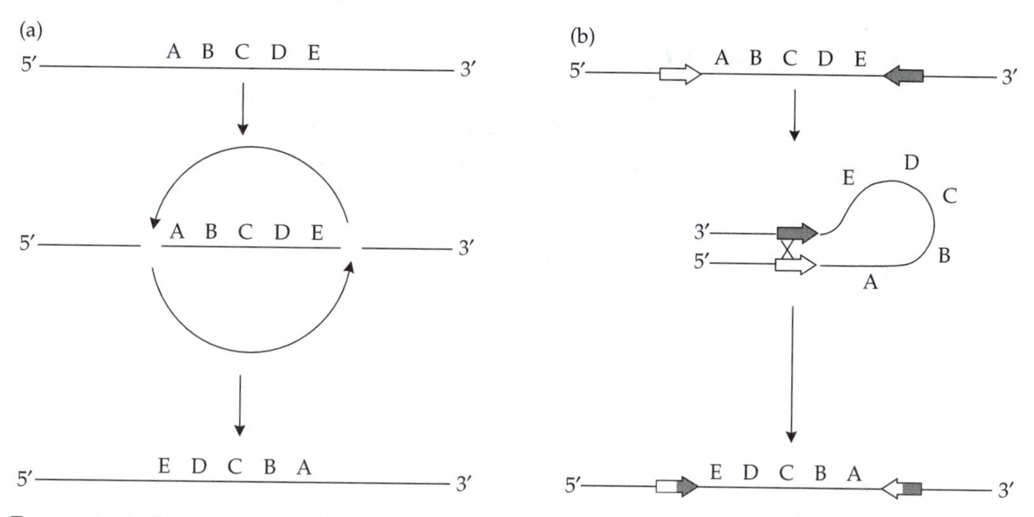

FIGURE 1.20 Mechanisms of inversion. (a) Chromosome breakage and rejoining. (b) Crossing over (×) between homologous segments (arrows) on the same chromosome that are oriented in opposite directions results in an inversion involving the DNA sequence between the homologous inverted repeats.

10^{-2} mutations per nucleotide site per year, i.e., approximately 2 million times higher than the rate of mutation in the nuclear DNA of vertebrates. The rate of mutation in the Rous sarcoma virus may be even higher, as 9 mutations were detected out of 65,250 replicated nucleotides, i.e., a rate of 1.4×10^{-4} mutations per nucleotide per replication cycle (Leider et al. 1988).

There are as yet only few direct estimates of the rates of deletion, insertion, recombination, and inversion in the literature (e.g., Sommer 1995).

Spatial distribution of mutations

Mutations do not occur randomly throughout the genome. Some regions are more prone to mutate than others, and they are called **hotspots** of mutation. One such hotspot is the dinucleotide 5′—CG—3′ (often denoted as CpG), in which the cytosine is frequently methylated in many animal genomes, changing it to 5′—TG—3′. The dinucleotide 5′—TT—3′ is a hotspot of mutation in prokaryotes but usually not in eukaryotes. In bacteria, regions within the DNA containing short palindromes (i.e., sequences that read the same on the complementary strand, such as 5′—GCCGGC—3′, 5′—GGCGCC—3′, and 5′—GGGCCC—3′) were found to be more prone to mutate than other regions. In eukaryotic genomes, short tandem repeats are often hotspots for deletions and insertions, probably as a result of slipped-strand mispairing. Sequences containing runs of alternating purine–pyrimidine dimers, such as GC, AT, and GT, are capable of adopting a left-handed DNA conformation called Z-DNA, which constitutes a hotspot for deletions involving even numbers of base pairs (Freund et al. 1989).

Patterns of mutation

The direction of mutation is nonrandom. In particular, transitions were found to occur more frequently than transversions. In animal nuclear DNA, transitions were found to account for about 60–70% of all mutations, whereas the proportion of transitions under random mutation is expected to be only 33%. Thus, in animal nuclear genomes, transitional mutations occur twice as frequently as transversions. In animal mitochondrial genomes, the ratio of transitions to transversions is about 15 to 20. Some nucleotides are more mutable than others. For example, in nuclear DNA of mammals, G and C tend to mutate more frequently than A and T.

Are mutations random?

Mutations are commonly said to occur "randomly." However, as we have seen, mutations do not occur at random with respect to genomic location, nor do all types of mutations occur with equal frequency. So, what aspect of mutation is random? Mutations are claimed to be random in respect to their effect on the fitness of the organism carrying them (Chapter 2). That is, any given mutation is expected to occur with the same frequency under conditions in which this mutation confers an advantage on the organism carrying it, as under conditions in which this mutation confers no advantage or is deleterious. "It may seem a deplorable imperfection of nature," said Dobzhansky (1970), "that mutability is not restricted to changes that enhance the adeptness of their carriers." And indeed, the issue of whether mutations are random or not with respect to their effects on fitness is periodically debated in the literature, sometimes with fierce intensity (see, e.g., Hall 1990; Lenski and Mittler 1993; Rosenberg et al. 1994; and Sniegowski 1995).

FURTHER READINGS

Alberts B., D. Bray, J. Lewis, M. Raff, K. Roberts, and J. D. Watson. 1995. *Molecular Biology of the Cell*, 3rd Ed. Garland, New York.

Bränden, C.-I. and J. Tooze. 1991. *Introduction to Protein Structure*. Garland, New York.

Griffiths, A. J. F., J. H. Miller, and D. T. Suzuki. 1996. *An Introduction to Genetic Analysis*, 6th Ed. Freeman, New York.

Lewin, B. 1997. *Genes VI*. Oxford University Press, New York.

Lodish, H., D. Baltimore, A. Berk. S. L. Zipursky, P. Matsudaira, and J. Darnell. 1995. *Molecular Cell Biology*, 3rd Ed. Freeman, New York.

Portin, P. 1993. The concept of the gene: Short history and present status. Q. Rev. Biol. 68: 173–223.

Sinden, R. R., C. E. Pearson, V. N. Potaman, and D. W. Ussery. 1998. DNA: Structure and function. Adv. Genome Biol. 5A: 1–141.

Stryer, L. 1995. *Biochemistry*, 4th Ed. Freeman, New York.

Chapter 2

Dynamics of Genes in Populations

Evolution is the process of change in the genetic makeup of populations. Consequently, the most basic component of the evolutionary process is the change in gene frequencies with time. Population genetics deals with genetic changes that occur within populations. In this chapter we review some basic principles of population genetics that are essential for understanding molecular evolution.

A basic problem in evolutionary population genetics is to determine how the frequency of a mutant allele will change in time under the effect of various evolutionary forces. In addition, from the long-term point of view, it is important to determine the probability that a new mutant variant will completely replace an old one in the population, and to estimate how fast and how frequent the replacement process will be. Unlike morphological changes, many molecular changes are likely to have only a small effect on the phenotype and, consequently, on the fitness of the organism. Thus, the frequencies of molecular variants are subject to strong chance effects, and the element of chance should be taken into account when dealing with molecular evolution.

Toward the end of this chapter, we introduce three types of evolutionary explanations—mutationism, neutralism, and selectionism—and discuss briefly the issue of polymorphism within populations. Finally, we use some of the predictions of the neutral mutation hypothesis concerning the pattern of distribution of polymorphic sites to characterize the driving forces in molecular evolution.

CHANGES IN ALLELE FREQUENCIES

The chromosomal or genomic location of a gene is called a **locus**, and alternative forms of the gene at a given locus are called **alleles**. In a population, more than one allele may be present at a locus. The relative proportion of an allele is referred to as the **allele frequency** or **gene frequency**. For example, let us assume that in a haploid population of size N individuals, two alleles, A_1 and A_2, are present at a certain locus. Let us further assume that the number of copies of allele A_1 in the population is n_1, and the number of copies of allele A_2 is n_2. Then the allele frequencies are equal to n_1/N and n_2/N for alleles A_1 and A_2, respectively. Note that $n_1 + n_2 = N$, and $n_1/N + n_2/N = 1$. The set of all alleles existing in a population at all loci is called the **gene pool**.

Evolution is the process of change in the genetic makeup of populations. For a new mutation to become significant from an evolutionary point of view, it must increase in frequency and ultimately become fixed in the population (i.e., all the individuals in a subsequent generation will share the same mutant allele). If it does not increase in frequency, a mutation will have little effect on the evolutionary history of the species. For a mutant allele to increase in frequency, factors other than mutation must come into play. The major factors affecting the frequency of alleles in populations are natural selection and random genetic drift. In classical evolutionary studies involving morphological traits, natural selection has been considered as the major driving force in evolution. In contrast, random genetic drift is thought to have played an important role in evolution at the molecular level.

There are two mathematical approaches to studying genetic changes in populations: deterministic and stochastic. The **deterministic model** is simpler. It assumes that changes in the frequencies of alleles in a population from generation to generation occur in a unique manner and can be unambiguously predicted from knowledge of initial conditions. Strictly speaking, this approach applies only when two conditions are met: (1) the population is infinite in size, and (2) the environment either remains constant with time or changes according to deterministic rules. These conditions are obviously never met in nature, and therefore a purely deterministic approach may not be sufficient to describe the temporal changes in allele frequencies. Random or unpredictable fluctuations in allele frequencies must also be taken into account, and dealing with random fluctuations requires a different mathematical approach. **Stochastic models** assume that changes in allele frequencies occur in a probabilistic manner. That is, from knowledge of the conditions in one generation, one cannot predict unambiguously the allele frequencies in the next generation; one can only determine the probabilities with which certain allele frequencies are likely to be attained.

Obviously, stochastic models are preferable to deterministic ones, since they are based on more realistic assumptions. However, deterministic models are much easier to treat mathematically and, under certain circumstances, they yield sufficiently accurate approximations. The following discussion deals with natural selection in a deterministic fashion.

NATURAL SELECTION

Natural selection is defined as the differential reproduction of genetically distinct individuals or **genotypes** within a population. Differential reproduction is caused by differences among individuals in such traits as mortality, fertility, fecundity, mating success, and the viability of the offspring. Natural selection is predicated on the availability of genetic variation among individuals in characters related to reproductive success. When a population consists of individuals that do not differ from one another in such traits, it is not subject to natural selection. Selection may lead to changes in allele frequencies over time. However, a mere change in allele frequencies from generation to generation does not necessarily indicate that selection is at work. Other processes, such as random genetic drift (discussed later in this chapter), can bring about temporal changes in allele frequencies as well. Interestingly, the opposite is also true: a lack of change in allele frequencies does not necessarily indicate that selection is absent.

The **fitness** of a genotype, commonly denoted as w, is a measure of the individual's ability to survive and reproduce. However, since the size of a population is usually constrained by the carrying capacity of the environment in which the population resides, the evolutionary success of an individual is determined not by its **absolute fitness**, but by its **relative fitness** in comparison to the other genotypes in the population. In nature, the fitness of a genotype is not expected to remain constant for all generations and under all environmental circumstances. However, by assigning a constant value of fitness to each genotype, we are able to formulate simple theories or models, which are useful for understanding the dynamics of change in the genetic structure of a population brought about by natural selection. In the simplest class of models, we assume that the fitness of the organism is determined solely by its genetic makeup. We also assume that all loci contribute independently to the fitness of the individual (i.e., that the different loci do not interact with one another in any manner that affects the fitness of the organism), so that each locus can be dealt with separately.

Mutations occurring in the genic fraction of the genome may or may not alter the organism's phenotype. In case they do, they may or may not affect the fitness of the organism that carries the mutation. Most new mutations arising in a population reduce the fitness of their carriers. Such mutations are called **deleterious**, and they will be selected against and eventually removed from the population. This type of selection is called **negative** or **purifying selection**. Occasionally, a new mutation may be as fit as the best allele in the population. Such a mutation is selectively **neutral**, and its fate is not determined by selection. In exceedingly rare cases, a mutation may arise that increases the fitness of its carriers. Such a mutation is called **advantageous**, and it will be subjected to **positive** or **advantageous selection**.

In the following, we shall consider the case of one locus with two alleles, A_1 and A_2. Each allele can be assigned an intrinsic fitness value; it can be advantageous, deleterious, or neutral. However, this assignment is only applicable to haploid organisms. In diploid organisms, fitness is ultimately determined by the interaction between the two alleles at a locus. With two alleles, there are three possible diploid genotypes: A_1A_1, A_1A_2, and A_2A_2, and their fitnesses can be denoted by w_{11}, w_{12}, and w_{22}, respectively. Given that the frequency of allele A_1 in a population is p, and the frequency of the complementary allele, A_2, is $q = 1 - p$, we can show that, under random mating, the frequencies of the A_1A_1, A_1A_2, and A_2A_2 genotypes are p^2, $2pq$, and q^2, respectively. A population in which such genotypic ratios are maintained is said to be at **Hardy-Weinberg equilibrium**. Note that $p = p^2 + 1/2(2pq) = p^2 + pq$, and $q = 1/2(2pq) + q^2 = q^2 + pq$. In the general case, the three genotypes are assigned the following fitness values and initial frequencies:

Genotype	A_1A_1	A_1A_2	A_2A_2
Fitness	w_{11}	w_{12}	w_{22}
Frequency	p^2	$2pq$	q^2

Let us now consider the dynamics of allele frequency changes following selection. The relative contribution of each genotype to the subsequent generation is the product of its initial frequency and its fitness. Given the frequencies of the three genotypes and their fitnesses as above, the relative contributions of the three genotypes to the next generation will be p^2w_{11}, $2pqw_{12}$ and q^2w_{22}, for A_1A_1, A_1A_2 and A_2A_2, respectively. Since half of the alleles carried by A_1A_2 individuals and all of the alleles carried by A_2A_2 individuals are A_2, the frequency of allele A_2 in the next generation (q_{t+1}) will become

$$q_{t+1} = \frac{pqw_{12} + q^2w_{22}}{p^2w_{11} + 2pqw_{12} + q^2w_{22}} \tag{2.1}$$

The extent of change in the frequency of allele A_2 per generation is denoted as Δq. We can show that

$$\Delta q = q_{t+1} - q_t = \frac{pq\left[p(w_{12} - w_{11}) + q(w_{22} - w_{12})\right]}{p^2w_{11} + 2pqw_{12} + q^2w_{22}} \tag{2.2}$$

In the following, we shall assume that A_1 is the original or "old" allele in the population. We shall also assume that the population is diploid, and therefore the initial population consists of only one genotype, A_1A_1. We then consider the dynamics of change in allele frequencies following the appearance of a new mutant allele, A_2, and the consequent creation of two new genotypes, A_1A_2 and A_2A_2. For mathematical convenience, we shall assign a relative fitness value of 1 to the A_1A_1 genotype. The fitness of the newly created genotypes, A_1A_2 and A_2A_2, will depend on the mode of interaction between A_1 and A_2. For example, if A_2 is completely dominant over A_1, then w_{11}, w_{12}, and w_{22} will be written as 1, $1 + s$, and $1 + s$, respectively, where s is the difference between the fitness of an A_2-carrying genotype and the fitness of A_1A_1. A positive value of s denotes an increase in fitness in comparison with A_1A_1 (**selective advantage**), while a negative value denotes a decrease in fitness (**selective disadvantage**). When $s = 0$, the fitness of an A_2-carrying genotype will be the same as that of A_1A_1 (**selective neutrality**).

Five common modes of interaction will be considered: (1) codominance or genic selection, (2) complete recessiveness, (3) complete dominance, (4) overdominance, and (5) underdominance.

Codominance

In **codominant** or **genic selection**, the two homozygotes have different fitness values, whereas the fitness of the heterozygote is the mean of the fitnesses of the two homozygous genotypes. The relative fitness values for the three genotypes can be written as

Genotype	A_1A_1	A_1A_2	A_2A_2
Fitness	1	$1 + s$	$1 + 2s$

From Equation 2.2, we obtain the following change in the frequency of allele A_2 per generation under codominance:

$$\Delta q = \frac{spq}{1 + 2spq + 2sq^2} \tag{2.3}$$

By iteration, Equation 2.3 can be used to compute the frequency of A_2 in any generation. However, the following approximation leads to a much more convenient solution. Note that if s is small, as is usually the case, the denominator in Equation 2.3 is approximately 1, and the equation is reduced to $\Delta q = spq$, which can be approximated by the differential equation

$$\frac{dq}{dt} = spq = sq(1 - q) \tag{2.4}$$

The solution of Equation 2.4 is given by

$$q_t = \frac{1}{1 + \left(\frac{1 - q_0}{q_0}\right) e^{-st}} \tag{2.5}$$

where q_0 and q_t are the frequencies of A_2 in generations 0 and t, respectively.

In Equation 2.5, the frequency q_t is expressed as a function of time t. Alternatively, t can be expressed as a function of q as

$$t = \frac{1}{s} \ln \frac{q_t(1 - q_0)}{q_0(1 - q_t)} \tag{2.6}$$

where ln denotes the natural logarithmic function. From this equation, one can calculate the number of generations required for the frequency of A_2 to change from one value (q_0) to another (q_t).

Figure 2.1a illustrates the increase in the frequency of allele A_2 for $s = 0.01$. We see that codominant selection always increases the frequency of one allele at the expense of the other, regardless of the relative allele frequencies in the population. Therefore, genic selection is a type of **directional selection**. Note, however, that at low frequencies, selection for a codominant allele is not very efficient (i.e., the change in allele frequencies is slow). The reason is that at low

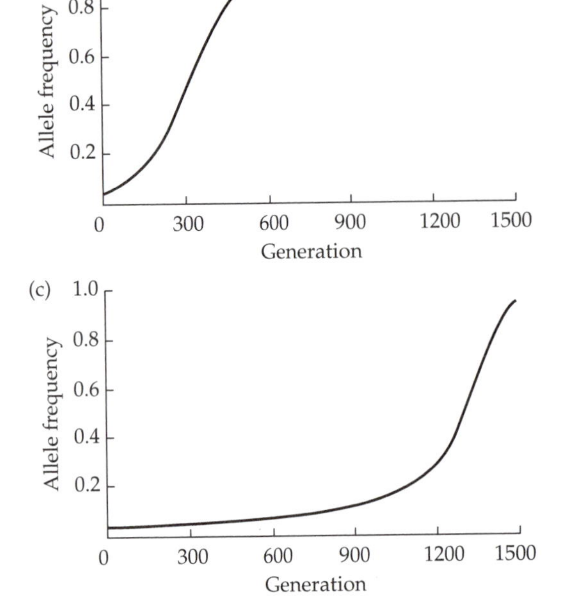

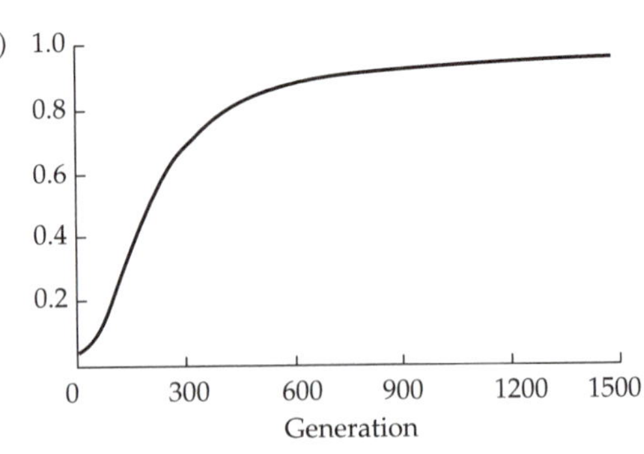

FIGURE 2.1 Changes in the frequency of an advantageous allele A_2 under (a) codominant selection, (b) dominant selection where A_2 is dominant over A_1, and (c) dominant selection where A_1 is dominant over A_2. In all cases, the frequency of A_2 at time 0 is $q = 0.04$, and the selective advantage is $s = 0.01$.

frequencies, the proportion of A_2 alleles residing in heterozygotes is large. For example, when the frequency of A_2 is 0.5, 50% of the A_2 alleles are carried by heterozygotes, whereas when the frequency of A_2 is 0.01, 99% of all A_2 alleles reside in heterozygotes. Because heterozygotes, which contain both alleles, are subject to weaker selective pressures than are A_2A_2 homozygotes (s versus $2s$), the overall change in allele frequencies at low values of q will be small.

Dominance

In **dominant selection**, the two homozygotes have different fitness values, whereas the fitness of the heterozygote is the same as the fitness of one of the two homozygous genotypes. In the following, we distinguish between two cases. In the first case, the new allele, A_2, is **dominant** over the old allele, A_1, and the fitness of the heterozygote, A_1A_2, is identical to the fitness of the A_2A_2 homozygote. The relative fitness values of the three genotypes are

Genotype	A_1A_1	A_1A_2	A_2A_2
Fitness	1	$1+s$	$1+s$

From Equation 2.2, we obtain the following change in the frequency of the A_2 allele per generation:

$$\Delta q = \frac{p^2 qs}{1 - s - p^2 s} \tag{2.7}$$

In the second case, A_1 is dominant over A_2, and the fitness of the heterozygote, A_1A_2, is identical to that of the A_1A_1 homozygote. That is, the new allele, A_2, is **recessive**. The relative fitness values are now

Genotype	A_1A_1	A_1A_2	A_2A_2
Fitness	1	1	$1 + s$

From Equation 2.2, we obtain the following change in the frequency of A_2 per generation:

$$\Delta q = \frac{pq^2 s}{1 + q^2 s} \tag{2.8}$$

Figures 2.1b–c illustrate the increase in the frequency of allele A_2 for $s = 0.01$ for dominant and recessive A_2 alleles, respectively. Both types of selection are also directional because the frequency of A_2 increases at the expense of A_1, regardless of the allele frequencies in the population.

In Figure 2.1b, the new allele is advantageous and dominant over the old one, so selection is very efficient and the frequency of A_2 increases rapidly. If the new dominant allele is deleterious, it will rapidly be eliminated from the population; this is why dominant deleterious alleles are seldom seen in natural populations. (A phenotypically dominant pathological allele such as Huntington chorea is neither dominant nor deleterious from an evolutionary point of view, since the age of onset for the disease is in the postreproductive period, and therefore the fitness of carriers of the Huntington allele is not significantly lower than that of individuals who do not carry it.)

Again we note that the less frequent allele A_2 is, the larger the proportion of A_2 alleles in a heterozygous state will be. Thus, as far as selection against recessive homozygotes is concerned, at low frequencies only very few alleles in each generation will be subject to selection, and their frequency in the population will not change much. In other words, selection will not be very efficient in increasing the frequency of beneficial alleles or, conversely, in reducing the frequency of deleterious ones, as long as the frequency of the allele in a population is low. For this reason, it is almost impossible to rid a population of recessive deleterious alleles, even when such alleles are lethal in a homozygous state (i.e., $s = -1$). The persistence of such genetic diseases as Tay-Sachs syndrome and cystic fibrosis in human populations attests to the inefficiency of directional selection at low allele frequencies. As we shall see in the next section, however, some recessive deleterious alleles may be maintained in the population because they have a selective advantage in the heterozygous state.

Overdominance and underdominance

In **overdominant selection**, the heterozygote has the highest fitness. Thus,

Genotype	A_1A_1	A_1A_2	A_2A_2
Fitness	1	$1 + s$	$1 + t$

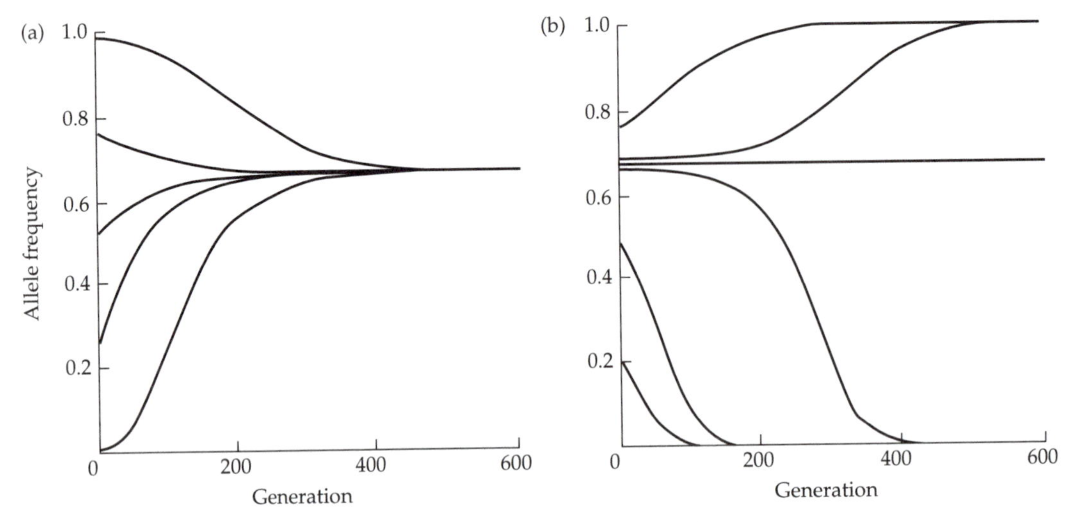

FIGURE 2.2 (a) Changes in the frequency of an allele subject to overdominant selection. Initial frequencies from top to bottom curves: 0.99, 0.75, 0.50, 0.25, and 0.01; $s = 0.04$ and $t = 0.02$. Since the s and t values are exceptionally large, the change in allele frequency is rapid. Note that there is a stable equilibrium at $q = 0.667$. (b) Changes in the frequency of an allele subject to underdominant selection. Initial frequencies from top to bottom curves: 0.75, 0.668, 0.667, 0.666, 0.50, and 0.20; $s = -0.02$ and $t = -0.01$. Again, because of the large values of s and t, the change in allele frequency is rapid. Note that there is an equilibrium at $q = 0.667$. This equilibrium, however, is unstable, since even the slightest deviation from it will cause one of the alleles to be eliminated from the population.

In this case, $s > 0$ and $s > t$. Depending on whether the fitness of A_2A_2 is higher than, equal to, or lower than that of A_1A_1, t can be positive, zero, or negative, respectively. The change in allele frequencies is expressed as

$$\Delta q = \frac{pq(2sq - tq - s)}{1 + 2spq + tq^2} \quad \textbf{(2.9)}$$

Figure 2.2a illustrates the changes in the frequency of allele A_2.

In contrast to the codominant or dominant selection regimes, in which allele A_1 is eventually eliminated from the population, under overdominant selection the population sooner or later reaches a **stable equilibrium** in which alleles A_1 and A_2 coexist. The equilibrium is stable because in case of a deviation from equilibrium, selection will quickly restore the equilibrium frequencies. After equilibrium is reached, no further change in allele frequencies will be observed (i.e., $\Delta q = 0$). Thus, overdominant selection belongs to a class of selection regimes called **balancing** or **stabilizing selection**.

The frequency of allele A_2 at equilibrium, $\hat{q}$, is obtained by solving Equation 2.9 for $\Delta q = 0$:

$$\hat{q} = \frac{s}{2s - t} \quad \textbf{(2.10)}$$

When $t = 0$ (i.e., both homozygotes have identical fitness values), the equilibrium frequencies of both alleles will be 50%.

In **underdominant selection**, the heterozygote has the lowest fitness, i.e., $s < 0$ and $s < t$. As in the case of overdominant selection, here too the change in the frequency of A_2 is described by Equation 2.9 and the frequency of allele A_2 at equilibrium by Equation 2.10. However, in this case the population reaches an **unstable equilibrium**, i.e., any deviation from the equilibrium frequencies in Equation 2.10 will drive the allele below the equilibrium frequency to extinction, while the other allele will be fixed (Figure 2.2b). An allele may thus be eliminated from the population even if the fitness of its homozygote (A_2A_2) is much higher than that of the prevailing homozygote (A_1A_1). Since a new allele arising in the population by mutation always has a very low frequency (i.e., much below the equilibrium frequency), new underdominant mutants will always be eliminated from the population. Underdominant selection is thus a powerful illustration of the fallacy of intuitive evolutionary thinking, according to which an allele that increases the fitness of its carriers will always replace one that does not.

RANDOM GENETIC DRIFT

As noted above, natural selection is not the only factor that can cause changes in allele frequencies. Allele frequency changes can also occur by chance, in which case the changes are not directional but random. An important factor in producing random fluctuations in allele frequencies is the random sampling of gametes in the process of reproduction (Figure 2.3). Sampling occurs because, in the vast majority of cases in nature, the number of gametes available in any generation is much larger than the number of adult individuals produced in the next generation. In other words, only a minute fraction of gametes succeeds in developing into adults. In a diploid population subject to Mendelian segregation, sampling can still occur even if there is no excess of gametes, i.e., even if each individual contributes exactly two gametes to the next generation. The reason is that heterozygotes can produce two types of gametes, but the two gametes passing on to the next generation may by chance be of the same type.

To see the effect of sampling, let us consider an idealized situation in which all individuals in the population have the same fitness and selection does not operate. We further simplify the problem by considering a population with nonoverlapping generations (e.g., a group of individuals that reproduce simultaneously and die immediately afterwards), such that any given generation can be unambiguously distinguished from both previous and subsequent generations. Finally, we assume that the population size does not change from generation to generation.

The population under consideration is diploid and consists of N individuals, so that at any given locus, the population contains $2N$ genes. Let us again consider the simple case of one locus with two alleles, A_1 and A_2, with frequencies p and $q = 1 - p$, respectively. When $2N$ gametes are sampled from

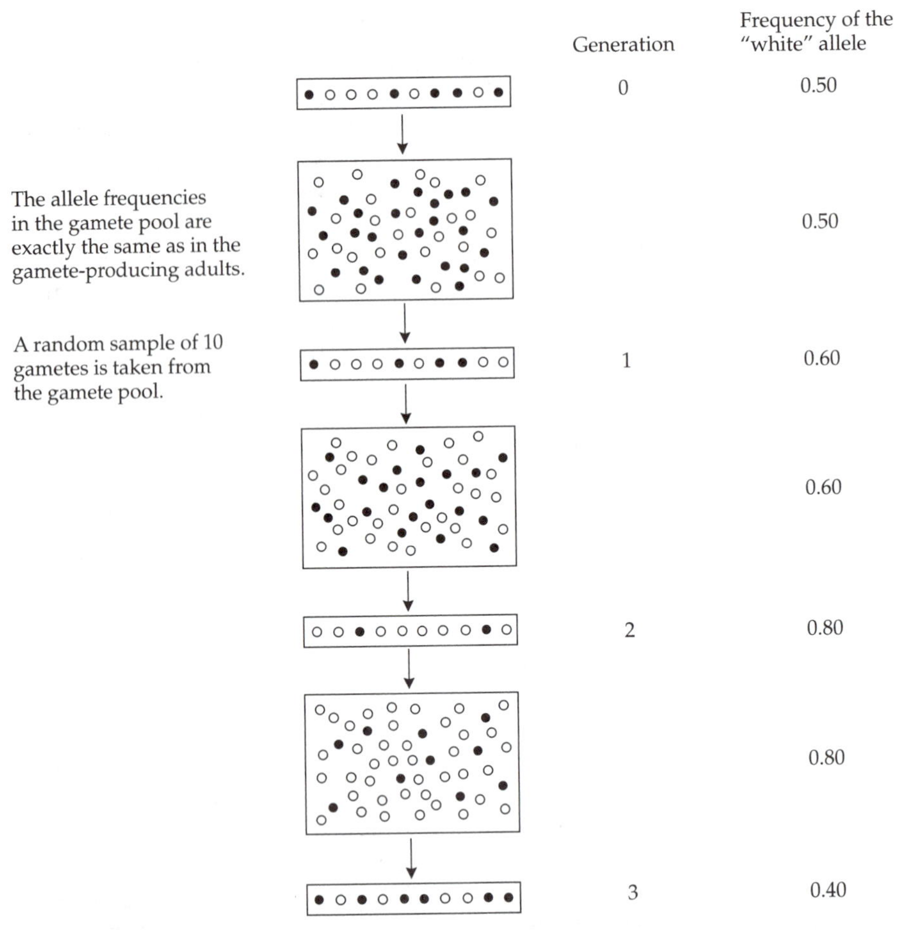

FIGURE 2.3 **Random sampling of gametes. Allele frequencies in the gamete pools (large boxes) in each generation are assumed to reflect exactly the allele frequencies in the adults of the parental generation (small boxes). Since the population size is finite, allele frequencies fluctuate up and down. Modified from Bodmer and Cavalli-Sforza (1976).**

the infinite gamete pool, the probability, P_i, that the sample contains exactly i alleles of type A_1 is given by the binomial probability function

$$P_i = \frac{(2N)!}{i!(2N-i)!} p^i q^{2N-i} \quad \textbf{(2.11)}$$

where ! denotes the factorial, and $(2N)! = 1 \times 2 \times 3 \times ... \times (2N)$. Since P_i is always greater than 0 for populations in which the two alleles coexist (i.e., $0 < p < 1$), the allele frequencies may change from generation to generation without the aid of selection.

The process of change in allele frequency due solely to chance effects is called **random genetic drift**. One should note, however, that random genetic

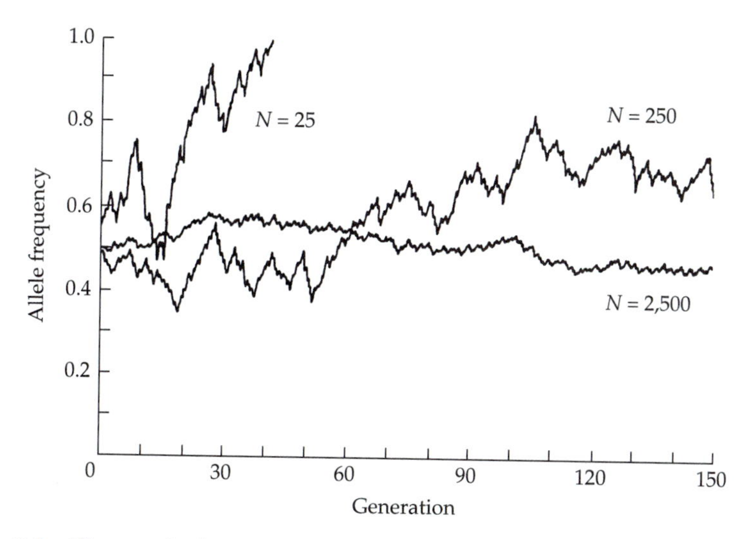

FIGURE 2.4 Changes in frequencies of alleles subject to random genetic drift in populations of different sizes. The smallest population (N = 25) reached fixation after 42 generations. The other two populations were still polymorphic after 150 generations, but will ultimately reach fixation if the experiment is continued long enough. From Bodmer and Cavalli-Sforza (1976).

drift can also be caused by processes other than the sampling of gametes. For example, stochastic changes in selection intensity can also bring about random changes in allele frequencies (Gillespie 1991).

In Figure 2.4 we illustrate the effects of random sampling on the frequencies of alleles in populations of different sizes. The allele frequencies change from generation to generation, but the direction of the change is random at any point in time. The most obvious feature of random genetic drift is that fluctuations in allele frequencies are much more pronounced in small populations than in larger ones. In Figure 2.5 we show two possible outcomes of random genetic drift in a population of size $N = 25$. In one replicate, allele A_2 is lost in generation 27; in the other replicate, allele A_1 is lost in generation 49.

Let us follow the dynamics of change in allele frequencies due to the process of random genetic drift in succeeding generations. The frequencies of allele A_1 are written as $p_0, p_1, p_2, ..., p_t$, where the subscripts denote the generation number. The initial frequency of allele A_1 is p_0. In the absence of selection, we expect p_1 to be equal to p_0, and so on for all subsequent generations. The fact that the population is finite, however, means that p_1 will be equal to p_0 only on the average (i.e., when repeating the sampling process an infinite number of times). In reality, sampling occurs only once in each generation, and p_1 is usually different from p_0. In the second generation, the frequency p_2 will no longer depend on p_0 but only on p_1. Similarly, in the third generation, the frequency p_3 will depend on neither p_0 nor p_1 but only on p_2. Thus, the most important property of random genetic drift is its cumulative behavior:

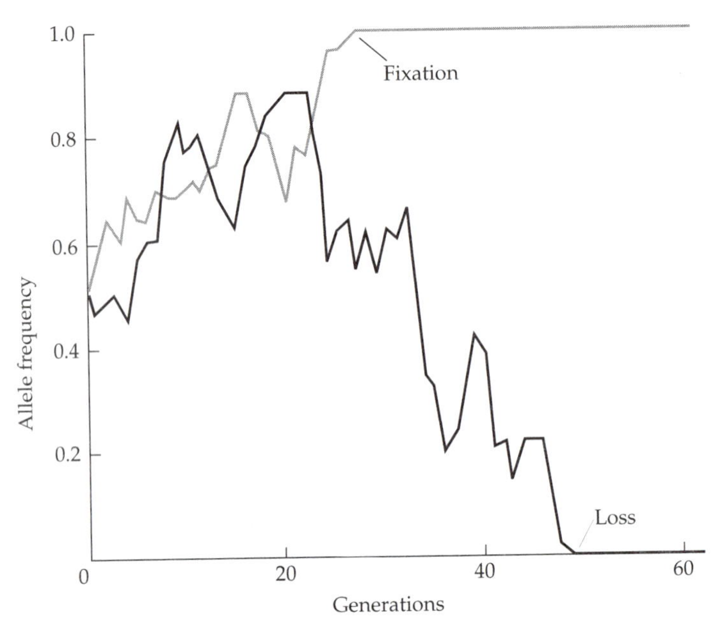

FIGURE 2.5 Two possible outcomes of random genetic drift in populations of size 25 and $p_0 = 0.5$. In each generation, 25 alleles were sampled with replacement from the previous generation. In the population represented by the gray line, the allele becomes fixed in generation 27; in the other population, the allele is lost in generation 49. Modified from Li (1997).

from generation to generation, the frequency of an allele will tend to deviate more and more from its initial frequency.

In mathematical terms, the mean and variance of the frequency of allele A_1 at generation t, denoted by $\bar{p}_t$ and $V(p_t)$, respectively, are given by

$$\bar{p}_t = p_0 \tag{2.12}$$

and

$$V(p_t) = p_0(1-p_0)\left[1-\left(1-\frac{1}{2N}\right)^t\right] \approx p_0(1-p_0)\left[1-e^{-t/(2N)}\right] \tag{2.13}$$

where p_0 denotes the initial frequency of A_1. Note that although the mean frequency does not change with time, the variance increases with time; that is, with each passing generation the allele frequencies will tend to deviate further and further from their initial values. However, the change in allele frequencies will not be systematic in its direction.

To see the cumulative effect of random genetic drift, let us consider the following numerical example. A certain population is composed of five diploid individuals in which the frequencies of the two alleles at a locus, A_1 and A_2, are each 50%. We now ask, What is the probability of obtaining the same allele frequencies in the next generation? By using Equation 2.11, we obtain a probability of 25%. In other words, in 75% of the cases the allele frequencies in the second generation will be different from the initial allele frequencies. Moreover, the probability of retaining the initial allele frequencies in subsequent generations will no longer be 0.25, but will become progressively smaller. For example, the probability of having an equal number of alleles A_1 and A_2 in the population in the third generation (i.e., $p_3 = q_3 = 0.5$) is about 18%. The probability drops to only about 5% in the tenth generation (Figure 2.6). Concomitantly, the probability of either A_1 or A_2 being lost increases with time, because in every generation there is a finite probability that all the chosen gametes happen to carry the same allele. In the above example, the probability of one of the alleles being lost is already about 0.1% in the first generation, and this probability increases dramatically in subsequent generations.

Once the frequency of an allele reaches either 0 or 1, its frequency will not change in subsequent generations. The first case is referred to as **loss** or **extinction**, and the second **fixation**. If the process of sampling continues for long periods of time, the probability of such an eventuality reaches certainty. Thus, the ultimate result of random genetic drift is the fixation of one allele and the loss of all the others. This will happen unless there is a constant input of alleles into the population by such processes as mutation or migration, or unless polymorphism is actively maintained by a balancing type of selection.

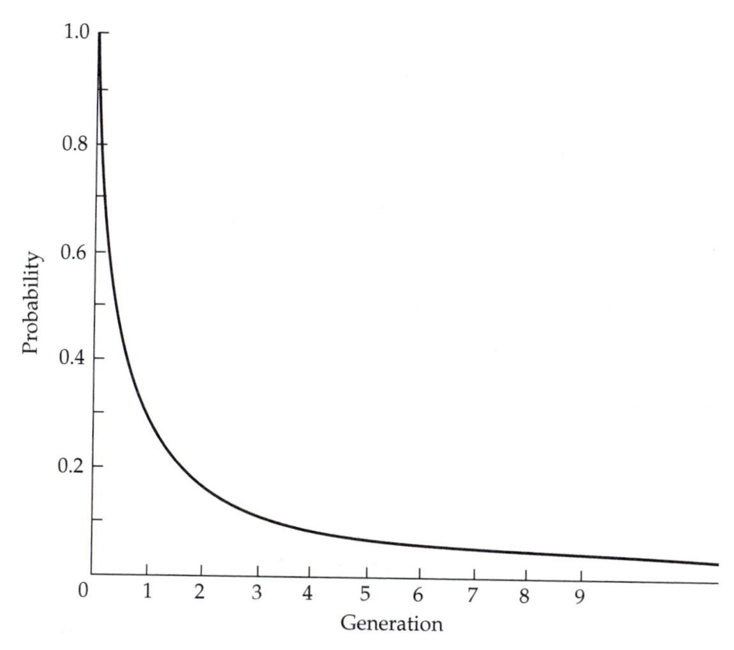

FIGURE 2.6 Probability of maintaining the same initial allele frequencies over time for two selectively neutral alleles. $N = 5$ and $p = 0.5$.

EFFECTIVE POPULATION SIZE

A basic parameter in population biology is the **census population size**, N, defined as the total number of individuals in a population. From the point of view of population genetics and evolution, however, the relevant number of individuals to be considered consists of only those individuals that actively participate in reproduction. Since not all individuals take part in reproduction, the population size that matters in the evolutionary process is different from the census size. This part is called the **effective population size** and is denoted by N_e. Wright (1931) introduced the concept of effective population size, which he rigorously defined as the size of an idealized population that would have the same effect of random sampling on allele frequencies as that of the actual population.

Consider, for instance, a population with a census size N, and assume that the frequency of allele A_1 at generation t is p. If the number of individuals taking part in reproduction is N, then the variance of the frequency of allele A_1 in the next generation, p_{t+1}, may be obtained from Equation 2.13 by setting $t = 1$.

$$V(p_{t+1}) = \frac{p_t(1-p_t)}{2N} \tag{2.14}$$

In practice, since not all individuals in the population take part in the reproductive process, the observed variance will be larger than that obtained from Equation 2.14. The effective population size is the value that is substituted for N in order to satisfy Equation 2.14, i.e.,

$$V(p_{t+1}) = \frac{p_t(1-p_t)}{2N_e} \tag{2.15}$$

In general, N_e is smaller, sometimes much smaller, than N. For example, the effective population size of the mosquito *Anopheles gambiae* in Kenya has been estimated to be about 2,000—about six orders of magnitude smaller than the census population size (Lehmann et al. 1998).

Various factors can contribute to this difference. For example, in a population with overlapping generations (and in particular if the variation in the number of offspring among individuals is great), at any given time part of the population will consist of individuals in either their prereproductive or postreproductive stage. Due to this age stratification, the effective size can be considerably smaller than the census size. For example, according to Nei and Imaizumi (1966), in humans, N_e is only slightly larger than $N/3$.

Reduction in the effective population size in comparison to the census size can also occur if the number of males involved in reproduction is different from the number of females. This is especially pronounced in polygamous species, such as social mammals and territorial birds, or in species in which a nonreproductive caste exists (e.g., social bees, ants, termites, and naked mole rats). If a population consists of N_m males and N_f females ($N = N_m + N_f$), N_e is given by

$$N_e = \frac{4N_m N_f}{N_m + N_f} \quad (2.16)$$

Note that unless the number of females taking part in the reproductive process equals that of males, N_e will always be smaller than N. For an extreme example, let us assume that in a population of size N in which the sexes are equal in number, all females ($N/2$) but only one male take part in the reproductive process. From Equation 2.16, we get $N_e = 2N/(1 + N/2)$. If N is considerably larger than 1, such that $N/2 + 1 \approx N/2$, then N_e becomes 4, regardless of the census population size.

The effective population size can also be much reduced due to long-term variations in the population size, which in turn are caused by such factors as environmental catastrophes, cyclical modes of reproduction, and local extinction and recolonization events. The long-term effective population size in a species for a period of n generations is given by

$$N_e = \frac{n}{\frac{1}{N_1} + \frac{1}{N_2} + \cdots + \frac{1}{N_n}} \quad (2.17)$$

where N_i is the population size of the ith generation. In other words, N_e equals the harmonic mean of the N_i values, and consequently it is closer to the smallest value of N_i than to the largest one. Similarly, if a population goes through a bottleneck, the long-term effective population size is greatly reduced even if the population has long regained its pre-bottleneck census size. Many estimates of the long-term (i.e., ~2 million years) effective population size in humans have been published. Most estimates converge on an N_e value of about 10,000 (Li and Sadler 1991; Takahata 1993; Hammer 1995; Takahata et al. 1995; Harding et al. 1997; Sherry et al. 1997; Clark et al. 1998).

GENE SUBSTITUTION

Gene substitution is defined as the process whereby a mutant allele completely replaces the predominant or **wild type** allele in a population. In this process, a mutant allele arises in a population as a single copy and becomes fixed after a certain number of generations. Not all mutants, however, reach fixation. In fact, the majority of them are lost after a few generations. Thus, we need to address the issue of **fixation probability** and discuss the factors affecting the chance that a new mutant allele will reach fixation in a population.

The time it takes for a new allele to become fixed is called the **fixation time**. In the following we shall identify the factors that affect the speed with which a new mutant allele replaces an old one in a population.

New mutations arise continuously within populations. Consequently, gene substitutions occur in succession, with one allele replacing another and being itself replaced in time by a newer allele. Thus, we can speak of the **rate of gene substitution**, i.e., the number of fixations of new alleles per unit time.

Fixation probability

The probability that a particular allele will become fixed in a population depends on (1) its frequency, (2) its selective advantage or disadvantage, s, and (3) the effective population size, N_e. In the following, we shall consider the case of genic selection and assume that the relative fitness of the three genotypes A_1A_1, A_1A_2, and A_2A_2 are 1, $1 + s$, and $1 + 2s$, respectively.

Kimura (1962) showed that the probability of fixation of A_2 is

$$P = \frac{1 - e^{-4N_e sq}}{1 - e^{-4N_e s}} \quad \textbf{(2.18)}$$

where q is the initial frequency of allele A_2. Since $e^{-x} \approx 1 - x$ for small values of x, Equation 2.18 reduces to $P \approx q$ as s approaches 0. Thus, for a neutral allele, the fixation probability equals its frequency in the population. For example, a neutral allele with a frequency of 40% will become fixed in 40% of the cases and will be lost in 60% of the cases. This is intuitively understandable because in the case of neutral alleles, fixation occurs by random genetic drift, which favors neither allele.

We note that a new mutant arising as a single copy in a diploid population of size N has an initial frequency of $1/(2N)$. The probability of fixation of an individual mutant allele, P, is thus obtained by replacing q with $1/(2N)$ in Equation 2.18. When $s \neq 0$,

$$P = \frac{1 - e^{-(2N_e s/N)}}{1 - e^{-4N_e s}} \quad \textbf{(2.19)}$$

For a neutral mutation, i.e., $s = 0$, Equation 2.19 becomes

$$P = \frac{1}{2N} \quad \textbf{(2.20)}$$

If the population size is equal to the effective population size, Equation 2.19 reduces to

$$P = \frac{1 - e^{-2s}}{1 - e^{-4Ns}} \quad \textbf{(2.21)}$$

If the absolute value of s is small, we obtain

$$P = \frac{2s}{1 - e^{-4Ns}} \quad \textbf{(2.22)}$$

For positive values of s and large values of N, Equation 2.22 reduces to

$$P \approx 2s \quad \textbf{(2.23)}$$

Thus, if an advantageous mutation arises in a large population and its selective advantage over the rest of the alleles is small, say up to 5%, the probability of its fixation is approximately twice its selective advantage. For example, if a new codominant mutation with $s = 0.01$ arises in a population, the probability of its eventual fixation is 2%.

Let us now consider a numerical example. A new mutant arises in a population of 1,000 individuals. What is the probability that this allele will become fixed in the population if (1) it is neutral, (2) it confers a selective advantage of 0.01, or (3) it has a selective disadvantage of 0.001? For simplicity, we assume that $N = N_e$. For the neutral case, the probability of fixation as calculated by using Equation 2.20 is 0.05%. From Equations 2.23 and 2.21, we obtain probabilities of 2% and 0.004% for the advantageous and deleterious mutations, respectively. These results are noteworthy, since they essentially mean that an advantageous mutation does not always become fixed in the population. In fact, 98% of all the mutations with a selective advantage of 0.01 will be lost by chance. This theoretical finding is of great importance, since it shows that the perception of adaptive evolution as a process in which advantageous mutations arise in populations and invariably take over the population in subsequent generations is a naive concept. Moreover, even deleterious mutations have a finite probability of becoming fixed in a population, albeit a small one. However, the mere fact that a deleterious allele may become fixed in a population at the expense of "better" alleles illustrates in a powerful way the importance of chance events in determining the fate of mutations during evolution.

As the population size becomes larger, the chance effects become smaller. For instance, in the above example, if the effective population size is 10,000 rather than 1,000, then the fixation probabilities become 0.005%, 2%, and $\sim 10^{-20}$, for neutral, advantageous, and deleterious mutations, respectively. Thus, while the fixation probability for the advantageous mutation remains approximately the same, that for the neutral mutation becomes smaller, and that for the deleterious allele becomes indistinguishable from zero.

Fixation time

The time required for the fixation or loss of an allele depends on (1) the frequency of the allele, (2) its selective advantage or disadvantage, and (3) the size of the population. The mean time to fixation or loss becomes shorter as the frequency of the allele approaches 1 or 0, respectively.

When dealing with new mutations, it is more convenient to treat fixation and loss separately. In the following, we deal with the mean fixation time of those mutants that will eventually become fixed in the population. This variable is called the **conditional fixation time**. In the case of a new mutation whose initial frequency in a diploid population is by definition $q = 1/(2N)$, the mean conditional fixation time, $\bar{t}$, was calculated by Kimura and Ohta (1969). For a neutral mutation, it is approximated by

$$\bar{t} = 4N \text{ generations} \qquad \textbf{(2.24)}$$

and for a mutation with a selective advantage of s, it is approximated by

$$\bar{t} = (2/s)\ln(2N) \text{ generations} \qquad \textbf{(2.25)}$$

To illustrate the difference between different types of mutation, let us assume that a mammalian species has an effective population size of about 10^6 and a mean generation time of 2 years. Under these conditions, it will take a neutral mutation, on average, 8 million years to become fixed in the population. In comparison, a mutation with a selective advantage of 1% will become fixed in the same population in only about 5,800 years. Interestingly, the conditional fixation time for a deleterious allele with a selective disadvantage $-s$ is the same as that for an advantageous allele with a selective advantage s (Maruyama and Kimura 1974). This is intuitively understandable given the high probability of loss for a deleterious allele. That is, for a deleterious allele to become fixed in a population, fixation must occur very quickly.

In Figure 2.7, we present in a schematic manner the dynamics of gene substitution for advantageous and neutral mutations. We note that advantageous mutations are either rapidly lost or rapidly fixed in the population. In contrast, the frequency changes for neutral alleles are slow, and the fixation time is much longer than that for advantageous mutants.

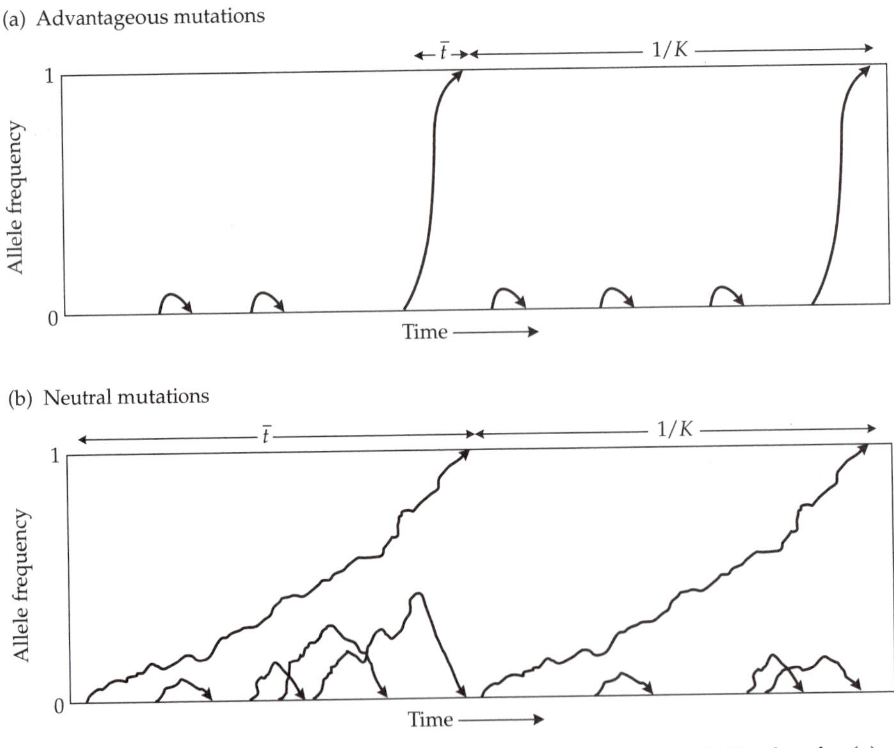

FIGURE 2.7 Schematic representation of the dynamics of gene substitution for (a) advantageous and (b) neutral mutations. Advantageous mutations are either quickly lost from the population or quickly fixed, so that their contribution to genetic polymorphism is small. The frequency of neutral alleles, on the other hand, changes very slowly by comparison, so that a large amount of transient polymorphism is generated. The conditional fixation time is $\bar{t}$, and $1/K$ is the mean time between two consecutive fixation events. Modified from Nei (1987).

Rate of gene substitution

Let us now consider the **rate of gene substitution**, defined as the number of mutants reaching fixation per unit time. We shall first consider neutral mutations. If neutral mutations occur at a rate of u per gene per generation, then the number of mutants arising at a locus in a diploid population of size N is $2Nu$ per generation. Since the probability of fixation for each of these mutations is $1/(2N)$, we obtain the rate of substitution of neutral alleles by multiplying the total number of mutations by the probability of their fixation:

$$K = 2Nu\left(\frac{1}{2N}\right) = u \tag{2.26}$$

Thus, for neutral mutations, the rate of substitution is equal to the rate of mutation—a remarkably simple and important result (Kimura 1968b). This result can be intuitively understood by noting that, in a large population, the number of mutations arising every generation is high but the fixation probability of each mutation is low. In comparison, in a small population, the number of mutations arising every generation is low, but the fixation probability of each mutation is high. As a consequence, the rate of substitution for neutral mutations is independent of population size.

For advantageous mutations, the rate of substitution can also be obtained by multiplying the rate of mutation by the probability of fixation for advantageous alleles as given in Equation 2.23. For genic selection with $s > 0$, we obtain

$$K = 4Nsu \tag{2.27}$$

In other words, the rate of substitution for the case of genic selection depends on the population size (N) and the selective advantage (s), as well as on the rate of mutation (u).

The inverse of K (i.e., $1/K$) is the mean time between two consecutive fixation events (Figure 2.7).

GENETIC POLYMORPHISM

A population is **monomorphic** at a locus if there exists only one allele at the locus. A locus is said to be **polymorphic** if two or more alleles coexist in the population. However, if one of the alleles has a very high frequency, say 99% or more, then none of the other alleles are likely to be observed in a sample unless the sample size is very large. Thus, for practical purposes, a locus is commonly defined as polymorphic only if the frequency of the most common allele is less than 99%. This definition is obviously arbitrary, and in the literature one may find thresholds other than 99%.

Gene diversity

One of the simplest ways to measure the extent of polymorphism in a population is to compute the average proportion of polymorphic loci (P) by dividing the number of polymorphic loci by the total number of loci sampled. For example, if 4 of the 20 loci are polymorphic, then $P = 4/20 = 0.20$. This measure, however, is dependent on the number of individuals studied, because the smaller the sample size, the more difficult it is to identify polymorphic loci as such.

A more appropriate measure of genetic variability is the **mean expected heterozygosity**, or **gene diversity**. This measure (1) does not depend on an arbitrary delineation of polymorphism, (2) can be computed directly from knowledge of the allele frequencies, and (3) is less affected by sampling effects. Gene diversity at a locus, or **single-locus expected heterozygosity**, is defined as

$$h = 1 - \sum_{i=1}^{m} x_i^2 \tag{2.28}$$

where x_i is the frequency of allele i and m is the total number of alleles at the locus. For any given locus, h is the probability that two alleles chosen at random from the population are different from each other. The average of the h values over all the loci studied, H, can be used as an estimate of the extent of genetic variability within the population. That is,

$$H = \frac{1}{n}\sum_{i=1}^{n} h_i \tag{2.29}$$

where h_i is the gene diversity at locus i, and n is the number of loci.

As we have seen previously, random genetic drift is an anti-polymorphic force in evolution. Thus, gene diversity is expected to decrease under random genetic drift. Wright (1942) and Kimura (1955) have shown that, in the absence of mutational input, gene diversity will decrease by a fraction of $1/2N_e$ each generation, where N_e is the effective population size.

Nucleotide diversity

The gene diversity measures h and H are used extensively for electrophoretic and restriction enzyme data. However, they are mostly unsuitable for DNA sequence data, since the extent of genetic variation at the DNA level in nature is quite extensive. In particular, when long sequences are considered, each sequence in the sample is likely to be different by one or more nucleotides from the other sequences, and in most cases both h and H will be close to 1. Thus,

(a)
G **A** G G T G C A A C A G
G **C** G G T G C A A C A G
G **T** G G T G C A A C A G
G **G** G G T G C A A C A G

(b)
G **A** G G **T G C** A **A** C A **G**
G **A** G G **A C C** A **A** C A **G**
G **A** G G **T G C** A **T** C A **A**
G **G** G G **T G G** A **A** C A **G**

FIGURE 2.8 Two groups of four DNA sequences. In (a) each sequence differs from any other sequence at a single nucleotide site (shaded). In (b) each sequence differs from any other sequence at two or more nucleotide sites. However, since in both cases, each sequence is represented in its group only once, the values of the single-locus diversity measure will be the same for both groups.

these gene diversity measures will not discriminate among different loci or populations and will no longer be informative measures of polymorphism. Consider, for instance, the two groups of sequences in Figure 2.8. Intuitively, we would think that the sequences in Figure 2.8a should be less polymorphic than the ones in Figure 2.8b. However, the values of h and H will be the same for both groups of sequences.

For DNA sequence data, a more appropriate measure of polymorphism in a population is the average number of nucleotide differences per site between any two randomly chosen sequences. This measure is called **nucleotide diversity** (Nei and Li 1979) and is denoted by Π:

$$\Pi = \sum_{ij} x_i x_j \pi_{ij} \quad \textbf{(2.30)}$$

where x_i and x_j are the frequencies of the ith and jth type of DNA sequences, respectively, and π_{ij} is the proportion of different nucleotides between the ith and jth types. The Π values for the sequences in Figure 2.8 are 0.031 and 0.094, i.e., the nucleotide diversity measures are in agreement with our intuitive perception that group (a) is more variable than group (b). (Note that for the case of $\pi_{ij} = 1$, the value of Π will be the same as that of h in Equation 2.28.)

One of the first studies on nucleotide diversity at the DNA sequence level concerned the alcohol dehydrogenase (*Adh*) locus in *Drosophila melanogaster.* Eleven sequences spanning the *Adh* region were sequenced by Kreitman (1983). The aligned sequences were 2,379 nucleotides long. Disregarding deletions and insertions, there were nine different alleles, one of which was represented in the sample by three sequences (*8-F*, *9-F*, and *10-F*), while the rest were represented by one sequence each (Figure 2.9). Thus, the frequencies x_1–x_8 were each 1/11, and the frequency x_9 was 3/11.

Forty-three nucleotide sites were polymorphic. We first calculate the proportion of different nucleotides for each pair of alleles. For example, alleles *1-S* and *2-S* differ from each other by three nucleotides out of 2,379, or $\pi_{12} = 0.13\%$. The π_{ij} values for all the pairs in the sample are listed in Table 2.1. By using Equation 2.30, the nucleotide diversity is estimated to be $\Pi = 0.007$. Six of the alleles studied were slow-migrating electrophoretic variants (*S*), and five were fast (*F*). The products of *S* and *F* alleles are distinguished from each other by one amino acid replacement that confers a different electrophoretic mobility to the proteins. The nucleotide diversity for each of these electrophoretic classes was calculated separately. We obtain $\Pi = 0.006$ for the *S* class, and $\Pi = 0.003$ for *F*; i.e., the *S* alleles are twice as variable as the *F* alleles.

THE DRIVING FORCES IN EVOLUTION

Evolutionary explanations can be broadly classified into three types according to the relative importance assigned to random genetic drift versus the various forms of selection in determining a particular evolutionary outcome. **Mutationist** hypotheses are those theories in which an evolutionary phenomenon

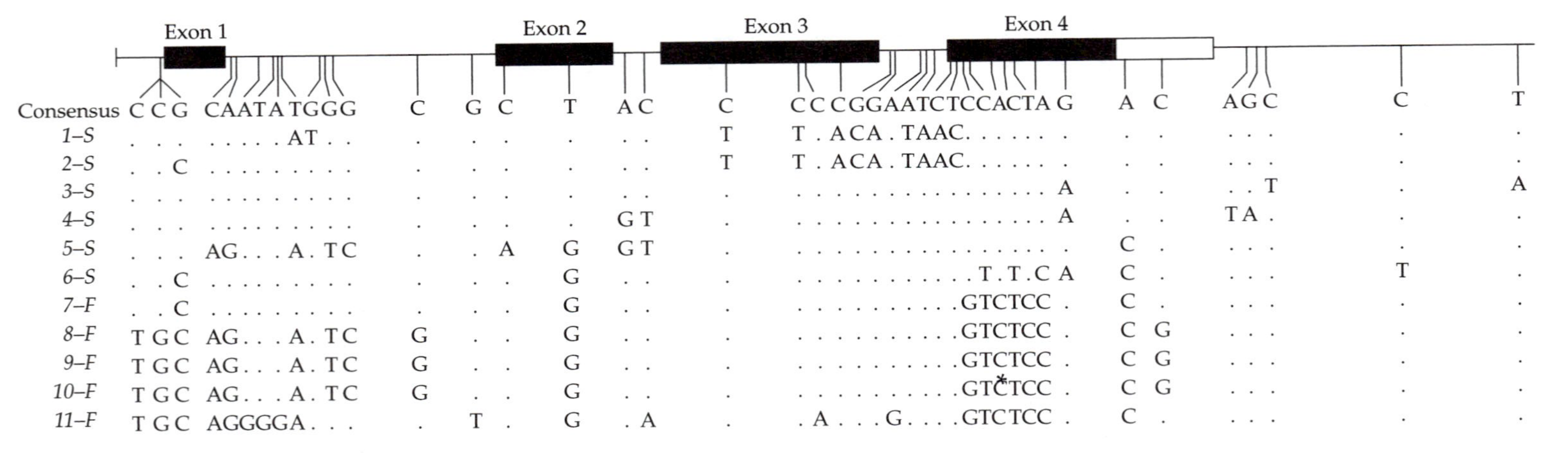

FIGURE 2.9 Polymorphic nucleotide sites among 11 sequences of the alcohol dehydrogenase gene in *Drosophila melanogaster*. Exons are shown as boxes; translated regions are in black. Only differences from the consensus sequence are shown. Dots indicate identity with the consensus sequence. The asterisk in exon 4 indicates the site of the lysine-for-threonine replacement that is responsible for the mobility difference between the fast (*F*) and slow (*S*) electrophoretic alleles. Modified from Hartl and Clark (1997).

TABLE 2.1 Pairwise percent nucleotide differences among 11 alleles of the alcohol dehydrogenase locus in *Drosophila melanogaster*[a]

Allele	*1-S*	*2-S*	*3-S*	*4-S*	*5-S*	*6-S*	*7-F*	*8-F*	*9-F*	*10-F*
1-S										
2-S	0.13									
3-S	0.59	0.55								
4-S	0.67	0.63	0.25							
5-S	0.80	0.84	0.55	0.46						
6-S	0.80	0.67	0.38	0.46	0.59					
7-S	0.84	0.71	0.50	0.59	0.63	0.21				
8-S	1.13	1.10	0.88	0.97	0.59	0.59	0.38			
9-S	1.13	1.10	0.88	0.97	0.59	0.59	0.38	0.00		
10-S	1.13	1.10	0.88	0.97	0.59	0.59	0.38	0.00	0.00	
11-S	1.22	1.18	0.97	1.05	0.84	0.67	0.46	0.42	0.42	0.42

From Nei (1987); data from Kreitman (1983).

[a]Total number of compared sites is 2,379. *S* and *F* denote the slow- and fast-migrating electrophoretic alleles, respectively.

is explained mainly by the effects of mutational input and random genetic drift. **Neutralist** hypotheses explain evolutionary phenomena by stressing the effects of mutation, random genetic drift, and purifying selection. **Selectionist** explanations emphasize the effects of the advantageous and balancing modes of selection as the main driving forces in the evolutionary process. The above distinctions provide a useful framework for understanding some of the most important controversies in the history of molecular evolution.

The neo-Darwinian theory and the neutral mutation hypothesis

Darwin proposed his theory of evolution by natural selection without knowledge of the sources of variation in populations. After Mendel's laws were rediscovered and genetic variation was shown to be generated by mutation, Darwinism and Mendelism were used as the framework of what came to be called the **synthetic theory of evolution**, or **neo-Darwinism**. According to this theory, although mutation is recognized as the ultimate source of genetic variation, natural (i.e., positive) selection is given the dominant role in shaping the genetic makeup of populations and in the process of gene substitution.

In time, neo-Darwinism became dogma in evolutionary biology, and selection came to be considered the only force capable of driving the evolutionary process. Factors such as mutation and random genetic drift were thought of as minor contributors at best. This particular brand of neo-Darwinism was called **pan-selectionism**.

According to the selectionist perception of the evolutionary process, gene substitution occurs as a consequence of selection for advantageous mutations.

Polymorphism, on the other hand, is maintained by balancing selection. Thus, selectionists regard substitution and polymorphism as two separate phenomena driven by different evolutionary forces. Gene substitution is the end result of a positive adaptive process whereby a new allele takes over future generations of the population if and only if it improves the fitness of the organism, while polymorphism is maintained when the coexistence of two or more alleles at a locus is advantageous for the organism or the population. Neo-Darwinian theories maintain that most genetic polymorphism in nature is stable, i.e., the same alleles are maintained at constant frequencies for long periods of evolutionary time.

The late 1960s witnessed a revolution in population genetics. The availability of protein sequence data removed the species boundaries in population genetics studies and for the first time provided adequate empirical data for examining theories pertaining to the process of gene substitution. In 1968, Kimura postulated that the majority of molecular changes in evolution are due to the random fixation of neutral or nearly neutral mutations (Kimura 1968a; see also King and Jukes 1969). This hypothesis, now known as the **neutral theory of molecular evolution**, contends that at the molecular level the majority of evolutionary changes and much of the variability within species are caused neither by positive selection of advantageous alleles nor by balancing selection, but by random genetic drift of mutant alleles that are selectively neutral (or nearly so). Neutrality, in the sense of the theory, does not imply strict equality in fitness for all alleles. It only means that the fate of alleles is determined largely by random genetic drift. In other words, selection may operate, but its intensity is too weak to offset the influences of chance effects. For this to be true, the absolute value of the selective advantage or disadvantage of an allele, $|s|$, must be smaller than $1/(2N_e)$, where N_e is the effective population size.

According to the neutral theory, the frequency of alleles is determined largely by stochastic rules, and the picture that we obtain at any given time is merely a transient state representing a temporary frame from an ongoing dynamic process. Consequently, polymorphic loci consist of alleles that are either on their way to fixation or are about to become extinct. Viewed from this perspective, all molecular manifestations that are relevant to the evolutionary process should be regarded as the result of a continuous process of mutational input and a concomitant random extinction or fixation of alleles. Thus, the neutral theory regards substitution and polymorphism as two facets of the same phenomenon. Substitution is a long and gradual process whereby the frequencies of mutant alleles increase or decrease randomly, until the alleles are ultimately fixed or lost by chance. At any given time, some loci will possess alleles at frequencies that are neither 0% nor 100%. These are the polymorphic loci. According to the neutral theory, most genetic polymorphism in populations is unstable and transient, i.e., allele frequencies fluctuate with time and the alleles are replaced continuously.

Interestingly, the neutral theory, even in its strictest form, does not preclude adaptation. According to Kimura (1983), a population that is free from selection can accumulate many polymorphic neutral alleles. Then if a change in ecological circumstances occurs, some of the neutral alleles will no longer be neutral but deleterious, against which purifying selection may operate. After these alleles are removed, the population will become more adapted to its new circumstances than before. Thus, theoretically at least, adaptive evolution may occur without positive selection.

The essence of the dispute between neutralists and selectionists concerns the distribution of fitness values of mutant alleles. Both theories agree that most new mutations are deleterious, and that these mutations are quickly removed from the population so that they contribute neither to the rate of substitution nor to the amount of polymorphism within populations. The difference concerns the relative proportion of neutral mutations among nondeleterious mutations. While selectionists claim that very few mutations are selectively neutral, neutralists maintain that the majority of nondeleterious mutations are effectively neutral.

The heated controversy over the neutral mutation hypothesis during the 1970s and the 1980s has had a significant impact on molecular evolution. First, it has led to the general recognition that the effect of random genetic drift cannot be neglected when considering the evolutionary dynamics of molecular changes. Second, the synthesis between molecular biology and population genetics has been greatly strengthened by the introduction of the concept that molecular evolution and genetic polymorphism are but two aspects of the same phenomenon (Kimura and Ohta 1971). Although the controversy continues, it is now recognized that any adequate theory of evolution should be consistent with both these aspects of the evolutionary process at the molecular level. In fact, without the neutral theory as a null hypothesis, the selectionist paradigm comes perilously close to being "a theory which explains nothing because it explains everything" (Lewontin 1974).

Testing the neutral mutation hypothesis

According to the neutral mutation hypothesis, both the variation within population and the differences between populations are due to neutral or nearly neutral mutations. In other words, polymorphism is a transient phase of molecular evolution, and the rate of evolution is positively correlated with the level of within-population variation (Kimura and Ohta 1971). One may, therefore, test the neutral mutation hypothesis by comparing the degree of DNA sequence variation within populations with the variation between populations. Many such tests have been developed (e.g., Kreitman and Aguadé 1986; Hudson et al. 1987; Sawyer and Hartl 1992). In the following, we present a simple method proposed by McDonald and Kreitman (1991).

Consider two samples of protein-coding sequences from species 1 and 2. A nucleotide site in the sequences is said to be polymorphic if it exhibits any variation in one or both species. A site is deemed to represent a fixed difference between the two species if it shows no intraspecific variation within either species but differs between the species. All other sites are monomorphic and are not used in the analysis.

TABLE 2.2 Synonymous and nonsynonymous fixed differences and polymorphisms at the glucose-6-phosphate dehydrogenase locus of *Drosophila melanogaster* and *D. simulans*[a]

Type of change	Fixed differences	Polymorphisms
Synonymous	26	36
Nonsynonymous	21	2

Data from Eanes et al. (1993).

[a]The comparisons are based on 32 sequences from *D. melanogaster* and 12 sequences from *D. simulans*, with an aligned length of 1,705 bp.

The polymorphic and fixed-site differences are further divided into two categories, synonymous and nonsynonymous. The McDonald-Kreitman method uses a 2×2 contingency table to test the independence of one classification (polymorphic versus fixed) from the other (synonymous versus nonsynonymous). The test is based on the following assumptions: (1) only nonsynonymous mutations may be adaptive, (2) synonymous mutations are always neutral, and (3) a selectively advantageous mutation will be fixed in the population much more rapidly than a neutral mutation, and hence is less likely to be found in a polymorphic state. Under the neutral mutation hypothesis, the expectation is that the ratio of fixed nonsynonymous differences to fixed synonymous differences will be the same as the ratio of nonsynonymous polymorphisms to synonymous polymorphisms. A significant difference between the two ratios can therefore be used to reject the neutral mutation hypothesis.

Table 2.2 shows the numbers of synonymous and nonsynonymous fixed differences and polymorphisms at the glucose-6-phosphate dehydrogenase (*G6PD*) gene between *Drosophila melanogaster* and *D. simulans* as an example of the use of the McDonald-Kreitman method to detect deviations from neutral evolution (Eanes et al. 1993). The ratio of fixed nonsynonymous differences to fixed synonymous differences is $21/26 = 0.81$, whereas the ratio of nonsynonymous polymorphisms to synonymous polymorphisms is only $2/36 = 0.06$. This highly significant difference implies a tenfold excess of nonsynonymous changes over that expected if the *G6PD* gene had evolved in a strictly neutral fashion.

Application of McDonald-Kreitman method and other tests have revealed important patterns in molecular evolution, including (1) positive directional selection at several nuclear loci in *Drosophila* (e.g., Tsaur et al. 1998) and absence at others (e.g., King 1998), (2) balancing selection at several loci in humans (e.g., Hughes and Nei 1989), (3) mildly deleterious alleles in animal mitochondrial DNA (e.g., Nachman 1998), and (4) a positive association between levels of nucleotide diversity and recombination rates (e.g., Begun and Aquadro 1992).

FURTHER READINGS

Caballero, A. 1994. Developments in the prediction of effective population size. Heredity 73: 657–679.

Christiansen, F. B. and M. W. Feldman. 1986. *Population Genetics*. Blackwell Scientific Publications, Cambridge, MA.

Crow, J. F. and M. Kimura. 1970. *An Introduction to Population Genetics*. Harper & Row, New York.

Gillespie, J. H. 1991. *The Causes of Molecular Evolution*. Oxford University Press, New York

Hartl, D. L. and A. G. Clark. 1997. *Principles of Population Genetics*, 3rd Ed. Sinauer Associates, Sunderland, MA.

Hedrick, P. W. 1983. *Genetics of Populations*. Science Books International, Portola Valley, CA.

Hey, J. 1999. The neutralist, the fly, and the selectionist. Trends Ecol. Evol. 14: 35–38.

Kimura, M. 1983. *The Neutral Theory of Molecular Evolution*. Cambridge University Press, Cambridge.

Nei, M. and D. Graur. 1984. Extent of protein polymorphism and the neutral mutation theory. Evol. Biol. 17: 73–118.

Roughgarden, J. 1996. *Theory of Population Genetics and Evolutionary Ecology: An Introduction*. Prentice-Hall, New York.

Chapter 3

Evolutionary Change in Nucleotide Sequences

A basic process in the evolution of DNA sequences is the substitution of one nucleotide for another during evolutionary time. The process deserves a detailed consideration because changes in nucleotide sequences are used in molecular evolutionary studies both for estimating the rate of evolution and for reconstructing the evolutionary history of organisms. The process of nucleotide substitution is usually extremely slow, so it cannot be dealt with by direct observation. To detect evolutionary changes in a DNA sequence, we need to compare two sequences that have descended from a common ancestral sequence. Such comparisons require statistical methods, several of which are discussed in this chapter.

Nucleotide Substitution in a DNA Sequence

In the previous chapter, we described the evolutionary process as a series of gene substitutions in which new alleles, each arising as a mutation in a single individual, progressively increase their frequency and ultimately become fixed in the population. We now look at the process from a different point of view. We note that an allele that becomes fixed is different in its sequence from the allele that it replaces. That is, the substitution of a new allele for an old one is the substitution of a new sequence for a previous sequence. If we use a time scale in which one time unit is larger than the time of fixation, then the DNA sequence at any given locus will appear to change continuously. For this reason, it is interesting to study how the nucleotides within a DNA sequence change with time. As explained later, the results of these studies can be used to develop methods for estimating the number of substitutions between two sequences.

To study the dynamics of nucleotide substitution, we must make several assumptions regarding the probability of substitution of one nucleotide by another. Numerous such mathematical schemes have been proposed in the literature (for a review, see Li 1997). We shall restrict our discussion to only the simplest and most frequently used ones: Jukes and Cantor's one-parameter model, and Kimura's two-parameter model. More complicated models of nucleotide substitution will be described only briefly.

Jukes and Cantor's one-parameter model

The substitution scheme of Jukes and Cantor's (1969) model is shown in Figure 3.1. This simple model assumes that substitutions occur with equal probability among the four nucleotide types. In other words, there is no bias in the direction of change. For example, if the nucleotide under consideration is A, it will change to T, C, or G with equal probability. Conversely, the probabilities of either T, C or G changing to A are also equal. In this model, the rate of substitution for each nucleotide is 3α per unit time, and the rate of substitution in each of the three possible directions of change is α. Because the model involves a single parameter, α, it is called the **one-parameter model**.

Let us assume that the nucleotide residing at a certain site in a DNA sequence is A at time 0. First, we ask, What is the probability that this site will be occupied by A at time t? This probability is denoted by $P_{A(t)}$.

Since we start with A, the probability that this site is occupied by A at time 0 is $P_{A(0)} = 1$. At time 1, the probability of still having A at this site is given by

$$P_{A(1)} = 1 - 3\alpha \quad \textbf{(3.1)}$$

in which 3α is the probability of A changing to T, C, or G, and $1 - 3\alpha$ is the probability that A has remained unchanged.

The probability of having A at time 2 is

$$P_{A(2)} = (1 - 3\alpha)P_{A(1)} + \alpha\left[1 - P_{A(1)}\right] \quad \textbf{(3.2)}$$

To derive this equation, we considered two possible scenarios: (1) the nucleotide has remained unchanged from time 0 to time 2, and (2) the nucleotide has changed to T, C, or G at time 1, but has subsequently reverted to A at time 2 (Figure 3.2). The probability of the nucleotide being A at time 1 is $P_{A(1)}$, and the probability that it has remained A at time 2 is $1 - 3\alpha$. The product of these two independent variables gives us the probability for the first scenario, which constitutes the first term in Equation 3.2. The probability of the nucleotide not being A at time 1 is $1 - P_{A(1)}$, and its probability of changing back to A at time 2 is α. The product of these two variables gives us the probability for the second scenario, and constitutes the second term in Equation 3.2.

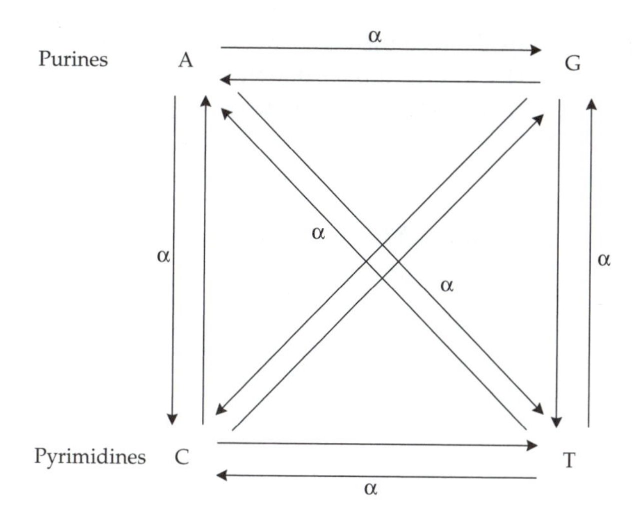

FIGURE 3.1 One-parameter model of nucleotide substitution. The rate of substitution in each direction is α.

Using the above formulation, we can show that the following recurrence equation applies to any t:

$$P_{A(t+1)} = (1-3\alpha)P_{A(t)} + \alpha\left[1-P_{A(t)}\right] \quad \textbf{(3.3)}$$

We note that Equation 3.3 will also hold for $t = 0$, because $P_{A(0)} = 1$, and hence $P_{A(0+1)} = (1 - 3\alpha)P_{A(0)} + \alpha[1 - P_{A(0)}] = 1 - 3\alpha$, which is identical with Equation 3.1. We can rewrite Equation 3.3 in terms of the amount of change in $P_{A(t)}$ per unit time as

$$\Delta P_{A(t)} = P_{A(t+1)} - P_{A(t)} = -3\alpha P_{A(t)} + \alpha\left[1-P_{A(t)}\right] = -4\alpha P_{A(t)} + \alpha \quad \textbf{(3.4)}$$

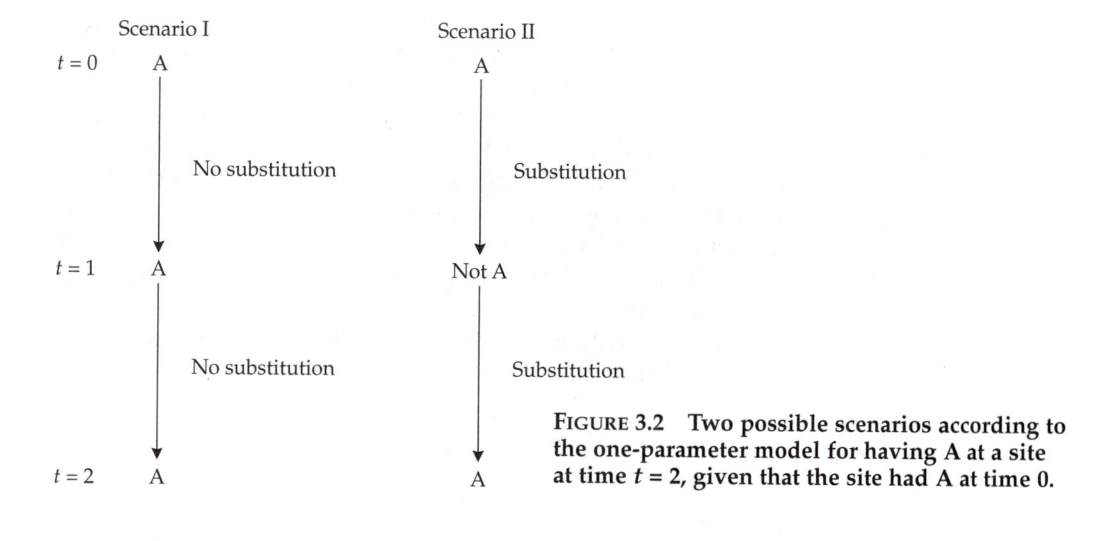

FIGURE 3.2 Two possible scenarios according to the one-parameter model for having A at a site at time $t = 2$, given that the site had A at time 0.

So far we have considered a discrete-time process. We can, however, approximate this process by a continuous-time model, by regarding $\Delta P_{A(t)}$ as the rate of change at time t. With this approximation, Equation 3.4 is rewritten as

$$\frac{dP_{A(t)}}{dt} = -4\alpha P_{A(t)} + \alpha \tag{3.5}$$

This is a first-order linear differential equation, and the solution is given by

$$P_{A(t)} = \frac{1}{4} + \left(P_{A(0)} - \frac{1}{4}\right)e^{-4\alpha t} \tag{3.6}$$

Since we started with A, the probability that the site has A at time 0 is 1. Thus, $P_{A(0)} = 1$, and consequently,

$$P_{A(t)} = \frac{1}{4} + \frac{3}{4}e^{-4\alpha t} \tag{3.7}$$

Actually, Equation 3.6 holds regardless of the initial conditions. For example, if the initial nucleotide is not A, then $P_{A(0)} = 0$, and the probability of having A at this position at time t is

$$P_{A(t)} = \frac{1}{4} - \frac{1}{4}e^{-4\alpha t} \tag{3.8}$$

Equations 3.7 and 3.8 are sufficient for describing the substitution process. From Equation 3.7, we can see that, if the initial nucleotide is A, then $P_{A(t)}$ decreases exponentially from 1 to ¼ (Figure 3.3). On the other hand, from Equation 3.8 we see that if the initial nucleotide is not A, then $P_{A(t)}$ will increase monotonically from 0 to ¼. This also holds true for T, C, and G.

Under the Jukes-Cantor model, the probability of each of the four nucleotides at equilibrium is ¼. After reaching equilibrium, there will be no further changes in probabilities, i.e., $P_{A(t)} = P_{T(t)} = P_{C(t)} = P_{G(t)} = ¼$ for all subsequent times.

So far, we have focused on a particular nucleotide site and treated $P_{A(t)}$ as a probability. However, $P_{A(t)}$ can also be interpreted as the frequency of A in a DNA sequence at time t. For example, if we start with a sequence made entirely of adenines, then $P_{A(0)} = 1$, and $P_{A(t)}$ is the expected frequency of A in the sequence at time t. The expected frequency of A in the sequence at equilibrium will be ¼, and so will the expected frequencies of T, C, and G. After reaching equilibrium, no further change in the nucleotide frequencies is expected to occur. However, the actual frequencies of the nucleotides will remain unchanged only in DNA sequences of infinite length. In practice, the lengths of DNA sequences are finite, and so fluctuations in nucleotide frequencies are likely to occur.

Equation 3.7 can be rewritten in a more explicit form to take into account the facts that the initial nucleotide is A and the nucleotide at time t is also A.

$$P_{AA(t)} = \frac{1}{4} + \frac{3}{4}e^{-4\alpha t} \tag{3.9}$$

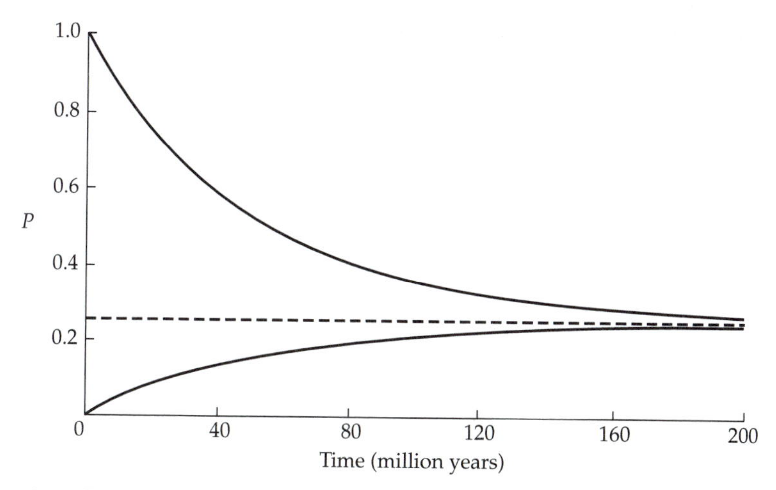

FIGURE 3.3 Temporal changes in the probability, P, of having a certain nucleotide at a position starting with either the same nucleotide (upper line) or with a different nucleotide (lower line). The dashed line denotes the equilibrium frequency (P = 0.25). $\alpha = 5 \times 10^{-9}$ substitutions per site per year.

If the initial nucleotide is G instead of A, then from Equation 3.8 we obtain

$$P_{\mathrm{GA}(t)} = \frac{1}{4} - \frac{1}{4}e^{-4\alpha t} \qquad \textbf{(3.10)}$$

Since all the nucleotides are equivalent under the Jukes-Cantor model, $P_{\mathrm{GA}(t)} = P_{\mathrm{CA}(t)} = P_{\mathrm{TA}(t)}$. In fact, we can consider the general probability, $P_{ij(t)}$, that a nucleotide will become j at time t, given that it was i at time 0. By using this generalized notation and Equation 3.9, we obtain

$$P_{ii(t)} = \frac{1}{4} + \frac{3}{4}e^{-4\alpha t} \qquad \textbf{(3.11)}$$

In a similar manner, from Equation 3.10 we obtain

$$P_{ij(t)} = \frac{1}{4} - \frac{1}{4}e^{-4\alpha t} \qquad \textbf{(3.12)}$$

where $i \neq j$.

Kimura's two-parameter model

The assumption that all nucleotide substitutions occur with equal probability, as in Jukes and Cantor's model, is unrealistic in most cases. For example, transitions (i.e., changes between A and G or between C and T) are generally more frequent than transversions. To take this fact into account, Kimura (1980) proposed a **two-parameter model** (Figure 3.4). In this scheme, the rate of transitional substitution at each nucleotide site is α per unit time, whereas the rate of each type of transversional substitution is β per unit time.

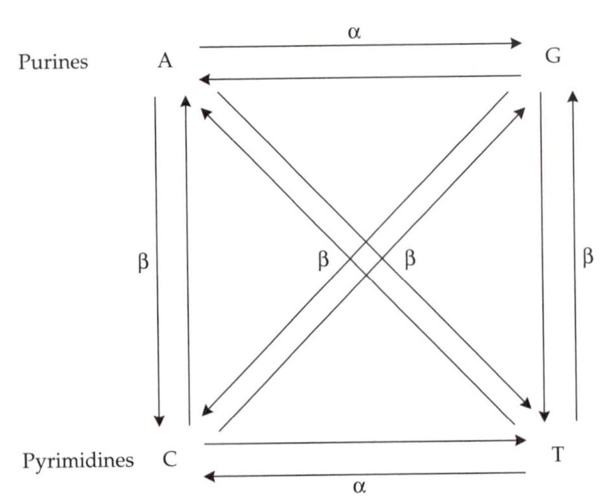

FIGURE 3.4 Two-parameter model of nucleotide substitution. The rate of transition (α) may not be equal to the rate of transversion (β).

Let us first consider the probability that a site that has A at time 0 will have A at time t. After one time unit, the probability of A changing into G is α, and the probability of A changing into either C or T is 2β. Thus, the probability of A remaining unchanged after one time unit is

$$P_{AA(1)} = 1 - \alpha - 2\beta \quad \textbf{(3.13)}$$

At time 2, the probability of having A at this site is given by the sum of the probabilities of four different scenarios: (1) A remained unchanged at $t = 1$ and $t = 2$; (2) A changed into G at $t = 1$ and reverted by a transition to A at $t = 2$; (3) A changed into C at $t = 1$ and reverted by a transversion to A at $t = 2$; and (4) A changed into T at $t = 1$ and reverted by a transversion to A at $t = 2$ (Figure 3.5). Hence,

$$P_{AA(2)} = (1 - \alpha - 2\beta)P_{AA(1)} + \beta P_{TA(1)} + \beta P_{CA(1)} + \alpha P_{GA(1)} \quad \textbf{(3.14)}$$

By extension we obtain the following recurrence equation for the general case:

$$P_{AA(t+1)} = (1 - \alpha - 2\beta)P_{AA(t)} + \beta P_{TA(t)} + \beta P_{CA(t)} + \alpha P_{GA(t)} \quad \textbf{(3.15)}$$

After rewriting this equation as the amount of change in $P_{AA(t)}$ per unit time, and after approximating the discrete-time model by the continuous-time model, we obtain the following differential equation:

$$\frac{dP_{AA(t)}}{dt} = -(\alpha + 2\beta)P_{AA(t)} + \beta P_{TA(t)} + \beta P_{CA(t)} + \alpha P_{GA(t)} \quad \textbf{(3.16)}$$

Similarly, we can obtain equations for $P_{TA(t)}$, $P_{CA(t)}$, and $P_{GA(t)}$, and from this set of four equations, we arrive at the following solution:

$$P_{AA(t)} = \frac{1}{4} + \frac{1}{4}e^{-4\beta t} + \frac{1}{2}e^{-2(\alpha+\beta)t} \quad \textbf{(3.17)}$$

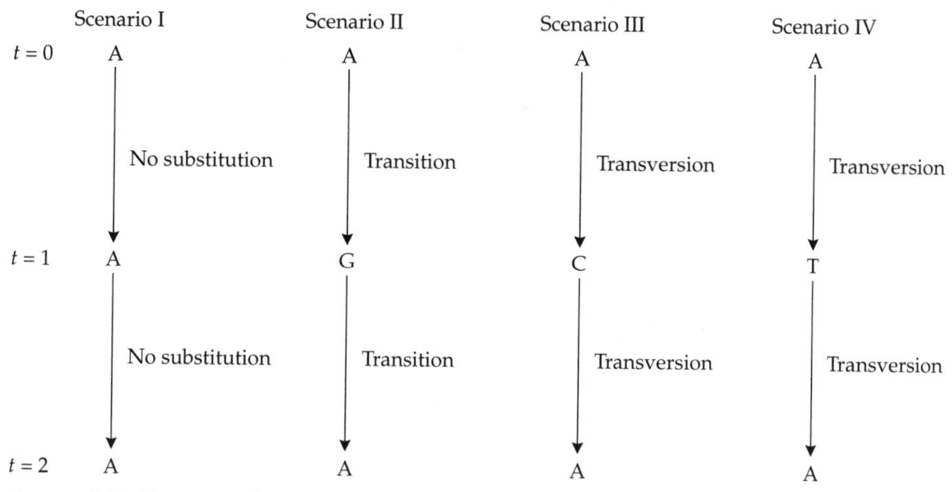

FIGURE 3.5 Four possible scenarios, according to Kimura's (1980) two-parameter model, for having A at a site at time t = 2, given that the site had A at time 0.

We note from Equation 3.11 that in the Jukes-Cantor model, the probability that the nucleotide at a site at time t is identical to that at time 0 is the same for all four nucleotides. In other words, $P_{AA(t)} = P_{GG(t)} = P_{CC(t)} = P_{TT(t)}$. Because of the symmetry of the substitution scheme, this equality also holds for Kimura's two-parameter model. We shall denote this probability by $X_{(t)}$. Therefore,

$$X_{(t)} = \frac{1}{4} + \frac{1}{4}e^{-4\beta t} + \frac{1}{2}e^{-2(\alpha+\beta)t} \tag{3.18}$$

At equilibrium, i.e., at $t = \infty$, Equation 3.18 reduces to $X_{(\infty)} = 1/4$. Thus, as in the case of Jukes and Cantor's model, the equilibrium frequencies of the four nucleotides are ¼.

Under the Jukes-Cantor model, Equation 3.12 holds regardless of whether the change from nucleotide i to nucleotides j is a transition or a transversion. In contrast, in Kimura's two-parameter model, we need to distinguish between transitional and transversional changes. We denote by $Y_{(t)}$ the probability that the initial nucleotide and the nucleotide at time t differ from each other by a transition. Because of the symmetry of the substitution scheme, $Y_{(t)} = P_{AG(t)} = P_{GA(t)} = P_{TC(t)} = P_{CT(t)}$. It can be shown that

$$Y_{(t)} = \frac{1}{4} + \frac{1}{4}e^{-4\beta t} - \frac{1}{2}e^{-2(\alpha+\beta)t} \tag{3.19}$$

The probability, $Z_{(t)}$, that the nucleotide at time t and the initial nucleotide differ by a specific type of transversion is given by

$$Z_{(t)} = \frac{1}{4} - \frac{1}{4}e^{-4\beta t} \tag{3.20}$$

Note that each nucleotide is subject to two types of transversion, but only one type of transition. For example, if the initial nucleotide is A, then the two possible transversional changes are A → C and A → T. Therefore, the probability that the initial nucleotide and the nucleotide at time t differ by one of the two types of transversion is twice the probability given in Equation 3.20. Note also that $X_{(t)} + Y_{(t)} + 2Z_{(t)} = 1$.

Number of Nucleotide Substitutions between Two DNA Sequences

The substitution of alleles in a population generally takes thousands or even millions of years to complete (Chapter 2). For this reason, we cannot deal with the process of nucleotide substitution by direct observation, and nucleotide substitutions are always inferred from pairwise comparisons of DNA molecules that share a common evolutionary origin. After two nucleotide sequences diverge from each other, each of them will start accumulating nucleotide substitutions. Thus, the number of nucleotide substitutions that have occurred since two sequences diverged is the most basic and commonly used variable in molecular evolution.

If two sequences of length N differ from each other at n sites, then the proportion of differences, n/N, is referred to as the **degree of divergence** or **Hamming distance**. Degrees of divergence are usually expressed as percentages ($n/N \times 100\%$). When the degree of divergence between the two sequences compared is small, the chance for more than one substitution to have occurred at any site is negligible, and the number of observed differences between the two sequences should be close to the actual number of substitutions. However, if the degree of divergence is substantial, then the observed number of differences is likely to be smaller than the actual number of substitutions due to **multiple substitutions** or **multiple hits** at the same site. For example, if the nucleotide at a certain site changed from A to C and then to T in one sequence, and from A to T in the other sequence, then the two sequences under comparison are identical at this site, despite the fact that three substitutions occurred in their evolutionary history (Figure 3.6). Many methods have been proposed in the literature to correct for multiple substitutions (e.g., Jukes and Cantor 1969; Holmquist 1972; Kimura 1980, 1981; Holmquist and Pearl 1980; Kaplan and Risko 1982; Lanave et al. 1984). In the following sections we shall review some of the most frequently used methods.

The number of nucleotide substitutions between two sequences is usually expressed in terms of the number of substitutions per nucleotide site rather than the total number of substitutions between the two sequences. This facilitates comparisons among sequence pairs that differ in length.

Protein-coding and noncoding sequences should be treated separately because they usually evolve at different rates. In the former case, it is advisable to distinguish between synonymous and nonsynonymous substitutions, since they are known to evolve at markedly different rates and because a distinction between the two types of substitution may provide us additional insight into the mechanisms of molecular evolution (Chapter 4).

Number of substitutions between two noncoding sequences

The results that we obtained earlier in this chapter for a single DNA sequence can be applied to study nucleotide divergence between two sequences that share a common origin. We assume that all the sites in a sequence evolve at

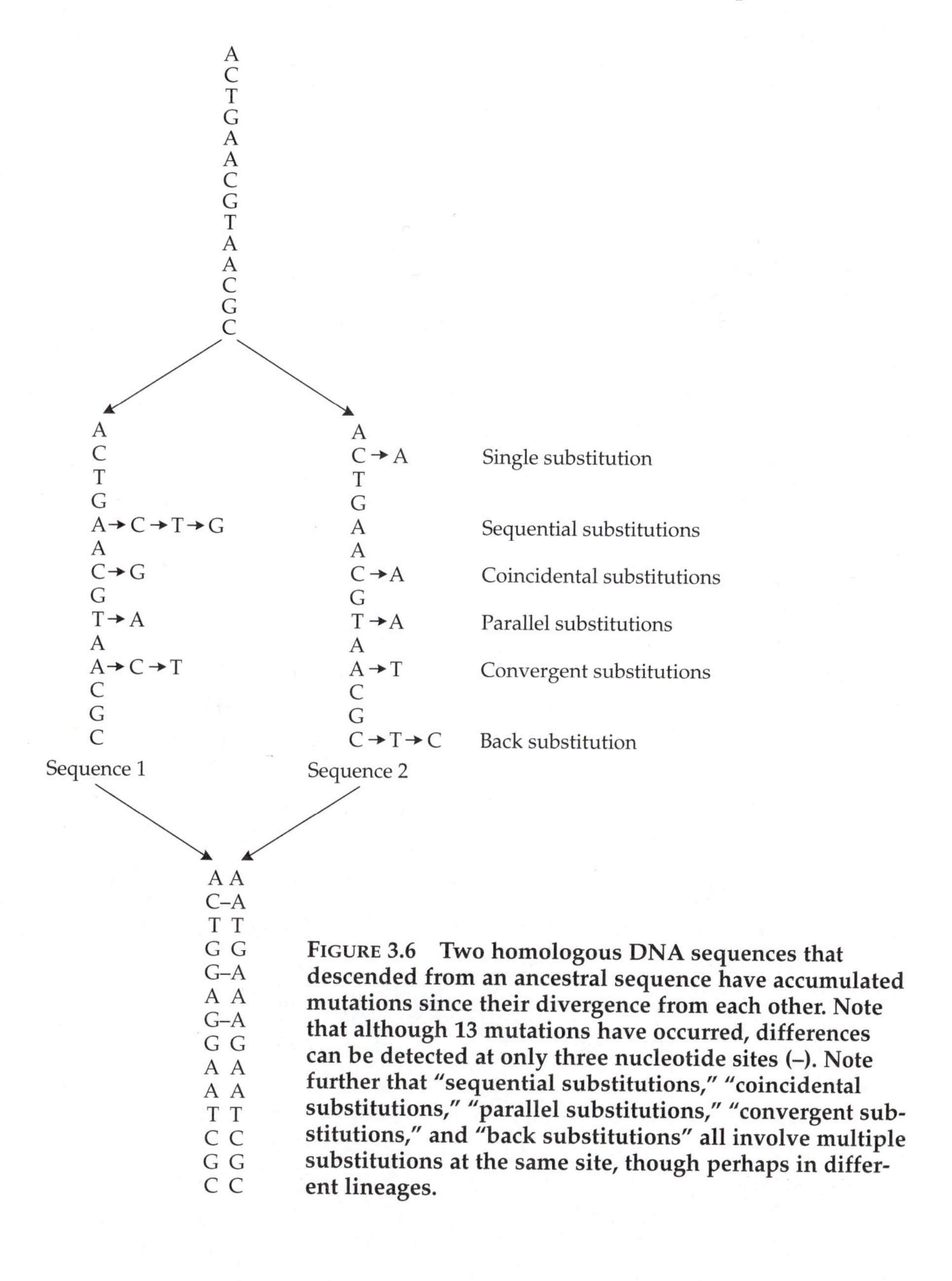

FIGURE 3.6 Two homologous DNA sequences that descended from an ancestral sequence have accumulated mutations since their divergence from each other. Note that although 13 mutations have occurred, differences can be detected at only three nucleotide sites (–). Note further that "sequential substitutions," "coincidental substitutions," "parallel substitutions," "convergent substitutions," and "back substitutions" all involve multiple substitutions at the same site, though perhaps in different lineages.

the same rate and follow the same substitution scheme. The number of sites compared between two sequences is denoted by L. Deletions and insertions are excluded from the analyses.

Let us start with the one-parameter model. In this model, it is sufficient to consider only $I_{(t)}$, which is the probability that the nucleotide at a given site at time t is the same in both sequences. Suppose that the nucleotide at a given site was A at time 0. At time t, the probability that a descendant sequence will have A at this site is $P_{AA(t)}$, and consequently the probability that two descendant sequences have A at this site is $P^2_{AA(t)}$. Similarly, the probabilities that both sequences have T, C, or G at this site are $P^2_{AT(t)}$, $P^2_{AC(t)}$, and $P^2_{AG(t)}$, respectively. Therefore,

$$I_{(t)} = P^2_{AA(t)} + P^2_{AT(t)} + P^2_{AC(t)} + P^2_{AG(t)} \quad (3.21)$$

From Equations 3.11 and 3.12, we obtain

$$I_{(t)} = \frac{1}{4} + \frac{3}{4}e^{-8\alpha t} \quad (3.22)$$

Equation 3.22 also holds for T, C, or G. Therefore, regardless of the initial nucleotide at a site, $I_{(t)}$ represents the proportion of identical nucleotides between two sequences that diverged t time units ago. Note that the probability that the two sequences are different at a site at time t is $p = 1 - I_{(t)}$. Thus,

$$p = \frac{3}{4}\left(1 - e^{-8\alpha t}\right) \quad (3.23)$$

or

$$8\alpha t = -\ln\left(1 - \frac{4}{3}p\right) \quad (3.24)$$

The time of divergence between two sequences is usually not known, and thus we cannot estimate α. Instead, we compute K, which is the number of substitutions per site since the time of divergence between the two sequences. In the case of the one-parameter model, $K = 2(3\alpha t)$, where $3\alpha t$ is the number of substitutions per site in a single lineage. By using Equation 3.24 we can calculate K as

$$K = -\frac{3}{4}\ln\left(1 - \frac{4}{3}p\right) \quad (3.25)$$

where p is the observed proportion of different nucleotides between the two sequences (Jukes and Cantor 1969). For sequences of length L, the sampling variance of K, $V(K)$, is approximately given by

$$V(K) = \frac{p - p^2}{L\left(1 - \frac{4}{3}p\right)^2} \quad (3.26)$$

(Kimura and Ohta 1972). Equation 3.26 is only applicable for large values of L.

In the case of the two-parameter model (Kimura 1980), the differences between two sequences are classified into transitions and transversions. Let P

and Q be the proportions of transitional and transversional differences between the two sequences, respectively. Then the number of nucleotide substitutions per site between the two sequences, K, is estimated by

$$K = \frac{1}{2}\ln\left(\frac{1}{1-2P-Q}\right) + \frac{1}{4}\ln\left(\frac{1}{1-2Q}\right) \tag{3.27}$$

Note that if we make no distinction between transitional and transversional differences, i.e., $p = P + Q$, then Equation 3.27 reduces to Equation 3.25, as in Jukes and Cantor's model. The sampling variance is approximately given by

$$V(K) = \frac{1}{L}\left[P\left(\frac{1}{1-2P-Q}\right)^2 + Q\left(\frac{1}{2-4P-2Q} + \frac{1}{2-4Q}\right)^2 - \left(\frac{P}{1-2P-Q} + \frac{Q}{2-4P-2Q} + \frac{Q}{2-4Q}\right)^2\right] \tag{3.28}$$

Let us now consider a hypothetical numerical example of two sequences of length 200 nucleotides that differ from each other by 20 transitions and 4 transversions. Thus, $L = 200$, $P = 20/200 = 0.10$, and $Q = 4/200 = 0.02$. According to the two-parameter model, we obtain $K \approx 0.13$. The total number of substitutions can be obtained by multiplying the number of substitutions per site, K, by the number of sites, L. In this case we obtain an estimate of about 26 substitutions, resulting in 24 observed differences between the two sequences. According to the one-parameter model, $p = 24/200 = 0.12$, and $K \approx 0.13$.

In the above example, the two models give essentially the same estimate because the degree of divergence is small enough that the corrected degree of divergence (i.e., the number of nucleotide substitutions) is only slightly larger than the uncorrected value (i.e., the number of nucleotide differences). In such cases, one may use Jukes and Cantor's model, which is simpler.

On the other hand, when the degree of divergence between two sequences is large, the estimates obtained by the two models may differ considerably. For example, consider two sequences with $L = 200$ that differ from each other by 50 transitions and 16 transversions. Thus, $P = 50/200 = 0.25$, and $Q = 16/200 = 0.08$. For the two-parameter model, $K \approx 0.48$. For the one-parameter model, $p = 66/200 = 0.33$, and $K \approx 0.43$, which is 10% smaller than the value obtained by using the two-parameter model. When the degree of divergence between two sequences is large, and especially in cases where there are prior reasons to believe that the rate of transition differs considerably from the rate of transversion, the two-parameter model tends to be more accurate than the one-parameter model.

Substitution schemes with more than two parameters

Since there are four types of nucleotides, and each of them can be substituted by any of the other three, there are 12 possible types of substitutions. Each of these substitution types has a certain probability to occur. These substitution probabilities can be written in the form of a matrix, the elements of which, α_{ij},

TABLE 3.1 General matrix of nucleotide substitution[a]

	A	T	C	G
A	$1 - \alpha_{12} - \alpha_{13} - \alpha_{14}$	α_{12}	α_{13}	α_{14}
T	α_{21}	$1 - \alpha_{21} - \alpha_{23} - \alpha_{24}$	α_{23}	α_{24}
C	α_{31}	α_{32}	$1 - \alpha_{31} - \alpha_{32} - \alpha_{34}$	α_{34}
G	α_{41}	α_{42}	α_{43}	$1 - \alpha_{41} - \alpha_{42} - \alpha_{43}$

[a]Values on the diagonal represent the probability of no change per unit time.

denote the probability of substitution of nucleotide *i* for nucleotide *j* per unit time (Table 3.1). For example, the substitution schemes in Figure 3.1 can also be represented in matrix form as in Table 3.2a.

The numerous mathematical models that have been used to study the dynamics of nucleotide substitution are different from each other in the particular probabilities assigned to the different types of substitution. In Table 3.2b we present Blaisdell's (1985) four-parameter substitution scheme. In this scheme, there are two different rates of transition, for A ↔ G and T ↔ C, respectively, and two different rates of transversion.

A general question arises: Which model should we use to calculate the number of substitutions between two sequences? Intuitively, it would seem that a model with a large number of parameters should perform better than one with fewer parameters. In practice, this is not necessarily the case, for several reasons. One reason is that in addition to the substitution scheme, it is usually necessary to make further assumptions in the formulation of each model. For example, to be able to estimate the number of substitutions between two sequences according to six-parameter models, we are required to assume that the common ancestral sequence was at equilibrium in terms of its nucleotide frequencies. Such an assumption is not necessary for the solution of either the one-parameter or two-parameter models. Obviously, the addition of assumptions (and parameters that need to be estimated) may greatly increase the estimation errors.

The second and probably more important reason is that sampling errors, which arise because the number of nucleotides compared is finite, can render a method inapplicable. This is because all estimators of K, such as Equations 3.25 and 3.27, involve logarithmic functions, and the argument can become zero or negative. This undesirable effect becomes more problematic with the increase in the number of parameters involved. The more two sequences are divergent from each other, the less applicable the multiple-parameter models will be. For example, when $K = 2.0$, four- and six-parameter methods may not be applicable to the vast majority of sequence comparisons. By contrast, the one-parameter method was found to be inapplicable in only very rare cases. The proportion of inapplicable cases decreases with the length of the sequence, L. However, even for large values of L (e.g., $L = 3{,}000$), four-parameter models are rendered inapplicable in about 25% of the cases when $K = 2.0$, and this proportion is even higher for six-parameter models.

TABLE 3.2 The one-parameter (Jukes and Cantor 1969) and four-parameter (Blaisdell 1985) schemes of nucleotide substitution in matrix form[a]

	A	T	C	G
One parameter				
A	$1-3\alpha$	α	α	α
T	α	$1-3\alpha$	α	α
C	α	α	$1-3\alpha$	α
G	α	α	α	$1-3\alpha$
Four parameters				
A	$1-\alpha-2\gamma$	γ	γ	α
T	δ	$1-\alpha-2\delta$	α	δ
C	δ	β	$1-\alpha-2\delta$	δ
G	β	γ	γ	$1-\beta-2\gamma$

[a]Values on the diagonals represent the probability of no change per unit time.

Finally, we note that for closely related sequences, the estimates obtained by the different methods are quite similar to one another.

Violation of assumptions

In the distance measures discussed so far, several assumptions have been made that are not necessarily met by the sequences under study. For example, the rate of nucleotide substitution was assumed to be the same for all sites. This assumption may not hold, as the rate may vary greatly from site to site (Chapter 4).

An additional assumption was that substitutions occur in an independent manner, i.e., that the probability of a certain substitution occurring at a site is not affected by (1) the context of the surrounding nucleotides, (2) the occurrence of a substitution at a different site, or (3) the history of substitutions at the site in question. The assumption of independence may not always apply in nature. For example, hairpin structures require nucleotide changes in one part of the sequence to be "compensated" by complementary changes elsewhere.

Finally, in the methods described above, the substitution matrix was assumed not to change in time, so that the nucleotide frequencies are maintained at a constant equilibrium value throughout their evolution. This may not be the case, especially in protein-coding sequences exhibiting extreme codon-usage biases (Chapter 4).

For each of these cases, special distance methods and corrections have been developed (e.g., Jin and Nei 1990; Tamura and Nei 1993; Lake 1994; Lockhart et al. 1994).

Number of substitutions between two protein-coding genes

Computing the number of substitutions between two protein-coding sequences is generally more complicated than computing the number of substitutions between two noncoding sequences, because a distinction should be made between **synonymous** and **nonsynonymous substitutions**. Several methods have been proposed in the literature for estimating the numbers of synonymous and nonsynonymous substitution (Perler et al. 1980; Miyata and Yasunaga 1980; Li et al. 1985b; Nei and Gojobori 1986; Li 1993; Pamilo and Bianchi 1993; Ina 1995; Comeron 1995).

In studying protein-coding sequences, we usually exclude the initiation and the termination codons from analysis because these two codons seldom change with time.

If the number of nucleotide substitutions between two DNA sequences is small, such that the number of nucleotide differences equals the number of substitutions, and if we are only interested in the absolute number of substitutions, then the numbers of synonymous and nonsynonymous substitutions can be obtained by simply counting synonymous and amino acid-altering nucleotide differences. Consider the following comparison:

Ser	Thr	Glu	Met	Cys	Leu
TCA	ACT	GAG	ATG	TGT	TTA
↕	↕		↕		↕
TCG	ACA	GAG	ATA	TGT	CTA
Ser	Thr	Glu	Ile	Cys	Leu

There are four codon differences between the sequences. Three of the differences are synonymous (TCA ↔ TCG, ACT ↔ ACA, and TTA ↔ CTA), and one is nonsynonymous (ATG ↔ ATA). Since 18 nucleotide sites are compared, the total number of substitutions per site is estimated as $4/18 = 0.22$.

However, if we want to compute the number of substitutions per site for synonymous and nonsynonymous substitutions separately, we must determine the appropriate denominators, i.e., the numbers of **synonymous** and **nonsynonymous sites**, respectively. In general, these numbers cannot be easily determined, for two reasons. The first is that the classification of a site changes with time. For example, the third position of CGG (Arg) is synonymous. However, if the first position changes to T, then the third position of the resulting codon, TGG (Trp), becomes nonsynonymous. The second reason is that many sites are neither completely synonymous nor completely nonsynonymous. For example, a transition in the third position of codon GAT (Asp) will be synonymous, while a transversion to either GAG or GAA will alter the amino acid. Consequently, many sites must be counted as part synonymous and part nonsynonymous. Note also that transitions result in synonymous substitutions more frequently than transversions do; therefore, the substitution scheme used to determine the number of synonymous and nonsynonymous substitutions must take into account that transitions occur with different frequencies than transversions.

One way to deal with this problem was proposed by Miyata and Yasunaga (1980) and Nei and Gojobori (1986). In their method, nucleotide sites are classified as follows. Consider a particular position in a codon. Let i be the number of possible synonymous changes at this site. Then this site is counted as $i/3$ synonymous and $(3 - i)/3$ nonsynonymous. For example, in the codon TTT (Phe), the first two positions are counted as nonsynonymous because no synonymous change can occur at these positions, and the third position is counted as one-third synonymous and two-thirds nonsynonymous because one of the three possible changes at this position is synonymous. As another example, the codon ACT (Thr) has two nonsynonymous sites (the first two positions) and one synonymous site (the third position), because all possible changes at the first two positions are nonsynonymous while all possible changes at the third position are synonymous. When comparing two sequences, one first counts the number of synonymous and the number of nonsynonymous sites in each sequence and then computes the averages between the two sequences. We denote the average number of synonymous sites by N_S and that of nonsynonymous sites by N_A.

Next, we classify nucleotide differences into synonymous and nonsynonymous differences. For two codons that differ by only one nucleotide, the difference is easily inferred. For example, the difference between the two codons GTC (Val) and GTT (Val) is synonymous, while the difference between the two codons GTC (Val) and GCC (Ala) is nonsynonymous. For two codons that differ by more than one nucleotide, the problem of estimating the numbers of synonymous and nonsynonymous substitutions becomes more complicated, because we need to determine the order in which the substitutions occurred.

Let us consider the case in which two codons differ from each other by two substitutions. For example, for the two codons CCC (Pro) and CAA (Gln), there are two possible pathways:

Pathway I: CCC (Pro) ↔ CCA (Pro) ↔ CAA (Gln)

Pathway II: CCC (Pro) ↔ CAC (His) ↔ CAA (Gln)

Pathway I requires one synonymous and one nonsynonymous change, whereas pathway II requires two nonsynonymous changes.

There are basically two approaches to deal with multiple substitutions at a codon. The first approach assumes that all pathways are equally probable (Nei and Gojobori 1986), so we average the numbers of the different types of substitutions for all the possible scenarios. This approach is called the **unweighted method**. For example, if we assume that the two pathways shown above are equally likely, then the number of nonsynonymous differences is $(1 + 2)/2 = 1.5$, and the number of synonymous differences is $(1 + 0)/2 = 0.5$.

The second approach, the **weighted method**, employs *a priori* criteria to decide which pathway is more probable. For instance, it is known that synonymous substitutions occur considerably more frequently than nonsynonymous substitutions (Chapter 4), and so, in the above example, pathway I is more probable that pathway 2. If we give a weight of 0 to pathway II, then the number of synonymous differences is 1 and the number of nonsynonymous differences is 1. Alternatively, we may give weights other than 0 and 1 to the two pathways. For example, if we assume a weight of 0.9 for pathway I and a weight of 0.1 for pathway II, then the number of nonsynonymous differences between the two codons is estimated as $(0.9 \times 1) + (0.1 \times 2) = 1.1$, and the number of synonymous differences as $(0.9 \times 1) + (0.1 \times 0) = 0.9$. The weights used here are hypothetical. To determine quantitatively and realistically the relative probabilities of the different pathways, adequate information on the relative likelihood of all possible codon changes is necessary. These values have been estimated empirically by Miyata and Yasunaga (1980) from protein sequence data and by Li et al. (1985b) from DNA sequence data.

In practice, the weighted and unweighted approaches usually yield quite similar estimates (Nei and Gojobori 1986), but they can be important for short genes coding for conservative protein-coding genes. One such example involves mammalian glucagon (Lopez et al. 1984). The arginine at the seventeenth amino acid position is encoded by CGC in hamster and by AGG in cow. There are two possible pathways between these two codons: (1) CGC (Arg) ↔ CGG (Arg) ↔ AGG (Arg), or (2) CGC (Arg) ↔ AGC (Ser) ↔ AGG (Arg). The first pathway requires two synonymous substitutions and the second, two nonsynonymous substitutions. Under the assumption that both pathways are equally probable, we would infer that one nonsynonymous substitution and one synonymous substitution have occurred at this position since the divergence between hamster and cow. Since glucagon consists of only 29 amino acids, this inference will greatly inflate the rate of nonsynonymous substitution at the expense of the synonymous substitution rate. Such errors may also occur when dealing with leucine codons belonging to different codon families, e.g., TTA and CTT.

Finally, we note that the situation is even more complex when two codons differ at all three positions. For example, if one sequence has CTT at a codon site and the second sequence has AGG, we have to consider six possible pathways:

Pathway I: CTT(Leu) ↔ ATT(Ile) ↔ AGT(Ser) ↔ AGG(Arg)
Pathway II: CTT(Leu) ↔ ATT(Ile) ↔ ATG(Met) ↔ AGG(Arg)
Pathway III: CTT(Leu) ↔ CGT(Arg) ↔ AGT(Ser) ↔ AGG(Arg)
Pathway IV: CTT(Leu) ↔ CTG(Leu) ↔ ATG(Met) ↔ AGG(Arg)
Pathway V: CTT(Leu) ↔ CGT(Arg) ↔ CGG(Arg) ↔ AGG(Arg)
Pathway VI: CTT(Leu) ↔ CTG(Leu) ↔ CGG(Arg) ↔ AGG(Arg)

The numbers of synonymous and nonsynonymous differences between two protein-coding sequences are denoted by M_S and M_A. We can therefore compute the number of synonymous differences per synonymous site as $p_S = M_S/N_S$ and the number of nonsynonymous differences per nonsynonymous site as $p_A = M_A/N_A$. These formulas obviously do not take into account the effect of multiple hits at the same site. We can make such corrections by using Jukes and Cantor's formula (Equation 3.25):

$$K_S = -\frac{3}{4}\ln\left(1-\frac{4M_S}{3N_S}\right) \tag{3.29}$$

and

$$K_A = -\frac{3}{4}\ln\left(1-\frac{4M_A}{3N_A}\right) \tag{3.30}$$

An alternative method for calculating K_S and K_A has been proposed by Li et al. (1985b). According to this method, we first classify the nucleotide sites into **nondegenerate**, **twofold degenerate**, and **fourfold degenerate** sites. A site is nondegenerate if all possible changes at this site are nonsynonymous, twofold degenerate if one of the three possible changes is synonymous, and fourfold degenerate if all possible changes at the site are synonymous. For example, the first two positions of codon TTT (Phe) are nondegenerate, the third position of TTT is twofold degenerate, and the third position of codon GTT (Val) is fourfold degenerate (see Table 1.3). In the universal genetic code, the third positions of the three isoleucine codons are treated for simplicity as twofold degenerate sites, although in reality the degree of degeneracy at these positions is threefold. In vertebrate mitochondrial genes, there are only two codons for isoleucine, and the third position in these codons is indeed a twofold degenerate site (see Table 1.4). Using the above rules, one first counts the numbers of the three types of sites for each of the two sequences, and then computes the averages, denoting them by L_0 (nondegenerate), L_2 (twofold degenerate), and L_4 (fourfold degenerate).

From the above classification of nucleotide sites, we can calculate the number of substitutions between two coding sequences for the three types of sites separately. The nucleotide differences in each class are further classified into transitional (S_i) and transversional (V_i) differences, where $i = 0$, 2, and 4 denote nondegeneracy, twofold degeneracy and fourfold degeneracy, respectively. Note that by definition, all the substitutions at nondegenerate sites are nonsynonymous. Similarly, all the substitutions at fourfold degenerate sites are synonymous. At twofold degenerate sites, transitional changes (C $\leftrightarrow$ T and A $\leftrightarrow$ G) are synonymous, whereas all the other changes, which are transversions, are nonsynonymous. There are no exceptions to this rule in the vertebrate mitochondrial genetic code. In the universal genetic code, on the other hand, there are two exceptions: (1) the first position of four arginine codons (CGA, CGG, AGA, and AGG), in which one type of transversion is synonymous while the other type is nonsynonymous; and (2) the last position in the three isoleucine codons (ATT, ATC, and ATA). In the first case, C $\leftrightarrow$ A transversions in the first codon position are included in S_2, and C $\leftrightarrow$ T and C $\leftrightarrow$ G changes are included in V_2. In the second case, T $\leftrightarrow$ C, T $\leftrightarrow$ A, and C $\leftrightarrow$ A changes in the third codon position are included in S_2, and T $\leftrightarrow$ G, C $\leftrightarrow$ G, and A $\leftrightarrow$ G changes are included in V_2. Similar methodological adjustments may be required for other genetic codes.

The proportion of transitional differences at i-fold degenerate sites between two sequences is calculated as

$$P_i = \frac{S_i}{L_i} \quad (3.31)$$

Similarly, the proportion of transversional differences at i-fold degenerate sites between two sequences is

$$Q_i = \frac{V_i}{L_i} \quad (3.32)$$

Kimura's (1980) two-parameter method is used to estimate the numbers of transitional (A_i) and transversional (B_i) substitutions per ith type site. The means are given by

$$A_i = \frac{1}{2}\ln(a_i) - \frac{1}{4}\ln(b_i) \quad (3.33)$$

and

$$B_i = \frac{1}{2}\ln(b_i) \quad (3.34)$$

The variances are given by

$$V(A_i) = \frac{a_i^2 P_i + c_i^2 Q_i - (a_i P_i + c_i Q_i)^2}{L_i} \quad (3.35)$$

and

$$V(B_i) = \frac{b_i^2 Q_i (1 - Q_i)}{L_i} \quad (3.36)$$

where L_i is the number of i-class degeneracy sites, $a_i = 1/(1 - 2P_i - Q_i)$, $b_i = 1/(1 - 2Q_i)$, and $c_i = (a_i - b_i)/2$. The total number of substitutions per ith type degenerate site, K_i, is given by

$$K_i = A_i + B_i \quad (3.37)$$

with an approximate sampling variance of

$$V(K_i) = \frac{a_i^2 P_i + d_i^2 Q_i - (a_i P_i + d_i Q_i)^2}{L_i} \quad (3.38)$$

where $d_i = b_i + c_i$. We note that A_2 and B_2 denote the numbers of synonymous and nonsynonymous substitutions per twofold degenerate site, respectively, $K_4 = A_4 + B_4$ denotes the number of synonymous substitutions per fourfold degenerate site, and $K_0 = A_0 + B_0$ denotes the numbers of nonsynonymous substitutions per nondegenerate site. These formulas can be used to compare the rates of substitution among the three different types of sites.

We denote by K_S the number of synonymous substitutions per synonymous site, and by K_A the number of nonsynonymous substitution per nonsynonymous site. Noting that one-third of a twofold degenerate site is synonymous and two-thirds are nonsynonymous, K_S and K_A may be obtained by

$$K_S = \frac{3(L_2 A_2 + L_4 K_4)}{L_2 + 3L_4} \quad (3.39)$$

and

$$K_A = \frac{3(L_2B_2 + L_0K_0)}{2L_2 + 3L_0} \quad (3.40)$$

Since transitional substitutions tend to occur more often than transversional substitutions, and since most transitional changes at twofold degenerate sites are synonymous changes, counting a twofold degenerate site as a one-third synonymous site will tend to overestimate the number of synonymous substitutions and to underestimate the number of nonsynonymous substitutions. To overcome these problems, Li (1993) and Pamilo and Bianchi (1993) proposed to calculate the number of synonymous substitutions by taking $(L_2A_2 + L_4A_4)/(L_2 + L_4)$, i.e., the weighted average of A_2 and A_4, as an estimate of the transitional component of nucleotide substitution at twofold and fourfold degenerate sites. Consequently, the number of (synonymous) substitutions per synonymous site is computed as

$$K_S = \frac{L_2A_2 + L_4A_4}{L_2 + L_4} + B_4 \quad (3.41)$$

Similarly, the weighted average $(L_0B_0 + L_2B_2)/(L_0 + L_2)$ is used as an estimate of the mean transversional number of substitutions at nondegenerate and twofold degenerate sites. The number of (nonsynonymous) substitutions per nonsynonymous site is given by

$$K_A = A_0 + \frac{L_0B_0 + L_2B_2}{L_0 + L_2} \quad (3.42)$$

The approximate variances of K_S and K_A are given by

$$V(K_S) = \frac{L_2^2V(A_2) + L_4^2V(A_4)}{(L_2 + L_4)^2} + V(B_4) - \frac{2b_4Q_4[a_4P_4 - c_4(1 - Q_4)]}{L_2 + L_4} \quad (3.43)$$

and

$$V(K_A) = V(A_0) + \frac{L_0^2V(B_0) + L_2^2V(B_2)}{(L_0 + L_2)^2} - \frac{2b_0Q_0[a_0P_0 - c_0(1 - Q_0)]}{L_0 + L_2} \quad (3.44)$$

Indirect estimations of the number of nucleotide substitutions

In estimating the number of nucleotide substitutions between two sequences (K), the highest resolution is obtained by comparing the nucleotide sequences themselves. However, K can also be estimated indirectly from other types of molecular data, such as those obtained by restriction enzyme mapping or DNA–DNA hybridization. Many methods have been developed to estimate K from such data (e.g., Britten et al. 1974; Upholt 1977; Nei and Li 1979; Engels 1981a; Kaplan 1983; Nei and Tajima 1983; Powell and Caccone 1990; Skiena and Sundaram 1994; Weigel and Scherba 1997). Indirect estimates of K values are subject to much larger sampling errors than those based on direct comparisons of nucleotide sequences.

NUMBER OF AMINO ACID REPLACEMENTS BETWEEN TWO PROTEINS

From the comparison of two amino acid sequences, we can calculate the observed proportion of different amino acids between the two sequences as

$$p = \frac{n}{L} \tag{3.45}$$

where n is the number of amino acid differences between the two sequences and L is the length of the aligned sequences.

A simple model that can be used to convert p into the number of amino acid replacements between two sequences is the Poisson process (Appendix II). The number of amino acid replacements per site, d, is estimated as

$$d = -\ln(1-p) \tag{3.46}$$

The variance of d is estimated as

$$V(d) = \frac{p}{L(1-p)} \tag{3.47}$$

ALIGNMENT OF NUCLEOTIDE AND AMINO ACID SEQUENCES

Comparison of two homologous sequences involves the identification of the location of deletions and insertions that might have occurred in either of the two lineages since their divergence from a common ancestor. The process is referred to as **sequence alignment**. We illustrate the process of alignment by using DNA sequences, but the same principles and procedures can be used to align amino acid sequences. As a matter of fact, one usually obtains more reliable alignments by using amino acid sequences than by using DNA sequences. There are two reasons for this: (1) amino acids change less frequently during evolution than nucleotides; and (2) there are 20 amino acids and only four nucleotides, so that the probability for two sites to be identical by chance is lower at the amino acid level than at the nucleotide level.

A DNA sequence alignment consists of a series of paired bases, one base from each sequence. There are three types of aligned pairs: **matches, mismatches**, and **gaps**. A matched pair is one in which the same nucleotide appears in both sequences, i.e., it is assumed that the nucleotide at this site has not changed since the divergence between the two sequences. A mismatched pair is a pair in which different nucleotides are found in the two sequences, i.e., at least one substitution has occurred in one of the sequences since their divergence from each other. A gap is a pair consisting of a base from one sequence and a **null base** from the other. Null bases are denoted by –. A gap indicates that a deletion has occurred in one sequence or an insertion has occurred in the other. However, the alignment itself does not allow us to distinguish between these two possibilities.

Consider the case of two DNA sequences, A and B, of lengths m and n, respectively. If we denote the number of matched pairs by x, the number of mismatched pairs by y and the total number of pairs containing a null base by z, we obtain

$$n + m = 2(x + y) + z \tag{3.48}$$

A distinction is sometimes made between **terminal** and **internal gaps**. For example, when aligning a partial sequence of a gene with a complete sequence of a homologous gene, it makes no sense to include the terminal gaps (i.e., the missing data from the first sequence) into the calculation. Internal gaps are sometimes also excluded from the calculation, for instance, in alignments between genomic sequences that contain introns and processed mRNA sequences that do not.

In evolutionary terms, each pair in an alignment represents an inference concerning **positional homology**, i.e., a claim to the effect that the two members of the pair descended from a common ancestral nucleotide. An error in an alignment means that an ancestral position has not been identified correctly, and consequently inferences of the number of substitutions will be incorrect.

Given that alignment is the first step in many evolutionary studies, and that errors in alignment tend to amplify in later computational stages, we must construct alignments very carefully. One should therefore remove all ambiguous parts of an alignment before any further analyses, even if the total aligned length decreases significantly and the sampling error associated with the estimate of the number of nucleotide substitutions between two sequences increases concomitantly.

Manual alignment by visual inspection

When there are few gaps and the two sequences are not too different from each other in any other respect, a reasonable alignment can be obtained by visual inspection using either specialized alignment editors or plain text editors. The advantages of this method include: (1) it uses the most powerful and trainable of all tools—the brain, and (2) it allows the direct integration of additional data, such as knowledge of domain structure (Chapter 6) or, more importantly, intuitive biological models that are not easily quantifiable. The main disadvantages of this method is that it is subjective and unscalable, i.e., its results cannot be compared to those derived from other methods.

The dot matrix method

In the **dot matrix method** (Gibbs and McIntyre 1970), the two sequences to be aligned are written out as column and row headings of a two-dimensional matrix (Figure 3.7). A dot is put in the **dot matrix plot** at a position where the

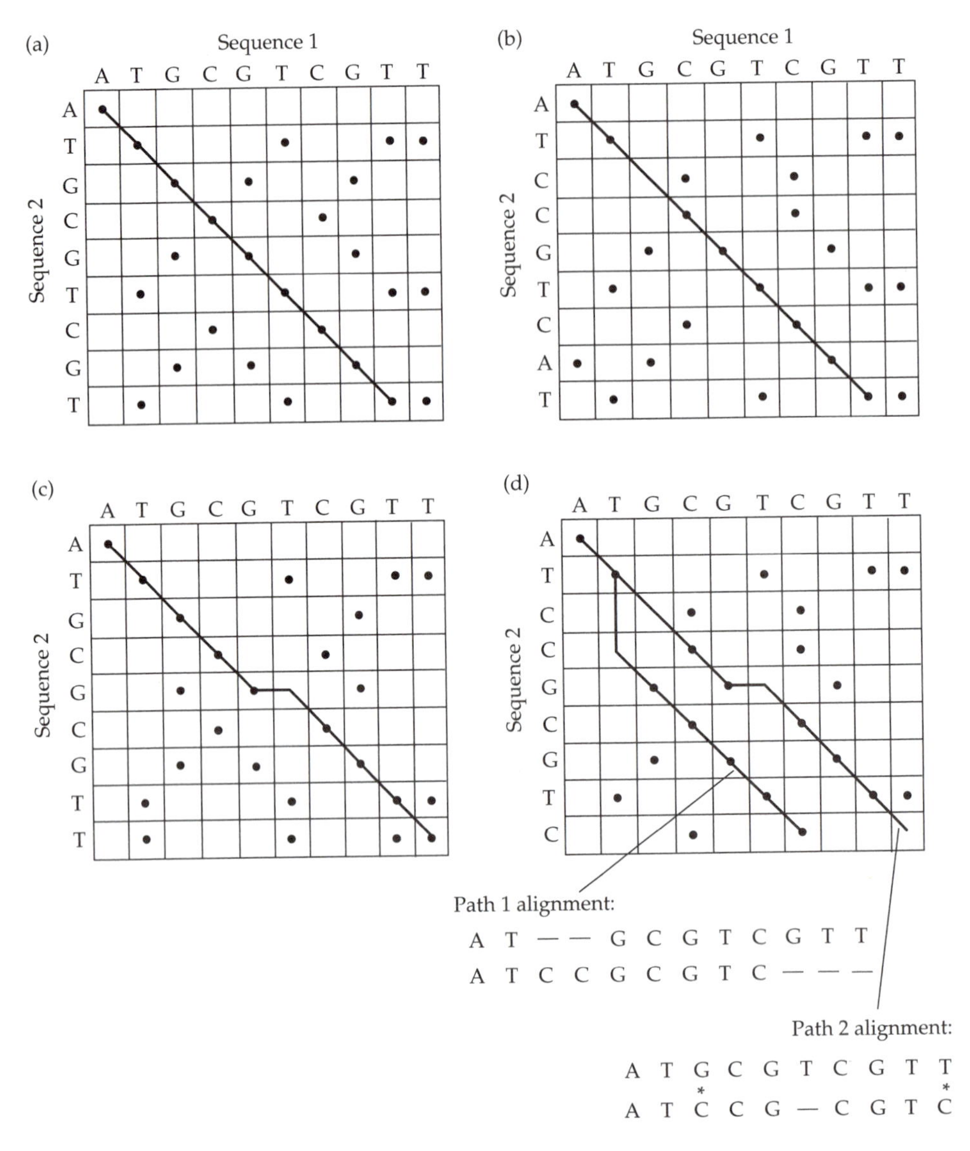

FIGURE 3.7 Dot matrices for aligning nucleotide sequences. (a) The two sequences are identical in the aligned part. (b) The two sequences differ from each other, the aligned part contains no gaps. (c) The alignment contains a gap, but otherwise the sequences are identical to each other. (d) Two possible alignment paths—containing both mismatches and gaps—are shown. Path 1 consists of six diagonal steps, none of which is empty, and two vertical steps. Path 2 contains eight diagonal steps, two of which are empty, and one horizontal step. The decision between the alignment in path 1 and that in path 2 depends on whether or not terminal gaps are taken into account, as well as on the type of gap penalty used, i.e., which evolutionary sequence of events is more probable: a two-nucleotide gap (—) as in path 1, or a one-nucleotide gap and two substitutions (*) as in path 2.

nucleotides in the two sequences are identical. That is, a dot plotted at point (x,y) indicates that the nucleotide at position x in the first sequence is the same as the nucleotide at position y in the second sequence.

The alignment is defined by a path through the matrix starting with the upper-left element and ending with the lower-right element. There are four possible types of steps in this path: (1) a diagonal step through a dot indicates a match, (2) a diagonal step through an empty element of the matrix indicates a mismatch, (3) a horizontal step indicates a null nucleotide in the sequence on the top of the matrix, and (4) a vertical step indicates a null nucleotide in the sequence on the left of the matrix.

If the two sequences are completely identical (or if a sequence is compared to itself), there will be dots in all the diagonal elements of the matrix (Figure 3.7a). If the two sequences differ from each other only by nucleotide substitutions, there will be dots in most of the matrix elements on the diagonal (Figure 3.7b). If an insertion occurred in one of the two sequences but there are no substitutions, there will be an area within the matrix where the alignment diagonal will be shifted either vertically or horizontally (Figure 3.7c). In each of these three cases, the alignment is self-evident. If the two sequences differ from each other by both gaps and nucleotide substitutions, however, it may be difficult to identify the location of the gaps and to choose between several alternative alignments (Figure 3.7d). In such cases, visual inspection and the dot matrix may not be reliable, so more rigorous methods are required.

As seen in Figure 3.8a, a dot matrix may become very cluttered. That is, many elements in the matrix, other than those representing the real alignment, are occupied by dots, thereby obscuring the alignment. The reason is that when comparing two DNA sequences, approximately 25% of the elements in the matrix will be occupied by dots by chance alone.

There are two parameters, **window size** and **stringency**, that determine the number of spurious matches and, hence, the resolution of the dot matrix plot. Thus, instead of using single-nucleotide sites, the sequences may be compared by using overlapping (sliding) fixed-length windows, and each comparison within the matrix is required to achieve a certain minimum threshold score summed over the window (stringency) to qualify as a match. For example, in Figure 3.8b, we use a window size that is three base pairs long, and a dot is put in the matrix only when at least two out of the three nucleotides show a match between the two sequences. Note that the alignment path is now more evident on a less cluttered background. For coding regions, one may use the amino acid sequences instead of the DNA sequences. The increase in the number of possible character states (letters) from 4 to 20 and the decrease in the lengths of the sequences from L to $L/3$ will greatly decrease the number of spurious dots (Figure 3.8c).

We note, however, that dot matrix methods also rely on the power of human cognition to recognize patterns indicative of similarity and to add gaps to the sequences to achieve an alignment. Moreover, it is very difficult to be sure that one has obtained the best possible alignment with the sequences at hand. For this reason, scientists developed criteria for quantifying

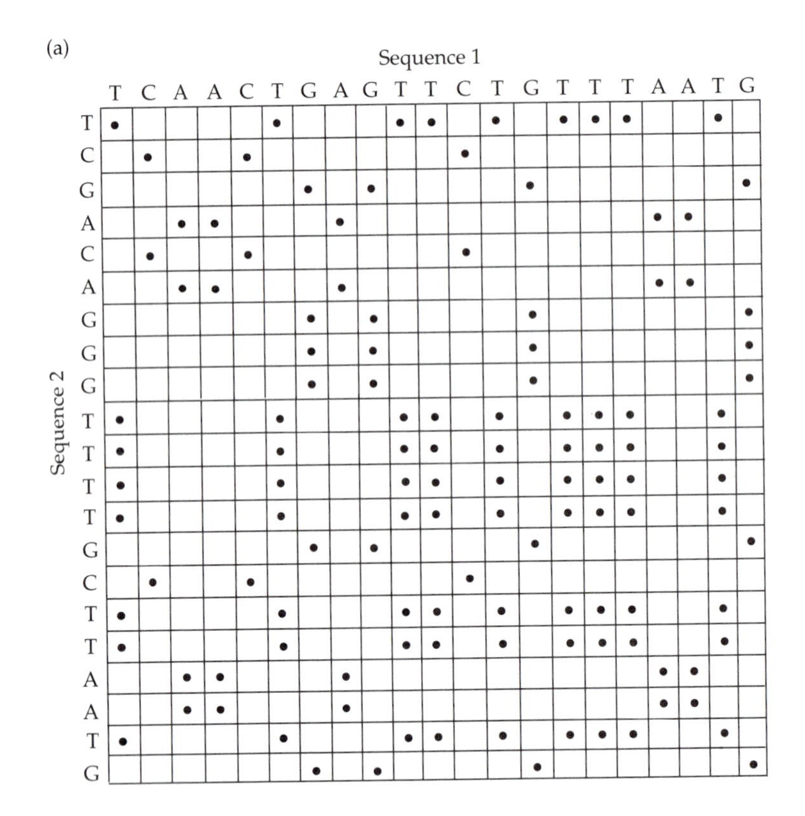

alignments and algorithms for identifying the best alignment according to the criteria chosen.

One should add, however, that dot matrix plots are very useful in their own right in unraveling important information on the evolution of sequences. For example, a dot matrix comparison between the amino acid sequences of human μ-crystallin (a lens protein) and glutamyl-tRNA reductase (an enzyme involved in the synthesis of the pigment porphyrin) from the bacterium *Salmonella typhimurium* indicates quite clearly that these two seemingly unrelated proteins share a common evolutionary origin (Figure 3.9).

Distance and similarity methods

The best possible alignment between two sequences, or the **optimal alignment**, is the one in which the numbers of mismatches and gaps are minimized according to certain criteria. Unfortunately, reducing the number of mismatches usually results in an increase in the number of gaps, and vice versa.

For example, consider the following two sequences, A and B:

A: T C A G A C G A T T G $L_A = 11$
B: T C G G A G C T G $L_B = 9$

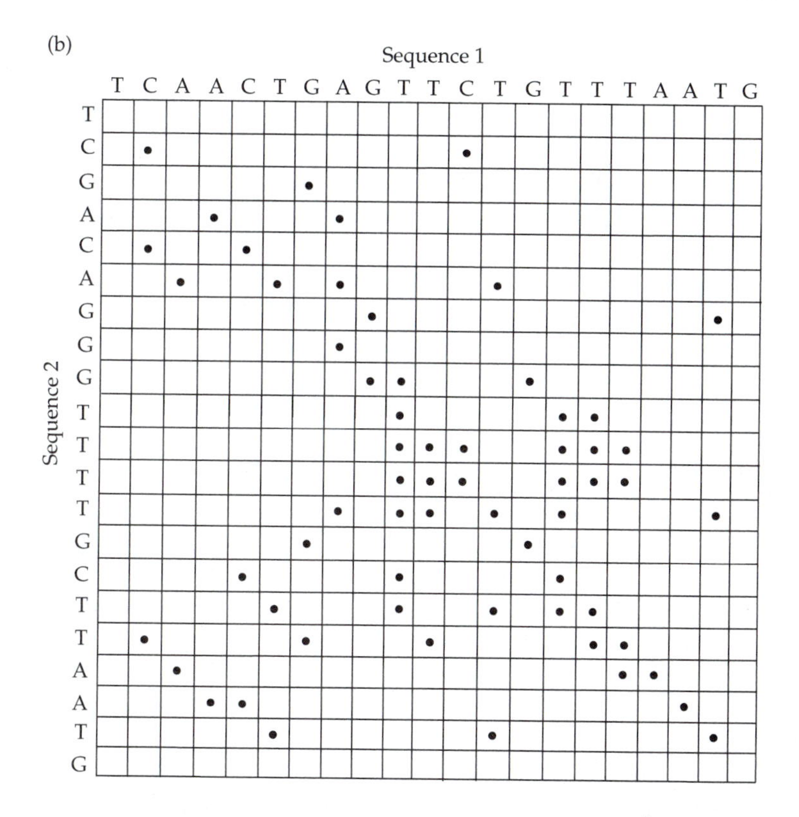

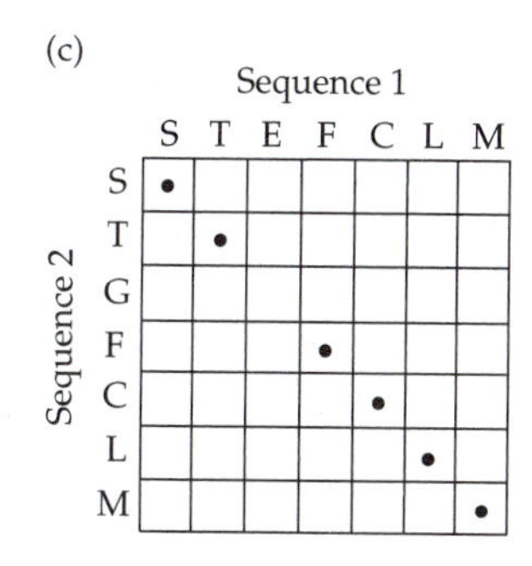

FIGURE 3.8 (a) Dot matrix for two nucleotide sequences. The alignment is obscured by the many spurious dots in the matrix. (b) Dot matrix for the two nucleotide sequences in (a), obtained by using a sliding window three nucleotides long and setting up a stringency threshold of two out of three matches for putting a dot in the relevant element. This filtering method clears many of the spurious dots, and the alignment now stands out on a less cluttered background. (c) Dot matrix for the two amino acid sequences obtained by translating the nucleotide sequences in (a). The alignment is now unambiguous.

We can reduce the number of mismatches to zero as follows:

(I)
```
T C A G – A C G – A T T G
T C – G G A – G C – T – G
```

The number of gaps in this case is 6. Conversely, the number of gaps can be reduced to a single gap having the minimum possible size of $|L_A - L_B| = 2$ nucleotides, with a consequent increase in the number of mismatches:

```
        T C A G A C G A T T G
(II)        *       *   *  *  *
        T C G G A G C T G - -
```

In this example, there is only one gap (two nucleotides in length), but the number of mismatches (indicated by asterisks) has increased to 5 (1 transition and 4 transversions).

Alternatively, we can choose an alignment that minimizes neither the number of gaps nor the number of mismatches. For example,

```
        T C A G - A C G A T T G
(III)                   *   *
        T C - G G A - G C T G -
```

In this case the number of mismatches is 2 (both transversions), and the number of gaps is 4.

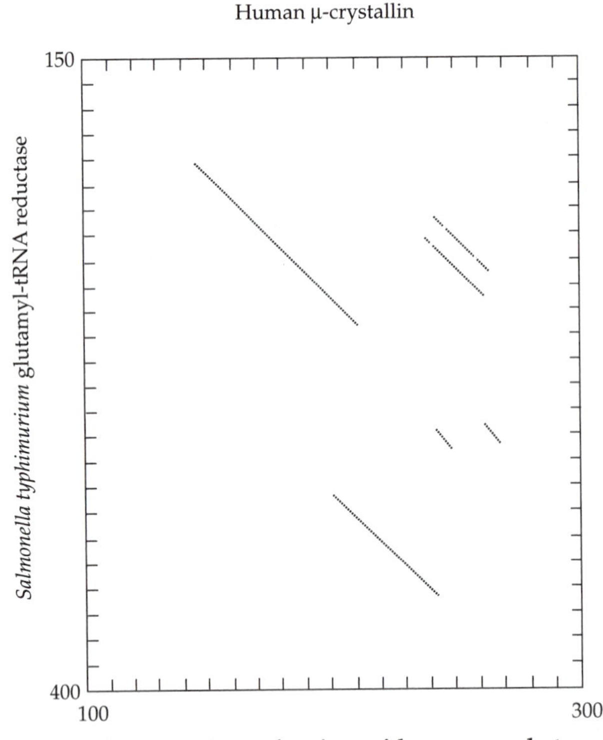

FIGURE 3.9 Dot matrix comparison of amino acid sequences between human μ-crystallin and glutamyl-tRNA reductase from *Salmonella typhimurium*. Only the C-terminal ends of the proteins were used. The window size was 60 amino acids and the stringency was 24 matches. Diagonals show regions of similarity. The vertical gap indicates that a coding region corresponding to approximately 75 amino acids has either been deleted from the human gene or inserted into the bacterial gene. The two diagonally oriented parallel lines in the upper right-hand corner of the plot most probably indicate that a small internal duplication has occurred in the bacterial gene. Modified from Segovia et al. (1997).

So, which of the three alignments is preferable? It is obvious that comparing mismatches with gaps is like comparing apples with oranges. As a consequence, we must find a common denominator with which to compare gaps and mismatches. The common denominator is called the **gap penalty** or **gap cost**. The gap penalty is a factor (or a set of factors) by which gap values (the numbers and lengths of gaps) are multiplied to make the gaps equivalent in value to the mismatches. The gap penalties are based on our assessment of how frequent different types of insertions and deletions occur in evolution in comparison with the frequency of occurrence of point substitutions. Of course, we must also assign **mismatch penalties**, i.e., an assessment of how frequently substitutions occur.

For any given alignment, we can calculate a **distance** or **dissimilarity index** (D) between the two sequences in the alignment as

$$D = \sum m_i y_i + \sum w_k z_k \qquad \textbf{(3.49)}$$

where y_i is the number of mismatches of type i, m_i is the mismatch penalty for an i-type of mismatch, z_k is the number of gaps of length k, and w_k is a positive number representing the penalty for gaps of length k.

Alternatively, the similarity between two sequences in an alignment may be measured by a **similarity index** (S). For any given alignment, the similarity between two sequences is

$$S = x - \sum w_k z_k \qquad \textbf{(3.50)}$$

where x is the number of matches, z_k is the number of gaps of length k, and w_k is a positive number representing the penalty for gaps of length k.

In the most frequently used gap penalty systems, it is assumed that the gap penalty has two components, a **gap-opening penalty** and a **gap-extension penalty**. There are many systems of assigning gap-extension penalties, depending on certain *a priori* notions on how frequently deletions and insertions of a certain length occur relative to nucleotide substitutions. In Figure 3.10, we present three systems. In the **fixed gap penalty system**, there are no gap-extension costs. In the **affine** or **linear gap penalty system**, the gap-extension cost is calculated by multiplying the gap length minus 1 by a constant representing the gap-extension penalty for increasing the gap by 1. For example, for a gap of length 1, the gap cost consists of the gap-opening penalty only; for a gap of length 3, the gap cost consists of the gap-opening penalty plus twice the gap extension penalty. Because the gap-extension cost for long gaps can become very large in the linear system, some researchers have proposed the **logarithmic gap penalty system** as a way of reducing the cost of long gaps (e.g., Gu and Li 1995). In this system, the gap-extension penalty increases more slowly with gap length.

Further complications in the gap penalty system may be introduced by distinguishing among different mismatches. In the case of amino acid sequences, for instance, we may classify mismatches by similarity between pairs of amino acids depending on the degree of conservation of chemical properties between them. For example, a mismatched pair consisting of Leu and Ile, which have very similar biochemical properties, may be given a lesser penalty

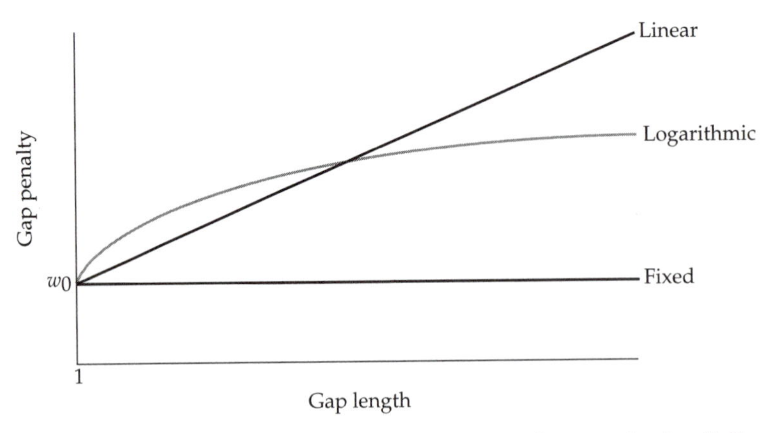

FIGURE 3.10 Three gap penalty systems. The gap-opening penalty in all three systems is w_0. In the linear gap penalty system, the gap-extension cost is the product of the gap length minus 1 and the gap-extension penalty for increasing the gap by 1. In the logarithmic gap penalty system, the gap-extension penalty increases with the logarithm of the gap length. In the fixed gap penalty system, there are no gap-extension costs.

than a mismatched pair consisting of Arg and Glu, which are very dissimilar from each other (Chapter 4).

Let us now compare alignments I, II, and III, by using a linear gap penalty system in which the mismatch penalty is 1, the gap-opening penalty is 2, and the gap-extension penalty is 6. The dissimilarity for alignment I is $D = (0 \times 1) + (6 \times 2) + 6(1 - 1) = 12$. The D values for alignments II and III are 12 and 10, respectively. Therefore, out of these three alignments, alignment III is judged to be the best. By using a different penalty system, in which the mismatch penalty is 1, the gap-opening penalty is 3, and the gap-extension penalty is 0, we obtain D values of 18, 8, and 14, for alignments I, II, and III, respectively. Thus, alignment II is judged to be the best. (Note that alignments II and III can be improved by visual inspection by aligning without mismatch the last two nucleotides from the two sequences.)

Alignment algorithms

The purpose of any alignment algorithm is to choose the alignment associated with the smallest D (or the largest S) from among all possible alignments. We note, however, that the number of possible alignments may be very large even for short sequences; for sequences usually encountered in molecular evolutionary studies, the number of possible alignments may literally be astronomical. For example, when two sequences 300 residues long each are compared, there are 10^{88} possible alignments, if any number of gaps of any length are allowed to occur. (In comparison, the number of elementary particles in the universe is only 10^{80}.) Fortunately, there are computer algorithms for searching for the optimal alignment between two sequences that do not require an exhaustive search of all the possibilities. Among the most frequently used methods are those of Needleman and Wunsch (1970) and Sellers (1974). Needleman and Wunsch's method uses similarity indices. In Sellers' method, distance indices are used. The two methods have been shown to be equivalent under certain conditions (Smith et al. 1981).

The **Needleman-Wunsch algorithm** uses **dynamic programming**, which is a general computational technique used in many fields of study. It is applicable when large searches can be divided into a succession of small stages such that (1) the solution of the initial search stage is trivial, (2) each partial solution in a later stage can be calculated by reference to only a small number of solutions in an earlier stage, and (3) the final stage contains the overall solution. Dynamic programming can be applied to alignment problems because similarity indices obey the following rule:

$$S_{1\to x,1\to y} = \max S_{1\to x-1,1\to y-1} + S_{x,y} \quad \textbf{(3.51)}$$

in which $S_{1\to x,1\to y}$ is the similarity index for the two sequences up to residue x in the first sequence and residue y in the second sequence, $\max S_{1\to x-1,1\to y-1}$ is the similarity index for the best alignment up to residues $x - 1$ in the first sequence and $y - 1$ in the second sequence, and $S_{x,y}$ is the similarity score for aligning residues x and y.

Alternatively, we may use dynamic programming with dissimilarity indices as follows:

$$D_{1\to x,1\to y} = \min D_{1\to x-1,1\to y-1} + D_{x,y} \quad \textbf{(3.52)}$$

in which $D_{1\to x,1\to y}$ is the dissimilarity index for the two sequences up to residue x in the first sequence and residue y in the second sequence, $\min D_{1\to x-1,1\to y-1}$ is the dissimilarity index for the best alignment up to residues $x - 1$ in the first sequence and $y - 1$ in the second sequence, and $D_{x,y}$ is the dissimilarity score for aligning residues x and y.

An alignment is calculated in two stages. First, the two sequences are arranged in the same way as in a dot matrix. For each element in the matrix, say x and y, the similarity index, $S_{1\to x,1\to y}$, is calculated. At the same time, the position of the best alignment score in the previous row or column is stored. This stored value is called a **pointer**. The relationship between the new $S_{1\to x,1\to y}$ value and the pointer is represented by an arrow. In the second stage the alignment is produced by starting at the highest similarity score in either the rightmost column or the bottom row, and proceeding from right to left by following the best pointers. This stage is called the **traceback**. The graph of pointers in the traceback is also referred to as the **path graph** because it defines the paths through the matrix that correspond to the optimal alignment.

Figure 3.11 shows a simple example of a dynamic programming alignment of the sequences ATGCG and ATCCGC. To make the example as simple as possible, we define $w_k = 0$, i.e., no penalty is applied for either opening gaps or extending them. The optimal alignment is therefore the alignment with the most matches. The matrix is filled from left to right and from top to bottom. In the first row, we have a single match, which receives a score of 1,

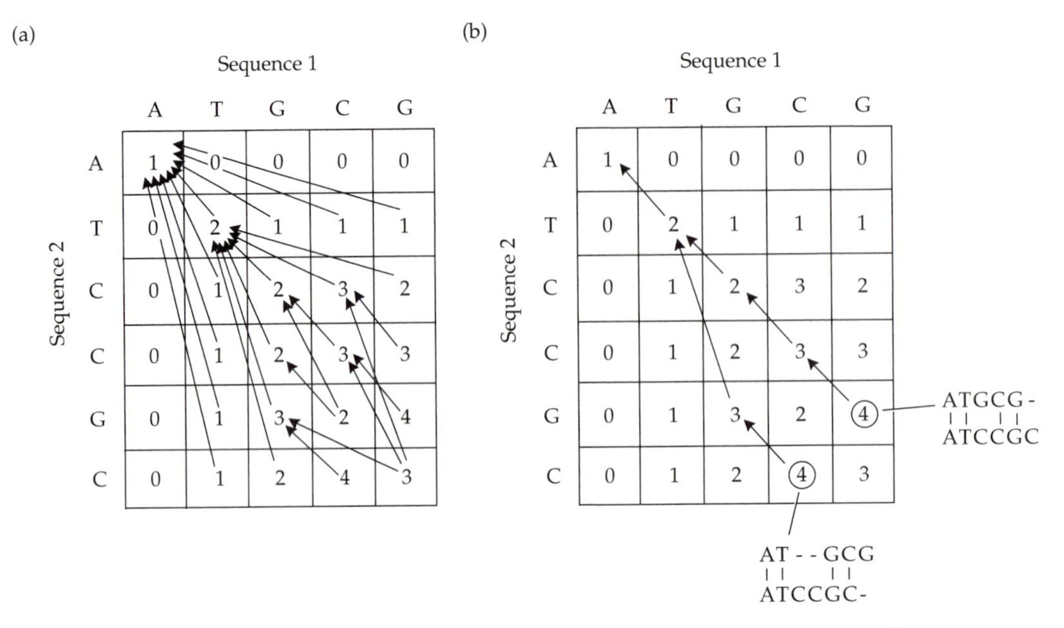

FIGURE 3.11 Calculation of a dynamic programming alignment. (a) The pointer values and paths connecting the pointers (arrows) are shown. In this calculation, matches receive a positive score of 1, and no penalty is applied to either mismatches or gaps. (b) Tracebacks start with the highest scores on the edges of the matrix (circled). Two traceback graphs and their corresponding optimal alignments are shown.

and four mismatches, each of which receives a score of 0. In the second row, we start by pairing A in the first sequences with T in the second. Since this is a mismatch and there can be no previous best alignments, the score is 0. For the second element in the second row, we pair T from the first sequence with T from the second. This is a match. Therefore, a score of 1 is added to the previous best score, i.e., 1 in row 1 column 1, and the total is written down. An arrow is drawn from this element to its pointer. We continue this process until the matrix is completed (Figure 3.11a). We note that an element may have more than a single pointer, and that arrows from many elements may lead to the same pointer. In the second stage, we start with the highest scores in the last row and column (circled numbers in Figure 3.11b), and complete the traceback by choosing in each step the pointer (or pointers) with the highest values. The path or paths defined by the traceback represent the optimal alignments.

The most important thing to remember is that the resulting alignment depends on the choices of gap penalties, which in turn depend on crucial assumptions about how frequently gap events occur in evolution relative to the frequency of point substitutions. For example, consider the alignment of human pancreatic hormone precursor and chicken pancreatic hormone (Fig-

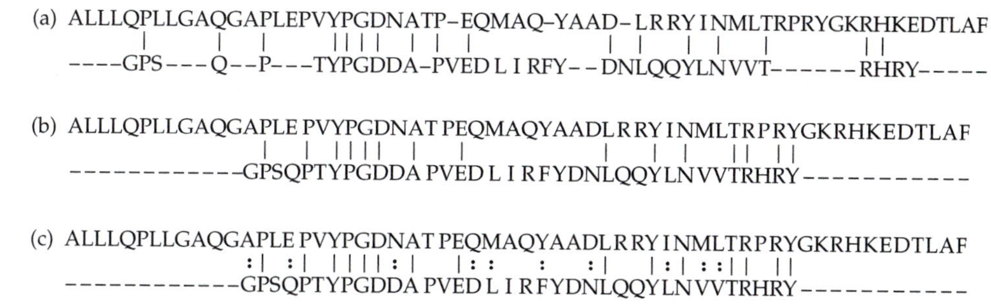

FIGURE 3.12 The effect of gap penalties on an amino acid alignment. The alignment of the human pancreatic hormone precursor and the chicken pancreatic hormone are shown. Perfect matches (identities) are indicated by vertical straight lines. (a) The penalty for gaps is 0. (b) The gap penalty for a gap of size k nucleotides was set at $wk = 1 + 0.1k$. (c) The same alignment as in (b), but the similarity between the two sequences is enhanced by showing pairs of biochemically similar amino acids (dots).

ure 3.12). When the two sequences are aligned with no gap penalties, the similarity between these homologous sequences is not evident (Figure 3.12a). However, when penalties are applied for the introduction of gaps ($w_k = 1 + 0.1k$, where k is the size of the gap), the similarity is clear (Figure 3.12b). The resemblance becomes even more enhanced if we consider not only perfect matches (**identities**) between the amino acids but also imperfect matches (**similarities**) between them (Figure 3.12c).

Multiple alignments

Multiple sequence alignment can be viewed as an extension of pairwise sequence alignment, but the complexity of the computation grows exponentially with the number of sequences being considered and, therefore, it is not feasible to search exhaustively for the optimal alignment. Several heuristic methods have been proposed in the literature, and computer programs are available for multiple alignment. The most popular such programs are MACAW (Schuler et al. 1991), CLUSTAL (Higgins and Sharp 1988, 1989), and MASH (Chappey et al. 1991). Most these programs use some sort of incremental or progressive algorithm (e.g., Feng and Doolittle 1987), in which a new sequence is added to a group of already aligned sequences in order of decreasing similarity. The process starts by calculating all possible pairwise alignments among the sequences under study, then identifying the pair with the highest similarity score. To improve the quality of multiple alignments, one may use auxiliary data, such as tertiary structure (Barton and Sternberg 1987) and phylogenetic data (Hein 1989). It is usually advisable to take a good look at the final multiple alignment, as such alignments can be frequently improved by visual inspection.

FURTHER READINGS

Doolittle, R. F. (ed.) 1990. *Molecular Evolution: Computer Analyses of Protein and Nucleic Acid Sequences*. Academic Press, San Diego.

Gribskov, M. and J. Devereux (eds.). 1991. *Sequence Analysis Primer*. Stockton Press, New York.

Harvey, P. H. and M. D. Pagel. 1991. *The Comparative Method in Evolutionary Biology*. Oxford University Press, Oxford.

McClure, M. A., T. K. Vasi, and W. M. Fitch. 1994. Comparative analysis of multiple-sequence alignment methods. Mol. Biol. Evol. 11: 571–592.

Nei, M. 1987. *Molecular Evolutionary Genetics*, Columbia University Press, New York.

Pearson, W. R. and W. Miller. 1992. Dynamic programming algorithms for biological sequence comparison. Methods Enzymol. 210: 575–601.

Sankoff, D. and J. B. Kruskal (eds.). 1983. *Time Warps, String Edits, and Macromolecules: The Theory and Practice of Sequence Comparison*. Addison-Wesley, Reading, MA.

Taylor, W. R. 1996. Multiple protein sequence alignment: Algorithms and gap insertion. Methods Enzymol. 266: 343–367.

Chapter 4

Rates and Patterns of Nucleotide Substitution

The theory developed in the preceding chapter can be used to calculate the rate of nucleotide substitution, which is one of the most basic quantities in the study of molecular evolution. Indeed, in order to characterize the evolution of a DNA sequence, the first thing we need to know is how fast it evolves. It is also interesting to compare the substitution rates among genes or among different DNA regions, because this can help us understand the mechanisms responsible for the different rates of nucleotide substitution during evolution. In this chapter, we present data on the rates and patterns of nucleotide substitution and discuss three factors affecting them: (1) functional constraint, (2) positive selection, and (3) mutational input. We also dissect the substitution rate into its constituent parts in order to infer the pattern of substitution, in particular the pattern of spontaneous mutation.

Knowing the rate of nucleotide substitution may also enable us to date evolutionary events, such as the divergence between species or higher taxa. This raises the issue of how variable the rate is among different evolutionary lineages. We investigate this variation and attempt to identify the factors affecting it. The rates of evolution of nuclear, organelle, and RNA genomes are also examined.

RATES OF NUCLEOTIDE SUBSTITUTION

The **rate of nucleotide substitution**, r, is defined as the number of substitutions per site per year. The mean rate of substitution can be calculated by dividing the number of substitutions, K, between two homologous sequences by $2T$, where T is the time of divergence between the two sequences (Figure 4.1). That is,

$$r = \frac{K}{2T} \tag{4.1}$$

T is assumed to be the same as the time of divergence between the two species from which the two sequences were taken, and is usually inferred from paleontological and biogeographical data. Equation 4.1 only holds when dealing with distantly related species. When dealing with closely related species, such as humans and chimpanzees, we must take into account the allelic divergence (polymorphism) that occurred within the ancestral population prior to the divergence event (e.g., Takahata and Satta 1997).

In this section we shall deal with the issue of rate variation among genes and among different regions in a gene. For this purpose, it is advisable to use the same species pair for all the genes under consideration. The reason is twofold. First, there are usually considerable uncertainties about paleontological estimates of divergence times. By using the same pair of species we can compare rates of substitution among genes without knowledge of the divergence time. Second, the rate of substitution may vary considerably among lineages. In this case, differences in rates between two genes may be due to dif-

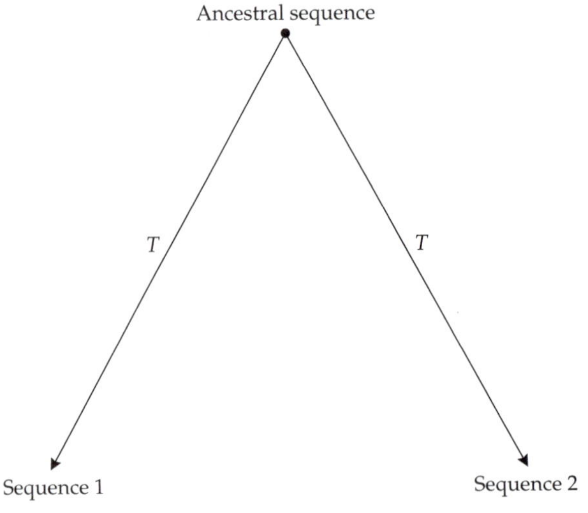

FIGURE 4.1 Divergence of two homologous sequences from a common ancestral sequence T years ago.

ferences between lineages rather than to differences that are attributable to the genes themselves.

Obtaining a reliable estimate of the rate of nucleotide substitution requires that the degree of sequence divergence be neither too small nor too large. If it is too small, the rate estimate will be influenced by large chance effects, whereas if it is too large, the estimate may be unreliable due to the difficulties in correcting for multiple substitutions at the same site (Chapter 3).

Coding regions

The protein-coding regions of genes have attracted the most attention from both molecular and evolutionary biologists because of their functional and medical importance. As a consequence, a large amount of sequence data has become available for these regions, and many comparative studies of nucleotide sequences in these regions have been published (e.g., Ohta 1995). In dealing with protein-coding sequences, it is important to discriminate between nucleotide changes that affect the primary structure of the encoded protein, i.e., nonsynonymous substitutions, and changes that do not affect the protein, i.e., synonymous changes.

In Table 4.1 we list the rates of synonymous and nonsynonymous substitution for 47 protein-coding genes. The genes in each group are arranged in order of increasing rate of nonsynonymous substitution. The rates were obtained from comparisons between human and murid (rat or mouse) homologous genes, by using the method of Li (1993) and Pamilo and Bianchi (1993), and by setting the time for the human–murid divergence event at 80 million years ago.

We note that the rate of nonsynonymous substitution is extremely variable among genes. It ranges from effectively zero in actin α to about 3.1×10^{-9} substitutions per nonsynonymous site per year in interferon γ. Nonsynonymous nucleotide substitutions are, of course, reflected in the rates of protein evolution, which may vary by as much as three orders of magnitude. As is well known, certain proteins (e.g., histones, actins, and ribosomal proteins) are extremely conservative. A very extreme case is ubiquitin, which is completely conserved between human and *Drosophila*, and which differs among animals, plants, and fungi by only 2 or 3 out of 76 amino acid residues. Some peptide hormones (e.g., somatostatin-28, glucagon, and insulin) are also extremely conservative, but others have evolved at intermediate rates (e.g., parathyroid hormone and erythropoietin) or at high rates (e.g., relaxin). The insulin C-peptide has often been used as an example of rapid evolution, but it actually evolves considerably more slowly than relaxin. Hemoglobins, myoglobin, and some carbonic anhydrases have evolved at intermediate rates, while apolipoproteins, immunoglobulins, interferons and interleukins have evolved very rapidly. Apolipoprotein B is a huge protein (4,536 amino acids) and the relatively high nonsynonymous rate in the region included in Table 4.1, which is the most conservative part of the protein, implies that the whole

TABLE 4.1 Rates of synonymous and nonsynonymous nucleotide substitutions (± standard errors) in various mammalian protein-coding genes[a]

Gene	Number of codons compared	Nonsynonymous rate	Synonymous rate
Ribosomal proteins			
S14	150	0.02 ± 0.02	2.16 ± 0.42
S17	134	0.06 ± 0.04	2.69 ± 0.53
Contractile system proteins			
Actin α	376	0.01 ± 0.01	2.92 ± 0.34
Myosin β heavy chain	1933	0.10 ± 0.01	2.15 ± 0.13
Activators, factors, and receptors			
Glucagon	29	0.00 ± 0.00	2.36 ± 1.08
Somatostatin-28	28	0.00 ± 0.00	3.10 ± 1.98
Translation elongation factor 2	857	0.02 ± 0.01	4.37 ± 0.36
Insulin	51	0.20 ± 0.10	3.03 ± 1.02
Prion protein	224	0.29 ± 0.06	3.89 ± 0.63
β1 adrenergic receptor	412	0.45 ± 0.06	3.07 ± 0.35
Insulin-like growth factor II	179	0.57 ± 0.11	2.01 ± 0.37
Atrial natriuretic factor	149	0.72 ± 0.13	3.38 ± 0.60
Erythropoietin	191	0.77 ± 0.12	3.56 ± 0.53
Tumor necrosis factor	231	0.76 ± 0.11	2.91 ± 0.45
Parathyroid hormone	90	1.00 ± 0.20	3.47 ± 0.87
Luteinizing hormone	140	1.05 ± 0.17	2.90 ± 0.54
Insulin C-peptide	31	1.07 ± 0.37	4.78 ± 2.14
Urokinase-plasminogen activator	430	1.34 ± 0.11	3.11 ± 0.35
Growth hormone	189	1.34 ± 0.17	3.79 ± 0.63
Interleukin-1	265	1.50 ± 0.15	3.27 ± 0.46
Relaxin	53	2.59 ± 0.51	6.39 ± 3.75
Blood proteins			
α-globin	141	0.56 ± 0.11	4.38 ± 0.77
Myoglobin	153	0.57 ± 0.11	4.10 ± 0.85
Fibrinogen γ	411	0.58 ± 0.07	4.13 ± 0.46
β-globin	146	0.78 ± 0.14	2.58 ± 0.49
Albumin	590	0.92 ± 0.07	5.16 ± 0.48
Apolipoproteins			
E	291	1.10 ± 0.12	3.72 ± 0.51
β-low-density lipoprotein receptor binding domain	273	1.32 ± 0.14	3.64 ± 0.50
A-I	235	1.64 ± 0.17	3.97 ± 0.63
Immunoglobulins			
Ig V_H	100	1.10 ± 0.20	4.76 ± 1.12
Ig κ	106	2.03 ± 0.30	5.56 ± 1.18

TABLE 4.1 Continued

Gene	Number of codons compared	Nonsynonymous rate	Synonymous rate
Interferons			
α1	166	1.47 ± 0.19	3.24 ± 0.66
β1	159	2.38 ± 0.27	5.33 ± 1.24
γ	136	3.06 ± 0.37	5.50 ± 1.45
Enzymes			
Aldolase A	363	0.09 ± 0.03	2.78 ± 0.33
Hydroxanthine phosphoribosyltransferase	217	0.12 ± 0.04	1.57 ± 0.31
Creatine kinase M	380	0.15 ± 0.03	2.72 ± 0.34
Lipoprotein lipase	437	0.19 ± 0.04	2.95 ± 0.33
Lactate dehydrogenase A	331	0.19 ± 0.04	4.06 ± 0.49
Glyceraldehyde-3-phosphate dehydrogenase	332	0.20 ± 0.04	2.30 ± 0.30
Glutamine synthetase	371	0.23 ± 0.04	2.95 ± 0.33
Thymidine kinase	232	0.43 ± 0.08	3.93 ± 0.59
Amylase	506	0.63 ± 0.06	3.42 ± 0.38
Adenine phosphoribosyltransferase	179	0.68 ± 0.11	3.56 ± 0.59
Carbonic anhydrase I	260	0.84 ± 0.11	3.22 ± 0.47
Average[b]		0.74 (0.67)	3.51 (1.01)

Modified from Li (1997).

[a]All rates are based on comparisons between human and mouse or rat genes. The time of divergence was set at 80 million years ago. Rates are in units of substitutions per site per 10^9 years.

[b]The average is the arithmetic mean, and values in parentheses are the standard deviations computed over all genes.

protein has evolved at an even higher rate. In contrast, another large protein, the myosin β heavy chain (1,933 amino acids), has evolved at a low rate.

The rate of synonymous substitution also varies considerably from gene to gene, though much less than the nonsynonymous rate. (The coefficient of variation for the nonsynonymous rates in Table 4.1 is 91%, whereas that for the synonymous rates is only 29%.)

In the vast majority of genes, the synonymous substitution rate greatly exceeds the nonsynonymous rate. An extreme example illustrating the difference between synonymous and nonsynonymous substitution rates is shown in Figure 4.2. From this comparison, we can determine that the synonymous substitution rate is at least 25 times higher than the nonsynonymous rate. The mean rate of nonsynonymous substitution for the genes in Table 4.1 is

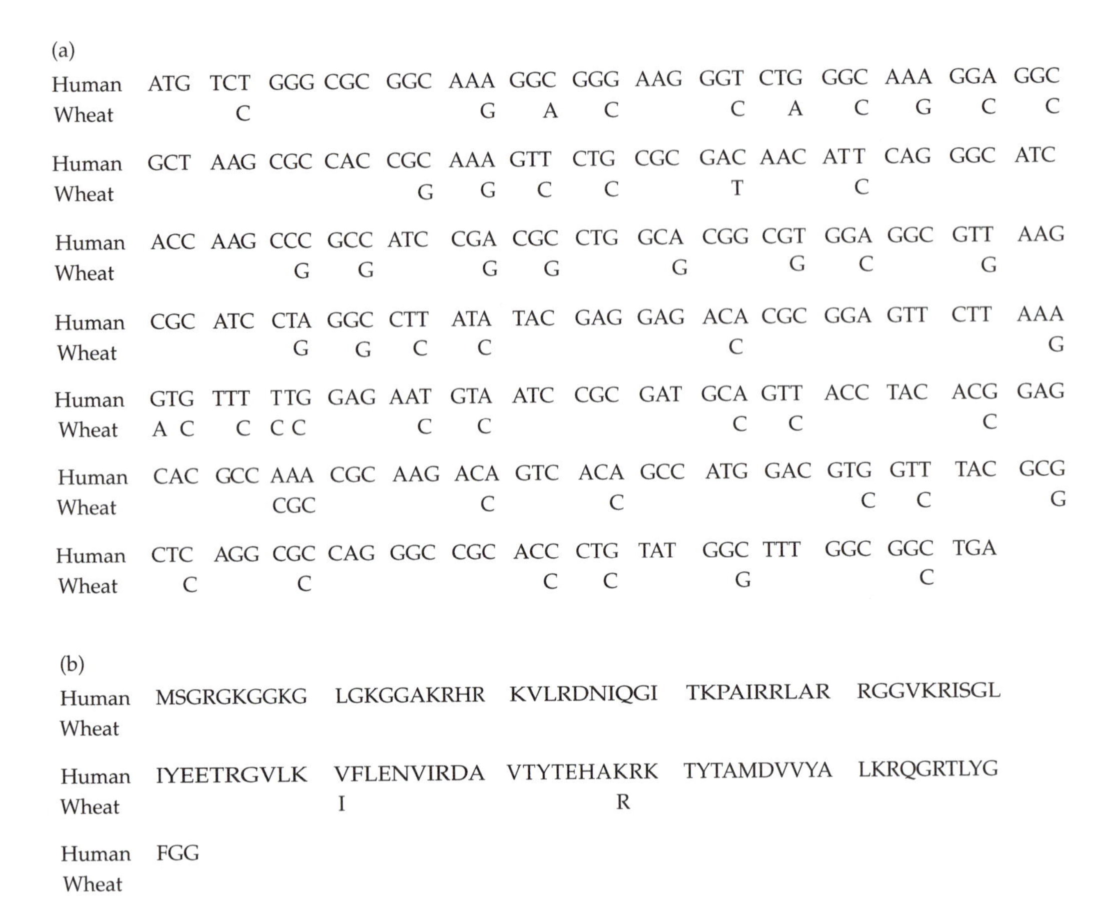

FIGURE 4.2 Preponderance of synonymous substitutions over nonsynonymous substitutions as revealed from the alignment of histone H4 genes (a) and proteins (b) from human and wheat. The human sequences are shown in their entirety; for the wheat sequences, only sites that differ from the human sequences are shown. The two genes differ from each other at 55 nucleotide positions (a), but at only two amino acid positions (b).

0.74×10^{-9} substitutions per nonsynonymous site per year, whereas the mean rate of synonymous substitution is 3.51×10^{-9} substitutions per synonymous site per year, i.e., about five times higher. We note, however, that some extreme values may be due to sampling errors. For example, in the relaxin gene, the synonymous rate of substitution is as high as 6.39×10^{-9} substitutions per synonymous site per year, but this estimate probably represents an extreme random deviate because it has a large standard error.

At fourfold degenerate sites it is possible to compare the rate of transitional substitution with that of transversional substitution (Table 4.2), since both types of substitution are synonymous. The rate of transition (2.24×10^{-9}) tends to be higher than that of transversion (1.47×10^{-9}), although at each fourfold degenerate site two types of transversional change and only one type of transitional change can occur. This observation can be largely explained by the fact that transitional mutations occur more frequently than transversional ones (Chapter 1). At twofold degenerate sites, the rate of transitional substitution is on average similar to that at fourfold degenerate sites, but the rate of transversional substitution is usually considerably lower than the corresponding rate at fourfold degenerate sites; the averages for the two rates at twofold degenerate sites are 1.86×10^{-9} and 0.38×10^{-9}, respectively. The latter rate is low because all transversional changes at twofold degenerate sites are nonsynonymous. At nondegenerate sites, at which all changes are nonsynonymous, the rates of transitional and transversional substitution are on the average about the same (0.40×10^{-9} and 0.38×10^{-9}, respectively), and are usually considerably lower than the corresponding values at fourfold degenerate sites. Therefore, the rates of nucleotide substitution are lowest at nondegenerate sites, intermediate at twofold degenerate sites, and highest at fourfold degenerate sites; the average rates are 0.78×10^{-9}, 2.24×10^{-9}, and 3.71×10^{-9}, respectively.

Noncoding regions

Data from noncoding regions are much less abundant than data from coding regions, and so only a limited comparative analysis can be done at the present time. (Note that in order to estimate the rate of substitution in a sequence we must have data from at least two species.) Since most published sequences are cDNA sequences derived from mRNAs, and since these sequences do not include introns and flanking regions, the 5′ and 3′ untranslated regions are the only noncoding regions that can be studied in detail. Table 4.3 shows the substitution rates in these two regions for 16 genes based on comparisons between humans and murids. Within both regions, the rates vary greatly among

TABLE 4.2 Rates of transitional and transversional substitutions (per site per 10^9 years) at nondegenerate, twofold degenerate, and fourfold degenerate codon sites[a]

Type of substitution	Nondegenerate	Twofold degenerate	Fourfold degenerate
Transition	0.40	1.86	2.24
Transversion	0.38	0.38	1.47
Total	0.78	2.24	3.71

From Li (1997)

[a]The rates are averages over the genes in Table 4.1.

TABLE 4.3 Rates of nucleotide substitution (± standard errors) in 5′ and 3′ untranslated regions and at fourfold degenerate sites of protein-coding genes, based on comparisons between human and mouse or rat genes[a]

Gene	5' untranslated		3' untranslated		Fourfold degenerate	
	L[b]	Rate	L	Rate	L	Rate
Corticotropin-β-lipoprotein precursor	99	1.87 ± 0.41	97	2.32 ± 0.49	275	2.78 ± 0.34
Aldolase A	124	1.08 ± 0.26	154	1.73 ± 0.32	195	3.16 ± 0.48
Apolipoprotein A-IV	83	3.06 ± 0.68	134	1.73 ± 0.33	160	3.38 ± 0.50
Apolipoprotein E	23	1.27 ± 0.69	84	1.70 ± 0.42	153	4.00 ± 0.60
Na,K-ATPase β	118	2.45 ± 0.45	1,117	0.57 ± 0.06	118	2.87 ± 0.54
Creatine kinase M	70	1.71 ± 0.46	168	1.79 ± 0.30	178	2.81 ± 0.41
α-fetoprotein	47	3.64 ± 1.13	144	2.79 ± 0.49	225	4.14 ± 0.54
α-globin	34	1.56 ± 0.65	90	2.21 ± 0.50	81	4.47 ± 0.98
β-globin	50	1.30 ± 0.46	126	2.85 ± 0.49	78	2.42 ± 0.56
Glyceraldehyde-3-phosphate dehydrogenase	70	1.34 ± 0.38	121	1.74 ± 0.36	170	2.43 ± 0.39
Growth hormone	21	1.79 ± 0.85	91	1.83 ± 0.41	83	3.82 ± 0.78
Insulin	56	2.92 ± 0.80	53	3.09 ± 0.81	62	4.19 ± 1.00
Interleukin I	59	1.09 ± 0.38	1,046	2.02 ± 0.14	105	2.97 ± 0.60
Lactate dehydrogenase A	95	2.79 ± 0.55	470	2.48 ± 0.23	152	3.64 ± 0.60
Metallothionein II	61	1.88 ± 0.52	111	2.57 ± 0.48	23	2.37 ± 1.00
Parathyroid hormone	84	1.79 ± 0.43	228	2.21 ± 0.30	38	3.85 ± 1.21
Average[c]		1.96 (0.78)		2.10 (0.61)		3.33 (0.69)

[a]Rates are in units of substitutions per site per 10^9 years. As in Table 4.1, the time of divergence has been set at 80 million years ago.

[b]L = number of sites.

[c]Average is arithmetic mean, and values in parentheses are the standard deviations, computed over all genes.

genes, but this variation may largely represent sampling effects due to the fact that both these regions are usually very short. An exceptional case is the gene for Na,K-ATPase β, in which the rate in the 3′ untranslated region is extremely low, though this region is long. The reason is that the Na,K-ATPase β gene encodes multiple forms of the protein by using alternative polyadenylation sites, and therefore not all the 3′ untranslated region is really noncoding.

In almost all genes, the substitution rates in the 5′ and 3′ untranslated regions are lower than those at fourfold degenerate sites (i.e., sites at which all possible nucleotide substitutions are synonymous). The average rates for the 5′ and 3′ untranslated regions are 1.96×10^{-9} and 2.10×10^{-9} substitutions per site per year, respectively, which are both about 55% of the average rate at fourfold degenerate sites (3.71×10^{-9} substitutions per site per year).

TABLE 4.4 Numbers of nucleotide substitutions per site (K) between cow and goat β- and γ-globin genes and between cow and goat β-globin pseudogenes

Region	K^a
5′ flanking region	5.3 ± 1.2
5′ untranslated region	4.0 ± 2.0
Fourfold degenerate sites	8.6 ± 2.5
Introns	8.1 ± 0.7
3′ untranslated region	8.8 ± 2.2
3′ untranslated region	8.0 ± 1.5
Pseudogenes	9.1 ± 0.9

[a]Means and standard errors.

Pseudogenes are DNA sequences that were derived from functional genes but have been rendered nonfunctional by mutations that prevent their proper expression (Chapters 1, 6, and 7). Table 4.4 shows a comparison between the rate of substitution in cow and goat $\psi\beta^X$ and $\psi\beta^Z$ globin pseudogenes and the rates in the noncoding regions and fourfold degenerate sites in the β- and γ-globin genes. The rate in these pseudogenes is slightly higher than that in the other regions. This seems to be generally true for pseudogenes.

In Figure 4.3 we present a schematic comparison of the rates of substitution in different regions of the gene, as well as in pseudogenes. The rates for the 5′ and 3′ untranslated regions, nondegenerate sites, twofold degenerate sites, and fourfold degenerate sites are the average rates for the genes in Table 4.2. The rate for the 5′ flanking region was computed by assuming that the ratio of this rate to that at fourfold degenerate sites is 5.3/8.6 = 0.62 (as suggested by the values in Table 4.4), and that the average rate at fourfold degenerate sites is 3.71×10^{-9} substitutions per site per year (Table 4.2). The rates for introns, 3′ flanking regions, and pseudogenes were computed in the same manner. Since these are rough estimates based on limited data, and since the rates vary from gene to gene, the values shown in Figure 4.3 may not be applicable to any particular gene but are meant to provide a rough general comparison of the substitution rates in different DNA regions. With this precaution, we note that the substitution rate in a gene is highest at fourfold degenerate sites; slightly lower in introns and the 3′ flanking region; intermediate for the 3′ untranslated region, the 5′ flanking and untranslated regions, and twofold degenerate sites; and lowest at nondegenerate sites. Pseudogenes have on average the highest rate of substitution, although only slightly higher than that at the fourfold degenerate sites of functional genes.

Similarity profiles

A quick and visually powerful method of detecting variation in rates of substitution can be achieved by means of a **similarity profile** (Figure 4.4). A simi-

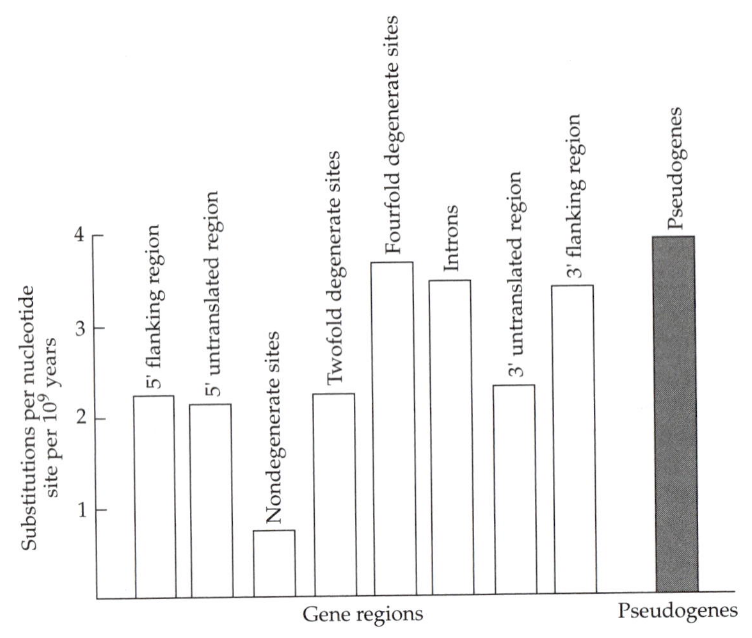

FIGURE 4.3 Average rates of substitution in different parts of genes (white) and in pseudogenes (gray). From Li (1997).

larity profile is a graph of similarity along an alignment of two homologous sequences. Similar (i.e., slowly evolving) parts emerge as peaks. This method can provide instant insight into evolutionary history, by revealing different levels of sequence conservation, as well as functional information, by revealing the location of such slowly evolving regions as protein-coding exons.

CAUSES OF VARIATION IN SUBSTITUTION RATES

To infer the causes underlying the observed variation in substitution rates among DNA regions, we note that the rate of substitution is determined by two factors: (1) the rate of mutation, and (2) the probability of fixation of a mutation (Chapter 2). The latter depends on whether the mutation is advantageous, neutral, or deleterious. Since the rate of mutation is unlikely to vary much within a gene but may vary among genes, we shall discuss the rate variation among different regions of a gene and the variation among genes separately.

Functional constraints

The intensity of purifying selection is determined by the degree of intolerance characteristic of a site or a genomic region towards mutations. This **functional**

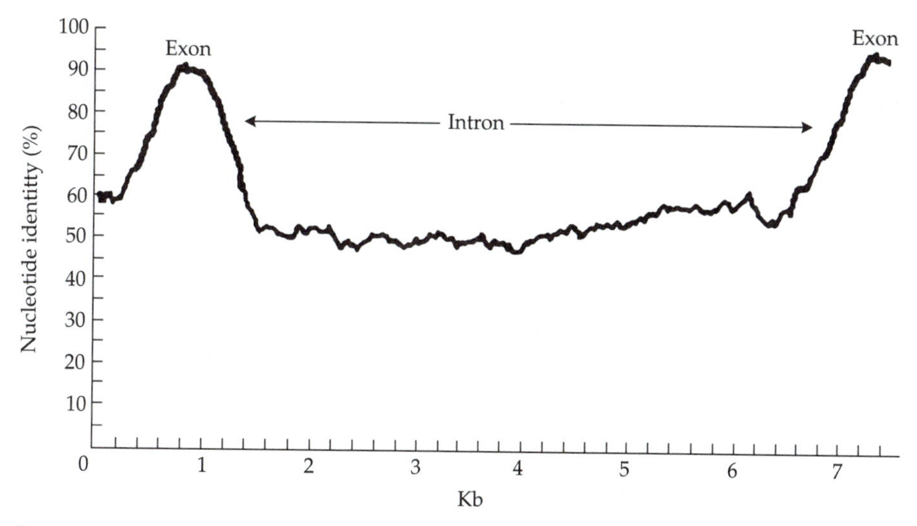

FIGURE 4.4 A similarity profile for two aligned DNA sequences. Conserved regions show up as peaks, which in protein-coding genes may allow us to identify the junctions between exons and introns.

or **selective constraint** (Miyata et al. 1980; Jukes and Kimura 1984) defines the range of alternative nucleotides that is acceptable at a site without affecting negatively the function or structure of the gene or the gene product. DNA regions (e.g., protein-coding regions or regulatory sequences), in which a mutation is likely to affect function, have a more stringent functional constraint than regions devoid of function. The stronger the functional constraints on a macromolecule are, the slower the rate of substitution will be.

Kimura (1977, 1983) has illustrated this principle by means of a simple model. Suppose that a certain fraction, f_0, of all mutations in a certain molecule are selectively neutral or nearly neutral and the rest are deleterious. (Advantageous mutations are assumed to occur only very rarely, such that their relative frequency is effectively zero, and they do not contribute significantly to the overall rate of molecular evolution.) If we denote by v_T the total mutation rate per unit time, then the rate of neutral mutation is

$$v_0 = v_T f_0 \tag{4.2}$$

According to the neutral theory of molecular evolution, the rate of substitution is $K = v_0$ (Chapter 2). Hence,

$$K = v_T f_0 \tag{4.3}$$

From Equation 4.3, we see that the highest rate of substitution is expected to occur in a sequence that does not have any function, such that all mutations in it are neutral (i.e., $f_0 = 1$). Indeed, pseudogenes, which are devoid of function, seem to have the highest rate of nucleotide substitution (Table 4.4 and Figure 4.3). Thus, although the model is clearly oversimplified, it is helpful for explaining the rate differences among different DNA regions.

As far as protein-coding genes are concerned, there have been several attempts to quantify functional constraints independently of their rate of substitution. One such measure is the **functional density** (Zuckerkandl 1976b). The functional density of a gene, F, is defined as n_s/N, where n_s is the number of sites committed to specific functions and N is the total number of sites. F, therefore, is the proportion of amino acids that are subject to stringent functional constraints. The higher the functional density, the lower the rate of substitution is expected to be. Thus, a protein in which the active sites constitute only 1% of its sequence will be less constrained, and therefore will evolve more quickly than a protein that devotes 50% of its sequence to performing specific biochemical or physiological tasks. It is usually very difficult to compute the functional density of a protein. However, for those protein-coding genes for which F is known, a rough negative correlation exists between functional density and rates of amino acid-altering substitution.

Numerous comparative studies of both protein and DNA sequence data have led to the general conclusion that an inverse relationship exists between stringency of functional constraint, or importance, on the one hand and the rate of evolution on the other. Therefore, functionally less important molecules or parts of a molecule evolve faster than more important ones. This rule is frequently read backwards, and the rate of nucleotide substitution is used to infer the stringency of structural and functional constraints in a particular sequence. In fact, whenever a particularly conservative sequence is found, researchers start looking for a specific function in this region. This practice may have the unfortunate result of turning the argument into a circular one (Graur 1985).

Synonymous versus nonsynonymous rates

Because the rates of mutation (v_T) at synonymous and nonsynonymous sites within a gene should be the same, or at least very similar, the difference in substitution rates between synonymous and nonsynonymous sites may be attributed to differences in the intensity of purifying selection between the two types of sites. This is understandable in light of the neutral theory of molecular evolution (Chapter 2). Mutations that result in an amino acid replacement have a greater chance of causing deleterious effects on the function of the protein than do synonymous changes. Consequently, the majority of nonsynonymous mutations will be eliminated from the population by purifying selection. The result will be a reduction in the rate of substitution at nonsynonymous sites. In contrast, synonymous changes have a better chance of being neutral, and a larger proportion of them will be fixed in a population.

Of course, nonsynonymous substitutions may have a better chance of improving the function of a protein. Therefore, if advantageous selection plays a major role in the evolution of proteins, the rate of nonsynonymous substitution should exceed that of synonymous substitution (see page 119).

Since v_T is the same for both synonymous and nonsynonymous sites, the observation that synonymous substitution rates are higher than the rates of nonsynonymous substitution (e.g., Miyata et al. 1982) is a result of the f_0 values being higher for synonymous sites than for nonsynonymous ones. The contrast between synonymous and nonsynonymous rates in protein-coding genes serves as an illuminating demonstration of the inverse relationship between the intensity of the functional constraint and the rate of molecular evolution. That the average rate of synonymous substitution is somewhat lower than the rate of substitution in pseudogenes would seem to indicate that some synonymous mutations are selected against. One reason for this may be that not all synonymous substitutions are silent at the amino acid level (Chapter 1). Another reason may be that some codons are preferred over their synonyms (see page 132).

Variation among different gene regions

Within a protein-coding gene, we assume that the rates of mutation are the same in all its various parts. Therefore, the observation that 5′ and 3′ untranslated regions have lower substitution rates than the rate of synonymous substitution in coding regions leads us to conclude that these regions are functionally constrained. Indeed, untranslated regions are known to contain important signals concerned with the regulation of the translation process. Introns and 3′ untranscribed regions, on the other hand, evolve at about the same rate as fourfold degenerate sites and only slightly less rapidly than pseudogenes. This observation suggests that these regions are only slightly constrained and that most nucleotide substitutions in this regions do not affect the fitness of the organism one way or another.

Within a protein, the different structural or functional domains are likely to be subject to different functional constraints and to evolve at different rates. A classic example is provided by insulin, a dimeric hormone secreted by the β cells of the pancreatic islets of Langerhans. The precursor of insulin, preproinsulin, is a chain of 86 amino acids consisting of four segments: A, B, C, and a signal peptide (Figure 4.5). After the signal peptide is removed, the remaining 62-amino-acid-long proinsulin folds into a specific three-dimensional structure stabilized by two disulfide bonds. Internal cuts then excise the 31-amino-acid long C peptide, which resides in the middle of the proinsulin chain, to create the active insulin hormone out of the two remaining segments, A and B. Neither the signal peptide nor the C peptide take part in the hormonal activity of insulin. The signal peptide is thought to facilitate the secretion of preproinsulin, whereas the C peptide is necessary for the creation of the proper tertiary structure of the hormone. Presumably, therefore, the signal peptide and the C peptide are less constrained than the A and B chains. Indeed, the nonsynonymous substitution rate for the region coding for the C peptide is about 5 times higher than the nonsynonymous rate for the A and B chains, and the rate of nonsynonymous substitution for the signal peptide coding region is approximately 6 times higher than that for the A and B chains. However, considerable constraints must still operate on the C peptide and the signal peptide, because their nonsynonymous rates are rather low in comparison with other protein-coding genes.

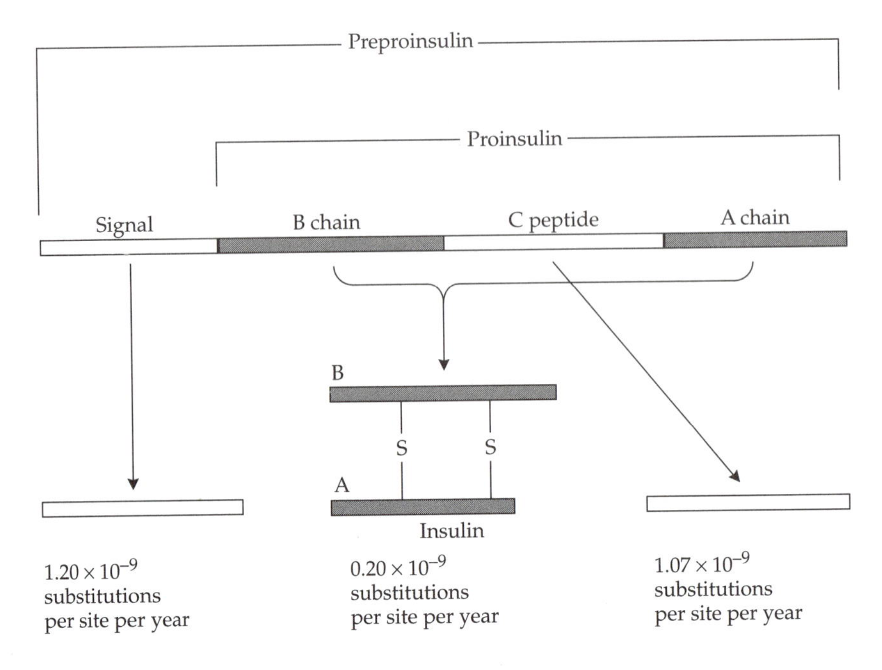

FIGURE 4.5 Comparison among the rates of nonsynonymous nucleotide substitution for DNA regions coding for functional insulin, the C peptide, and the signal peptide. A mature insulin consists of A and B chains, linked by two disulfide (S—S) bonds. The rates are based on comparisons between human and rat genes. The time of divergence was set at 80 million years ago.

An additional example involves hemoglobin, a tetrameric protein composed of two α and two β chains. The surface of the hemoglobin molecule performs no specific function and is constrained only by the requirement that it must be hydrophilic. On the other hand, the internal residues, especially the amino acids lining the heme pocket, play an important role in the normal function of the molecule. Indeed, the rates of substitution on the surfaces of the α and β chains are 1.35×10^{-9} and 2.73×10^{-9} amino acid substitutions per site per year, respectively, while the rates of substitution in the interior are only 0.17×10^{-9} and 0.24×10^{-9} for α and β, respectively. Purifying selection seems to be the mechanism responsible for the fact that in both the α and the β chains, the residues on the surface of the molecule evolve 8–11 times faster than the residues in the interior. Indeed, mutations affecting the interior of hemoglobin were shown to cause particularly harmful abnormalities, whereas replacements of amino acids on the surface of the molecule often do not exhibit any clinical effects (see Perutz 1983).

Many genes are initially translated into long proteins, which are post-translationally cleaved to produce smaller active molecules. Let us consider, for example, the fibrinogens, a group of elongated proteins that are the precursors of fibrin, an essential polymer in the clotting process. The highly soluble fibrinogens are converted into insoluble fibrins by the proteolytic action of thrombin, which cleaves apart short polypeptides called fibrinopeptides. Fibrinopeptides have little known biological activity on their own, and although they do affect certain physiological functions, this action has been attributed to a minute portion of the peptide. Consequently, in fibrinopeptides virtually any amino acid change that will still allow them to be cleaved off will be acceptable. As a result of this relative lack of functional constraint, f_0 is probably close to 1, and the rate of substitution is consequently expected to be only slightly lower than the mutation rate. Indeed, fibrinopeptides are among the fastest evolving proteins, whereas the biologically active fibrins evolve at relatively low rates.

Variation among genes

To explain the large variation in the rates of nonsynonymous substitution among genes, we must consider two possible culprits: (1) the rate of mutation, and (2) the intensity of selection. The assumption of equal rates of mutation for different genes may not hold in this case, because different regions of the genome may have different propensities to mutate. However, the difference in mutation rates among different genomic regions is far too small to account for the approximately 1,000-fold range in nonsynonymous substitution rates. Thus the most important factor in determining the rates of nonsynonymous substitution seems to be the intensity of purifying selection, which in turn is determined by functional constraints.

To illustrate the effect of functional constraints on the evolution of different genes, let us consider apolipoproteins and histone H3, which exhibit markedly different rates of nonsynonymous substitution. Apolipoproteins are the major carriers of lipids in the blood of vertebrates, and their lipid-binding domains consist mostly of hydrophobic residues. Comparative analyses of apolipoprotein sequences from various mammalian orders suggest that in these domains, exchanges among hydrophobic amino acids (e.g., valine for leucine) are acceptable at many sites (Luo et al. 1989). This lax structural requirement may explain the fairly high nonsynonymous rate in these genes .

At the other extreme, we have histone H3. Since most amino acids in H3 interact directly with either the DNA or other core histones in the formation of the nucleosome (Figure 4.6), it is reasonable to assume that there are very few possible substitutions that can occur without hindering the function of the protein. In addition, H3 must retain its strict compactness and high alkalinity, which are necessary for its intrastrand location and its interaction with the acidic DNA molecule. As a consequence, H3 is very intolerant of most amino acid changes. Indeed, this protein is one of the slowest evolving proteins known, evolving more than 1,000 times more slowly than the apolipoproteins.

An illuminating example of the importance of functional constraint can be derived from comparing vertebrate hemoglobins and cytochrome *c*. As an

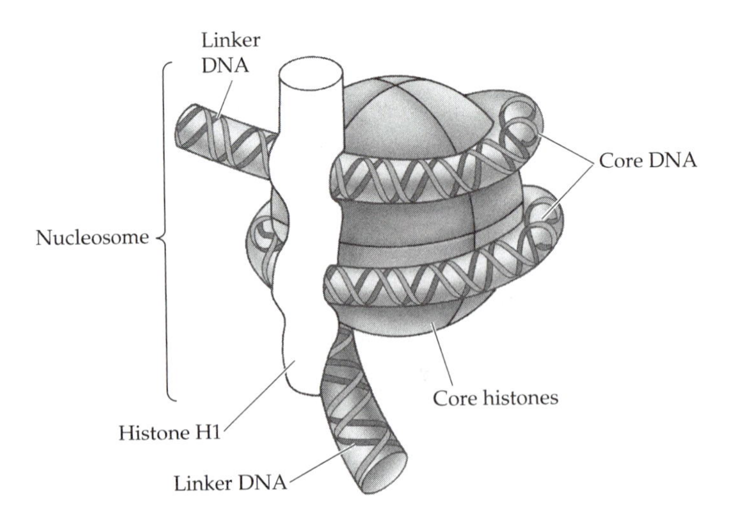

FIGURE 4.6 **Schematic diagram of a nucleosome. The DNA double helix is wound around the core histones (two each of histones H2A, H2B, H3, and H4). Histone H1 binds to the outside of this core particle and to the linker DNA.**

oxygen carrier, hemoglobin requires the attachment of heme prosthetic groups, and has the capability to respond structurally to changes in pH and CO_2 concentration. However, most of its functional requirements are restricted to the interior of the molecule, and as we have seen in the previous section, many amino acid-altering mutations, especially on the surface of this globular protein, are acceptable. Cytochrome *c* also carries oxygen, binds heme, and responds structurally to changes in physiological conditions, but in addition to these hemoglobin-like functions, this protein also interacts at its surface with two very large enzymes: cytochrome oxidase and cytochrome reductase. Thus, a higher proportion of the amino acids in cytochrome *c* take part in specific functions, and its rate of amino acid substitution is consequently lower than that of hemoglobin.

Why the rate of synonymous substitution also varies from gene to gene is less clear. There may be two reasons for this variation. First, the rate of mutation may differ among different regions of the genome, and the variation in the rates of synonymous substitution may in part reflect the chromosomal position of the gene (see Wolfe et al. 1989a). The second reason may be that, in some genes, not all synonymous codons have an equivalent effect on fitness. As a result, some synonymous substitutions may be selected against. Such purifying selection will create variation in the rate of synonymous substitution among genes. However, although purifying selection has been shown to affect the synonymous substitution rate, as well as the pattern of usage of synonymous codons in the genomes of bacteria, yeast, and *Drosophila*, the intensity of this type of selection in mammals is not clear (see page 137).

It has also been noted that there is a positive correlation between synonymous and nonsynonymous substitution rates in a gene (Graur 1985; Li et al. 1985b). This may be explained by assuming either that the rate of mutation varies among genes (and hence some genes will have both high synonymous and nonsynonymous rates of substitution), or that the extent of selection at synonymous sites is affected by the nucleotide composition at adjacent positions (Ticher and Graur 1989).

Acceleration of nucleotide substitution rates following partial loss of function

If the rate of substitution is indeed inversely related to the stringency of the functional constraint as claimed by the neutral theory, then we should observe an increase in the rate of nucleotide substitution in genes that lost their function. We have already seen that pseudogenes, in which all constraints have presumably been removed, are the fastest evolving sequences in the mammalian nuclear genome. Let us now examine what happens when selection constraints are only partially, rather than entirely, removed. Such a phenomenon is called **relaxation of selection**.

An illustration of such a case is provided by the evolution of the single-copy αA-crystallin gene in the subterranean blind mole rat (genus *Spalax*). In vertebrates, the crystallins function mainly as structural components of the eye lens. In the blind mole rat, however, αA-crystallin has long lost this functional role, since *Spalax* became subterranean and presumably lost the use of its eyes more than 25 million years ago. Since αA-crystallin is usually a slowly evolving protein, an increase in the rate of substitution should be readily detectable.

Hendriks et al. (1987) sequenced the αA-crystallin gene in the blind mole rat and compared its rate of substitution with those in other rodents, such as mouse, rat, gerbil, hamster, and squirrel, which possess fully functional eyes. The blind mole rat αA-crystallin gene turned out to possess all the prerequisites for normal function and expression, including the proper signals for alternative splicing, but its nonsynonymous rate of substitution was found to be exceptionally high, almost 20 times faster than the rate in rat. Nevertheless, the nonsynonymous rate was still slightly lower than that in pseudogenes, indicating that functional constraints may still be operating.

Hendriks et al. (1987) suggested that the αA-crystallin gene may not have lost all of its vision-related functions, such as photoperiod perception and adaptation to seasonal changes. However, the fact that the atrophied eye of *Spalax* does not respond to light argues against this explanation. An alternative explanation for the slower rate in the αA-crystallin gene than in pseudogenes may be that the blind mole rat lost its vision more recently than 25 million years ago. Consequently, the rate of nonsynonymous substitution after nonfunctionalization may have been underestimated. This argument, however, fails to explain why the αA-crystallin gene is still an intact gene as far as the essential molecular structures for its expression are concerned.

The most likely explanation at this point is that the αA-crystallin gene product may also serve a function unrelated to that of the eye. This possibility is supported by the fact that αA crystallin has been found in tissues other than the lens. Such multiple roles are known for several proteins, including many crystallins (Chapter 6). Support for this hypothesis was provided by the discovery that αA crystallin also functions as a molecular chaperone that can bind denaturing proteins and prevent their aggregation (Bova et al. 1997; Mornon et al. 1998). Moreover, the regions within the αA crystallin responsible for this activity have been identified. Interestingly, the sites involved in chaperone activity are conserved in the mole rat. Therefore, the functional constraints on αA-crystallin in the mole rat cannot be assumed to have been completely removed with the atrophication of the eye, but only to have been relaxed considerably. αA crystallin most probably represents a case of partial loss of function resulting in an increase in the rate of nucleotide substitution due to relaxation of selection.

Estimating the intensity of purifying selection in protein-coding genes

Ophir et al. (1999) proposed a simple method by which the intensity of purifying selection on a functional protein-coding gene can be quantified. Their method requires three homologous sequences: a pseudogene (ψA), a functional homologous gene from the same species (A), and a functional homolog from a different species (B). For each such trio (Figure 4.7) and by using the relative rate test (see page 142), it is possible to calculate the numbers of nucleotide substitutions along the branches leading to ψA and A, i.e., $K_{\psi A}$ and K_A. Following Kimura's (1977) model in Equation 4.3, the numbers of nucleotide substitutions along the two branches leading to ψA and A are given by

$$K_{\psi A} = v_{\psi A} f_{\psi A} \tag{4.4}$$

and

$$K_A = v_A f_A \tag{4.5}$$

where v is the total mutation rate per unit time, f is the fraction of mutations that are selectively neutral or nearly so, and subscripts identify the branches.

If we assume that the mutation rate is the same in the gene and the pseudogene, i.e., $v_A = v_{\psi A}$, and if we further assume that mutations occurring in a pseudogene do not affect the fitness of the organism, i.e., $f_{\psi A} = 1$, we obtain

$$f_A = \frac{K_A}{K_{\psi A}} \tag{4.6}$$

By definition, the fraction of deleterious mutations that are subject to purifying selection (or the intensity of selection) is $1 - f_A$.

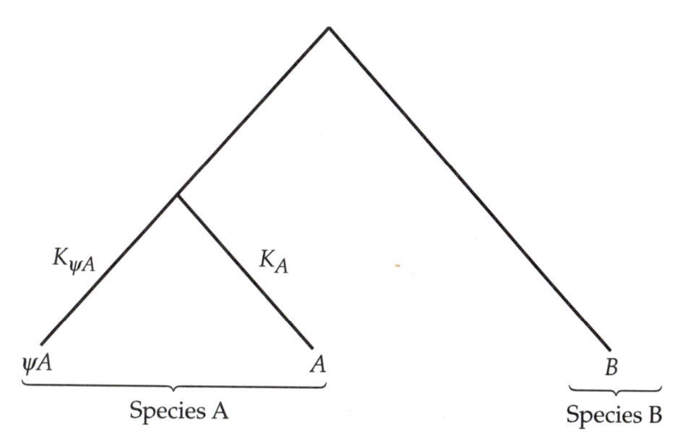

FIGURE 4.7 Phylogenetic tree for three homologous sequences used to quantify the intensity of purifying selection on a protein-coding gene. The sequences are: a pseudogene (ψA), a functional homologous gene (A) from the same species, and a functional homolog (B) from a different species. $K\psi_A$ and K_A denote the rates of substitution along the branches leading to ψA and A, respectively.

As expected, Ophir et al. (1998) found that the least constrained codon position is the third, followed by the first and second codon positions. In the second codon position, in which all possible mutations are nonsynonymous, they found that in humans approximately 7% of all mutations were neutral. The corresponding value for the first codon position was 21%. Interestingly, only about 42% of all the mutations occurring in the third codon position of protein-coding genes were found to be neutral in their sample of human genes. Given that about 70% of all possible mutations in this position are synonymous (see Table 1.5), these results indicate that a significant fraction of synonymous mutations are selected against.

Mutational input: Male-driven evolution

From Equation 4.3 it is clear that the rate of substitution should be influenced not only by functional constraints, but also by the rate of mutation. It is possible, therefore, that the observed variation in the rates of synonymous substitution among genes partly reflects mutational variation. An interesting example of the effect of the mutation rate on the rate of substitution is the so-called **male-driven evolution.**

Because mammalian oogenesis (egg production) differs fundamentally from the process of spermatogenesis (sperm production), the number of germ cell divisions from one generation to the next in males, n_m, is usually much larger than that in females, n_f. In humans, n_f was estimated to be about 33, whereas n_m is approximately 200 if the mean reproductive age is 20 years (Chang et al. 1994). In mice, $n_m = 57$ and $n_f = 28$, and in rats, $n_m = 58$ and $n_f = 29$. Haldane (1947) stated: "If mutation is due to faulty copying of genes at the nuclear division, we might expect it to be commoner in males than in females." He then, however, added: "It is difficult to see how this could be proved or disproved for many years to come." Fortunately, a simple and elegant approach proposed by Miyata et al. (1987) provides us with a means to test Haldane's hypothesis.

Let u_m and u_f be the mutation rates in males and females, respectively, and α be the ratio of male to female mutation rates. That is,

$$\alpha = \frac{u_m}{u_f} \quad (4.7)$$

Since an autosomal sequence is derived from the father or the mother with equal probabilities, the mutation rate per generation for an autosomal sequence is

$$A = \frac{u_m + u_f}{2} \quad (4.8)$$

An X-linked sequence is carried two-thirds of the time by females and one-third of the time by males. Therefore, the mutation rate per generation for a sequence located on the X chromosome is

$$X = \frac{u_m + 2u_f}{3} \quad (4.9)$$

A Y-linked sequence is only carried by males, so its rate of mutation per generation is

$$Y = u_m \quad (4.10)$$

From these three equations it is easy to see that the ratio of Y to A is

$$Y/A = \frac{2\alpha}{1+\alpha} \quad (4.11)$$

Similarly,

$$X/A = \frac{2(2+\alpha)}{3(1+\alpha)} \quad (4.12)$$

and

$$Y/X = \frac{3\alpha}{2+\alpha} \quad (4.13)$$

The zinc finger protein-coding genes are a good case for studying the ratio of male to female mutation rates because in all eutherian mammals studied to date there are at least two homologous genes, an X-linked gene (*Zfx*) and a Y-linked one (*Zfy*). Shimmin et al. (1993) sequenced the last intron of the human, orangutan, baboon, and squirrel monkey *Zfx* and *Zfy* genes, which are highly similar. As seen previously, there are very few functional constraints on introns, and therefore we may disregard selective forces in this case. For all pairwise comparisons, Shimmin et al. (1993) found that the Y sequences were more divergent, i.e., have evolved faster, than their X-linked homologs. The mean Y/X ratio was 2.25, which by using Equation 4.12 translates into an estimate of $\alpha \approx 6$. This indicates that in primates the mutation rate is considerably higher in the male germline than in the female germline, i.e., evolution is "male-driven." Interestingly, the ratio of the number of germ cell divisions from one generation to the next in males, n_m, to that in females, n_f, is also approximately $200/33 \approx 6$.

In rat, mouse, hamster, and fox, the mutation rate in males was found to be twice as large as that in females (Lanfear and Holland 1991; Chang et al. 1994), which agrees with the $n_m/n_f = 2$ ratio in these species.

McVean and Hurst (1997) raised the possibility that the higher rate of nucleotide substitution in Y-linked sequences than in X-linked ones is not due to male-driven evolution but due to a reduction in the mutation rate in the X chromosome in comparison to the mutation rates in Y and the autosomes. They claimed that selection will often favor a lower mutation rate on the X chromosome than on autosomes, owing to the exposure of deleterious recessive mutations on hemizygous chromosomes (i.e., chromosomes that exist in a single copy in one sex but in two copies in the other). To test this possibility against the hypothesis of male-driven evolution, Ellegren and Fridolfsson (1997) studied rates of mutation in birds. As opposed to the situation in mammals, in which the females are homogametic (XX) and the males are heterogametic (XY), in birds males are homogametic (WW) and females are heterogametic (WZ). They found that the male-to-female ratio in mutation rates ranged from 4 to 7. Under McVean and Hurst's (1997) hypothesis, this high ratio would have to be explained by a reduction in the mutation rate on the Z chromosome, which is homologous to the Y chromosome in mammals, in comparison with that on the W chromosome. This explanation, however, would amount to special pleading, and it is much simpler to assume that male-driven evolution occurs in both mammals and birds.

POSITIVE SELECTION

The rates of nucleotide substitution in the vast majority of genes and nongenic regions of the genome can be explained by a combination of (1) mutational input, (2) random genetic drift of neutral or nearly neutral alleles, and (3) purifying selection against deleterious alleles. In a few cases, however, positive selection was found to play an important role in the molecular evolution of genes.

Detecting positive selection

Nonsynonymous changes are far more likely than synonymous changes to improve the function of a protein. Since advantageous mutations undergo fixation in a population much more rapidly than neutral mutations (Chapter 2), the rate of nonsynonymous substitution should exceed that of synonymous substitution if advantageous selection plays a major role in the evolution of a protein. Therefore, one way to detect positive Darwinian selection is to show that the number of substitutions per nonsynonymous site is significantly greater than the number of substitutions per synonymous site. Several methods have been proposed in the literature to test for this difference. In the following we present two simple tests.

In the first test, we use the number of substitutions per nonsynonymous site (K_A) and the number of substitutions per synonymous site (K_S), and their variances, which have been calculated by Equations 3.29 and 3.30 (Chapter 3). The test statistic is Student's t

$$t = \frac{K_A - K_S}{\sqrt{V(K_A) + V(K_S)}} \tag{4.14}$$

where V denotes the variance. Assuming that the statistic follows the t distribution with an infinite number of degrees of freedom, we may perform a one-tailed test of the null hypothesis that $K_A = K_S$ against the alternative hypothesis that $K_A > K_S$. Rejection of the null hypothesis would mean that positive selection has played a major role in the evolution of the sequences under study.

When the numbers of synonymous and nonsynonymous substitutions are small, the statistic is unlikely to follow the t distribution and we are likely to reject the null hypothesis more often than expected by chance. To overcome this problem, Zhang et al. (1997) proposed the following test. In this test, we use the numbers of synonymous and nonsynonymous differences between the two protein-coding sequences, M_S and M_A, and the average numbers of synonymous and nonsynonymous sites, N_S and N_A (Chapter 3). Under the null hypothesis of neutral evolution, i.e., no positive selection, we expect $M_S/N_S = M_A/N_A$. We can therefore use a 2×2 table as follows:

	Nonsynonymous	Synonymous	Total
Changes	M_A	M_S	$M_A + M_S$
No changes	$N_A - M_A$	$N_S - M_S$	$L - (M_A + M_S)$
Total	N_A	N_S	L

where L is the length of the aligned sequence. We can test the null hypothesis by using the χ^2 test or the exact binomial distribution test (Fisher's exact test).

This test and others have been used extensively in the literature to detect departures from the neutral mode of molecular evolution. In some immunoglobulin genes, the nonsynonymous rate in the complementarity-determining regions was higher than the synonymous rate. The higher rate has been attributed to overdominant selection for antibody diversity (Tanaka and Nei 1989). Nevertheless, when the entire immunoglobulin gene is considered, the nonsynonymous rate is considerable lower than the synonymous rate (Table 4.1). This result indicates that, even in immunoglobulins, most nonsynonymous mutations are detrimental and are eliminated from the population. Hughes and Nei (1989) reported a similar situation in certain regions of the major histocompatibility complex genes, i.e., the rate of nonsynonymous substitution exceeds the rate of synonymous substitution. They, too, attributed the higher rates of nonsynonymous substitution to overdominant selection.

Parallelism and convergence: The evolution of lysozymes in ruminants, langurs, and hoatzins

Parallelism at the molecular level is defined as the independent occurrence of two or more nucleotide substitutions (or amino acid replacements) of the same type at homologous sites in different evolutionary lineages. Molecular **convergence** is the occurrence of two or more nucleotide substitutions (or amino acid replacements) at homologous sites in different evolutionary lineages resulting in the same outcome (see Figure 3.6). Given the number of substitutions that have occurred during the evolution of a gene, a limited degree of parallelism and convergence is expected to be observed purely by chance. However, if the numbers of parallel and convergent substitutions significantly exceed the chance expectation, then it is unlikely that they have occurred by random genetic drift, and we must invoke positive selection to explain their existence. Thus, parallel and convergent changes may be taken as evidence for positive selection. One such example is discussed below.

Lysozyme is a 130 amino-acid-long enzyme whose catalytic function is to cleave the β(1-4) glycosidic bonds between *N*-acetyl glucoseamine and *N*-acetyl muramic acid in the cell walls of eubacteria, thereby depriving the bacteria of their protection against osmotic pressures and subsequent lysis. By virtue of its catalytic function and its expression in body fluids such as saliva, serum, tears, avian egg white, and mammalian milk, lysozyme usually serves as a first-line defense against bacterial invasion. In foregut fermenters (i.e., in animals in which the anterior part of the stomach functions as a chamber for bacterial fermentation of ingested plant matter), lysozyme is also secreted in the posterior parts of the digestive system and is used to free nutrients from within the bacterial cells.

Foregut fermentation has arisen independently at least twice in the evolution of placental mammals and at least once in birds. In this section, we shall deal with foregut fermentation in two mammalian taxa, the ruminants (e.g., cows, deer, sheep, giraffes) and the colobine monkeys (e.g., langurs). Foregut fermentation in birds will be represented by the hoatzin (pronounced Watson), *Opisthocomus hoazin*, an enigmatic South American bird most probably related to the cuckoos. In all these three cases, lysozyme, which in other animals is not normally secreted in the digestive system, has been recruited to degrade the walls of bacteria that carry on the fermentation in the foregut. Stewart and Wilson (1987) and Kornegay et al. (1994) identified five amino acid positions that have experienced parallel and convergent replacements (Figure 4.8), which may be regarded as adaptations enabling the lysozyme to function in the hostile environment of the stomach .

A phylogenetic reconstruction by Kornegay et al. (1994) indicated that the foregut lysozymes evolved from two branches of the lysozyme gene family.

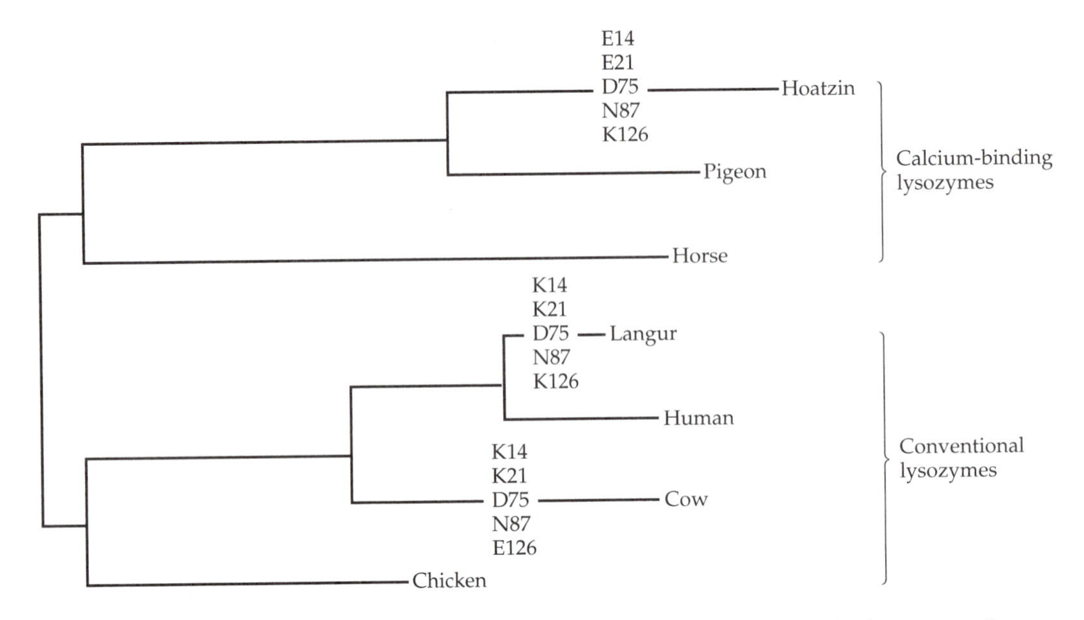

FIGURE 4.8 Parallel and convergent amino acid replacements in lysozymes from the foregut of cow, langur, and hoatzin. The lengths of the branches are proportional to the total numbers of amino acid replacements along them. Only convergent replacements are shown, denoted by a one-letter abbreviation of the resultant amino acid (see Table 1.2) followed by the position number at which the replacement occurred. Modified from Kornegay et al. (1994).

The hoatzin lysozyme is a calcium-binding lysozyme, whereas the ruminant and langur lysozymes originated from the conventional lysozyme branch of the gene family (Figure 4.8). The functional convergence of the lysozyme in the three lineages is the result of parallel amino acid replacements that occurred independently in each lineage. For example, position 75 changed to aspartic acid independently in three lineages. It is unlikely that the parallel amino acid replacements in the three evolutionary lineages occurred by chance, and we must therefore assume that adaptive selection played a major role in the evolution of lysozyme.

In all three lineages, fewer basic residues seem to be required for the optimal function of lysozyme in the stomach, resulting in the stomach proteins having a much lower isoelectric point (6.0–7.7) than the other lysozymes (10.6–11.2). For example, arginine was found to be heavily selected against at the surface of the foregut lysozymes, probably because this amino acid is susceptible to inactivation by digestive products such as diacyl. Indeed, six out of the ten arginines in the common ancestor of pigeon and hoatzin have been replaced by less basic amino acids in the hoatzin lineage. Moreover, it has been determined that these replacements contribute to a better performance of lysozyme at low pH values and confer protection against the proteolytic activity of pepsin in the stomach.

Clearly, these three lysozymes share adaptations to conditions prevalent in the digestive system. Conversely, these lysozymes perform less efficiently at higher pH values than human or chicken lysozymes. In conclusion, it seems safe to deduce that we are dealing here with a case of parallel occurrence of advantageous substitutions in different evolutionary lines resulting in parallel adaptations to similar selective agents. We note, however, that there is some disagreement in the literature on the exact number of adaptive substitutions during the evolution of stomach lysozymes (e.g., Zhang et al. 1997; Yang 1998).

Prevalence of positive selection

According to a survey by Endo et al. (1996), positive selection affecting entire protein-coding sequences is suspected in only very few cases. In their study of 3,595 groups of homologous sequences, they found only 17 gene groups (about 0.45%) in which the ratio of nonsynonymous to synonymous substitution was significantly larger than 1, i.e. in which positive selection may have played an important role in their evolution. Interestingly, 9 out of the 17 gene groups consisted of genes encoding surface antigens of parasites and viruses. We note, however, that in this survey only whole genes were used, so positive selection affecting only parts of genes or individual sites might have gone undetected.

Interestingly, elevated K_A/K_S ratios are frequently found in sex-related genes, e.g., genes involved in mating behavior, fertilization, spermatogenesis, ejaculation, or sex determination (Tsaur and Wu 1997; Civetta and Singh 1998; Tsaur et al. 1998). Indeed, the highest ratio of nonsynonymous to synonymous substitution ($K_A/K_S = 5.15$) for a full-length protein was found in the 18-kilodalton protein in the acrosomal vesicle at the anterior of the sperm cell of several abalone species (Vacquier et al. 1997). It is thought that sex-related genes are subject to positive selection for short periods of time during speciation as a means of erecting reproductive barriers that restrict gene flow between the speciating populations (Civetta and Singh 1998).

PATTERNS OF SUBSTITUTION AND REPLACEMENT

There are 12 possible types of nucleotide substitution (e.g., from A to G, from A to C, from T to A). For calculating the number of nucleotide substitution between two DNA sequences, we pool all 12 types together. However, it is sometimes interesting to determine separately the frequency with which each type of nucleotide substitution occurs. The **pattern of nucleotide substitution** is defined as the relative frequency with which a certain nucleotide changes into another during evolution. The pattern is usually shown in the form of a 4 × 4 matrix, in which each of the 12 elements of the matrix (excluding the four diagonal elements which represent the case of no substitution) denote the number of changes from a certain nucleotide to another.

Let P_{ij} be the proportion of base changes from the ith type to the jth type of nucleotide (i, j = A, T, C or G, and $i \neq j$). This proportion is calculated as

$$P_{ij} = \frac{n_{ij}}{n_i} \quad \textbf{(4.15)}$$

where n_{ij} is the number of substitutions from i to j, and n_i is the number of the i nucleotides in the ancestral sequence. To be able to compare the patterns of nucleotide substitution between sequences, we define f_{ij}, the relative substitution frequency from nucleotide i to nucleotide j, as

$$f_{ij} = \frac{P_{ij}}{\sum_i \sum_{j \neq i} P_{ij}} \times 100 \quad \textbf{(4.16)}$$

Thus, f_{ij} represents the expected number of base changes from the ith type nucleotide to the jth type among every 100 substitutions in a random sequence, (i.e., in a sequence in which the four bases are equally frequent).

Pattern of spontaneous mutation

Because point mutation is one of the most important factors in the evolution of DNA sequences, molecular evolutionists have long been interested in knowing the **pattern of spontaneous mutation** (e.g., Beale and Lehmann 1965; Fitch 1967; Zuckerkandl et al. 1971; Vogel and Kopun 1977; Sinha and Haimes 1980). This pattern can serve as the standard for inferring how far the observed frequencies of interchange between nucleotides in any given DNA sequence have deviated from the values expected under the assumption of no selection, or strict selective neutrality.

One way to study the pattern of point mutation is to examine the pattern of substitution in regions of DNA that are subject to no selective constraint. Pseudogenes are particularly useful in this respect. Since they are devoid of function, all mutations occurring in pseudogenes are selectively neutral and become fixed in the population with equal probability. Thus, the rate of nucleotide substitution in pseudogenes is expected to equal the rate of mutation. Similarly, the pattern of nucleotide substitution in pseudogenes is expected to reflect the pattern of spontaneous point mutation.

Figure 4.9 shows a simple method for inferring the nucleotide substitutions in a pseudogene sequence (Gojobori et al. 1982). Sequence 1 is a pseudogene, sequence 2 is a functional counterpart from the same species, and sequence 3 is a functional sequence that diverged before the emergence of the pseudogene. Suppose that at a certain nucleotide site, sequences 1 and 2 have A and G, respectively. Then we can assume that the nucleotide in the pseudogene sequence has changed from G to A if sequence 3 has G, but that the nucleotide in sequence 2 has changed from A to G if sequence 3 has A. However, if sequence 3 has T or C, then we cannot decide the direction of change, and in this case the site is excluded from the comparison. In general, when the ancestral nucleotide at a site cannot be determined uniquely, the site should be

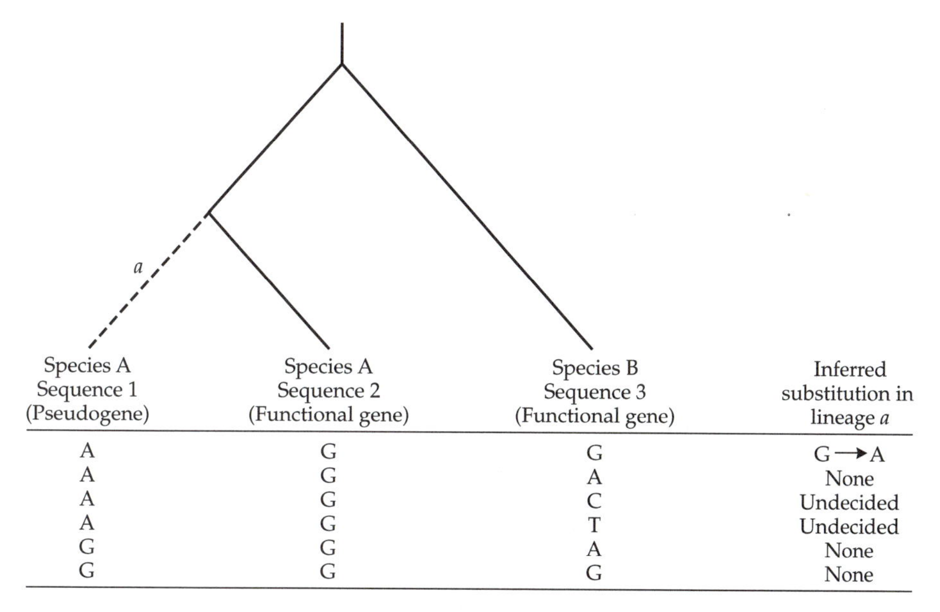

Species A Sequence 1 (Pseudogene)	Species A Sequence 2 (Functional gene)	Species B Sequence 3 (Functional gene)	Inferred substitution in lineage *a*
A	G	G	G → A
A	G	A	None
A	G	C	Undecided
A	G	T	Undecided
G	G	A	None
G	G	G	None

FIGURE 4.9 A tree for inferring the pattern of nucleotide substitution in a pseudogene sequence. The dashed line *a* implies "nonfunctional." In cases where the nucleotides occupying homologous sites in sequences 2 and 3 are identical, but different from the nucleotide in sequence 1, the type of substitution in lineage *a* can be unambiguously inferred.

excluded from the analysis. Similarly, deletions and insertions should also be excluded. Since the rate of substitution is usually much higher in pseudogenes than that in the homologous functional genes, differences in the nucleotide sequence between a gene and a pseudogene are explained in the majority of cases by substitutions that have occurred in the pseudogene rather than by substitutions in the functional gene.

In a study of 105 mammalian processed pseudogenes, Ophir and Graur (unpublished data) found that the f_{ij} values vary greatly from one pseudogene to another. This variation was attributed largely to chance effects. Thus, to obtain a reliable estimate of the mutation pattern from the pattern of substitution in pseudogenes we need to use many sequences.

The matrix in Table 4.5 represents the combined pattern of substitution inferred from 55 processed pseudogene sequences from humans. We note that the direction of mutation is nonrandom. For example, A changes more often to G than to either T or C. The four elements from the upper right corner to the lower left corner are the f_{ij} values for transitions, while the other eight elements represent transversions. All transitions, and in particular C → T and G → A, occur more often than transversions. The sum of the relative frequencies of transitions is 67.5% (66.2% if CG dinucleotides are excluded; see below). Because there are four types of transitions and eight types of transversions, the expected proportion of transitions under the assumption that all

TABLE 4.5 Pattern of substitution in pseudogenes[a]

From	To: A	T	C	G	Row totals
A	—	3.4 ± 0.7	4.5 ± 0.8	12.5 ± 1.1	20.3
		(3.6 ± 0.70)	(4.8 ± 0.9)	(13.3 ± 1.1)	(21.6)
T	3.3 ± 0.6	—	13.8 ± 1.9	3.3 ± 0.6	20.4
	(3.5 ± 0.6)		(14.7 ± 2.0)	(3.5 ± 0.6)	(21.7)
C	4.2 ± 0.5	20.7 ± 1.3	—	4.6 ± 0.6	29.5
	(4.2 ± 0.5)	(16.4 ± 1.3)		(4.4 ± 0.6)	(25.1)
G	20.4 ± 1.4	4.4 ± 0.6	4.9 ± 0.7	—	29.7
	(21.9 ± 1.5)	(4.6 ± 0.6)	(5.2 ± 0.8)		(31.6)
Column totals	27.9 (29.5)	28.5 (24.6)	23.2 (23.2)	20.5 (21.3)	

Courtesy of Dr. Ron Ophir.

[a]Table entries are the inferred percentages (f_{ij}) of nucleotide changes from i to j based on 105 processed pseudogene sequences from humans. Values in parentheses were obtained by excluding all CG dinucleotides from the comparison.

possible mutations occur with equal frequencies is 33.3%. The observed proportion (67.4%) is about twice the expected value.

Some nucleotides are more mutable than others. In the last column of Table 4.5, we list the relative frequencies of all mutations from A, T, C, and G to any other nucleotide. Were all the four nucleotides equally mutable, we would expect a value of 25% in each of the column's elements. In practice, we see that G mutates with a relative frequency of 29.7% (31.6% if CG dinucleotides are excluded), i.e., G is a highly mutable nucleotide, while A mutates with a relative frequency of 20.3% (i.e., it is not as mutable). In the bottom row, we list the relative frequencies of all mutations that result in A, T, C, or G. We note that 56.4% of all mutations result in either A or T, while the expectation for the case of equal probabilities of mutation in all directions is 50%. Since there is a tendency for C and G to change to A or T, and since A and T are not as mutable as C and G, pseudogenes are expected to become rich in A and T. This should also be true for other noncoding regions that are subject to no functional constraint. Indeed, noncoding regions are generally AT-rich.

It is known that, in addition to base mispairing, the transition from C to T can also arise from conversion of methylated C residues to T residues upon deamination (Coulondre et al. 1978; Razin and Riggs 1980). The effect will elevate the frequencies of C:G → T:A and G:C → A:T; i.e., f_{CT} and f_{GA}. Since about 90% of methylated C residues in vertebrate DNA occur at 5′–CG–3′ dinucleotides (Razin and Riggs 1980), this effect should be expressed mainly as changes of the CG dinucleotides to TG or CA. After a gene becomes a pseudogene, such changes would no longer be subject to any functional constraint and can contribute significantly to C → T and G → A transitions, if the frequency of CG is relatively high before the silencing of the gene (i.e., its loss of function). The substitution pattern obtained by excluding all nucleotides sites where the CG dinucleotides occur is given in parentheses in Table 4.5. This pattern is probably more suitable for predicting the pattern of mutations in a sequence that has not been subject to functional constraints for a long time (e.g., some parts of an intron), because in such a sequence few CG dinucleotides would exist to begin with. The pattern obtained after excluding the CG dinucleotides is somewhat different from that obtained otherwise. In particular, the relative frequencies of the transition C → T is lower by about 20%.

Strand inequalities: Pattern of substitution in human mitochondrial DNA

Initial studies on primate mitochondrial DNA (mtDNA) revealed an extreme transitional bias in the pattern of nucleotide substitution (Brown et al. 1982). Tamura and Nei (1993) studied the pattern of substitution in the control region of mtDNA using extensive data from humans and chimpanzees. With the exception of a conservative central portion, which was excluded from the analysis, the control region is thought to be devoid of functional constraints, and as such its pattern of substitution may reflect the pattern of spontaneous mutation in mtDNA. The estimated pattern is shown in Table 4.6. Transversions were found to occur with very low frequencies and the average transition/transversion ratio was 15.7, much larger than the ratio of 2 in nuclear DNA. Moreover the relative frequency of transition between pyrimidines (C ↔ T) is almost twice as large as that between purines (G ↔ A), violating the two equalities, $f_{CT} = f_{GA}$ and $f_{TC} = f_{AG}$, and suggesting that the patterns and rates of mutation may be different between the two strands (Chapter 8). In comparison, no evidence for strand inequality is obtained from the nuclear pattern in Table 4.5, where the relative frequency of transition between pyrimidines (34.5%) is about the same as that between purines (32.9%).

TABLE 4.6 Pattern of substitution in the control region of human mitochrondrial DNA[a]

	To				
From	**A**	**T**	**C**	**G**	**Row totals**
A	—	0.4	1.1	14.1	15.6
T	0.3	—	33.8	0.3	34.4
C	1.1	25.8	—	0.5	27.4
G	20.0	1.1	1.6	—	22.7
Column total	21.4	27.3	36.5	14.9	

From Tamura and Nei (1993).

[a]Table entries are the inferred percentages (f_{ij}) of nucleotide changes from *i* to *j* based on 95 sequences.

TABLE 4.7 Physicochemical distances between pairs of amino acids[a]

Arg	Leu	Pro	Thr	Ala	Val	Gly	Ile	Phe	Tyr	Cys	His	Gln	Asn	Lys	Asp	Glu	Met	Trp	
110	145	74	58	99	124	56	142	155	144	112	89	68	46	121	65	80	135	177	**Ser**
	102	103	71	112	96	125	97	97	77	180	29	43	86	26	96	54	91	101	**Arg**
		98	92	96	32	138	**5**	22	36	198	99	113	153	107	172	138	15	61	**Leu**
			38	27	68	42	95	114	110	169	77	76	91	103	108	93	87	147	**Pro**
				58	69	59	89	103	92	149	47	42	65	78	85	63	81	128	**Thr**
					64	60	94	113	112	195	86	91	111	106	126	107	84	148	**Ala**
						109	29	50	55	192	84	96	133	97	152	121	21	88	**Val**
							135	153	147	159	98	87	80	127	94	98	127	184	**Gly**
								21	33	198	94	109	149	102	168	134	10	61	**Ile**
									22	205	100	116	158	102	177	140	28	40	**Phe**
										194	83	99	143	85	160	122	36	37	**Tyr**
											174	154	139	202	154	170	196	**215**	**Cys**
												24	68	32	81	40	87	115	**His**
													46	53	61	29	101	130	**Gln**
														94	23	42	142	174	**Asn**
															101	56	95	110	**Lys**
																45	160	181	**Asp**
																	126	152	**Glu**
																		67	**Met**

From Grantham (1974).

[a]Mean distance is 100. The largest and smallest distances are emphasized with shading.

Patterns of amino acid replacement

There are many measures in the literature aimed at quantifying the similarity or dissimilarity between two amino acids (e.g., Sneath 1966; Grantham 1974; Miyata et al. 1979). These so-called **physicochemical distances** are based on such properties of the amino acids as polarity, molecular volume, and chemical composition. Grantham's (1974) physicochemical distances are shown in Table 4.7. A replacement of an amino acid by a similar one (e.g., leucine to isoleucine or leucine to methionine; Figure 4.10a) is called a **conservative replacement**, and a replacement of an amino acid by a dissimilar one (e.g., glycine to tryptophan or cysteine to tryptophan; Figure 4.10b) is called a **radical replacement**. Some amino acids, such as leucine, isoleucine, glutamine, and methionine are **typical amino acids**, since they have a number of similar alternative amino acids with which they can be replaced through a single nonsynonymous substitution. Other amino acids, such as cysteine, tryptophan, tyrosine, and glycine, are **idiosyncratic amino acids**; they have few similar alternative amino acids with which they can be replaced. Graur (1985) devised a **stability index**, which is the mean physicochemical distance between an amino acid and its mutational derivatives that can be produced through a single nucleotide substitution. The stability index can be used to predict the evolutionary propensity of amino acids to undergo replacement.

It has been known since the early work of Zuckerkandl and Pauling (1965) that conservative replacements occur more frequently than radical replacements in protein evolution. The conservative nature of amino acid re-

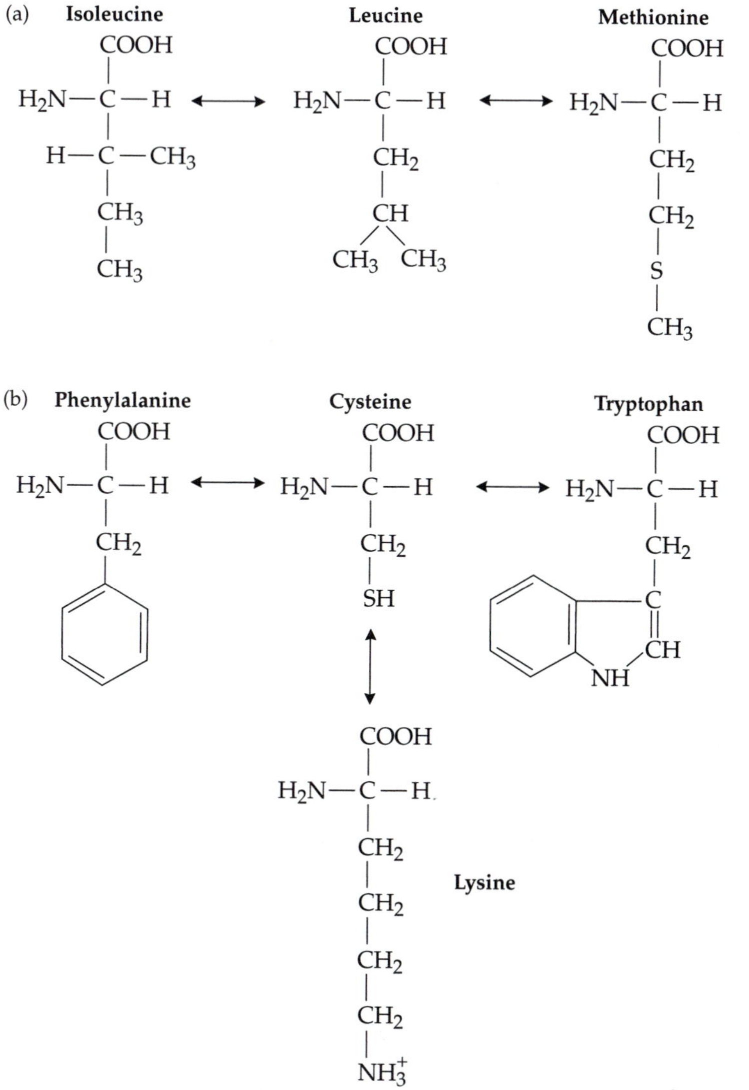

(a) Isoleucine ⟷ Leucine ⟷ Methionine

(b) Phenylalanine ⟷ Cysteine ⟷ Tryptophan; Cysteine ⟷ Lysine

FIGURE 4.10 **(a) The most similar amino acid pairs are leucine and isoleucine (Grantham's distance = 5), and leucine and methionine (15). (b) The most dissimilar amino acid pairs are cysteine and tryptophan (215), cysteine and phenylalanine (205), and cysteine and lysine (202). Note that all three most dissimilar pairs involve the idiosyncratic amino acid cysteine.**

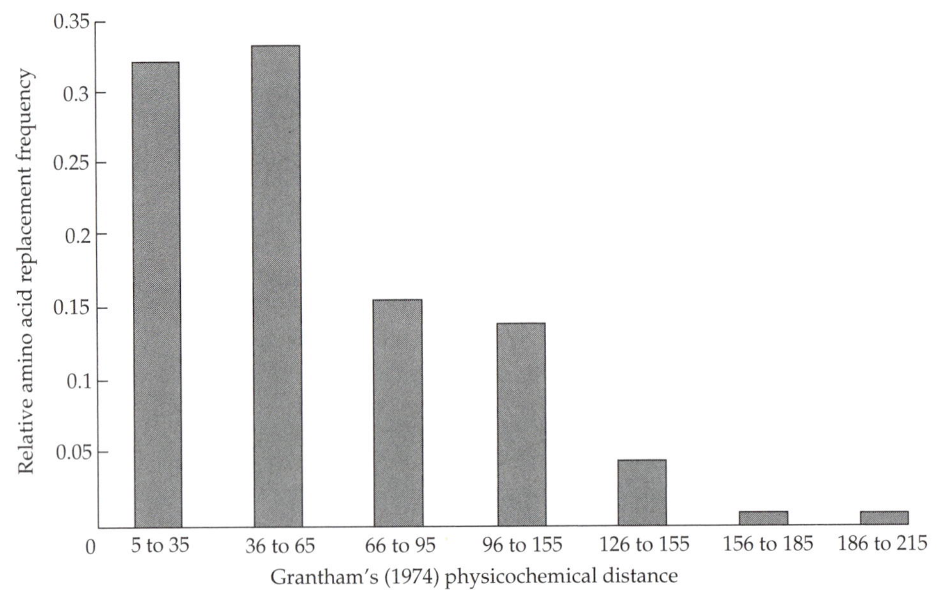

FIGURE 4.11 Relationship between physicochemical distance and relative amino acid replacement frequency in 20 mammalian proteins (ryanodine receptor, dystrophin, ataxia-telangiectasia locus protein, coagulation factors VIII and IX, cystic fibrosis transmembrane conductance regulator, α glucosidase, low-density lipoprotein receptor, pyruvate kinase, butyrylcholinesterase, hexoseaminidases A and B, glucocerebrosidase, phenylalanine hydroxylase, fumarylacetate hydrolase, galactose-1-phosphate uridyltransferase, peripherin, uroporphyrinogene III synthase, CD40 ligand, and von Hippel-Lindau factor). Courtesy of Ms. Inna Gur-El.

placement is evident when plotting the relative frequencies of amino acid replacements against the physicochemical distance (Figure 4.11). From the standpoint of the neutral theory, this phenomenon can be easily explained by using Equation 4.3. Conservative replacements are likely to be less disruptive than radical ones; hence, the probability of a mutational change being selectively neutral (as opposed to deleterious) is greater if the amino acid replacement occurs between two similar amino acids than if it occurs between two dissimilar ones. However, in some cases the codons of similar amino acids differ by more than one nucleotide, and so a conservative amino acid replacement may be less probable than a more radical replacement.

Argyle (1980) devised a circular graphical representation of amino acid exchangeability. A modified version by Pieber and Tohá (1983) is shown in Figure 4.12. Depending on the protein, 60–90% of observed amino acid replacements involve the nearest or second-nearest neighbors on the ring.

What protein properties are conserved in evolution?

The evolution of each protein-coding gene is constrained by the functional requirements of the specific protein it produces. However, it may be interesting

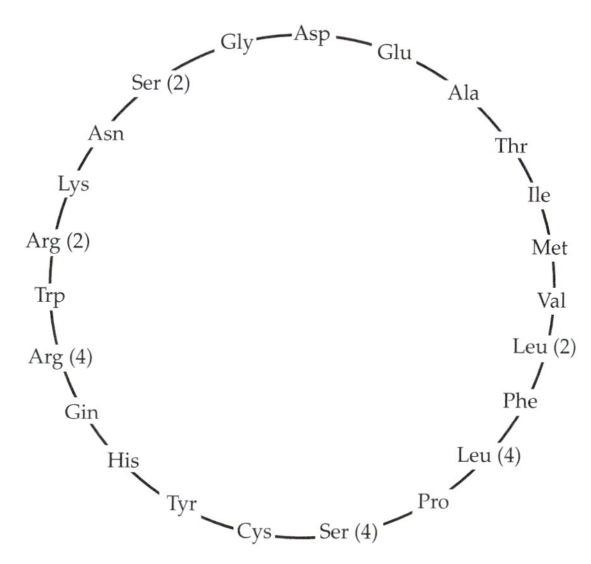

FIGURE 4.12 A circular graphical representation of amino acid exchangeability according to Argyle's (1980) method. Numbers in parentheses denote number of codons in a family in cases where an amino acid is encoded by two codon families. About 60–90% of the observed amino acid replacements involve the nearest or second-nearest neighbors on the ring. Data from Pieber and Tohá (1983).

to find out whether or not there are also some general properties that are constrained during evolution in all proteins. The answer seems to be that several properties are indeed conserved during protein evolution (Soto and Tohá 1983). The two most highly conserved properties are bulkiness (volume) and refractive index (a measure of protein density). Hydrophobicity and polarity also seem to be moderately well conserved, whereas optical rotation seems to be an irrelevant property in the evolution of proteins. Surprisingly, the distribution of charged amino acids, which might be expected to be an important factor determining protein evolution, is one of the least conserved properties during protein evolution (Leunissen et al. 1990).

As a consequence of the conservation of bulkiness and hydrophobicity, some amino acids tend to be relatively impervious to replacement during evolution. Indeed, glycine, the smallest amino acid, tends to be extremely conserved during evolution regardless of its proximity to functionally active sites. Conceivably, replacements at sites occupied by glycine will invariably introduce into the polypeptide chain a much bulkier amino acid. Since such a structural disruption has a high probability to have an adverse effect on the function of the protein regardless of the location of glycine relative to the active site, such mutations are frequently selected against. Consequently, genes that encode proteins that contain a large proportion of glycine residues will tend to evolve more slowly than proteins that are glycine-poor (Graur 1985). In addition to glycine, other amino acids (e.g., lysine, cysteine, and proline) are also consistently conserved (Naor et al. 1996). Lysine and cysteine are most probably conserved because of their involvement in crosslinking between polypeptide chains, while proline is conserved because of its unique contribution to the contorted bending of proteins.

NONRANDOM USAGE OF SYNONYMOUS CODONS

Because of the degeneracy of all the genetic codes, most of the 20 amino acids are encoded by more than one codon (Chapter 1). Since synonymous mutations do not cause any change in amino acid sequence and since natural selection is thought to operate predominantly at the protein level, synonymous mutations were proposed as candidates for selectively neutral mutations (Kimura 1968b; King and Jukes 1969). However, if all synonymous mutations are indeed selectively neutral, and if the pattern of mutation is symmetrical, then the synonymous codons for an amino acid should be used with more or less equal frequencies. As DNA sequence data accumulated, however, it became evident that the usage of synonymous codons is distinctly nonrandom in both prokaryotic and eukaryotic genes (Grantham et al. 1980). In fact, in many yeast and *Escherichia coli* genes, biased usage of synonymous codons is highly conspicuous. For example, 21 of the 23 leucine residues in the *E. coli* outer membrane protein II (*ompA*) are encoded by the codon CUG, although five other codons for leucine are available. How to explain the widespread phenomenon of nonrandom codon usage became a controversial issue, to which, fortunately, some clear answers seem to be emerging.

Measures of codon-usage bias

A simple measure of nonrandom usage of synonymous codons in a gene is the **relative synonymous codon usage** (***RSCU***). The *RSCU* value for a codon is the number of occurrences of that codon in the gene divided by the number of occurrences expected under the assumption of equal codon usage.

$$RSCU_i = \frac{X_i}{\frac{1}{n}\sum_{i=1}^{n} X_i} \quad \textbf{(4.17)}$$

where n is the number of synonymous codons ($1 \leq n \leq 6$) for the amino acid under study, i is the codon, and X_i is the number of occurrences of codon i (Sharp et al. 1986). If the synonymous codons of an amino acid are used with equal frequencies, their *RSCU* values will equal 1.

If certain codons are used preferentially in an organism, it is useful to devise a measure with which to distinguish between genes that use the preferred codons and those that do not. Sharp and Li (1987b) devised a **codon adaptation index** (***CAI***). In the first step of computing *CAI* values, a reference table of *RSCU* values for highly expressed genes is compiled. From this table, it is possible to identify the codons that are most frequently used for each amino acid. The **relative adaptiveness** of a codon, w_i, is computed as

$$w_i = \frac{RSCU_i}{RSCU_{max}} \quad \textbf{(4.18)}$$

where $RSCU_{max}$ is the *RSCU* value for the most frequently used codon for an amino acid.

The *CAI* value for a gene is calculated as the geometric mean of w_i values for all the codons used in that gene:

$$CAI = \left(\prod_{i=1}^{L} w_i \right)^{\frac{1}{L}} \quad \textbf{(4.19)}$$

where L is the number of codons. *CAI* values are frequently used to identify genomic regions that have been horizontally transferred among species (Chapter 8).

Another measure of codon bias is the **effective number of codons** (*ENC*). It is estimated by

$$ENC = 2 + \frac{9}{F_2} + \frac{1}{F_3} + \frac{5}{F_4} + \frac{3}{F_6} \quad \textbf{(4.20)}$$

where F_i (i = 2, 3, 4, or 6) is the average probability that two randomly chosen codons for an amino acid with i codons will be identical (Wright 1990). (Note the analogy with nucleotide diversity presented in Chapter 2.) In genes translated by the universal genetic code, *ENC* values may range from 20 (the number of amino acids), which means that the bias is at a maximum and only one codon is used from each synonymous-codon group, to 61 (the number of sense codons), which indicates no codon-usage bias.

Universal and species-specific patterns of codon usage

An observation that has been helpful for understanding the phenomenon of nonrandom codon usage is that genes in an organism generally show the same pattern of choices among synonymous codons (Grantham et al. 1980). The choices of biased codon usage, however, differ from one organism to another. Thus, mammalian, *Escherichia coli*, and yeast genes fall into distinct classes of codon usage. In light of this fact, Grantham et al. (1980) proposed the **genome hypothesis**. According to this hypothesis, the genes in any given genome use the same strategy with respect to choices among synonymous codons; in other words, the bias in codon usage is species-specific. The genome hypothesis turned out to be true in general, although there is considerable heterogeneity in codon usage among genes within a single genome.

There are, however, a few features of codon usage shared among all organisms. For example, in a comparison of codon-usage patterns among *E. coli*, yeast, *Drosophila*, and primates, Zhang et al. (1991) discovered that codons that contain the CG dinucleotide are universally avoided (**low-usage codons**). This phenomenon is particularly notable as far as the arginine codons CGA and CGG are concerned. The reason for this avoidance may be the previously mentioned propensity of CG dinucleotides to mutate and, hence, disappear from genomic sequences (see pages 126–127).

Codon usage in unicellular organisms

Studies of codon usage in *E. coli* and yeast have greatly increased our understanding of the factors that affect the choice of synonymous codons. Post et al. (1979) found that *E. coli* ribosomal protein-coding genes preferentially use synonymous codons that are recognized by the most abundant tRNA species. They suggested that the preference resulted from natural selection because using a codon that is translated by an abundant tRNA species will increase translational efficiency and accuracy. Their finding prompted Ikemura (1981, 1982) to gather data on the relative abundances of tRNAs in *E. coli* and the yeast *Saccharomyces cerevisiae*. He showed that, in both species, a positive correlation exists between the relative frequencies of the synonymous codons in a gene and the relative abundances of their cognate tRNA species. The correlation is very strong for highly expressed genes.

Figure 4.13 shows schematically the correspondence between the frequencies of the six leucine codons and the relative abundances of their cognate tRNAs. In *E. coli*, $\text{tRNA}_1^{\text{Leu}}$ is the most abundant species, and indeed in highly expressed genes, CUG (the codon recognized by this tRNA) is much more frequently used than the other five codons. In contrast, $\text{tRNA}_{\text{CUA}}^{\text{Leu}}$ is relatively rare, and indeed CUA appears only very rarely in highly expressed genes. In yeast, the most abundant leucine-tRNA species is $\text{tRNA}_3^{\text{Leu}}$, and the codon recognized by this tRNA (UUG) is the predominant one. The least abundant leucine-tRNA species in yeast is $\text{tRNA}_2^{\text{Leu}}$, and the codons recognized by this species (CUC and CUU) are seldom encountered in highly expressed protein-coding genes. In contrast, in genes with low levels of expression, the correspondence between tRNA abundance and the use of the respective codons is much weaker in both species (Figure 4.13).

The importance of translational efficiency in determining the codon-usage pattern in highly expressed genes is further supported by the following observation (Ikemura 1981). It is known that codon–anticodon pairing involves **wobbling** at the third position. For example, U in the first position of anticodons can pair with both A and G. Similarly, G can pair with both C and U. On the other hand, C in the first anticodon position can only pair with G at the third position of codons, and A can only pair with U. Wobbling is also made possible by the fact that some tRNAs contain modified bases at the first anticodon position, and these can recognize more than one codon. For example, inosine (a modified adenine) can pair with any of the three bases U, C, and A. Interestingly, most tRNAs that can recognize more than one codon exhibit differential preferences for one of them. For example, 4-thiouridine (S^4U) in the wobble position of an anticodon can recognize both A and G in the wobble position; however, it has a marked preference for A-terminated codons over G-terminated ones. Such a preference should be reflected in highly expressed genes. The two codons for lysine in *E. coli* are recognized by a tRNA molecule that has S^4U in the wobble position of the anticodon and indeed, in the *E. coli ompA* gene, 15 of the 19 lysine codons are AAA, and only four are AAG.

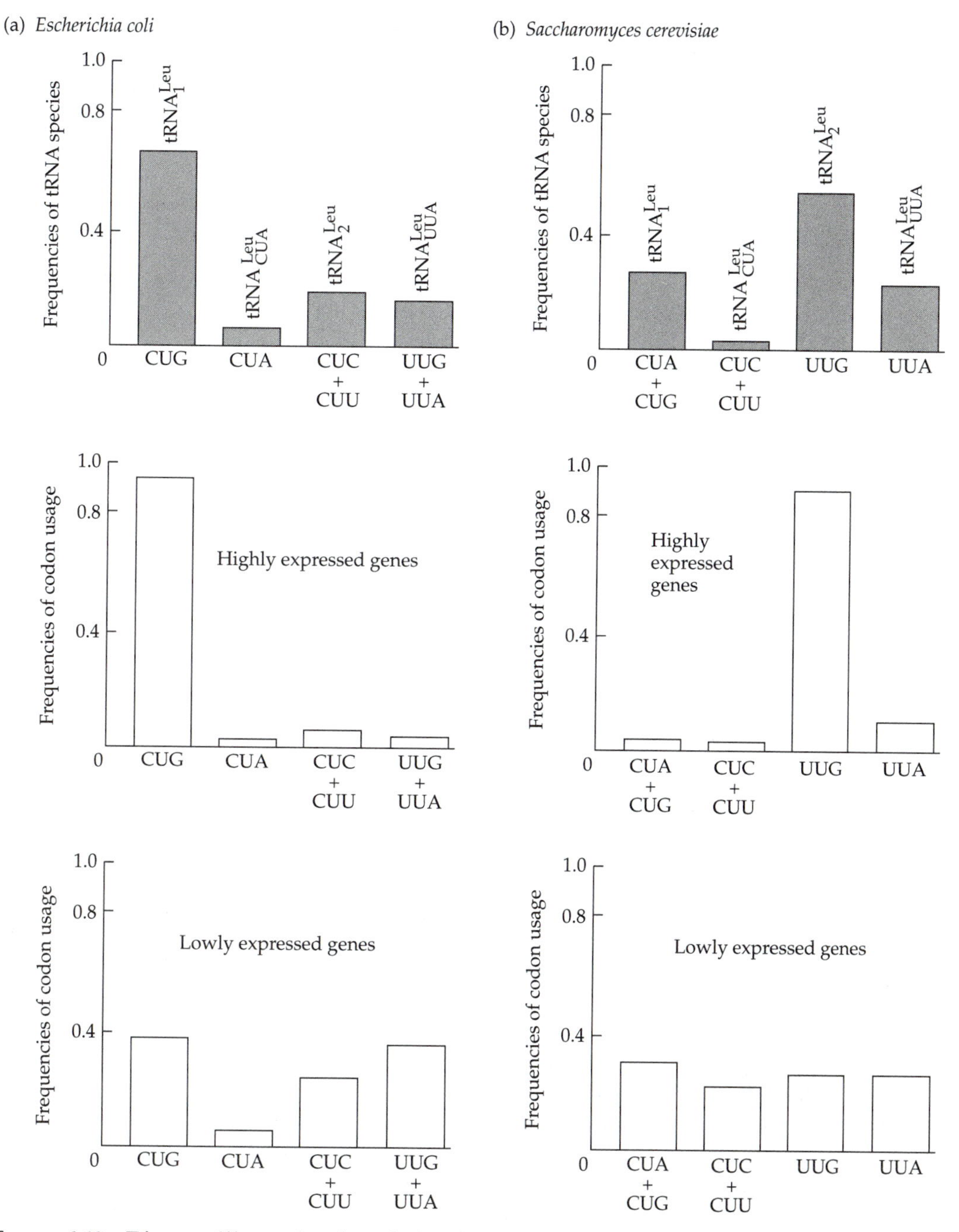

FIGURE 4.13 Diagram illustrating the relationship between the relative frequency of codon usage for leucine (open bars) and the relative abundance of the corresponding cognate tRNA species (gray bars) in (a) *Escherichia coli* and (b) the yeast *Saccharomyces cerevisiae*. The plus signs between codons indicate that the two condons are recognized by a single tRNA species (e.g., both CUC and CUU in *E. coli* are recognized by tRNA$_2^{Leu}$).

TABLE 4.8 Codon usage in *Escherichia coli* and *Saccharomyces cerevisiae*[a]

		E. coli		*S. cerevisiae*	
Amino acid	**Codon**	**High**	**Low**	**High**	**Low**
Leu	UUA	0.06	1.24	0.49	1.49
	UUG	0.07	0.87	5.34	1.48
	CUU	0.13	0.72	0.02	0.73
	CUC	0.17	0.65	0.00	0.51
	CUA	0.04	0.31	0.15	0.95
	CUG	5.54	2.20	0.02	0.84
Val	GUU	2.41	1.09	2.07	1.13
	GUC	0.08	0.99	1.91	0.76
	GUA	1.12	0.63	0.00	1.18
	GUG	0.40	1.29	0.02	0.93
Ile	AUU	0.48	1.38	1.26	1.29
	AUC	2.51	1.12	1.74	0.66
	AUA	0.01	0.50	0.00	1.05
Phe	UUU	0.34	1.33	0.19	1.38
	UUC	1.66	0.67	1.81	0.62

From Sharp et al. (1988)

[a]For each group of synonymous codons, the sum of the relative frequencies equals the number of codons in the group. For example, there are six codons for leucine, and so the sum of the relative frequencies for these six codons is 6. Under equal codon usage, the relative frequencies for each codon in a group should be 1, and so the degree of deviation from 1 indicates the degree of bias in codon usage. "High" and "Low" denote genes with high and low expression levels, respectively.

Table 4.8 shows part of a compilation of codon usage by Sharp et al. (1988). For each group of synonymous codons, if the usage is equal, the relative frequency of each codon should be 1. This is clearly not so in the majority of cases. Moreover, in both *E. coli* and yeast, the codon-usage bias is much stronger in highly expressed genes than in lowly expressed ones. For instance, in highly expressed *E. coli* genes, the ratio of the most frequently used to the least frequently used Leu codon is about 140. In lowly expressed genes, the ratio is only 7. A simple explanation for this difference is that in highly expressed genes, purifying selection for translational efficiency and accuracy is strong, so the codon-usage bias is pronounced. In lowly expressed genes, on the other hand, selection is relatively weak, so the usage pattern is mainly affected by mutation pressure and random genetic drift, and is consequently less skewed (Sharp and Li 1986).

Translational selective constraints on codon usage will result in purifying selection, thus slowing down the rate of synonymous substitution (Ikemura 1981; Kimura 1983; Sharp and Li 1987a). Indeed, it has been shown that the

TABLE 4.9 Factors determining the choice of optimal codons in unicellular organisms

1. tRNA availability.
2. Preference for A over G when thiolated uridine or 5-carboxymethyl are at the anticodon wobble position.
3. Preference for T and C over A when inosine is at the anticodon wobble position.
4. Preference for C in the third position of codons AAN, ATN, TAN, and TTN

After Ikemura and Ozeki (1983).

rate of synonymous substitution in *E. coli* and other related enterobacteria is negatively correlated with the degree of codon-usage bias (Sharp and Li 1986). Therefore, the phenomenon of nonrandom usage of synonymous codons may not be taken as evidence against the neutral theory of molecular evolution, since it can be explained in terms of the principle that stronger selective constraints result in lower rates of evolution (see page 109 and Kimura 1983). Additional support for the role of translational efficiency has been obtained from a study of ribosomal protein-coding genes from three organisms. Moriyama and Powell (1998) found a positive correlation between gene length and codon-usage bias. They suggested that energetically costly longer genes have a higher codon-usage bias to maximize translational efficiency.

In conclusion, in unicellular organisms, the choice of synonymous codons appears to be constrained mainly by tRNA availability and other factors related to translational efficiency. Mutation patterns as well as large-scale aspects of chromatin structure might also influence codon usage to a certain extent (Dujon et al. 1994). Ikemura and Ozeki (1983) identified a set of four factors that determine the choice of optimal codons in unicellular organisms (Table 4.9).

Codon usage in multicellular organisms

Figure 4.14 shows the distribution of the effective numbers of codons (*ENC*) for 1,117 *Drosophila melanogaster* genes, as well as for a subset of 28 ribosomal protein-coding genes (Moriyama and Powell 1997). The vast majority of genes show codon-usage biases, as indicated by a mean *ENC* value of 46. In the ribosomal protein-coding genes, the codon-usage bias is much stronger (mean *ENC* = 35), indicating that selection for translational efficiency plays an important role in determining the choice of synonymous codons in *Drosophila*. We note, however, that in multicellular organisms, different cells produce different proteins, and therefore a simple relationship between codon usage and tRNA abundance is not expected. In some cases, there is evidence for tissue-specific adaptations of the number of isoacceptor tRNA species and codon frequencies. For example, the middle part of the silk gland in the silk moth *Bombyx mori* secretes a protein called sericin (31% serine), while the posterior part of the gland produces fibroin (46% glycine, 29% alanine, and 12% serine). Interestingly, the

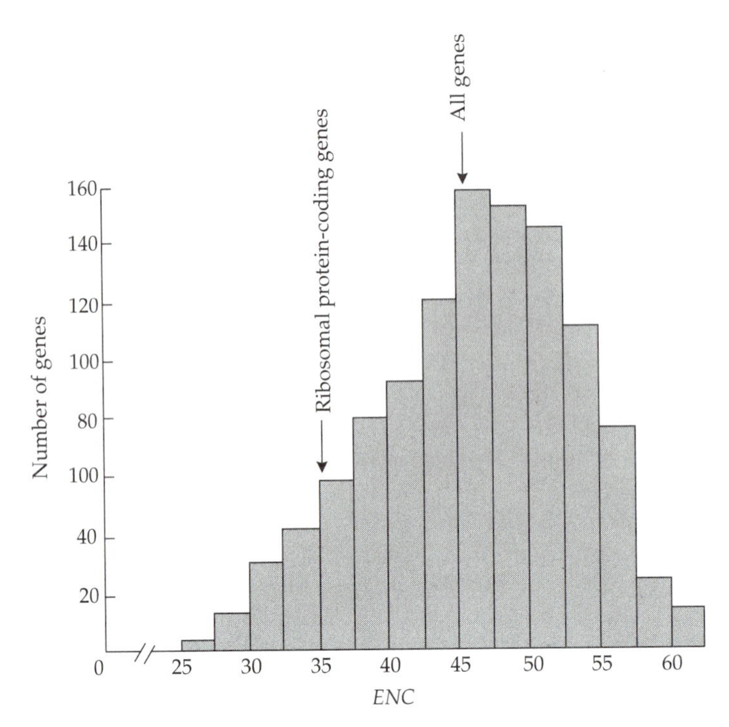

FIGURE 4.14 Distribution of effective number of codons (*ENC*) for 1,117 *Drosophila melanogaster* genes Mean *ENC* values for all these genes, as well as for a subset of 28 ribosomal protein-coding genes are shown. Modified from Moriyama and Powell (1997).

population of isoaccepting $tRNA^{Ser}$ species is different between the two parts of the gland. While cells in the middle part contain mainly $tRNA_1^{Ser}$, which recognizes AGU and AGC codons, the posterior part contains mainly $tRNA_2^{Ser}$, which recognizes UCA (Hentzen et al. 1981). This example notwithstanding, in most cases the correlation between the relative quantity of isoacceptor tRNA species and codon frequencies is far from being convincing.

In most mammalian genes, codons tend to end in either G or C (i.e., to have a high GC content at the third codon position). In other genes, the third position of codons tends to be GC-poor. A possible interpretation of these findings would be that the former group of genes is highly expressed, whereas the latter is not. There are, however, several reasons to believe that these contrasting biases in codon usage may not be related to gene expression. First, the α- and β-globin genes have different GC contents at the third position of codons (high and low, respectively), although they are both expressed in the same tissue (erythrocytes) in approximately equal quantities and therefore should have the same level of expression. Second, in chicken genes, the frequency with which codons are used does not correlate well with tRNA availability (Ouenzar et al. 1988), although this observation may not be directly applicable to mammalian genes. Finally, the GC content at the third position is strongly correlated with the GC level in both flanking regions and in introns. For example, the α-globin is high in GC and resides in a high-GC region, whereas the β-globin gene is low in GC and resides in a low-GC region (Bernardi et al. 1985). Thus, it appears that the codon-usage bias in a mammalian gene is largely determined by the GC content in the region that contains the gene. As will be discussed in Chapter 8, whether the GC content in a chromosomal region is determined by natural selection or mutational bias is still a controversial question.

Codon usage and population size

If codon usage is affected by selection, the strength of such selection (either positive or negative) ought to be very weak. In fact, it may be so weak that random genetic drift would dominate the evolutionary dynamics of codon substitution in species with a small effective population size, whereas selection would be the dominant force in the evolution of codon bias in species with large effective population sizes (Chapter 2). Indeed, in a comparison of codon biases between two *Drosophila* species, Akashi (1997) found that *D. simulans*, which has a larger effective population size than *D. melanogaster*, also has a stronger codon bias.

MOLECULAR CLOCKS

In their comparative studies of hemoglobin and cytochrome *c* protein sequences from different species, Zuckerkandl and Pauling (1962, 1965) and Margoliash (1963) first noticed that the rates of amino acid replacement were approximately the same among various mammalian lineages (Figure 4.15). Zuckerkandl and Pauling (1965) therefore proposed that for any given protein, the rate of molecular evolution is approximately constant over time in all lineages or, in other words, that there exists a **molecular clock**. The proposal immediately stimulated a great deal of interest in the use of macromolecules in evolutionary studies. Indeed, if proteins evolve at constant rates, they can be used to determine dates of species divergence and to reconstruct phylogenetic relationships among organisms. This practice would be analogous to the dating of geological times by measuring the decay of radioactive elements.

Under the molecular clock assumption, Equation 4.1 can be used to estimate the rate of substitution, r, for any given protein or DNA region by using a species pair for which the date of divergence can be established on the basis of paleontological data. This estimated rate could then be used to date the divergence time between two species for which paleontological data on their divergence time is lacking. For example, let us assume that the rate of nonsynonymous substitution for the α chain of hemoglobin is 0.56×10^{-9} substitutions per site per year, and that α-globins from rat and human differ by 0.093 substitutions per site. Then, under the molecular clock hypothesis, the

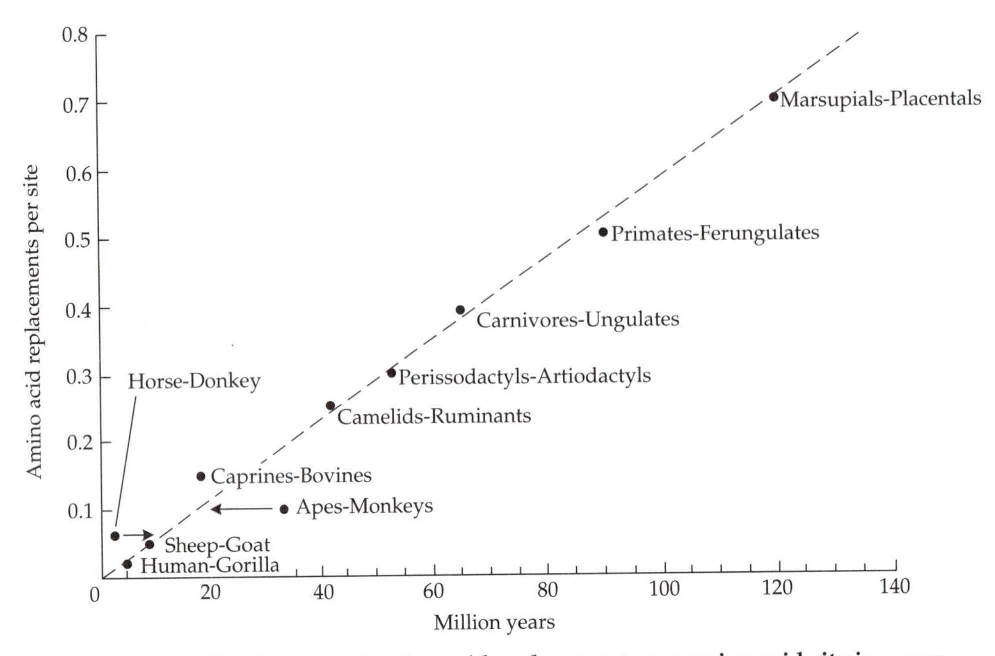

FIGURE 4.15 **Number of amino acid replacements per amino acid site in a combined sequence consisting of hemoglobins α and β, cytochrome *c*, and fibrinopeptide A among various mammalian groups plotted against geological estimates of divergence times. The dashed line represents the molecular clock expectation of equal rates of amino acid replacement in all evolutionary lineages. There are two large deviations of the observed values from the expected line. These deviations indicate a slowdown in evolution following the divergence between apes and monkeys, and an acceleration following the divergence between horse and donkey. However, these inferences are based on specific paleontological estimates of divergence times (33 million years for the ape-monkey split and 2 million years for the horse-donkey split), and if these time estimates are inaccurate (arrows), the deviation of these lineages from a strict molecular clock may not be significant. Modified from Langley and Fitch (1974).**

divergence time between the human and rat lineages is estimated to be approximately $0.093/(2 \times 0.56 \times 10^{-9}) = 80$ million years ago.

In the 1970s, protein sequence data became available and many statistical analyses were conducted to test the molecular clock hypothesis (Dickerson 1971; Ohta and Kimura 1971; Langley and Fitch 1974; Wilson et al. 1977). A general finding was that a rough linear relationship exists between the estimated number of amino acid replacements and divergence time. This finding led to the proposal of the following evolutionary rule by Kimura (1983): "For each protein, the rate of evolution in terms of amino acid substitutions is approximately constant per year per site for various lines, as long as the function and tertiary structure of the molecule remain essentially unaltered." The condition appended to the rule was meant to take care of instances in which, in a particular evolutionary lineage, a gene loses some or all of its functions (see page 115), or acquires a new biological role (Chapter 6). In such cases, the functional constraints operating on the gene are altered, and as a consequence the gene will no longer evolve at the same rate as the homologous genes in other organisms, which did retain their original function.

It should be noted, however, that from the beginning, proponents of the molecular clock hypothesis recognized the existence of exceptional cases. Thus, while a given protein usually evolves at a characteristic rate regardless of the organismic lineage, in some particular lineage it may evolve at a markedly different rate. A well-known case was that of insulin, which has evolved much faster in the evolutionary lineage leading to the guinea pig than along other lines (King and Jukes 1969). It has also been recognized that amino acid replacement does not usually follow a simple Poisson process (see Appendix II); the observed variance in amino acid replacement rates among different lineages is about two times larger than the expected Poisson variance (Ohta and Kimura 1971; Langley and Fitch 1974; Kimura 1983).

The molecular clock hypothesis stimulated a great deal of controversy. Classical evolutionists, for instance, argued against it because the suggestion of rate constancy did not sit well with the erratic tempo of evolution at the morphological and physiological levels. The hypothesis met with particularly heated opposition when the assumption of rate constancy was used to obtain an estimate of 5 million years for the divergence between humans and the African apes (Sarich and Wilson 1967), in sharp contrast to the then-prevailing view among paleontologists that humans and apes diverged at least 15 million years ago. Further controversy ensued when the approximate rate constancy observed in protein sequence data was used to support the neutral mutation hypothesis (Kimura 1969).

Many molecular evolutionists have also challenged the validity of the molecular clock hypothesis. In particular, Goodman (1981a,b) and his associates (Czelusniak et al. 1982) contended that the rate of evolution often accelerates following gene duplication. For example, they claimed that extremely high rates of amino acid replacement occurred following the gene duplication that gave rise to the α- and β-hemoglobins, and that the high rates were due to advantageous mutations that improved the function of these globin chains. Moreover, Goodman and his associates also objected to the molecular clock on grounds that protein sequence evolution often proceeds much more rapidly at times of adaptive radiation than during periods in which no speciation occurs. For instance, Goodman (1981b) and Goodman et al. (1974, 1975) claimed that an increase in the rate of amino acid replacement in cytochrome *c* and in the α- and β-globins occurred at least three times during the evolution of these proteins: in the period immediately after the teleost–tetrapod divergence (~400 million years ago), in the early phases of divergence among eutherian orders (~80 million years ago), and during the initial stages of divergence among primates (less than 55 million years ago).

Although the rate constancy assumption has always been controversial, it has been widely used in the estimation of divergence times and the reconstruction of phylogenetic trees (e.g., Dayhoff 1972; Nei 1975; Wilson et al. 1977). Thus, the question of the validity of the molecular clock is a vital issue in molecular evolution. The rapid accumulation of DNA sequence data in recent years affords an unprecedented opportunity for testing the hypothesis. Such data allow a closer examination of the hypothesis than do protein sequences, and can be interpreted more directly than DNA–DNA hybridization data or immunological distances.

RELATIVE RATE TESTS

The controversy over the molecular clock hypothesis often involves disagreements on dates of species divergence. For if we are allowed to move any point horizontally in Figure 4.15, then we will be able to fit any set of data to a straight line. To avoid this problem, several tests that do not require knowledge of divergence times have been developed. In the following, we present three such tests. Many other methods have been suggested in the literature (e.g., Felsenstein 1988; Muse and Weir 1992; Goldman 1993; Takezaki et al. 1995; Robinson et al. 1998).

Margoliash, Sarich, and Wilson's test

The **relative rate test** of Margoliash (1963) and Sarich and Wilson (1973) is illustrated in Figure 4.16. Suppose that we want to compare the rates in lineages A and B. Then, we use a third species, C, as an outgroup reference. We should be certain that the reference species branched off earlier than the divergence of species A and B. For example, to compare the rates in the human and chimpanzee lineages, we can use orangutan as reference (Chapter 5).

From Figure 4.16, it is easy to see that the number of substitutions between species A and B, K_{AB}, is equal to the sum of substitutions that have occurred from point O to point A (K_{OA}) and from point O to point B (K_{OB}). That is,

$$K_{AB} = K_{OA} + K_{OB} \tag{4.21}$$

Similarly,

$$K_{AC} = K_{OA} + K_{OC} \tag{4.22}$$

and

$$K_{BC} = K_{OB} + K_{OC} \tag{4.23}$$

Since K_{AC}, K_{BC}, and K_{AB} can be directly estimated from the nucleotide sequences (Chapter 3), we can easily solve the three equations to find the values of K_{OA}, K_{OB}, and K_{OC}:

$$K_{OA} = \frac{K_{AC} + K_{AB} - K_{BC}}{2} \tag{4.24}$$

$$K_{OB} = \frac{K_{AB} + K_{BC} - K_{AC}}{2} \tag{4.25}$$

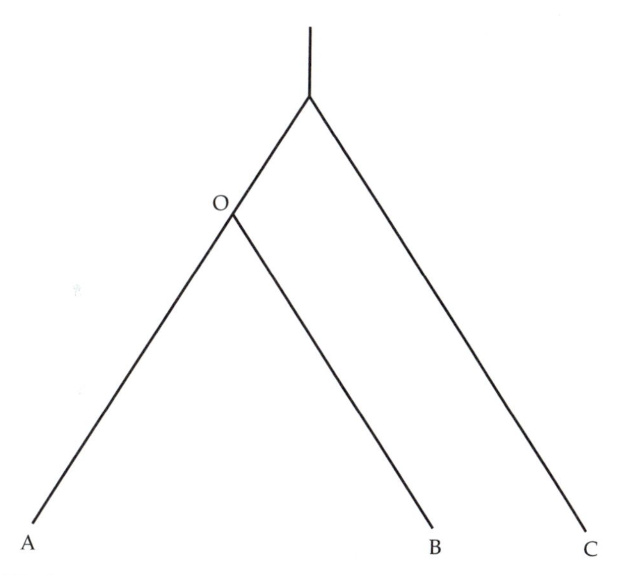

FIGURE 4.16 Phylogenetic tree used in the relative rate test. O denotes the common ancestor of species A and B. C is the outgroup.

$$K_{OC} = \frac{K_{AC} + K_{BC} - K_{AB}}{2} \quad \textbf{(4.26)}$$

We can now decide whether the rates of substitution are equal in lineages A and B by comparing the value of K_{OA} with that of K_{OB}. The time that has passed since species A and B last shared a common ancestor is by definition equal for both lineages. Thus, according to the molecular clock hypothesis, K_{OA} and K_{OB} should be equal, i.e., $K_{OA} - K_{OB} = 0$. From the above equations, we note that $K_{OA} - K_{OB} = K_{AC} - K_{BC}$. Therefore, we can compare the rates of substitution in A and B directly from K_{AC} and K_{BC}. In statistical terminology, the above approach means that we use $K_{AC} - K_{BC}$ as an estimator of $K_{OA} - K_{OB}$. $K_{OA} - K_{OB}$ represents the difference in branch length between the two lineages leading from node O to species A and B. We denote this difference by d. A positive d value means that the molecule has evolved faster in lineage A than in lineage B, whereas a negative d value means that the molecule evolved faster in lineage B.

The variance of d is given by

$$V(d) = V(K_{AC}) + V(K_{BC}) - 2V(K_{OC}) \quad \textbf{(4.27)}$$

Under the one-parameter model, $V(K_{AC})$ and $V(K_{BC})$ can be obtained from Equation 3.27 (Chapter 3). $V(K_{OC})$ can be obtained by putting

$$p = \frac{3}{4}\left(1 - e^{-4/3K_{OC}}\right) \quad \textbf{(4.28)}$$

into Equation 3.27. For a more detailed explanation of the derivation of the formulas, see Wu and Li (1985).

A simple way to test whether an observed d value is significantly different from 0 is to compare it with its standard error. For example, an absolute d value larger than two times the standard error may be considered significantly different from 0 at the 5% level. Similarly, an absolute d value larger than 2.7 times the standard error may be considered significantly different from 0 at the 1% level.

The relative rate test can be used even if the order of divergence among species A, B and C is not known for certain. We note that Equations 4.21–4.26 also hold for the two alternative cases: (1) A and C diverged from each other after the divergence of B, and (2) B and C diverged from each other after the divergence of A. In other words, even when the order of divergence among three species is unknown, two of the species must be more closely related to each other than either is to the third. In this case we can compare the values of K_{OA}, K_{OB}, and K_{OC}, and find out whether any two values are identical. Failure to find even one identity among the three comparisons would cast doubt on the molecular clock hypothesis for the three species under consideration.

Tajima's 1D method

Although Sarich and Wilson's (1973) relative rate test does not rely on knowledge of divergence times, it does assume (1) that the substitution model (e.g., Jukes and Cantor one-parameter method) is known, and (2) that the substitution rates among different sites vary according to some prespecified distribution, e.g., the Γ distribution (Jin and Nei 1990).

Tajima (1993) suggested several simple, albeit less powerful, methods for overcoming these difficulties. Here, we shall only discuss the **one degree of freedom (1D) method**. We start with three aligned nucleotide sequences, 1, 2, and 3. Let n_{ijk} be the observed number of sites where sequences 1, 2, and 3 have nucleotides i, j, and k, respectively, where i, j, and k can be nucleotides A, G, C, or T. If sequence 3 is the outgroup, then the expectation of n_{ijk} should be equal to that of n_{jik}, i.e.,

$$E(n_{ijk}) = E(n_{jik}) \tag{4.29}$$

This equality holds regardless of the substitution model or the pattern of variation in substitution rates among sites.

We define m_1 as follows:

$$\begin{aligned} m_1 &= \sum n_{ijj} \\ &= n_{AGG} + n_{ACC} + n_{ATT} + n_{GAA} + n_{GCC} + n_{GTT} \\ &\quad + n_{CAA} + n_{CGG} + n_{CTT} + n_{TAA} + n_{TGG} + n_{TCC} \end{aligned} \tag{4.30}$$

Similarly, we define m_2 as

$$\begin{aligned} m_2 &= \sum n_{jij} \\ &= n_{AGA} + n_{ACA} + n_{ATA} + n_{GAG} + n_{GCG} + n_{GTG} \\ &\quad + n_{CAC} + n_{CGC} + n_{CTC} + n_{TAT} + n_{TGT} + n_{TCT} \end{aligned} \tag{4.31}$$

Note that only sites in which exactly two types of nucleotides exist in the three sequences are used in this analysis.

When sequence 3 is the outgroup, the expectation of m_1 is equal to that of m_2 under the molecular clock:

$$E(m_1) = E(m_2) \tag{4.32}$$

The equality can be tested by using χ^2 with one degree of freedom, namely,

$$\chi^2 = \frac{(m_1 - m_2)^2}{m_1 + m_2} \tag{4.33}$$

Tests involving comparisons of duplicate genes

The molecular clock hypothesis can also be tested by comparing homologous genes that originated through a gene duplication event (Chapter 6). If a gene duplication occurred prior to (but not too long before) the split of two species (Figure 4.17), then we can determine whether or not the homologous genes evolve at the same rate in the two lineages by comparing the number of substitutions between the two duplicate genes in one lineage with the number of substitutions between the duplicate genes in the second lineage (Wu and Li 1985).

If we denote the number of substitutions between sequence i and sequence j by K_{ij}, we see from Figure 4.17 that the number of substitutions between A_1 and B_1 ($K_{A_1B_1}$) can be calculated as

$$K_{A_1B_1} = K_{AA_1} + K_{OA} + K_{OB} + K_{BB_1} \tag{4.34}$$

Similarly, the number of substitutions between A_2 and B_2 is

$$K_{A_2B_2} = K_{AA_2} + K_{OA} + K_{OB} + K_{BB_2} \tag{4.35}$$

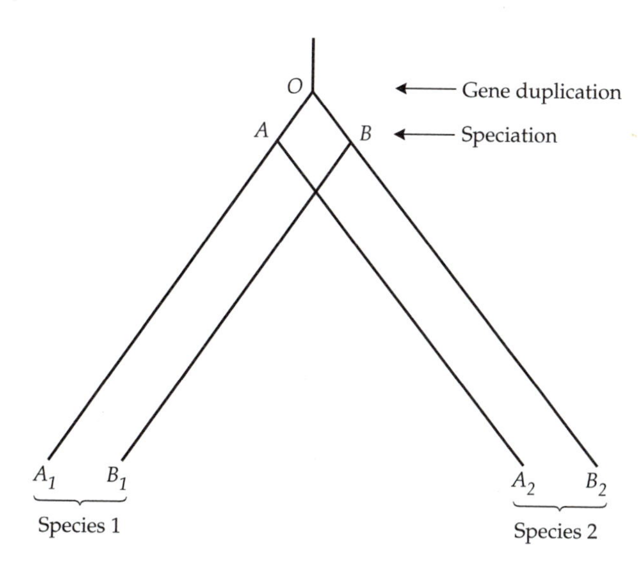

FIGURE 4.17 Phylogenetic tree used to test constancy of rates among lineages A and B by using duplicate genes. *O* denotes the common ancestor of all four genes. *A* is the ancestor of the duplicated genes A_1 and A_2, *B* is the ancestor of the duplicated genes B_1 and B_2 and the subscripts denote the species.

From Equations 4.34 and 4.35 we obtain

$$K_{A_1B_1} - K_{A_2B_2} = K_{AA_1} + K_{BB_1} - K_{AA_2} - K_{BB_2} \quad \textbf{(4.36)}$$

If A_1 evolves at the same rate as A_2 and B_1 evolves at the same rate as B_2 (i.e., if the assumption of rate constancy holds for both genes), then $K_{AA_1} = K_{AA_2}$ and $K_{BB_1} = K_{BB_2}$. Therefore, under the molecular clock hypothesis, $K_{A_1B_1} - K_{A_2B_2} = 0$. Note that genes A_1 and A_2 serve as outgroups for comparing the rates between the lineages leading to B_1 and B_2, and vice versa.) This test requires neither knowledge of the divergence time between the two species under study, nor does it assume that the rates of substitution of A_1 and A_2 are equal to those of B_1 and B_2. The drawback is that gene conversion may occur between the duplicated genes (Chapter 6) and distort the results. Thus, this method must be used with caution.

LOCAL CLOCKS

Sometimes we are interested in whether or not a molecular clock exists for a particular group of organisms. Such a clock, if found, would be applicable only to the group of organisms under study and to no other. Such clocks are called **local clocks**. In the following we shall review the question of local clocks in several mammalian taxa by using the methods presented above.

Nearly equal rates in mice and rats

Table 4.10 shows a comparison of the rates of synonymous and nonsynonymous substitution in mice and rats using the relative rate test. In all cases, species A is the mouse, species B is the rat, and species C is hamster. A positive sign for the value of $K_{AC} - K_{BC}$ means that the rate in mice is higher than that in rats, whereas a negative sign indicates the opposite. The sequence data consists of 28 genes with a combined total aligned length of 11,295 nucleotides (O'hUigin and Li 1992). Neither $K_{AC} - K_{BC}$ value is significantly different from 0, and therefore the null hypothesis of equal substitution rates in

TABLE 4.10 Differences in the number of substitutions per 100 sites between mice (species 1) and rats (species 2), with hamsters (species 3) as a reference

Type of substitution	Number of sites compared	Number of substitutions[a]			
		K_{12}[b]	K_{13}	K_{23}	$K_{13} - K_{23}$
Synonymous	4,855	19.9 ± 0.7	31.1 ± 0.9	32.4 ± 1.0	–1.3 ± 7.9
Nonsynonymous	17,440	1.9 ± 0.1	2.9 ± 0.1	2.7 ± 0.1	0.3 ± 1.3

Modified from O'hUigin and Li (1992).

[a]Mean ± standard error. K_{ij} = number of substitutions per 100 sites between species *i* and *j*.

mice and rats cannot be rejected. Since the rate constancy holds not only for synonymous but also for nonsynonymous substitutions, the results are consistent with the neutral mutation hypothesis, which postulates that the majority of molecular changes in evolution are due to neutral or nearly neutral mutations.

Lower rates in humans than in African apes and monkeys

On the basis of immunological distances and protein sequence data, Goodman (1961) and Goodman et al. (1971) suggested that a rate slowdown occurred in hominoids (humans and apes) after their separation from the Old World monkeys. Wilson et al. (1977), however, contended that the slowdown is an artifact owing to the use of an erroneous estimate of the ape–human divergence time. They conducted relative rate tests using both immunological distance data and protein sequence data and concluded that there was no evidence for a hominoid slowdown.

DNA sequences data provide a better resolution of the controversy. In the following we shall use Tajima's (1993) 1D test on the η-globin pseudogene sequences from human, common chimpanzee (*Pan troglodytes*), pygmy chimpanzee (*P. paniscus*), gorilla (*Gorilla gorilla*), orangutan (*Pongo pygmaeus*), and common gibbon (*Hylobates lar*). In each pairwise comparison, we use as outgroup the closest possible reference species. For example, in the *Pan troglodytes/Pan paniscus* comparison, we use humans as outgroup. Similarly, in the *Pan/Homo* comparison we use *Gorilla* as outgroup. The results are shown in Table 4.11. In the comparisons between *Homo* (species 1) on the one hand and *P. paniscus*, *P. troglodytes*, or *Gorilla gorilla* (species 2) on the other, we see that the values of m_1 are significantly smaller than the m_2 values. There-

TABLE 4.11 Tajima's (1993) 1D test for unequal rates of substitution between pairs of hominoids[a]

Species 1	Species 2	Species 3	m_1	m_2	χ^2
Pan troglodytes	*Pan paniscus*	*Homo sapiens*	21	14	1.40
Homo sapiens	*Pan paniscus*	*Gorilla gorilla*	33	55	5.50*
Homo sapiens	*Pan troglodytes*	*Gorilla gorilla*	34	61	7.67**
Homo sapiens	*Gorilla gorilla*	*Pongo pygmaeus*	36	58	5.15*
Pan troglodytes	*Gorilla gorilla*	*Pongo pygmaeus*	63	56	0.41
Pan paniscus	*Gorilla gorilla*	*Pongo pygmaeus*	56	58	0.04
Homo sapiens	*Pongo pygmaeus*	*Hylobates lar*	91	105	1.00
Pan troglodytes	*Pongo pygmaeus*	*Hylobates lar*	120	104	1.14
Pan paniscus	*Pongo pygmaeus*	*Hylobates lar*	109	104	0.12
Gorilla gorilla	*Pongo pygmaeus*	*Hylobates lar*	106	112	0.17

*Significant at the 5% level; ** significant at the 1% level.

[a]Species 1 and 2 are the ingroups. Species 3 is the outgroup.

TABLE 4.12 Differences in the number of nucleotide substitutions per 100 sites in noncoding regions and the relative rates of substitution between the African monkey (species 1) and the human (species 2) lineages, with a New World monkey (species 3) as reference

Type of sequence	Sequence length	K_{12}[a]	K_{13}	K_{23}	$K_{13} - K_{12}$	Ratio[b]
Pseudogene	8,781	6.7	11.8	10.7	1.1 ± 0.3**	1.4
Introns	8,478	7.1	14.7	13.9	0.8 ± 0.3**	1.3
Flanking and untranslated regions	936	7.9	14.9	11.7	3.1 ± 1.1**	2.3

Data from Bailey et al. (1991), Porter et al. (1995), and Ellsworth et al. (1993).

**Significant at the 1% level.

[a] K_{ij} = number of substitutions per 100 sites between species i and j.

[b] The ratio of the rate in the African monkey region to that in the human lineage.

fore, the rate of substitution in humans is smaller than that in the African apes. In the comparison between human and orangutan with gibbon as outgroup, $m_1 < m_2$, but the difference is not statistically significant.

A study by Seino et al. (1992) has shown that the slowdown in the human lineage is not restricted to the η-globin pseudogene, but constitutes a general phenomenon probably affecting the entire genome. Li and Tanimura (1987a) estimated that the chimpanzee and gorilla lineages evolved approximately 1.5 times faster than the human lineage. DNA–DNA hybridization experiments have also provided strong evidence for the slowdown hypothesis (Sibley and Ahlquist 1984, 1987).

The divergence between hominoids and African monkeys (Cercopithecidae) is estimated to have occurred in the Oligocene (approximately 20–30 million years ago). The divergence between the Old World monkeys (Catarrhini), which include Hominoidea and Cercopithecidae, and the New World monkeys (Platyrrhini) is estimated to have occurred at least 10 million years earlier (Gingerich 1984; Pilbeam 1984; Fleagle et al. 1986). It is therefore possible to test for inequalities in the rates of substitution between humans and African monkeys by using a New World monkey as an outgroup. In Table 4.12, K_{13} and K_{23} are the distances between an African monkey and a New World monkey and between human and a New World monkey, respectively. In all cases, $K_{13} - K_{12}$ is significantly larger than 0 and we may, therefore, conclude that the rate of substitution in the noncoding regions, which presumably reflects the rate of mutation, is higher in the African monkeys than in humans.

Higher rates in rodents than in primates

Based on DNA–DNA hybridization data, Laird et al. (1969), Kohne (1970), Kohne et al. (1972), and Rice (1972) found that the rate of nucleotide substitution in rat and mouse is much higher than that in human and chimpanzee. Their findings were challenged by Sarich (1972), Sarich and Wilson (1973), and Wilson et al. (1977), who argued that some of the studies were based on questionable estimates of divergence time. However, some of the observed differences were too large to be explained by inaccuracies in the estimation of divergence times. For example, Kohne (1970) estimated the rates of evolution in two murid lineages (mouse and rat) to be 16 times higher than those in human and chimpanzee .

Gu and Li (1992) applied the relative rate test to compare the rates of substitution in the rodent and human lineages, using chicken as outgroup. They used amino acid sequences instead of DNA, because the chicken and mammalian lineages diverged about 300 million years ago, so that it is difficult to obtain reliable estimates of divergence at synonymous sites. We denote by N_R the number of residues at which the human and rodent sequences are different but the human and the chicken sequences are identical; this case represents an instance in which an amino acid replacement presumably occurred in the rodent lineage. We denote by N_H the number of residues at which the human and rodent sequences are different but the rodent and the chicken sequences are identical; this case represents an instance in which an amino acid replacement presumably occurred in the human lineage. Under the null assumption of equal rates in the human and rodent lineages, N_R is expected to be equal to N_H. For the 54 proteins used in Gu and Li's (1992) study, $N_R = 600$, and $N_H = 416$ ($p < 0.001$). Therefore, there is good evidence for an overall faster substitution rate in the lineage leading to mouse and rat than in the human lineage.

Gu and Li (1992) also compared the frequencies of insertions and deletions (gaps) in the human and rodent sequences using chicken as reference. The total length of gaps in the rodent lineage was 385 amino acids, as compared to only 108 in the human lineage ($p < 0.001$). The number of gaps was also higher in humans (44 gaps) than in rodents (31 gaps), but because of the small number of gaps the difference could not be shown to be statistically significant ($0.05 < p < 0.1$). Again, we may conclude that the rate of insertion/deletion is higher in the rodent lineage than in the human lineage.

The null hypothesis of equal rates in rodents and primates has also been tested by using comparisons between duplicated-gene pairs (e.g., Harlow et al. 1988). Table 4.13 shows the results from an analysis of the β-globin and aldolase duplicate genes in humans and mice (Li et al. 1987a). In the case of the globin genes, duplicate copies in the mouse are always more divergent than those in humans. This is true for both synonymous and nonsynonymous substitutions. For the three gene pairs considered, the ratios of the number of substitutions in mouse to those in human are 1.24, 1.56, and 1.71 for synonymous substitutions, and 1.13, 1.17, and 1.50 for nonsynonymous substitutions. In the case of the aldolase A and B genes, the degree of divergence at synonymous sites is 1.24 times higher in rat than in man, but the degrees of divergence at nonsynonymous sites are not significantly different between the two species.

Thus, the results of tests of the molecular clock hypothesis using duplicate genes also support the view that the rate of nucleotide substitution is higher in rodents than in humans. We must remember, however, that the ratios are underestimates, because the rates of substitution in the two species

TABLE 4.13 Numbers of nucleotide substitutions per synonymous site (K_S) and per nonsynonymous site (K_A) between duplicated genes in humans and rodents

Gene pair	K_S	K_A
β-like globin genes[a]		
Human adult–Human fetal	0.73	0.18
Mouse adult–Mouse fetal	0.90	0.21
Human adult–Human embryonic	0.62	0.16
Mouse adult–Mouse embryonic	0.97	0.18
Human fetal–Human embryonic	0.56	0.10
Mouse fetal–Mouse embryonic	0.96	0.15
Aldolase *A* and *B* genes		
Human *A*–Human *B*	1.55	0.21
Rat *A*–Rat *B*	1.92	0.21

From Li et al. (1987a)

[a]The adult globin genes are β in human and β_{maj} in mouse; the fetal genes are ${}^{A}\gamma$ in human and β_{H1} in mouse; and the embryonic genes are ε in human and $y2$ in mouse.

are not entirely independent of each other. For example, from Figure 4.17, we see that genes A_1 and A_2 share the same evolutionary history between the time of gene duplication (O) and the time of species divergence (A). Similarly genes B_1 and B_2 share the same evolutionary history between the time of gene duplication (O) and the time of species divergence (B). In order to obtain reliable estimates for the ratio of substitution rates in two lineages we must use genes that had been duplicated shortly before the time of speciation, i.e., we have to make sure that K_{OA} and K_{OB} are much smaller than K_{AA_1}, K_{AA_2}, K_{BB_1}, and K_{BB_2}.

Li et al. (1996) estimated the rate of substitution in rats and mice to be 4–6 times higher than that in humans. Other rodents, e.g., voles and lemmings (Arvicolidae), also evolve much faster than humans (Catzeflis et al. 1987).

EVALUATION OF THE MOLECULAR CLOCK HYPOTHESIS

In its most extreme form, the molecular clock hypothesis postulates that homologous stretches of DNA evolve at essentially the same rate along all evolutionary lineages for as long as they maintain their original function (e.g., Wilson et al. 1987). Analyses of DNA sequences from several orders of mammals indicate that no global molecular clock exists in the class Mammalia. In fact, significant variations in the rates of nucleotide substitution have been found both within and among the different orders of mammals. Thus, rodents seem to evolve considerably faster than artiodactyls, which in turn evolve faster than primates. Within the mammalian orders, the variation in the rates of substitution among species is somewhat smaller. However, there are cases in which substantial intraordinal differences have been shown to exist. For example, within the primates, the rate of synonymous substitution in the Old World monkeys is almost twice that in the apes (Vawter and Brown 1986; Li and Tanimura 1987a).

Since the assumption of rate constancy is violated even within Mammalia, a truly universal molecular clock that applies to all organisms cannot be assumed to exist. Indeed, the substitution rate among distantly related organisms has been invariably shown to vary to a greater or lesser extent. For example, it has been estimated that the rate of substitution in *Drosophila* species is 5–10 times faster than the rate in most vertebrates, and that this difference holds even for the most conservative portions of the genome (Caccone and Powell 1990). Therefore, extreme caution must be exercised when using the molecular clock assumption to infer times of divergence, more so when distantly related species are concerned (Chapter 5).

The above results should not, however, be taken as evidence that no molecular clocks exist. Rather, there are many local clocks that tick fairly regularly for many groups of closely related species, e.g., in murids. Moreover, one must note the remarkable fact that, while parameters related to the life histories of plants, bacteria, insects, and mammals are different from each other by many orders of magnitude, the mean synonymous substitution rates in the nuclear genomes of these organisms differ by no more than one or two orders of magnitude. Therefore, the molecular clock can still be used to estimate times of species divergence with a fair degree of confidence, provided appropriate corrections for the unequal rates of molecular evolution among lineages are made.

Causes of variation in substitution rates among evolutionary lineages

Many hypotheses have been proposed to account for the differences in the rate of nucleotide substitution among evolutionary lineages. The factors most commonly invoked to explain these differences are roughly divided into **replication-dependent factors**, such as generation time and DNA repair efficiency, and **replication-independent factors**, such as basal metabolic rate and body size.

The higher substitution rates in monkeys than in humans and the higher rates in rodents than in primates were explained by the so-called **generation time effect** (Laird et al. 1969; Kohne 1970). The generation time in rodents is much shorter than that in humans, whereas the numbers of germline DNA replication cycles per generation are similar. Therefore, the number of germline replications per year could be many times higher in rodents than in humans. Assuming that mutations accumulate chiefly during the process of DNA replication, the more cycles of replication there are, the more mutational errors will occur. In the absence of differences in selection intensities, rodents are expected to have higher substitution rates than humans. Similarly, monkeys have shorter generation time than humans and so are expected to have a higher substitution rate. In mammals, the rate of substitution in the nuclear genome was found to behave according to the expectations of the generation time effect hypothesis. This hypothesis is also supported by the finding that the rate of substitution in the chloroplast genome is more than five times higher in annual grasses than in perennial palms (Gaut et al. 1992). As opposed to the situation in nuclear genes, the rate of substitution in the mitochondrial genome of mammals does not support the generation time effect hypothesis. In particular, mitochondrial genes in some mammalian species with long generation times, such as humans and elephants, are known to evolve faster than their homologs in rodents, which have much shorter generation times (Stanhope et al. 1998).

The differences in substitution rates could also be partly accounted for by differences in the efficiency of the DNA repair system (Britten 1986). There are limited data indicating that rodents have a less efficient DNA repair system than humans and consequently accumulate more mutation per replication cycle. One such example involves the removal of hypoxanthine residues, a product of adenine deamination that constitutes an important source of transition mutations. The removal of hypoxanthine from DNA is carried on by a group of enzymes called 3-methyladenine DNA glucosylases. Saparbaev and Laval (1994) have shown that 3-methyladenine DNA glucosylases from humans are more efficient than their counterparts in rats.

It has also been suggested that metabolic rates (defined as amounts of O_2 consumed per weight unit per unit time) may influence the rate of nucleotide substitution (Martin and Palumbi 1993). For example, poikilotherms (cold-blooded organisms) seem to have lower rates of nucleotide substitution that homeotherms (warm-blooded organisms). According to Martin and Palumbi's (1993) suggestion, the correlation between metabolic rate and the rate of nucleotide substitution is mediated by (1) the mutagenic effects of oxygen radicals that are abundant by-products of aerobic respiration and whose production is proportional to the metabolic rate, and (2) increased rates of DNA turnover (i.e., DNA synthesis, repair, and degradation) in organisms with high metabolic rates. However, although there seems to be a correlation between metabolic and evolutionary rates in some groups of organisms, studies based on a large number of DNA sequences from many diverse taxa indicate that the correlation may not be of general application (e.g., Seddon et al. 1998).

It should be emphasized that the alternative hypotheses presented above are not mutually exclusive. Rather, all the factors may contribute to varying extents to the rate variation among lineages. Determining their relative contributions to the rate variation, however, requires additional research. Finally, all these factors correlate with one another, making it difficult to assess the relative contribution of each of them to the rate of molecular evolution. For example, large animals usually have longer generation times as well as lower metabolic rates than small animals.

All the factors presented above involve variations in mutation rates. However, rates of substitutions may vary among evolutionary lineages due to differences in selection intensities. For example, in large populations, a greater proportion of the mutations may be affected by purifying selection (Chapter 2). This will result in a decrease in the proportion of neutral mutations and a consequent decrease in the rates of substitution (e.g., Ohta 1995). Alternatively, one species may be subjected to stronger selective constraints than another species, and in this case it will evolve slower.

Are living fossils molecular fossils too?

Living fossils are defined as taxa that have not changed morphologically for long periods of time (usually at least 100 million years). It is of interest to find out whether the morphological stasis is also accompanied by molecular stasis. Quite early in the history of molecular evolution, it was noted that sharks, which have not changed to any conspicuous degree since the Devonian, evolved at about the same rate as other "nonfossil" organisms (Fisher and Thompson 1979; Kimura 1989). Furthermore, the mitochondrial DNA of the alligator (*Alligator mississippiensis*), which is also considered to be a living fossil, evolves much faster than that of birds, which presumably appeared on the evolutionary stage much more recently (Janke and Árnason 1997). Turtles, on the other hand, which have remained morphologically unchanged since Triassic times, seem to evolve at the molecular equivalent of a "turtle's pace" (Avise et al. 1992).

The most spectacular example of a lack of relationship between morphology and molecular evolution is most probably that of the horseshoe crab (class Merostomata), which despite their name are more closely related to scorpions, spiders, and mites than to the crustaceans. While the morphology of horseshoe crabs has changed little in the last 500 million years—indeed, one extant horseshoe crab, *Limulus polyphemus* is almost indistinguishable from its extinct Jurassic relatives—their rates of molecular evolution are unexceptional (e.g., Nguyen et al. 1986; Tokugana et al. 1993). Thus, there seems to be no obvious relationship between morphological and molecular change.

"Primitive" versus "advanced: A question of rates

In the literature one often encounters the adjective "primitive" attached to the name of an organism or a taxonomic group. For example, the flagellated protozoan *Giardia* is often said to be a "primitive eukaryote." In a similar vein, sponges are defined as "primitive metazoans," and the whisk fern, *Psilotum nudum*, is defined as a "primitive vascular plant." We note, however, that the terms "primitive" and "advanced" cannot be defined unambiguously. We further note that all the organisms mentioned above are extant organisms and as such they are as "primitive" or as "advanced" as other extant organisms, such as *Homo sapiens*. The ranking of organisms started with the Aristotelian "Great

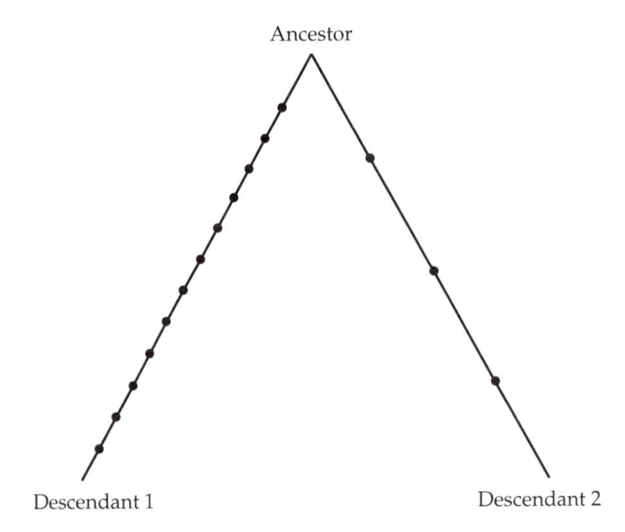

FIGURE 4.18 An ancestral taxon gives rise to two descendent taxa. The branch leading to descendant 1 has accumulated more substitutions (dots) than the branch leading to descendant 2.

Chain of Being," and was used by Linnaeus in his *Systema Naturae* to rank animals into Primates (humans and monkeys), Secundates (mammals), and Tertiates (all others).

The relative rate tests allow us to attach an objective meaning to the terms "primitive" and "advanced." In Figure 4.18, we show an ancestral taxon that gave rise to two descendant taxa. The lineage leading to descendant 1 evolved faster (i.e., accumulated more substitutions) than the lineage leading to descendant 2. Therefore, descendant 2 resembles the ancestral taxon more than descendant 1. Since, by temporal placement, the ancestor pre-dates the descendants (i.e., the ancestor is the primitive taxon), we may conclude that descendant 2 is more "primitive" than descendant 1. The question of "primitive" versus "advanced" is, therefore, a question of rates of evolution. The faster a lineage evolves, the less primitive its decendants are.

We now note that within the order of placental mammals (Eutheria), the lineage leading to humans has experienced the slowest average rate of nuclear gene evolution. Ergo, *Homo sapiens* may be the most primitive mammal!

Phyletic gradualism versus punctuated equilibria at the molecular level

In the literature, there are two conflicting models concerning long-term evolutionary rates. The first is **phyletic gradualism**, which regards evolution as a gradual process, albeit with some variation in rates. In contrast, the **punctuated equilibria hypothesis** (Eldredge and Gould 1972) maintains that evolution is mainly characterized by long periods of time during which very little change occurs (**stasis**). These lengthy "calm" periods are interrupted by short bursts of rapid evolutionary change (**revolutions**), mainly coinciding with periods of intense speciation. In its original formulation, the punctuated equilibria hypothesis was intended to explain phenotypic evolution, but its predictions can be tested at the molecular level.

As noted previously, the rate of evolution at the molecular level exhibits some variation, but most of the variation can be accommodated within the phyletic gradualist view (see Gillespie 1991). The mammalian growth hormone genes, however, exhibit a pattern of evolutionary change that seems more consistent with the punctuated equilibria hypothesis (Wallis 1994, 1996). In Figure 4.19, we see that growth hormone genes evolve quite slowly throughout most mammalian evolution. The mean evolutionary rate is approximately 0.3×10^{-9} replacements per amino acid site per year. There are, however, two independent bursts of rapid evolution, one prior to the divergence among ruminants, and one before primate divergence. Wallis (1994) estimated that during these short times, which constitute less than 10% of the total evolutionary time, there was a 20-fold and a 40-fold increase in the rate of evolution in ruminants and primates, respectively.

Three possible explanations could account for the increased evolutionary rates in ruminants and primates: (1) an increase in the mutation rate, (2) positive selection for altered biological properties, or (3) relaxation of purifying selection. To distinguish between these possibilities, Wallis (1996) computed the ratios of nonsynonymous (K_A) to synonymous (K_S) substitutions during the slow and rapid phases of evolution. The results are shown in Table 4.14. We see that during the rapid phases of evolution, there is a significant increase in the K_A/K_S values, indicative of either positive selection or relaxation of selection.

RATES OF SUBSTITUTION IN ORGANELLE DNA

The vast majority of eukaryotes possess at least one extranuclear genome, which is replicated independently of the nuclear genome. Organelle genomes,

TABLE 4.14 Rates of amino acid replacement ($\times 10^9 \pm$ standard error) and the ratio of nonsynonymous (K_A) to synonymous (K_S) substitution in growth hormones during mammalian evolution

Phase	Rate of amino acid replacement	K_A/K_S
Slow phase[a]	0.3 ± 0.1	0.03
Ruminant rapid phase	5.6 ± 1.4	0.30
Primate rapid phase	10.8 ± 1.3	0.49

From Wallis (1996).

[a]Based on all data excluding the rapid phases in primate and ruminant evolution.

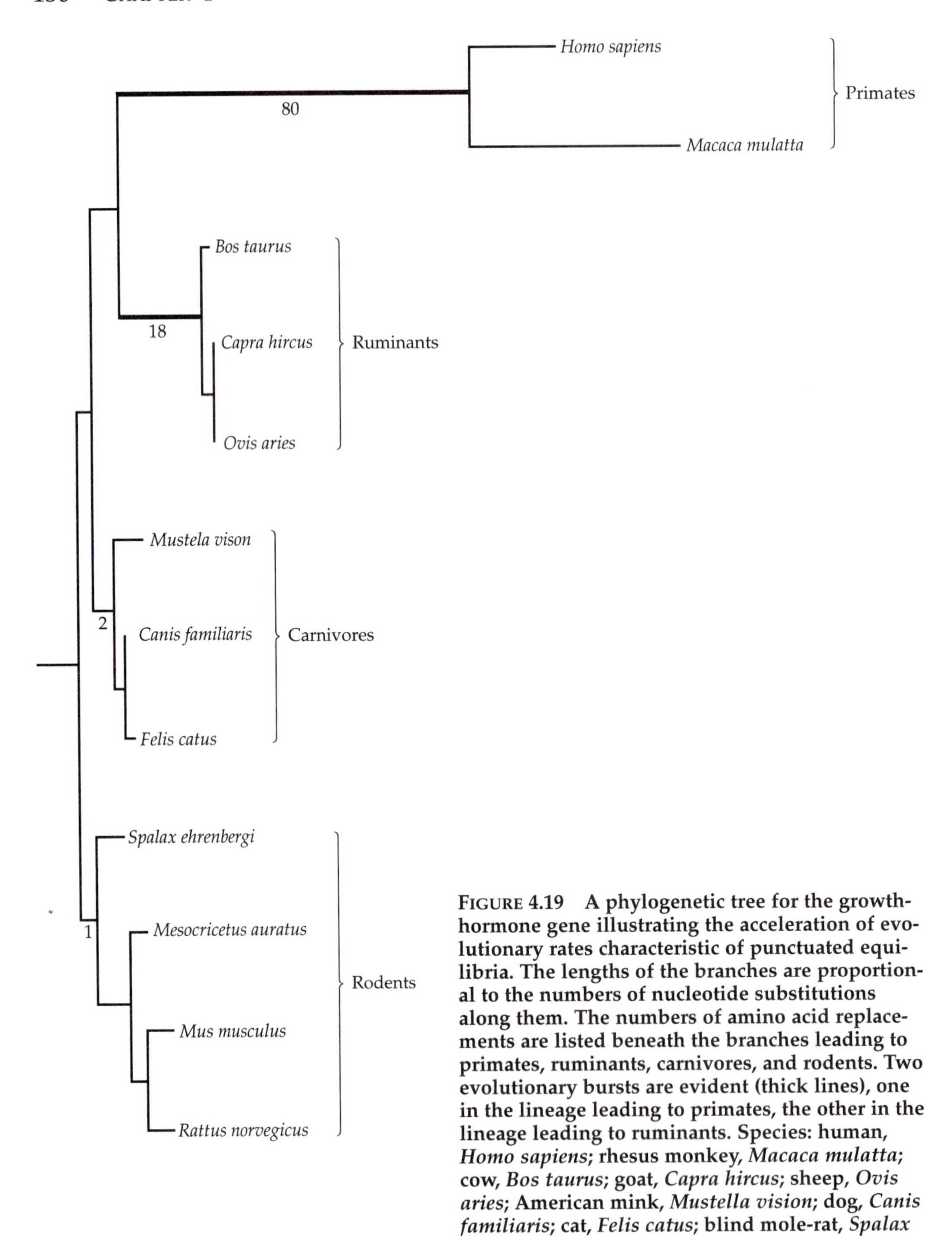

FIGURE 4.19 A phylogenetic tree for the growth-hormone gene illustrating the acceleration of evolutionary rates characteristic of punctuated equilibria. The lengths of the branches are proportional to the numbers of nucleotide substitutions along them. The numbers of amino acid replacements are listed beneath the branches leading to primates, ruminants, carnivores, and rodents. Two evolutionary bursts are evident (thick lines), one in the lineage leading to primates, the other in the lineage leading to ruminants. Species: human, *Homo sapiens*; rhesus monkey, *Macaca mulatta*; cow, *Bos taurus*; goat, *Capra hircus*; sheep, *Ovis aries*; American mink, *Mustella vision*; dog, *Canis familiaris*; cat, *Felis catus*; blind mole-rat, *Spalax ehrenberghi*; golden hamster, *Mesocricetus auratus*; mouse, *Mus musculus*; rat, *Rattus norvegicus*.

such as those of mitochondria and chloroplasts, are much smaller and easier to investigate experimentally than nuclear genomes. Organelles are almost always inherited uniparentally. Animal mitochondria are inherited maternally, although some paternal "leaking" has been detected. Plant chloroplasts and mitochondria are mostly inherited maternally too, but there are exceptions. In conifers, the chloroplasts are inherited paternally, and in *Chlamydomonas reinhartii*, the mitochondria are inherited paternally. In the following we shall discuss the rates of substitution in the mitochondrial DNA of mammals and in the mitochondrial and chloroplasmic genomes of vascular plants.

Mammalian mitochondrial genes

The mitochondrial genome of animals consists of a circular, double-stranded DNA about 15,000–17,000 base pairs long, approximately 1/10,000 of the smallest nuclear genome. It contains only unique (i.e., nonrepetitive) sequences: 13 protein-coding genes (7 subunits of NADH-ubiquinone oxireductase, 3 subunits of cytochrome *c* oxidase, 2 subunits of the H^+-ATP synthase, and cytochrome *b*), 22 tRNA genes (each specifying a distinct tRNA species), a control region that contains sites for replication and transcription initiation, and a few intergenic spacers. All the genes are intronless, and the genome is structurally and evolutionarily very stable (as is evident from the small variation in genome size and gene order among mammalian species).

The synonymous rate of substitution in mammalian mitochondrial protein-coding genes has been estimated to be 5.7×10^{-8} substitutions per synonymous site per year (Brown et al. 1979, 1982). This is about 10 times the value for synonymous substitutions in nuclear protein-coding genes. The rate of nonsynonymous substitution varies greatly among the 13 protein-coding genes, but is always much higher than the average nonsynonymous rate for nuclear genes. The principal reason for these high rates of substitution in mammalian mitochondrial DNA seems to be a high rate of mutation relative to the nuclear rate. The high mutation rate might be due to (1) a low fidelity of the DNA replication process in mitochondria, (2) an inefficient repair mechanism, and/or (3) a high concentration of mutagens (e.g., superoxide radicals, O_2^-) resulting from the metabolic functions performed by mitochondria. These mutational factors notwithstanding, a reduction in the intensity of purifying selection may also play a role in elevating the substitution rates in mammalian mitochondrial genomes (Lynch 1997).

Plant nuclear, mitochondrial, and chloroplast DNAs

Our current knowledge of the rates and mechanism of molecular evolution has been derived largely from comparative studies of genes and proteins of animals. Only recently has the study of the molecular biology of plants provided sufficient data to allow the evolution of plant genes to be investigated. Since the plant and animal kingdoms diverged about 1 billion years ago, the

TABLE 4.15 Genetic content of plant mitochondrial and chloroplast genomes

Species	Size (bp)	Open reading frames[a]	rRNAs	tRNAs	Introns
Mitochondria					
Chondrus crispus	25,836	36	3	23	1
Prototheca wickerhamii	55,328	36	3	26	5
Marchantia polymorpha	186,609	74	3	29	32
Arabidopsis thaliana	366,923	117	3	21	22
Chloroplasts					
Epiphagus virginiana	70,028	34	8	28	12
Odontella sinensis	119,704	140	6	29	0
Pinus thunbergii	119,707	156	4	36	15
Marchantia polymorpha	121,024	89	8	34	21
Oryza sativa	134,525	108	8	44	14
Zea mays	140,387	111	8	39	23
Euglena gracilis	143,172	65	8	43	146
Nicotiana tabacum	155,939	102	7	38	25

[a]Putative protein-coding genes larger than 100 codons. Some of the differences in protein-coding gene number among chloroplast genomes may be attributed to the presence of variable numbers of duplicated genes.

pattern of evolution in plants might have become very different from that in animals. Indeed, plants differ from animals in the organization of their organelle DNA by having a much larger and structurally more variable mitochondrial genome, and by having a third independent genome, the chloroplast (Palmer 1985).

Plant mitochondrial genomes exhibit much more structural variability than their animal counterparts. They undergo frequent rearrangements, duplications, and deletions (Palmer 1985). For this reason, the genome varies from about 26,000 bp to 2,500,000 bp. The mitochondrial genome in plants can be linear or circular, and in many cases the genetic information is divided into separate DNA molecules, referred to as **subgenomic circles**. The coding content of plant mitochondria has been fully determined in several species (Table 4.15). The mitochondrial genome of the liverwort *Marchantia polymorpha* is 186,609 bp long and contains 30 identified protein-coding genes (16 ribosomal proteins, 7 subunits of NADH-ubiquinone oxireductase, 3 subunits of cytochrome *c* oxidase, 3 subunits of H^+-ATP synthase, and cytochrome *b*), 3 rRNA-specifying genes, and 29 tRNA-specifying genes (for 27 tRNA species). The genome also contains several pseudogenes and about 40 unidentified reading frames, some of which may be protein-coding genes. In addition, 32 type I and type II introns have been identified (Oda et al. 1992).

Structural genes may appear in multiple copies in plant mitochondrial genomes. However, judging by the available data to date, the variation in number and types of genes seems to be quite limited despite the immense variability in genome size. For example, the mitochondrial genome of the chlorophyte alga *Prothoteca wickerhammi*, which is only 55,328 bp in length (i.e., about 30% the size of the *Marchantia* mitochondrial genome), contains essentially a very similar set of genes, the only differences being its containing 13 ribosomal protein-coding genes, 9 genes encoding subunits of NADH-ubiquinone oxireductase, and 26 tRNA-specifying genes, versus 16, 7, and 29, respectively, in *Marchantia* (Wolfe et al. 1994).

The largest mitochondrial genome sequenced to date is the 366,923-bp genome of *Arabidopsis thaliana* (Unseld et al. 1997). Identified and unidentified protein-coding genes account for about 20% of this genome. Introns, duplications, remnants of retrotransposons (Chapter 7), and integrated plasmid sequences account for an additional 20%. Therefore, 60% of the mitochondrial genome of *Arabidopsis thaliana* is unaccounted for.

The chloroplast genome in vascular plants is circular and varies in size from about 70,000 bp in some nonphotosynthetic plants to about 220,000 bp, with an average size of about 150,000 bp (Palmer 1985). Despite the much smaller variation in size in comparison to that in plant mitochondrial genomes, the genomic content of chloroplasts varies considerably among plants, especially in the numbers of protein-coding genes, tRNA-specifying genes, pseudogenes, and the various types of introns (Table 4.15).

Early studies based on a few gene sequences or on restriction enzyme mapping have suggested that chloroplast genes have lower rates of nucleotide substitution than mammalian nuclear genes (Curtis and Clegg 1984; Palmer 1985; Zurawski and Clegg 1987) and that plant mitochondrial DNA evolves slowly in nucleotide sequence, though it undergoes frequent rearrangements (Palmer and Herbon 1987). These results have been confirmed by more extensive analyses of DNA sequences (Wolfe et al. 1987, 1989b).

Table 4.16 shows a comparison of the substitution rates in the three genomes of vascular plants. The average numbers of substitution per nonsyn-

TABLE 4.16 Comparison of the number of nucleotide substitutions per site (± standard error) in plant chloroplast, mitochondrial, and nuclear genes[a]

Genome	K_S	L_S	K_A	L_A
Comparison between monocot and dicot species				
Chloroplast	0.58 ± 0.02	4,177	0.05 ± 0.00	14,421
Mitochondrial	0.21 ± 0.01	1,219	0.04 ± 0.00	4,380
Comparison between maize and wheat or barley				
Nuclear	0.71 ± 0.04	1,475	0.06 ± 0.00	5,098
Chloroplast	0.17 ± 0.01	2,068	0.01 ± 0.00	7,001
Mitochondrial	0.03 ± 0.01	413	0.01 ± 0.00	1,526

From Wolfe et al. (1987, 1989b).

[a] K_S, number of synonymous substitutions per synonymous site; K_A, number of nonsynonymous substitutions per nonsynonymous site; L_S and L_A are the number of synonymous and nonsynonymous sites, respectively.

onymous site (K_A) in the chloroplast and mitochondrial genomes are similar, but the average number of substitutions per synonymous site (K_S) in the chloroplast genome is almost three times that in the mitochondrial genome for the comparison between monocot and dicot species, and it is six times higher for the comparison between maize and wheat or barley. In the following, the former ratio will be used because it is based on a larger data set. The average synonymous substitution rate in plant nuclear genes is about four times that in the chloroplast genes. Thus, the synonymous substitution rates in plant mitochondrial, chloroplast, and nuclear genes are in the approximate ratio of 1:3:12 (Wolfe et al. 1989b). If we take the divergence time between maize and wheat to be 50–70 million years (Stebbins 1981; Chao et al. 1984), the nuclear data in Table 4.16 convert to an average synonymous rate of $5.1–7.1 \times 10^{-9}$ substitutions per site per year. This value is similar to the rates of synonymous substitution seen in mammalian nuclear genes. The average rates of synonymous substitution in plant chloroplast and mitochondrial genes are $1.2–1.7 \times 10^{-9}$ and $0.2–0.3 \times 10^{-9}$ substitutions per site per year, respectively.

It should be noted, however, that rates of molecular change vary quite widely among plant families and orders in a manner that violates the assumptions of a simple molecular clock. Thus, the above estimates should be regarded with caution (Clegg et al. 1994). Indeed, in a study of tRNA evolution, Lynch (1997) estimated that plant chloroplast DNA evolves more slowly than nuclear DNA, whereas plant mitochondrial DNA evolves at rates that are comparable to those in the nuclear genome.

Substitution and rearrangement rates

The rate of nucleotide substitution does not correlate well with the rate of structural changes in the genome of organelles. In mammals, the mitochondrial DNA evolves very rapidly in terms of nucleotide substitutions, but the spatial arrangement of genes and the size of the genome are fairly constant among species. In contrast, the mitochondrial genome of plants undergoes frequent structural changes in terms of size and gene order, but the rate of nucleotide substitution is extremely low. In chloroplast DNA, both the rates of nucleotide substitution and structural evolution are very low. The lack of correlation between the rates of substitution and the rates of structural evolution suggest that the two processes occur independently.

RATES OF SUBSTITUTION IN RNA VIRUSES

RNA viruses are known to evolve at exceptionally high rates (Holland et al. 1982), perhaps 1 million times faster than organisms with DNA genomes. Therefore, significant numbers of nucleotide substitutions accumulate over short time periods, and differences in nucleotide sequences between viral strains isolated at relatively short time intervals are easily detectable. This property allows for a different approach to estimating evolutionary rates than that used previously (Chapter 3).

Estimation models

Several methods have been proposed in the literature for estimating the rate of nucleotide substitution in RNA viruses (Buonagurio et al. 1986; Saitou and Nei 1986; Li et al. 1988). In all models it is assumed that the rate of substitution is constant over time and that no further mutation occurs after the virus is isolated. In the following we shall describe Li et al.'s (1988) method.

Figure 4.20 shows the model tree used in the calculations. Sequence 1 was isolated t years earlier than sequence 2, r is the rate of substitution per nucleotide site per year, and l_1 and l_2 are the expected numbers of substitutions per site from the ancestral node O to the time of isolation of sequences 1 and 2, respectively. Therefore

$$l_2 - l_1 = rt_2 - rt_1 = rt \qquad \textbf{(4.37)}$$

By using sequence 3 as an outgroup reference, we obtain

$$l_2 - l_1 = d_{23} - d_{13} \qquad \textbf{(4.38)}$$

where d_{ij} denotes the number of substitutions per site between sequences i and j. Combining Equations 4.37 and 4.38, we obtain

$$r = \frac{d_{23} - d_{13}}{t} \qquad \textbf{(4.39)}$$

Methods for calculating the variance of r can be found in Wu and Li (1985) and Li and Tanimura (1987a,b). Note that one may use multiple outgroup references, but the outgroup references should be closely related to sequences 1 and 2, otherwise the variance of r may become too large. The best

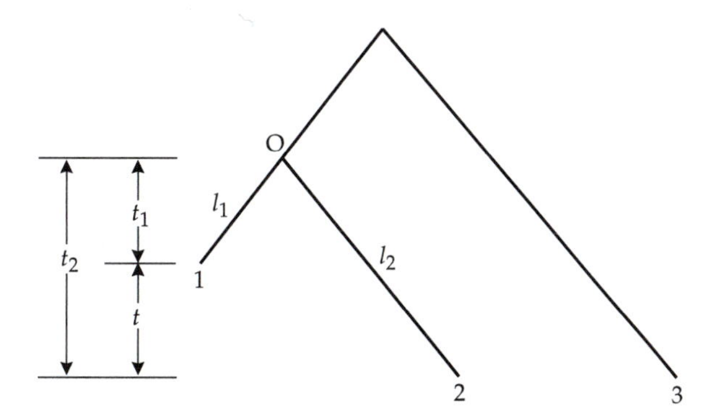

FIGURE 4.20 Model tree for estimating the rate of nucleotide substitution in RNA viruses. l_1 and l_2 denote the expected number of substitutions on the branches leading to isolates 1 and 2, respectively. Sequence 1, which was isolated at t_1, was collected t years earlier than sequence 2, which was isolated at t_2. Modified from Li et al. (1988).

situation in which to apply this model is when t is relatively large but t_1 is small. If t is small, $l_2 - l_1$ and $d_{23} - d_{13}$ would also be small, and then even a small error in the estimation of these variables would lead to a large error in the estimation of r. However, t should not be too large because it is difficult to obtain reliable estimates when the true values of d_{ij} are larger than 1. On the other hand, if t_1 is large, then l_1 will also be large, as will the variances of l_1, $l_2 - l_1$ and $d_{23} - d_{13}$. An example of the application of this method is given in the following section.

Human immunodeficiency viruses

The acquired immune deficiency syndrome (AIDS) is caused by a retrovirus called human immunodeficiency virus (HIV). Two major types of HIV, HIV-1 and HIV-2, are known. It is interesting to know the rates of nucleotide substitution in various regions of the HIV genome because, presumably, these rates determine the rate of change in viral pathogenicity and antigenicity (Rabson and Martin 1985; Gallo 1987). A number of authors (e.g., Hahn et al. 1986; Yokoyama and Gojobori 1987) have studied the rate of nucleotide substitution in HIV-1. The results below are from Li et al. (1988).

Three strains of the HIV-1 virus, denoted WMJ1, WMJ2, and WMJ3, were isolated from a two-year-old child on October 3, 1984, January 15, 1985, and May 3, 1985, respectively. The child was presumed to have been infected only once (perinatally by her mother) by a single strain of HIV. In the following, WMJ1 and WMJ2 are sequences 1 and 2, respectively, and $t = 3.4$ months (0.28 years). By using a number of reference outgroup isolates, Li et al. (1988) calculated the rate of nucleotide substitution in the WMJ strains.

The mean distances of the reference strains from WMJ1, d_{13}, was 0.0675, and from WMJ2, $d_{23} = 0.0675$. Therefore, $d_{23} - d_{13} = 0.0020$. From Equation 4.39, we obtain $r = 0.0020/0.28 = 7.1 \times 10^{-3}$ substitutions per site per year. Slightly different values are obtained when WMJ1 and WMJ3 or WMJ2 and WMJ3 are used, or when other HIV-1 sequences were selected as outgroup references.

Table 4.17 lists the synonymous (K_S) and nonsynonymous (K_A) rates of substitution for the different genes in HIV-1. The rates of synonymous substitution are always higher than the nonsynonymous rates. However, while the mean K_A/K_S ratio in mammalian nuclear genes is about 5 and in some highly conservative genes it can reach values in excess of 100 (see Table 4.1), the mean K_A/K_S ration in HIV is less than 2, and even in the two most conservative genes, *gag* and *pol*, the ratio is less than 7. Thus, the extent of purifying selection in HIV genes is very weak.

The estimates for the average synonymous and nonsynonymous rates for the HIV genome are more than 10^6 times greater than the corresponding rates in mammalian genes, and are similar to the values obtained for other RNA viruses (Holland et al. 1982; Gojobori and Yokoyama 1985; Hayashida et al. 1985; Buonagurio et al. 1986; Saitou and Nei 1986). The high substitution rates in HIV have been suggested to be mainly due to errors in the reverse tran-

TABLE 4.17 Rates of synonymous (K_S) and nonsynonymous (K_A) nucleotide substitution per site per year ($\times 10^{-3}$) in various coding regions of the HIV-1 genome

Coding region	Function	K_S (range)		K_A (range)	
gag	Group-specific antigen	9.7	(6.5–13.1)	1.7	(1.1–2.3)
pol	Polymerase	11.0	(7.4–14.8)	1.6	(1.0–2.1)
sor	Infectivity	9.1	(6.1–12.3)	4.7	(3.1–6.3)
tat (exon 2)	Regulatory	7.0	(4.7–9.5)	8.3	(5.6–11.2)
art (exon 3)	Regulatory	7.4	(5.0–10.0)	6.6	(4.5–8.9)
gp120	Outer membrane protein	8.1	(5.5–10.9)	3.3	(2.2–4.4)
envhv	Hypervariable region	17.2	(11.6–23.2)	14.0	(9.4–18.8)
gp41	Transmembrane protein	9.8	(6.6–13.2)	5.1	(3.5–6.9)
env	Envelope	9.2	(6.2–12.4)	5.1	(3.5–6.9)
p27	Capsid protein	7.9	(5.3–10.7)	5.9	(4.0–8.0)
Average[a]		9.64	(2.92)	5.63	(3.60)

Modified from Li et al. (1988).

[a]The average is the arithmetic mean, and values in parentheses are the standard deviations computed over all genes.

scription from RNA to DNA (Rabson and Martin 1985; Hahn et al. 1986; Coffin 1986). As pointed out by several authors, the high substitution rates may result in rapid modification of viral properties such as tissue tropism and sensitivity to antiviral drug therapy. In particular, the extremely high nonsynonymous rate in the hypervariable regions of the *env* gene may lead to extremely rapid changes in viral antigenicity.

FURTHER READINGS

Akashi, H. 1997. Codon bias evolution in *Drosophila*. Population genetics of mutation-selection drift. Gene 205: 269–278.

Easteal, S., C. C. Collet, and D. J. Betty. 1995. *The Mammalian Molecular Clock.* Springer & Landes, Austin, TX.

Ikemura, T. 1985. Codon usage and tRNA content in unicellular and multicellular organisms. Mol. Biol. Evol. 2: 13–34

Kimura, M. 1983. *The Neutral Theory of Molecular Evolution.* Cambridge University Press, Cambridge.

Miyata, T., K. Kuma, N. Iwabe, H. Hayashida, and T. Yasunaga. 1990. Different rates of evolution of autosome-, X chromosome- and Y chromosome-linked genes: Hypothesis of male-driven evolution. pp. 342–357. *In* N. Takahata and J. F. Crow (eds.), *Population Biology of Genes and Molecules.* Baifukan, Tokyo.

Morgan, G. J. 1998. Emile Zuckerkandl, Linus Pauling, and the molecular evolutionary clock, 1959–1965. J. Hist. Biol. 31: 155–178.

Nei, M. 1987. *Molecular Evolutionary Genetics.* Columbia University Press, New York.

Ohta, T. and J. H. Gillespie. 1996. Development of neutral and nearly neutral theories. Theor. Pop. Biol. 49: 128–142.
Sharp, P. M., M. Averof, A. T. Lloyd, G. Matassi, and J. F. Peden. 1995. DNA sequence evolution: The sounds of silence. Philos. Trans. Roy. Soc. London 349B: 241–247.
Sharp, P. M. and G. Matassi. 1994. Codon usage and genome evolution. Curr. Opin. Genet. Develop. 4: 851–860.
Takahata, N. 1996. Neutral theory of molecular evolution. Curr. Opin. Genet. Develop. 6: 767–772.
Xia, X. and W.-H. Li. 1998. What amino acid properties affect protein evolution? J. Mol. Evol. 47: 557–564.

Chapter 5

Molecular Phylogenetics

Molecular phylogenetics is the study of evolutionary relationships among organisms by using molecular data such as DNA and protein sequences, insertions of transposable elements, or other molecular markers. It is one of the areas of molecular evolution that have generated much interest in recent years, mainly because in many cases phylogenetic relationships are difficult to assess any other way. The objectives of phylogenetic studies are to reconstruct the correct genealogical ties among biological entities, to estimate the time of divergence between organisms (i.e., the time since they last shared a common ancestor), and to chronicle the sequence of events along evolutionary lineages. This chapter will (1) introduce the vocabulary of phylogenetics, (2) explain how to reconstruct a phylogenetic tree from molecular data, and (3) discuss some theoretical problems associated with molecular phylogenetic reconstruction. In the latter half of this chapter, we present a number of examples in which the molecular approach has been able to provide a much clearer resolution of longstanding phylogenetic issues than was possible with any nonmolecular approach.

Impacts of Molecular Data on Phylogenetic Studies

The study of molecular phylogeny began before the turn of the century, even before Mendel's laws were rediscovered in 1900. Immunochemical studies showed that serological cross-reactions were stronger for closely related organisms than for distantly related ones. The evolutionary implications of these findings were used by Nuttall (1902, 1904) to infer the phylogenetic relationships among various groups of animals, such as eutherians (placental mammals), primates, ungulates (hoofed mammals), and artiodactyls (even-hoofed ungulates). For example, he determined that the closest relatives of humans are the apes, followed in order of decreasing relatedness by the Old World monkeys, the New World monkeys, and the prosimians.

Since the late 1950s, various techniques have been developed in molecular biology, and this started the extensive use of molecular data in phylogenetic research. In particular, the study of molecular phylogeny progressed tremendously in the 1960s and 1970s as a result of the development of protein-sequencing methodologies. Less expensive and more expedient methods such as protein electrophoresis, DNA–DNA hybridization, and immunological methods, though less accurate than protein sequencing, were extensively used to study the phylogenetic relationships among populations or closely related species (Goodman 1962; Nei 1975; Ayala 1976; Wilson et al. 1977). The application of these methods also stimulated the development of measures of genetic distance and tree-making methods (e.g., Fitch and Margoliash 1967; Nei 1975; Felsenstein 1988; Miyamoto and Cracraft 1991; Swofford et al. 1996).

The rapid accumulation of DNA sequence data since the late 1970s has had a great impact on molecular phylogeny. DNA sequence data are more abundant and easier to analyze than protein sequence data. The advent of various molecular techniques, in particular the polymerase chain reaction (PCR), has led to an even more rapid accumulation of DNA sequence data and has resulted in an unprecedented level of activity in the field of molecular phylogenetics. Indeed, these data have been used on the one hand to infer the phylogenetic relationships among closely related populations or species, such as the relationships among human populations (Cann et al. 1987; Vigilant et al. 1991; Hedges et al. 1992; Templeton 1992; Horai et al. 1993; Torroni et al. 1993; Bailliet et al. 1994) or the relationships among apes (see page 217) and, on the other hand, they were used to study very ancient evolutionary occurrences, such as the origin of mitochondria and chloroplasts and the divergence of phyla and kingdoms (Woese 1987; Cedergren et al. 1988; Giovannovi et al. 1988; Lockhart et al. 1994). In the future, DNA sequences are likely to resolve many of the longstanding problems in phylogenetics, such as the evolutionary relationships among bacteria and unicellular eukaryotes (Sogin et al. 1986, 1989; Wainright et al. 1993; Doolittle et al. 1996), that have eluded resolution by traditional methods of evolutionary inquiry.

In a letter to Thomas Huxley, Charles Darwin wrote: "The time will come, I believe, though I shall not live to see it, when we shall have fairly true genealogical trees of each great kingdom of Nature." Molecular data have proved so powerful in the study of evolutionary history that we may be within grasp of Darwin's dream. Of course, we should not abandon traditional means of evolutionary inquiry, such as morphology, anatomy, physiology, and paleontology. Rather, different approaches provide complementary data. Indeed, taxonomy is based mainly on morphological and anatomical data, and paleontological information is one of the few types of data that can provide a time frame for evolutionary studies.

ADVANTAGES OF MOLECULAR DATA IN PHYLOGENETIC STUDIES

There are several reasons why molecular data, particularly DNA and amino acid sequence data, are much more suitable for evolutionary studies than morphological and physiological data. First, DNA and protein sequences are strictly heritable entities. This may not be true for many morphological traits that can be influenced to varying extents by environmental factors. Second, the description of molecular characters and character states is unambiguous. Thus, the third amino acid in the preproinsulin of the rabbit (*Oryctolagus cuniculus*) can be unambiguously identified as serine, and the homologous position in the preproinsulin of the golden hamster (*Mesocricetus auratus*) as leucine. In contrast, morphological descriptions frequently contain such ambiguous modifiers as "thin," "reduced," "slightly elongated," "partially enclosed," and "somewhat flattened." Third, molecular traits generally evolve in a much more regular manner than do morphological and physiological characters and therefore can provide a clearer picture of the relationships among organisms. Fourth, molecular data are often much more amenable to quantitative treatments than are morphological data. In fact, sophisticated mathematical and statistical theories have been developed for the quantitative analysis of DNA sequence data, whereas morphological studies retain a great deal of qualitative argumentation. Fifth, homology assessment is easier with molecular data than with morphological traits. Sixth, some molecular data can be used to assess evolutionary relationships among very distantly related organisms. For example, numerous protein and ribosomal RNA sequences can be used to reconstruct evolutionary relationships among such distantly related organisms as fungi, plants, and animals. In contrast, there are few morphological characters that can be used for such a purpose. Finally, molecular data are much more abundant than morphological data. This abundance is especially useful when working with organisms such as bacteria, algae, and protozoa, which possess only a limited number of morphological or physiological characters that can be used for phylogenetic studies.

TERMINOLOGY OF PHYLOGENETIC TREES

In phylogenetic studies, the evolutionary relationships among a group of organisms are illustrated by means of a **phylogenetic tree** (or **dendrogram**). A phylogenetic tree is a graph composed of **nodes** and **branches**, in which only one branch connects any two adjacent nodes (Figure 5.1). The nodes represent the taxonomic units. The taxonomic units represented by the nodes can be species (or higher taxa), populations, individuals, or genes. The branches define the relationships among the taxonomic units in terms of descent and ancestry. The branching pattern of a tree is called its **topology**.

We distinguish between **terminal** and **internal nodes**, and between **external branches** (branches that end in a tip) and **internal branches** (branches that do not end in a tip). For example, in Figure 5.1 nodes A, B, C, D, and E

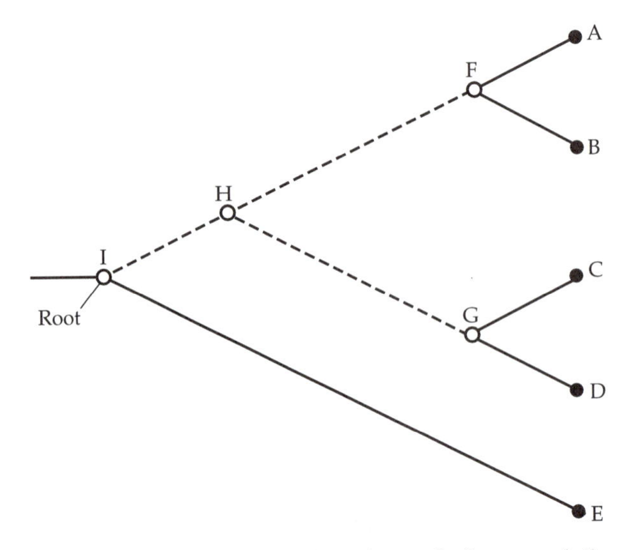

FIGURE 5.1 A phylogenetic tree illustrating the evolutionary relationships among five OTUs (A–E). Solid and white circles denote terminal and internal nodes, respectively. Solid and dashed lines denote terminal and internal branches, respectively. The internal nodes (F–H) represent the HTUs. I is the root.

are terminal, whereas all others are internal. Branches AF, BF, CG, DG, and EI in Figure 5.1 are external; all others are internal. Terminal nodes represent the extant taxonomic units under comparison, which are referred to as **operational taxonomic units (OTUs)**. Internal nodes represent inferred ancestral units, and since we have no empirical data pertaining to these taxa, they are sometimes referred to as **hypothetical taxonomic units (HTUs)**.

A node is **bifurcating** if it has only two immediate descendant lineages, but **multifurcating** if it has more than two immediate descendant lineages. In a strictly bifurcating tree, each internal node is incident to exactly three branches, two derived and one ancestral. In evolutionary studies we assume that the process of speciation is usually a binary one, i.e., that speciation results in the formation of not more than two species from a single stock at any one time. Thus, the common representation of phylogenies employs bifurcating trees, in which each ancestral taxon splits into two descendant taxa. There are two possible interpretations for a multifurcation (or **polytomy**) in a tree: either it represents the true sequence of events, whereby an ancestral taxon gave rise to three or more descendant taxa simultaneously, or it represents an instance in which the exact order of two or more bifurcations cannot be determined unambiguously with the available data. In the following, we assume that speciation is always a bifurcating process, and multifurcating trees will only be used for cases in which the exact temporal sequence of several bifurcations cannot be determined unambiguously.

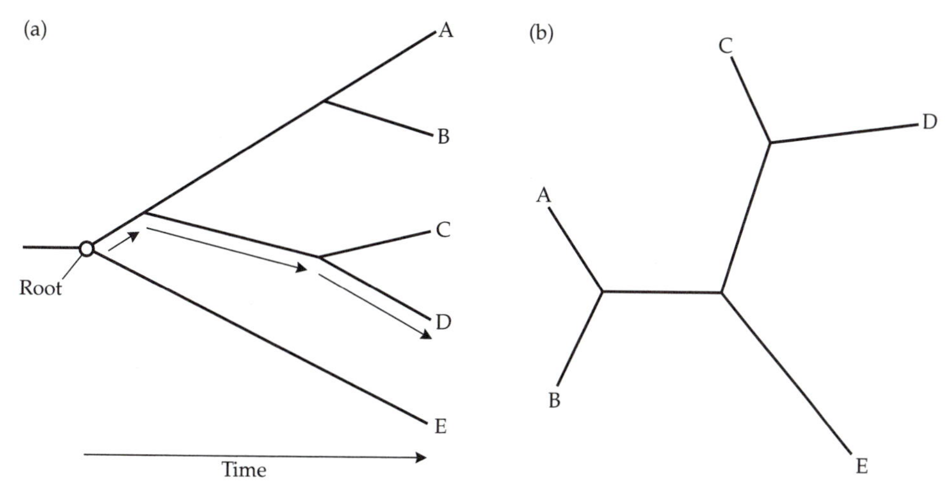

FIGURE 5.2 Rooted (a) and unrooted (b) trees. Arrows indicate the unique path leading from the root to OTU D.

Rooted and unrooted trees

Trees can be rooted or unrooted. In a **rooted tree** there exists a particular node, called the **root**, from which a unique path leads to any other node (Figure 5.2a). The direction of each path corresponds to evolutionary time, and the root is the most recent common ancestor of all the taxonomic units under study. An **unrooted tree** is a tree that only specifies the degree of kinship among the taxonomic units but does not define the evolutionary path (Figure 5.2b). Thus, strictly speaking, an unrooted tree may not in itself be considered a phylogenetic tree, since the time arrow is not specified. Unrooted trees neither make assumptions nor require knowledge about common ancestors.

An unrooted tree has n terminal nodes representing the OTUs and $n - 2$ internal nodes. Such a tree has $2n - 3$ branches, of which $n - 3$ are internal and n are external. In a rooted tree, there are n terminal nodes and $n - 1$ internal ones, as well as $2n - 2$ branches, of which $n - 2$ are internal and n are external. In an unrooted tree with four external nodes, the internal branch is frequently referred to as the **central branch.**

Scaled and unscaled trees

Figure 5.3 illustrates two common ways of drawing a phylogenetic tree. In Figure 5.3a, the branches are **unscaled**; their lengths are not proportional to the number of changes, which are indicated on the branches. This type of presentation allows us to line up the extant OTUs and to place the internal nodes representing divergence events on a time scale when the times of divergence are known or have been estimated. In Figure 5.3b, the branches are **scaled**, i.e., each **branch length** is proportional to the number of changes (e.g., nucleotide substitutions) that have occurred along that branch.

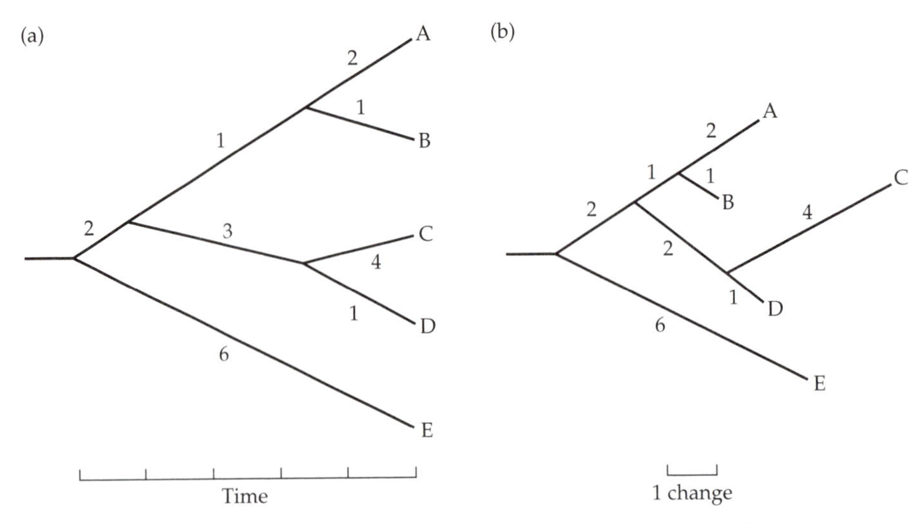

FIGURE 5.3 **Two alternative representations of a phylogenetic tree for 5 OTUs. (a) Unscaled branches: extant OTUs are lined up and nodes are positioned proportionally to times of divergence. (b) Scaled branches: lengths of branches are proportional to the numbers of molecular changes along them.**

The Newick format

In computer programs, trees are represented in a linear form by a series of nested parentheses, enclosing names and separated by commas. This type of representation is called the **Newick format**. The originator of this format was Cayley (1857). The Newick format for phylogenetic trees was adopted by an informal standards committee of the Society for the Study of Evolution in 1986. The Newick format currently serves as the standard employed by most phylogenetic computer packages. Unfortunately, it has yet to be described in a formal publication.

In the Newick format, the pattern of the parentheses indicates the topology of the tree by having each pair of parentheses enclose all members of a monophyletic group. For example, the rooted tree in Figure 5.4a may be written down as (((((A,B),C),D),E),F). Similarly, the unrooted tree in Figure 5.4b may be written down in the Newick format as ((A,B),(C,D),(E,F)). The three-way split of the unrooted tree is enclosed by the external (or bottommost) parentheses with two commas. Scaled trees are written down in the Newick format with the branch lengths placed immediately after the group descended from that branch and separated by a colon (Figure 5.4c).

Number of possible phylogenetic trees

For three species A, B, and C, there exists only one possible unrooted tree (Figure 5.5a). There are, however, 3 different rooted trees (Figure 5.5b). For 4 OTUs, there are 3 possible unrooted trees (Figure 5.5c) and 15 rooted ones

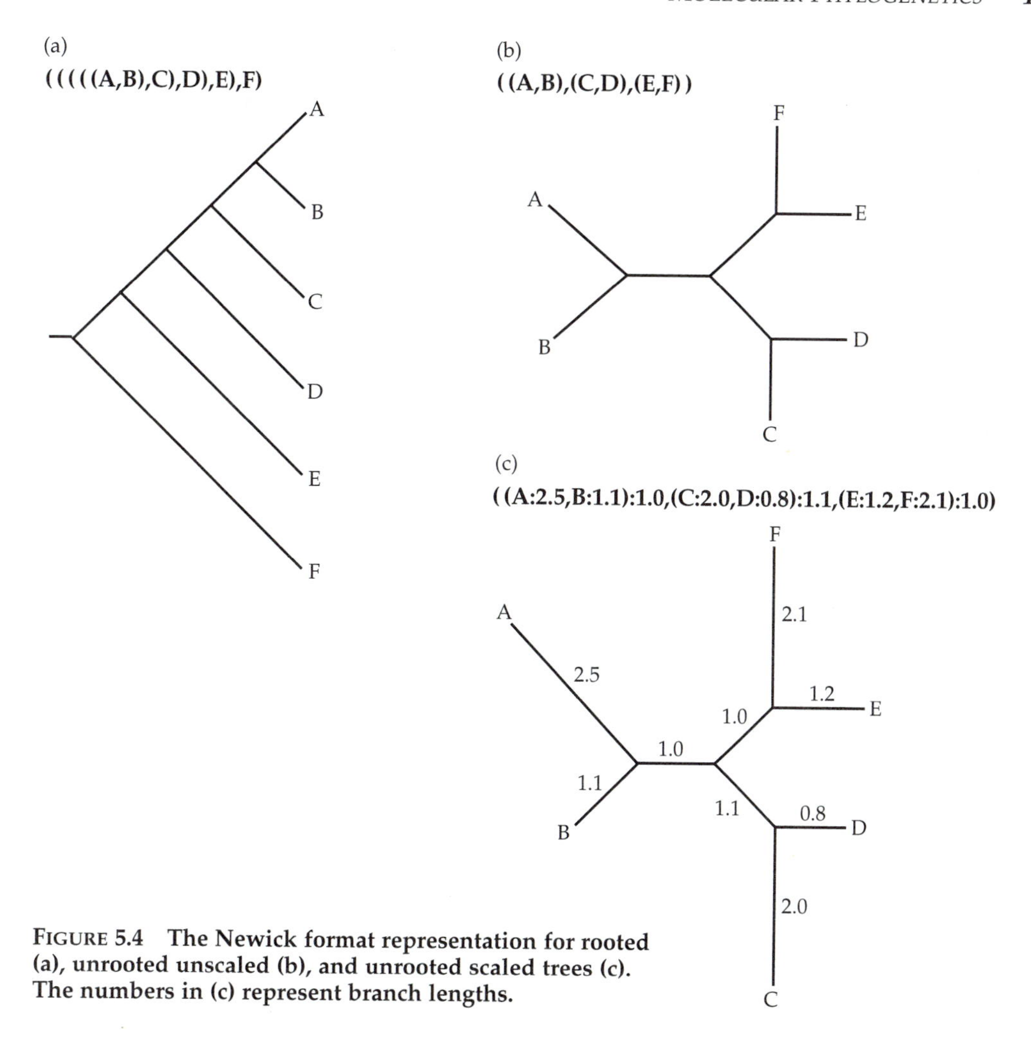

FIGURE 5.4 The Newick format representation for rooted (a), unrooted unscaled (b), and unrooted scaled trees (c). The numbers in (c) represent branch lengths.

(Figure 5.5d). The number of bifurcating rooted trees (N_R) for n OTUs is given by

$$N_R = \frac{(2n-3)!}{2^{n-2}(n-2)!} \tag{5.1}$$

when $n \geq 2$ (Cavalli-Sforza and Edwards 1967). The number of bifurcating unrooted trees (N_U) for $n \geq 3$ is

$$N_U = \frac{(2n-5)!}{2^{n-3}(n-3)!} \tag{5.2}$$

Note that the number of possible unrooted trees for n OTUs is equal to the number of possible rooted trees for $n - 1$ OTUs, i.e., rooting an unrooted tree

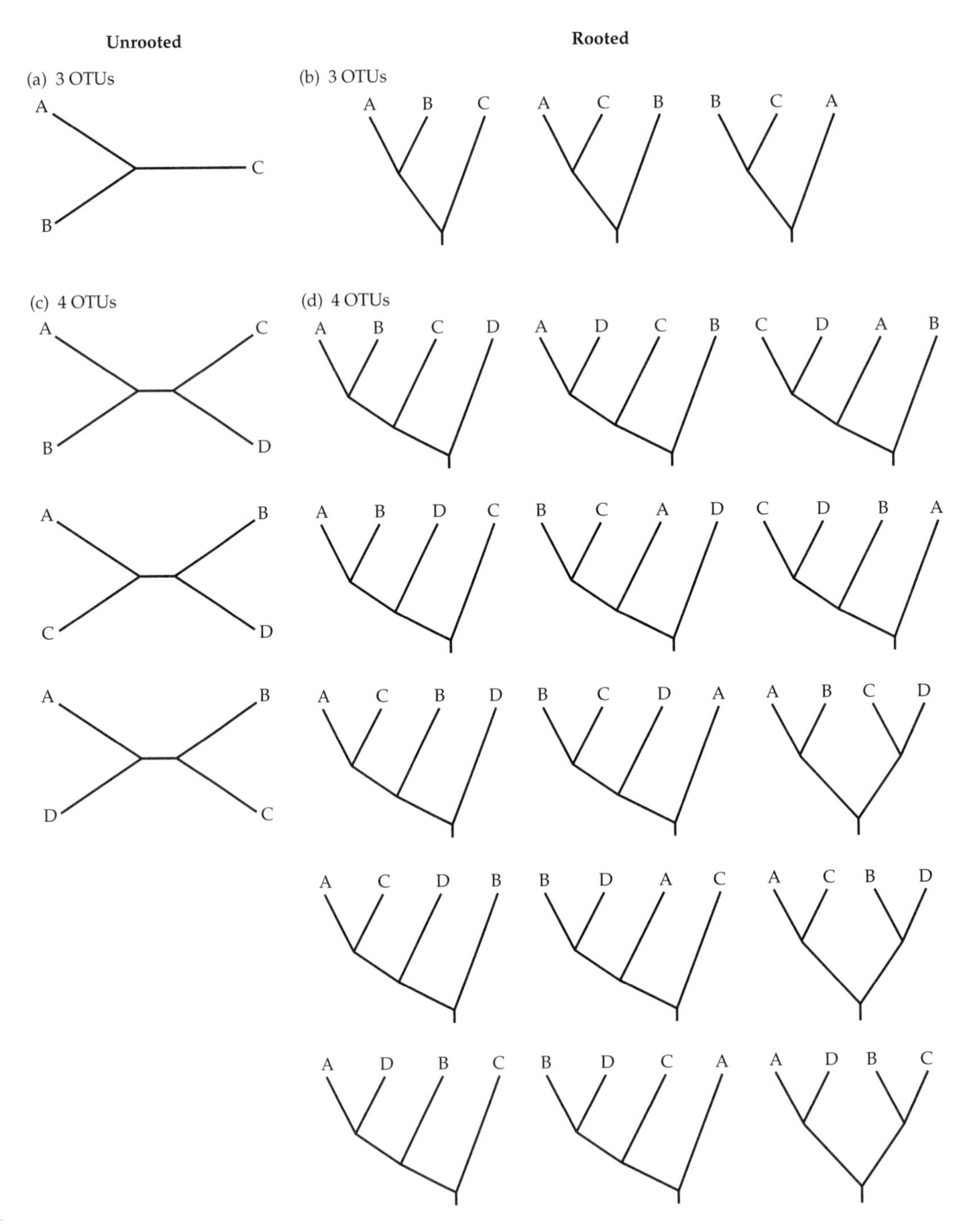

FIGURE 5.5 From three OTUs it is possible to construct only a single unrooted tree (a) but three different rooted ones (b). From four OTUs it is possible to construct three unrooted trees (c) and 15 rooted ones (d).

TABLE 5.1 Numbers of possible rooted and unrooted trees for up to 20 OTUs

Number of OTUs	Number of rooted trees	Number of unrooted trees
2	1	1
3	3	1
4	15	3
5	105	15
6	945	105
7	10,395	954
8	135,135	10,395
9	2,027,025	135,135
10	34,459,425	2,027,025
11	654,729,075	34,459,425
12	13,749,310,575	654,729,075
13	316,234,143,225	13,749,310,575
14	7,905,853,580,625	316,234,143,225
15	213,458,046,676,875	7,905,853,580,625
16	6,190,283,353,629,375	213,458,046,676,875
17	191,898,783,962,510,625	6,190,283,353,629,375
18	6,332,659,870,762,850,625	191,898,783,962,510,625
19	221,643,095,476,699,771,875	6,332,659,870,762,850,625
20	8,200,794,532,637,891,559,375	221,643,095,476,699,771,875

Data from Felsenstein (1978b).

is equivalent to adding one branch to each of its existing branches. The numbers of possible rooted and unrooted trees for up to 20 OTUs are given in Table 5.1. We see that both N_R and N_U increase very rapidly with n, and for 10 OTUs there are already more than 2 million bifurcating unrooted trees and close to 35 million rooted ones. For 20 OTUs there are close to 10^{22} rooted trees. Since only one of these trees correctly represents the true evolutionary relationships among the OTUs, it is usually very difficult to identify the true phylogenetic tree when n is large.

True and inferred trees

The sequence of speciation events that has led to the formation of any group of OTUs is historically unique. Thus, only one of all the possible trees that can be built with a given number of OTUs represents the true evolutionary history. Such a phylogenetic tree is called the **true tree**. A tree that is obtained by using a certain set of data and a certain method of tree reconstruction is called an **inferred tree**. An inferred tree may or may not be identical to the true tree.

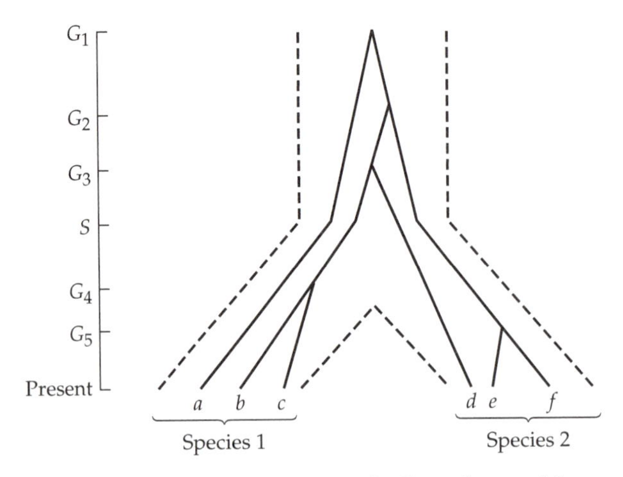

FIGURE 5.6 Diagram showing that in a genetically polymorphic population, gene splitting events (G_1–G_5) may occur before or after the speciation event (S). The evolutionary history of gene splitting resulting in the six alleles denoted *a–f* is shown in solid lines; speciation (i.e., population splitting), is shown by broken lines. Modified from Nei (1987).

Gene trees and species trees

Phylogeny is the representation of the branching history of the routes of inheritance of organisms. At every locus, if we trace back the history of any two alleles from any two populations, we will eventually reach a common ancestral allele from which both contemporary alleles have been derived. The routes of inheritance represent the passage of genes from parents to offspring, and the branching pattern depicts a **gene tree**. Different genes, however, may have different evolutionary histories, i.e., different routes of inheritance. We note, however, that the routes of inheritance are mostly confined by reproductive barriers—that is, gene flow occurs only within the species. A species is therefore like a bundle of genetic connections, in which many entangled parent–offspring lines form the ties that bundle individuals together into a species lineage (Figure 5.6). (Exceptional cases involving horizontal gene transfer will be dealt with in Chapter 7.)

Species are created by a process of **speciation** (or **cladogenesis**), i.e., the splitting of an ancestral species into two descendant ones. Thus, all life forms on earth, both extant and extinct, share a common origin, and their ancestries can be traced back to one or a few organisms that lived approximately 4 billion years ago. All animals, plants, and bacteria are related by descent to one another. Closely related organisms are descended from more recent common ancestors than are distantly related ones. The former are referred to as **recent taxa**, and the latter as **ancient taxa**. When we trace back the history of many genes from different species, we infer the routes of inheritance for the species, and in this case we obtain a phylogenetic tree for the species, or a **species tree**, representing the evolutionary relationships among species.

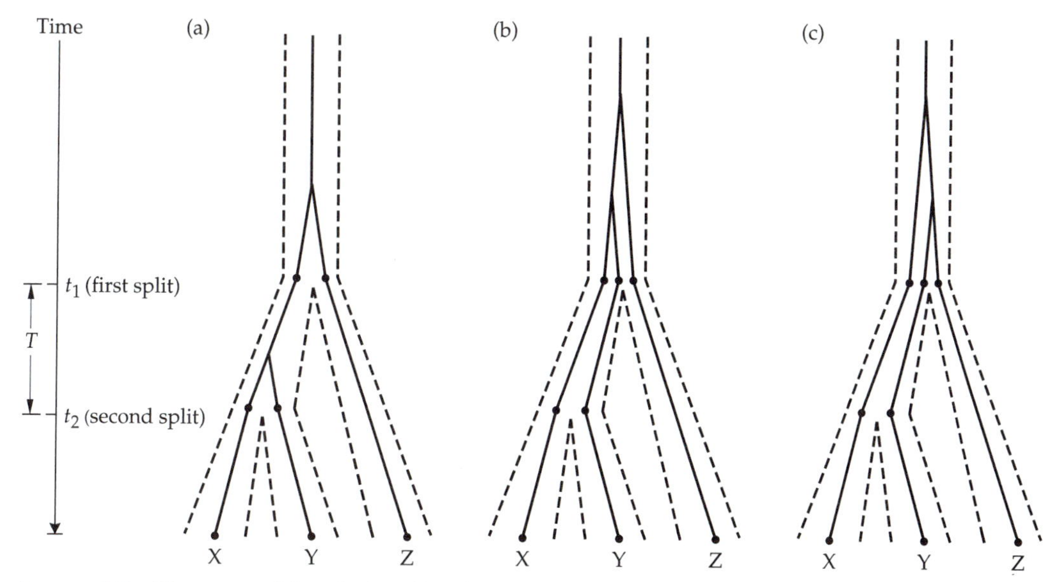

FIGURE 5.7 Three possible relationships between a species tree (broken lines) and a gene tree (solid lines). In (a) and (b), the topologies of the species trees are identical to those of the gene trees. Note that in (a) the time of divergence between the genes is roughly equal to the time of divergence between the populations. In (b), on the other hand, the time of divergence between genes X and Y greatly pre-dates the time of divergence between the respective populations. The topology of the gene tree in (c) is different from that of the species tree. Modified from Nei (1987).

In a species tree, a bifurcation represents the time of speciation, i.e., the time when the two species became distinct and reproductively isolated from one another. The gene tree can differ from the species tree in two respects. First, the divergence of two genes sampled from two different species may have pre-dated the divergence of the two species from each other (Figure 5.6). This will result in an overestimate of the branch length but will not represent a serious problem if we are concerned with long-term evolution, in which the component of divergence due to genetic polymorphism within each species may be ignored.

The second problem with gene trees is that the branching pattern of a gene tree (i.e., its topology) may be different from that of the species tree. The reason for this difference is genetic polymorphism in the ancestral species. Figure 5.7 shows three different possible relationships between the two trees. The topologies of gene trees (a) and (b) are identical with those of the corresponding species trees (e.g., OTUs X and Y form a cluster). Gene tree (c), however, is different from the true species tree, since Y and Z form a cluster in the gene tree, but not in the species tree. The probability of obtaining the erroneous topology (c) is quite high when the interval between the first and second splitting, $T = (t_1 - t_2)$, is short (Pamilo and Nei 1988), as is probably the case with the phylogenetic relationships among humans, chimpanzees, and gorillas. This probability cannot be substantially decreased by increasing the number of alleles sampled at a single locus (i.e., the number of individuals). To avoid this type of error, one needs to use many unlinked genes in the reconstruction of a phylogeny. A large amount of data is also required to avoid stochastic errors, which can occur because nucleotide substitutions occur randomly, so that, for instance, lineage Z in Figure 5.7a may have by chance accumulated fewer substitutions than lineages X and Y, despite the fact that it has branched off earlier in time.

When the gene under consideration belongs to a multigene family (see Chapter 6), it may be difficult to ascertain proper homology of the genes. We should therefore exercise great caution in inferring species trees from gene trees.

If the purpose of the phylogenetic reconstruction is to infer the evolutionary relationships among different members of a gene family, we must study gene trees.

Taxa and clades

A **taxon** is a species or a group of species (e.g., a genus, family, order, or class) that has been given a name; for example, *Homo sapiens* (the species name for modern humans) or Lepidoptera (an order of insects comprising the butterflies and moths). Codes of biological nomenclature attempt to ensure that every taxon has a single and stable name, and that every name is used for only one taxon.

One of the main aims of phylogenetic studies is to establish the evolutionary relationships among different taxa. In particular, we are interested in the identification of **natural clades** (or **monophyletic groups**). Strictly speaking, a clade is defined as a group of all the taxa that have been derived from a common ancestor, plus the common ancestor itself. In molecular phylogenetics, it is common to use the term "clade" for any group of taxa under study that share a common ancestor not shared by any species outside the group. If a clade is composed of two taxa, they are referred to as **sister taxa**. A taxonomic group whose common ancestor is shared by any other taxon is **paraphyletic**.

Figure 5.8 shows a possible evolutionary tree for the three classes of vertebrates: birds, reptiles, and mammals (Benton 1997). We see that the traditional taxonomic assignment of reptiles to a separate class does not fit the definition of a clade, since the three groups of reptiles share a common ancestor with another group, the birds, which is not included within the classical definition of reptiles. The reptiles are therefore paraphyletic. In this tree, birds and crocodiles constitute a natural clade, called the Archosauria, because they share a common ancestor not shared by any extant organism other than birds and crocodiles. Similarly, all birds and all reptiles taken together constitute a natural clade, nowadays called class Reptilia. An alternative internal arrangement of the Reptilia (inclusive of birds) has been proposed by Hedges and Poling (1999).

Named taxonomic groups that do not constitute a natural clade, such as fishes, protozoa, or insectivores—groups that are not monophyletic—are called **convenience taxa**.

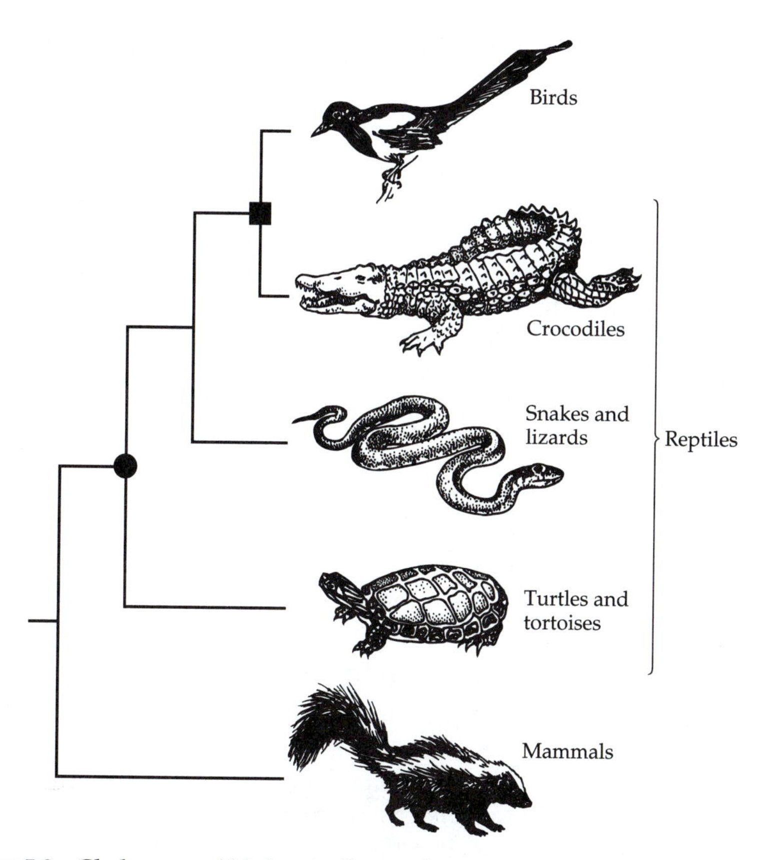

FIGURE 5.8 Cladogram of birds, reptiles, and mammals. The reptiles do not constitute a natural clade, since their most recent common ancestor (black circle) also gave rise to the birds, which are not included in the original definition of reptiles. Birds and crocodiles, on the other hand, constitute a natural clade (Archosauria), since they share a common ancestor (black box) that is not shared by any non-archosaurian organism.

TYPES OF DATA

Molecular data fall into one of two categories: characters and distances. A character provides information about an individual OTU. A distance represents a quantitative statement concerning the dissimilarity between two OTUs.

Character data

A **character** is a well-defined feature that in a taxonomic unit can assume one out of two or more mutually exclusive **character states**. In other words, a character is an independent variable, such as "height" or "the ninety-eighth amino acid position in cytochrome *c*," and the character state is the value of the character in a particular OTU, e.g., "1.68 cm" or "alanine."

Characters are either **quantitative** or **qualitative**. The character states of a quantitative character (e.g., height) are usually **continuous** and are measured on an interval scale. The character states of a qualitative character (e.g. amino acid positions in a protein) are **discrete**. Discrete characters can be assigned two or more values. When a character can only have two character states, it is referred to as **binary**. When three or more states are possible, the character is referred to as **multistate**.

Molecular data provide many binary characters that are useful in phylogenetic studies, usually taking the form of the presence or absence of a molecular marker. For example, the presence or absence of a retrotransposon at a certain genomic location can be used as a phylogenetic character.

In DNA and protein sequences, the qualitative multistate characters are the positions in the aligned sequences, and the character states are the particular nucleotide or amino acid residues at these positions in each of the taxa under study. For example, if nucleotide A is observed at position 139 of a mitochondrial sequence from pig, then the OTU is pig mitochondrial DNA, position 139 is the character, and A is the character state assigned to the character in this OTU.

Assumptions about character evolution

Methods of phylogenetic reconstruction require that we make explicit assumptions about (1) the number of discrete steps required for one character state to change into another, and (2) the probability with which such a change may occur. A character is designated as **unordered** if a change from one character state to another occurs in one step (Fitch 1971). Nucleotides in DNA sequences are usually assumed to be unordered, i.e., a change from any nucleotide to any other nucleotide is possible in one step.

Often a character state cannot change directly into another character state, but must pass through one or more intermediate states. A character is designated as **ordered** if the number of steps from one state to another equals the absolute value of the difference between their state number (Farris 1970; Swofford and Maddison 1987). Thus, a change from state 1 to state 5 is assumed to occur in four steps through the intermediate steps 2, 3, and 4. The process is assumed to be symmetrical in both directions, so a change from state 5 to state 1 is also assumed to require four steps. Perfectly ordered characters are rarely encountered in molecular data unless one is willing to make very strong assumptions concerning the mechanism via which a certain character state may change into another. For example, if the number of copies of a certain repetitive sequence in a genome is assumed to increase or decrease in a stepwise fashion, then we may treat the character "number of repeats" as ordered, so that, for instance, a change from two copies to four copies is assumed to require two steps.

Partially ordered characters are those characters in which the number of steps varies for the different pairwise combinations of character states, but for which no definite relationship exists between the number of steps and the

(a)

	A	C	T	G
A	0	1	1	1
C	1	0	1	1
T	1	1	0	1
G	1	1	1	0

(b)

	A	C	D	E	F	G	H	I	K	L	M	N	P	Q	R	S	T	V	W	Y
A	0	2	1	1	2	1	2	2	2	2	2	2	1	2	2	1	1	1	2	2
C	2	0	2	3	1	1	2	2	3	2	3	2	2	3	1	1	2	2	1	1
D	1	2	0	1	2	1	1	2	2	2	3	1	2	2	2	2	2	1	3	1
E	1	3	1	0	3	1	2	2	1	2	2	2	2	1	2	2	2	1	2	2
F	2	1	2	3	0	2	2	1	3	1	2	2	2	3	2	1	2	1	2	1
G	1	1	1	1	2	0	2	2	2	2	2	2	2	2	1	1	2	1	1	2
H	2	2	1	2	2	2	0	2	2	1	3	1	1	1	1	2	2	2	3	1
I	2	2	2	2	1	2	2	0	1	1	1	1	2	2	1	1	1	1	3	2
K	2	3	2	1	3	2	2	1	0	2	1	1	2	1	1	2	1	2	2	2
L	2	2	2	2	1	2	1	1	2	0	1	2	1	1	1	1	2	1	1	2
M	2	3	3	2	2	2	3	1	1	1	0	2	2	2	1	2	1	1	2	3
N	2	2	1	2	2	2	1	1	1	2	2	0	2	2	2	1	1	2	3	1
P	1	2	2	2	2	2	1	2	2	1	2	2	0	1	1	1	1	2	2	2
Q	2	3	2	1	3	2	1	2	1	1	2	2	1	0	1	2	2	2	2	2
R	2	1	2	2	2	1	1	1	1	1	1	2	1	1	0	1	1	2	1	2
S	1	1	2	2	1	1	2	1	2	1	2	1	1	2	1	0	1	2	1	1
T	1	2	2	2	2	2	2	1	1	2	1	1	1	2	1	1	0	2	2	2
V	1	2	1	1	1	1	2	1	2	1	1	2	2	2	2	2	2	0	2	2
W	2	1	3	2	2	1	3	3	2	1	2	3	2	2	1	1	2	2	0	2
Y	2	1	1	2	1	2	1	2	2	2	3	1	2	2	2	1	2	2	2	0

FIGURE 5.9 Step matrices. The elements in each matrix represent the number of steps (minimal number of nucleotide substitutions) required for a change between a character state in the column to a state in the row. (a) A step matrix for a nucleotide character. It is assumed that such a case can be suitably represented as a four-state unordered character. (b) A step matrix for amino acids encoded by the universal genetic code. An amino acid position in a protein can be represented as a twenty-state, partially ordered character.

character-state number (Swofford and Olsen 1990). Amino acid sequences are the most commonly encountered examples of partially ordered characters in molecular evolution. An amino acid cannot change into all other amino acids in a single step; sometimes two or three steps are required. For example, a tyrosine may only change into a leucine through an intermediate state, i.e., phenylalanine or histidine (see Table 1.3). The number of steps is written down in a matrix format called a **step matrix**, the elements of which indicate the number of steps required between any two character states. Two step matrices are shown in Figure 5.9.

Most discrete characters encountered in molecular evolution are **reversible**, i.e., they are assumed to change back and forth with equal probability. However, we sometimes come across characters in which it is reasonable to impose constraints on reversibility. The most common ones are binary characters in which one character state can change into the other quite easily, but the reverse occurs only rarely. For **irreversible** characters (Camin and Sokal 1965), it is assumed that changes in character state may only occur in one direction. The presence or absence of a retrosequence (Chapter 7) at a certain location in the genome is one example of an irreversible character. The reason is that retrosequences frequently insert themselves into the genome but are almost never excised precisely.

In addition to the number of steps between two characters, we may also consider the different probabilities with which different one-step changes occur. For example, we may assign different probabilities of occurrence to transitions and transversions.

Polarity and taxonomic distribution of character states

In terms of temporal appearance during evolution, the character states within a character of interest may be ranked by antiquity. A primitive or ancestral character state is called a **plesiomorphy** (literally, close to the original form), while the derived state representing an evolutionary novelty relative to the ancestral state is called an **apomorphy** (i.e., away from the original form) A primitive state that is shared by several taxa is a **symplesiomorphy**. A derived state that is shared by several taxa is a **synapomorphy**. A derived character state unique to a particular taxon is called an **autapomorphy**. A character state that has arisen in several taxa independently (through convergence, parallelism, and reversals) rather than being inherited from a common ancestor is called a **homoplasy**. Some methods of phylogenetic reconstruction rely solely on synapomorphies for the identification of monophyletic clades.

Distance data

Unlike character data, in which values are assigned to individual taxonomic units, **distances** involve pairs of taxa. Some experimental procedures, such as DNA–DNA hybridization, directly yield pairwise distances. Distance data cannot be converted into character data. In such cases, distance methods provide the only means of reconstructing phylogenetic trees. Much of the primary data produced by molecular studies, including sequences and restriction maps, consist of character data. These characters, however, can be transformed into distances, e.g., the number of substitutions per site between two nucleotide sequences (Chapter 3).

Swofford and Olsen (1990) outlined three possible reasons for converting characters into distances. First, a long list of character states, such as a DNA sequence, is in itself meaningless in an evolutionary context. On the other hand, if we can say that the similarity between two sequences is 93%, whereas the similarity between one of these sequences and a third one is only 50%, we evoke an intuitive (and often correct) image of a specific evolutionary relationship. Second, as pointed out in Chapter 3, one must take into account multiple substitutions at a site. By making reasonable assumptions about the nature of the evolutionary process, we are able to estimate the number of "unseen" events. These corrections apply to distances, such as the number of substitutions between two sequences, but not to the sequences themselves. Third, numerous methods exist for inferring phylogenetic trees from distance data. Most of these methods are very fast and efficient, and can be used even when the number of OTUs is so large as to preclude the use of many methods that are based on characters (see page 194).

Distance data can be additive, ultrametric, or neither. Distances are additive if the distance between any two OTUs is equal to the sum of the lengths of all the branches connecting them. A tree in which all the distances are additive is called an **additive tree**. For example, if additivity holds, then the distance between OTUs A and C in Figure 5.3a should be equal to $2 + 1 + 3 + 4 = 10$. The distance between two OTUs is calculated directly from molecular data (e.g., DNA sequences), while the branch lengths are estimated from the distances between the OTUs according to certain rules (see page 202). Additivity usually does not hold strictly if multiple substitutions have occurred at any nucleotide sites (see Figure 3.6). Distances are **ultrametric** if all the OTUs are equidistant from the root. This requires all the OTUs under study to evolve at the same rate (Chapter 4).

METHODS OF TREE RECONSTRUCTION

Inferring a phylogeny is an estimation procedure, in which a "best estimate" of the evolutionary history is made on the basis of incomplete information. In the context of molecular phylogenetics, we usually do not have information about the past; we only have access to contemporary sequences derived from contemporary organisms. Because many different phylogenetic trees can be produced from any set of OTUs (see Table 5.1), we must specify criteria for selecting one or a few trees as representing our best estimate of the true evolutionary history. Most phylogenetic inference methods seek to accomplish this goal by defining a criterion for comparing alternative phylogenies and deciding which tree is better. A phylogenetic reconstruction, therefore, consists of two steps: (1) definition of an **optimality criterion**, or **objective function**, i.e., the value that is assigned to a tree and is subsequently used for comparing one tree to another; and (2) design of specific **algorithms** to compute the value of the objective function and to identify the tree (or set of trees) that have the best values according to this criterion.

Several methods of tree reconstruction employ a specific sequence of steps (i.e., an algorithm) for constructing the best tree. This class of methods combines tree inference and the definition of the optimality criterion for selecting the preferred tree into a single statement. One must note, however, that an inferred tree is only as good as the assumptions on which the method of phylogenetic reconstruction is based.

Numerous tree-making methods have been proposed in the literature. For a detailed treatment, readers may consult Sneath and Sokal (1973), Nei (1987), and Felsenstein (1982, 1988). Here we describe several methods that are frequently used in molecular phylogenetic studies. For simplicity, we consider nucleotide sequence data, but the methods are equally applicable to other types of molecular data, such as amino acid sequences.

A longstanding controversy in phylogenetics has been the often acrimonious dispute between "cladists" and "pheneticists." **Cladistics** can be defined as the study of the pathways of evolution. In other words, cladists are interested in such questions as: How many branches are there among a group of organisms? Which branch connects to which other branch? and, What is the branching order? A tree that expresses such ancestor–descendant relationships is called a **cladogram**. To put it another way, a cladogram refers to the topology of a rooted phylogenetic tree.

On the other hand, **phenetics** is the study of relationships among a group of organisms on the basis of the degree of similarity between them, be that similarity molecular, phenotypic, or anatomical. A tree expressing phenetic relationships is called a **phenogram**. While a phenogram may be taken as an indicator of cladistic relationships, it is not necessarily identical with the cladogram. If there is a linear relationship between evolutionary time and the degree of genetic divergence, the two types of tree will be identical.

Among the methods discussed below, the maximum parsimony method is a typical representative of the cladistic approach, whereas UPGMA is a typical phenetic method. The other methods, however, cannot be easily classified according to the above criteria. For example, the transformed distance method and the neighbor-joining method have often been said to be phenetic methods, but this is not an accurate description. Although these methods use distance measures, they do not assume a direct connection between similarity and evolutionary kinship, nor are they intended to infer phenetic relationships.

In molecular phylogeny, a better classification of methods would be to distinguish between **distance matrix** and **character state approaches**. Methods belonging to the former approach are based on distance measures, such as the number of nucleotide substitutions or amino acid replacements, while methods belonging to the latter approach rely on character states, such as the nucleotide or amino acid at a particular site, or the presence or absence of a deletion or an insertion at a certain locus. According to this classification, UPGMA, the transformed distance method, the neighbors-relation method, and the neighbor-joining method are distance methods, while maximum parsimony is a character state method. Likelihood methods use both character states and distances, and cannot be assigned unequivocally to either of these two categories.

DISTANCE MATRIX METHODS

In the distance matrix methods, evolutionary distances (usually the number of nucleotide substitutions or amino acid replacements between two taxonomic units) are computed for all pairs of taxa, and a phylogenetic tree is constructed by using an algorithm based on some functional relationships among the distance values.

Unweighted pair-group method with arithmetic means (UPGMA)

This is the simplest method for tree reconstruction. It was originally developed for constructing taxonomic phenograms, i.e., trees that reflect the phenotypic similarities among OTUs (Sokal and Michener 1958), but it can also be used to construct phylogenetic trees if the rates of evolution are approximately constant among the different lineages so that an approximate linear relation exists between evolutionary distance and divergence time (Nei 1975). For such a relation to hold, linear distance measures such as the number of nucleotide substitutions should be used.

UPGMA employs a sequential clustering algorithm, in which local topological relationships are identified in order of decreasing similarity, and the phylogenetic tree is built in a stepwise manner. In other words, we first identify from among all the OTUs (or **simple OTUs**) the two that are most similar to each other and treat these as a new single OTU. Such an OTU is referred to as a **composite OTU**. For the new group of OTUs we compute a new distance matrix and identify the pair with the highest similarity. This procedure is repeated until we are left with only two OTUs.

To illustrate the method, let us consider a case of four OTUs, A, B, C, and D. The pairwise evolutionary distances are given by the following matrix:

OTU	A	B	C
B	d_{AB}		
C	d_{AC}	d_{BC}	
D	d_{AD}	d_{BD}	d_{CD}

In this matrix, d_{ij} stands for the distance between OTUs i and j. The first two OTUs to be clustered are the ones with the smallest distance. Let us assume that d_{AB} is the smallest. Then, OTUs A and B are the first to be clustered, and the branching point, l_{AB}, is positioned at a distance of $d_{AB}/2$ substitutions (Figure 5.10a).

Following the first clustering, A and B are considered as a single composite OTU (AB), and a new distance matrix is computed.

OTU	(AB)	C
C	$d_{(AB)C}$	
D	$d_{(AB)D}$	d_{CD}

In this matrix, $d_{(AB)C} = (d_{AC} + d_{BC})/2$, and $d_{(AB)D} = (d_{AD} + d_{BD})/2$. In other words, the distance between a simple OTU and a composite OTU is the average of the distances between the simple OTU and the constituent simple OTUs of the composite OTU. If $d_{(AB)C}$ turns out to be the smallest distance in

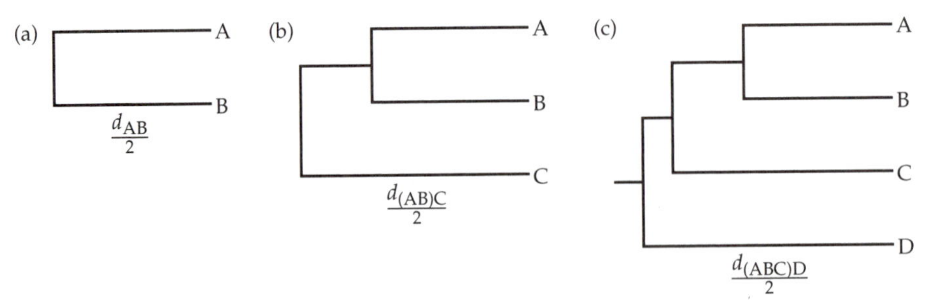

FIGURE 5.10 Diagram illustrating the stepwise construction of a phylogenetic tree for four OTUs by using UPGMA (see text).

the new matrix, then OTU C will be joined to the composite OTU (AB) with a branching node at $l_{(AB)C} = d_{(AB)C}/2$ (Figure 5.10b).

The final step consists of clustering the last OTU, D, with the new composite OTU, (ABC). The root of the entire tree is positioned at $l_{(ABC)D} = d_{(ABC)D}/2 = [(d_{AD} + d_{BD} + d_{CD})/3]/2$. The final tree inferred by using UPGMA is shown in Figure 5.10c.

In UPGMA, the branching point between two simple OTUs, *i* and *j*, is positioned at half the distance between them.

$$l_{ij} = \frac{d_{ij}}{2} \tag{5.3}$$

The branching point between a simple OTU, *i*, and a composite OTU, (*jm*), is positioned at half the arithmetic mean of the distances between the simple OTU and the constituent simple OTUs of the composite OTU.

$$l_{(i)(jm)} = \frac{(d_{ij} + d_{im})/2}{2} \tag{5.4}$$

The branching point between two composite OTUs is positioned at half the arithmetic mean of the distances between the constituent simple OTUs in each composite OTU. For example, the position of the branching point between a composite OTU, (*ij*), and a composite OTU, (*mn*), is

$$l_{(ij)(mn)} = \frac{(d_{im} + d_{in} + d_{jm} + d_{jn})/4}{2} \tag{5.5}$$

In the case of a tripartite composite OTU, (*ijk*), and a bipartite composite OTU, (*mn*), the position of the branching point is

$$l_{(ijk)(mn)} = \frac{(d_{im} + d_{in} + d_{jm} + d_{jn} + d_{km} + d_{kn})/6}{2} \tag{5.6}$$

UPGMA is one of the very few methods of phylogenetic reconstruction that yields a rooted tree. Note also that by using UPGMA one obtains the topology of the tree and the branch lengths simultaneously.

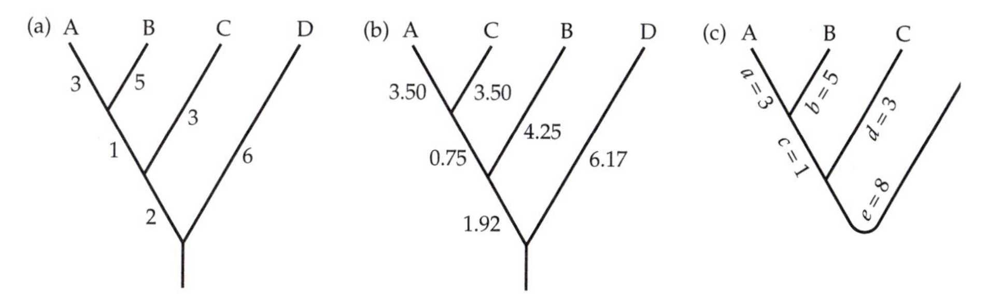

FIGURE 5.11 (a) The true phylogenetic tree. (b) The erroneous phylogenetic tree reconstructed by using UPGMA, which does not take into account the possibility of unequal substitution rates along the different branches. (c) The tree inferred by the transformed distance method. The root must be on the branch connecting OTU D and the node of the common ancestor of OTUs A, B, and C, but its exact location cannot be determined by the transformed distance method.

Transformed distance method

If the assumption of rate constancy among lineages does not hold, UPGMA may give an erroneous topology. For example, suppose that the phylogenetic tree in Figure 5.11a is the true tree. Under the assumption of additivity, the pairwise evolutionary distances are given by the following matrix:

OTU	A	B	C
B	8		
C	7	9	
D	12	14	11

By using UPGMA, we obtain an inferred tree (Figure 5.11b) that differs in topology from the true tree (Figure 5.11a). For example, OTUs A and C are grouped together in the inferred tree, whereas in the true tree A and B are sister OTUs. (Note that additivity does not hold in the inferred tree. For example, the true distance between A and B is 8, whereas the sum of the lengths of the estimated branches connecting A and B is 3.50 + 0.75 + 4.25 = 8.50.)

These topological errors might be remedied, however, by using a correction called the **transformed distance method** (Farris 1977; Klotz et al. 1979). In brief, this method uses an outgroup as a reference in order to make corrections for the unequal rates of evolution among the lineages under study and then applies UPGMA to the new distance matrix to infer the topology of the tree. An **outgroup** is an OTU or a group of several OTUs for which we have external knowledge, such as taxonomic or paleontological information, that clearly shows them to have diverged from the common ancestor prior to all the other OTUs under consideration (the **ingroup** taxa).

In the present case, let us assume that taxon D is an outgroup to all other taxa. D can then be used as a reference to transform the distances by the following equation:

$$d'_{ij} = \frac{d_{ij} - d_{iD} - d_{jD}}{2} + \bar{d}_D \quad \textbf{(5.7)}$$

where d'_{ij} is the transformed distance between OTUs i and j, and $\bar{d}_D$ is a correction term. It is calculated as

$$\bar{d}_D = \frac{\sum_{k=1}^{n} d_{kD}}{n} \quad \textbf{(5.8)}$$

where n is the number of ingroup OTUs. The term $\bar{d}_D$ was introduced to assure that all values are positive. This is done because a distance cannot be negative.

In our example, $\bar{d}_D = 37/3$ and the new distance matrix for taxa A, B, and C is

OTU	A	B
B	10/3	
C	13/3	13/3

Since d'_{AB} has the smallest value, OTUs A and B are the first to be clustered together, and C is added to the tree subsequently. By definition, the outgroup D determines the root of the tree, and is the last to be added. This gives the correct topology (Figure 5.11c). In the above example, we considered only three taxa with one outgroup, but the method can be easily extended to more taxa and more outgroups.

In many instances, it is impossible to decide *a priori* which of the taxa under consideration is an outgroup. To overcome this difficulty, a two-stage approach has been proposed (Li 1981). In the first step, one infers the root of the tree by using UPGMA. After that, the taxa on one side of the root are used as references (outgroups) for correcting the unequal rates of evolution among the lineages on the other side of the root (ingroups), and vice versa. In our example, this approach also identifies the correct tree.

If the purpose of the study is to infer an unrooted tree, then it is not necessary to know which of the taxa under study is an outgroup to the others, because any OTU can be used as a reference. If additivity holds, the true topology will be recovered regardless of the OTU chosen as reference. If additivity does not hold, the topologies inferred with different reference outgroups may be different from one another.

The transformed distance method does not provide branch lengths. However, after the topology is inferred, one can compute the branch lengths by using one of the methods that will be discussed later.

Sattath and Tversky's neighbors-relation method

In an unrooted bifurcating tree, two OTUs are said to be **neighbors** if they are connected through a single internal node. For example, in Figure 5.12a, A and B are neighbors, as are C and D. In contrast, A and C, A and D, B and C, and B and D are not neighbors because they are connected by two nodes. In Figure 5.12b, neither A and C nor B and C are neighbors; however, if we combine OTUs A and B into one composite OTU, then the composite OTU (AB) and the simple OTU C become a new pair of neighbors. (Note that neighbor taxa may or may not be sister taxa, depending on the position of the root.)

Let us now assume that the tree in Figure 5.12a is the true tree. Then if additivity holds, we should have

$$d_{AC} + d_{BD} = d_{AD} + d_{BC} = a + b + c + d + 2x = d_{AB} + d_{CD} + 2x \quad \textbf{(5.9)}$$

where a, b, c, and d are the lengths of the terminal branches and x is the length of the central branch. Therefore, the following two conditions hold:

$$d_{AB} + d_{CD} < d_{AC} + d_{BD} \quad \textbf{(5.10)}$$

and

$$d_{AB} + d_{CD} < d_{AD} + d_{BC} \quad \textbf{(5.11)}$$

These two conditions are collectively known as the **four-point condition** (Buneman 1971; Fitch 1981). They may hold even if additivity holds only approximately.

For four OTUs with unknown phylogenetic relationships, the four-point condition provides a simple way for inferring the topology of the tree because the condition can be used to identify the neighbors (A and B, or C and D), and once a pair of neighbors is identified, the topology of the phylogenetic tree is determined unambiguously. This approach to phylogenetic reconstruction is called the **neighborliness approach**. Note, however, that the four-point condition does not always lead to the true topology because in practice additivity may not hold.

Sattath and Tversky (1977) proposed the following method for dealing with more than four OTUs. First, compute the distance matrix as in UPGMA. For every possible combination of four OTUs, say OTUs i, j, m, and n, compute the following three values: (1) $d_{ij} + d_{mn}$, (2) $d_{im} + d_{jn}$, and (3) $d_{in} + d_{jm}$. Suppose that the first sum is the smallest; then, assign a score of 1 to both the pair i and j and the pair m and n. Pairs i and m, j and n, i and n, and j and m are

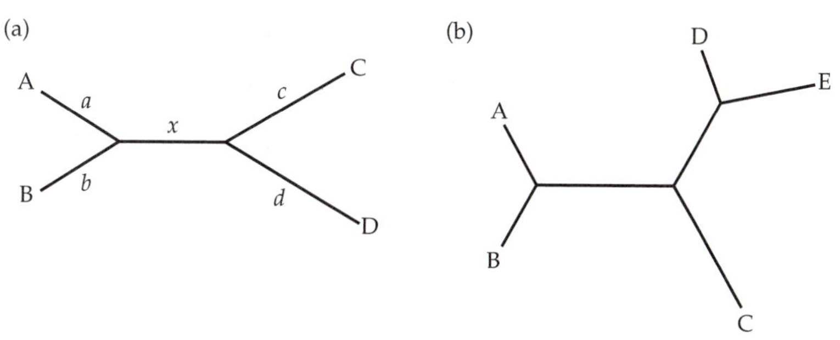

FIGURE 5.12 Bifurcating unrooted trees with (a) four OTUs and (b) five OTUs.

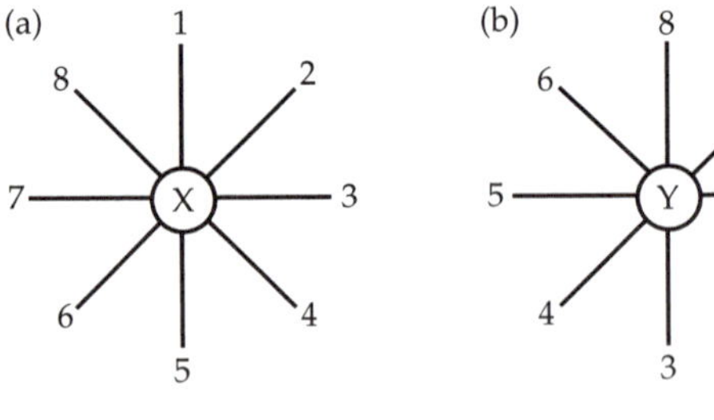

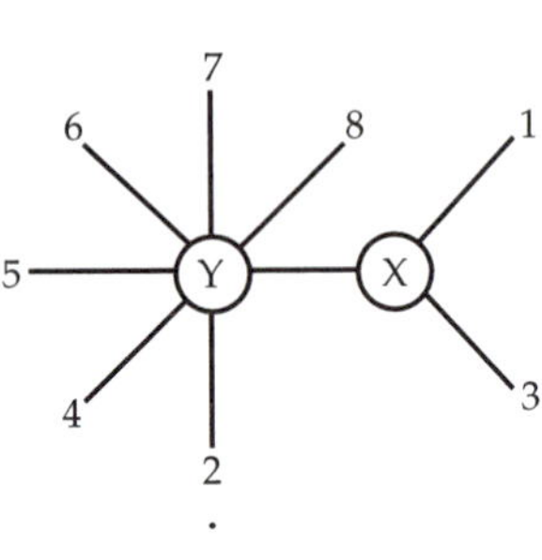

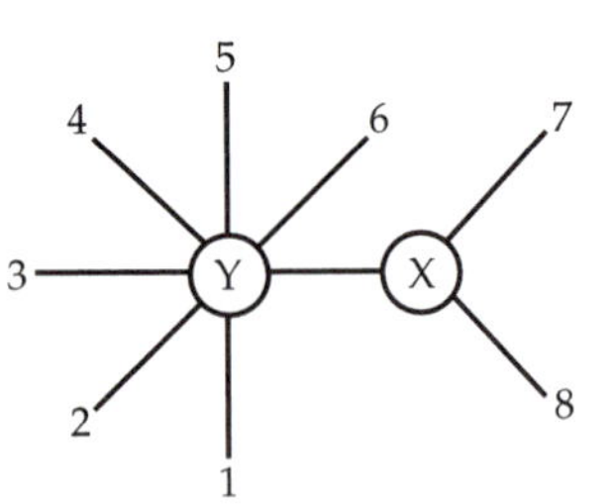

FIGURE 5.13 (a) A starlike tree for eight OTUs with no hierarchical structure. (b) Trees in which two of the OTUs are clustered at node X, and a single internal branch connects nodes X and Y. There are $N(N-1)/2$ ways of choosing pairs of OTUs. Three such examples are shown. Modified from Saitou and Nei (1987).

each assigned a score of 0. If, on the other hand, $d_{im} + d_{jn}$ has the smallest value, then we assign the pair i and m and the pair j and n each a score of 1, and assign the other four possible pairs a score of 0. When all the possible combinations of quadruples are considered, the pair with the highest total score is selected as the first pair of neighbors, and subsequently is treated as a new single composite OTU. Next, we compute a new distance matrix as in the case of UPGMA and repeat the process to select the second pair of neighbors. The process is repeated until we are left with three OTUs, at which time the topology of the tree is unambiguously inferred. A detailed illustration of this method will be given later in the chapter, when we consider the phylogeny of apes and humans.

Sattath and Tversky's method does not provide branch lengths. However, the inferred topology can be used to compute branch lengths separately (see page 202). Another method based on the neighborliness approach has been proposed by Fitch (1981).

Saitou and Nei's neighbor-joining method

The **neighbor-joining method** (Saitou and Nei 1987) is also a neighborliness method. It provides an approximate algorithm for finding the shortest (**minimum evolution**) tree. This is accomplished by sequentially finding neighbors that minimize the total length of the tree. The method starts with a starlike tree with N OTUs like the one shown in Figure 5.13a. The first step is to consider a tree that is of the form given in Figure 5.13b. In this tree, there is only one internal branch connecting nodes X and Y, where X is the node connecting OTUs 1 and 2, and Y is the node connecting all the other OTUs (3, 4, ..., N). For this tree, the sum of all the branch lengths is

$$S_{12} = \frac{1}{2(N-2)}\sum_{k=3}^{N}(d_{1k} + d_{2k}) + \frac{1}{2}d_{12} + \frac{1}{N-2}\sum_{3 \le i < j \le N}^{N} d_{ij} \qquad \textbf{(5.12)}$$

where d_{ij} is the distance between OTUs i and j. Any pair of OTUs can take positions 1 and 2 in the tree, and there are $N(N-1)/2$ ways of choosing the pairs. Among these possible pairs of OTUs (Figure 5.13b), the one that gives the smallest sum of branch lengths is chosen as the first pair of neighbors. This pair of OTUs is then regarded as a single composite OTU, and arithmetic mean distances between OTUs are computed to form a new distance matrix as in UPGMA. The next pair of OTUs that gives the smallest sum of branch length is again chosen. This procedure is continued until all $N-3$ internal branches are found. Saitou and Nei (1987) have shown that in the case of four OTUs, the necessary condition for this method to obtain the correct tree topology is also given by the four-point condition (Equations 5.10 and 5.11).

MAXIMUM PARSIMONY METHODS

The principle of **maximum parsimony** involves the identification of a topology that requires the smallest number of evolutionary changes (e.g., nucleotide substitutions) to explain the observed differences among the OTUs under study. It is often said that the principle of maximum parsimony abides by William of Ockham's razor, according to which the best hypothesis is the one requiring the smallest number of assumptions. In maximum parsimony methods, we use discrete character states, and the shortest pathway leading to these character states is chosen as the best tree. Such a tree is called a **maximum parsimony tree**. Often two or more trees with the same minimum number of changes are found, so that no unique tree can be inferred. Such trees are said to be **equally parsimonious**.

Several different parsimony methods have been developed for treating different types of data (see Felsenstein 1982). The method discussed below was first developed for amino acid sequence data (Eck and Dayhoff 1966), and was later modified for use on nucleotide sequences (Fitch 1977).

We start with the classification of sites. A site is defined as **invariant** if all the OTUs under study possess the same character state at this site. **Variable sites** may be **informative** or **uninformative**. A nucleotide site is phylogenetically informative only if it favors a subset of trees over the other possible trees. To illustrate the distinction between informative and uninformative sites, consider the following four hypothetical sequences:

	Site								
Sequence	**1**	**2**	**3**	**4**	**5**	**6**	**7**	**8**	**9**
1	A	A	G	A	G	T	T	C	A
2	A	G	C	C	G	T	T	C	T
3	A	G	A	T	A	T	C	C	A
4	A	G	A	G	A	T	C	C	T
					*		*		*

There are three possible unrooted trees for four OTUs (Figure 5.14). Site 1 is not informative because all sequences at this site have A, so that no change is required in any of the three possible trees. At site 2, sequence 1 has A, while all other sequences have G, and so a simple assumption is that the nucleotide has changed from G to A in the lineage leading to sequence 1. Thus, this site is also not informative, because each of the three possible trees requires 1 change. As shown in Figure 5.14a, for site 3 each of the three possible trees requires 2 changes, and so this site is also not informative. Note that if we assume that the nucleotide at the node connecting OTUs 1 and 2 in tree I in Figure 5.14a is C instead of G, the number of changes required for the tree remains 2. Figure 5.14b shows that for site 4, each of the three trees requires 3 changes; thus site 4 is also uninformative. For site 5, tree I requires only 1 change, whereas trees II and III require 2 changes each (Figure 5.14c). This site is, therefore, informative. The same is true for site 7. Site 9 is also informative, but in contrast to the previous two informative sites, this site favors tree II, which requires only 1 change, whereas trees I and III require 2 changes each.

From these examples, we see that a site is informative only when there are at least two different kinds of nucleotides at the site, each of which is represented in at least two of the sequences under study. In the above sequences, the informative sites (sites 5, 7, and 9) are indicated by asterisks.

To infer a maximum parsimony tree, we first identify all the informative sites. Next, for each possible tree we calculate the minimum number of substitutions at each informative site. In the above example, there are three informative sites. For sites 5, 7, and 9, tree I requires 1, 1, and 2 changes, respectively, tree II requires 2, 2, and 1 changes, and tree III requires 2, 2, and 2 changes. In the final step, we sum the number of changes over all the informative sites for each possible tree and choose the tree associated with the smallest number of changes. In our case, tree I is chosen because it requires only 4 changes at the informative sites, whereas trees II and III require 5 and 6 changes, respectively.

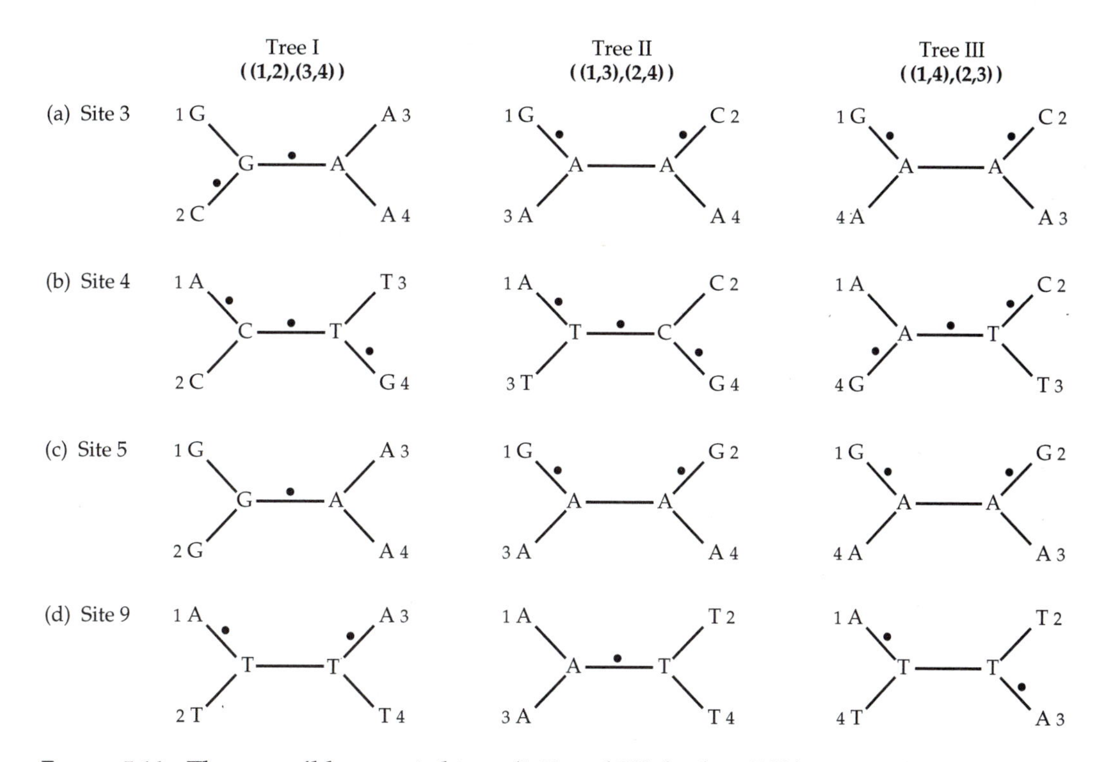

FIGURE 5.14 Three possible unrooted trees (I, II, and III) for four DNA sequences (1, 2, 3, and 4) that have been used to choose the most parsimonious tree. The possible phylogenetic relationships among the four sequences are shown in Newick format. The terminal nodes are marked by the sequence number and the nucleotide type at homologous positions in the extant species. Each dot on a branch means a substitution is inferred on that branch. Note that the nucleotides at the two internal nodes of each tree represent one possible reconstruction from among several alternatives. For example, the nucleotides at both the internal nodes of tree III(d) (bottom right) can be A instead of T. In this case, the two substitutions will be positioned on the branches leading to species 2 and 4. Alternatively, other combinations of nucleotides can be placed at the internal nodes. However, these alternatives will require three substitutions or more. The minimum number of substitutions required for site 9 is two.

In the case of four OTUs, an informative site can favor only one of the three possible alternative trees. For example, site 5 favors tree I over trees II and III, and is thus said to support tree I. It is easy to see that the tree supported by the largest number of informative sites is the most parsimonious tree. For instance, in the above example, tree I is the maximum parsimony tree because it is supported by two sites, whereas tree II is supported by only one site, and tree III by none. In cases where more than 4 OTUs are involved, an informative site may favor more than one tree, and the maximum parsimony tree may not necessarily be the one supported by the largest number of informative sites.

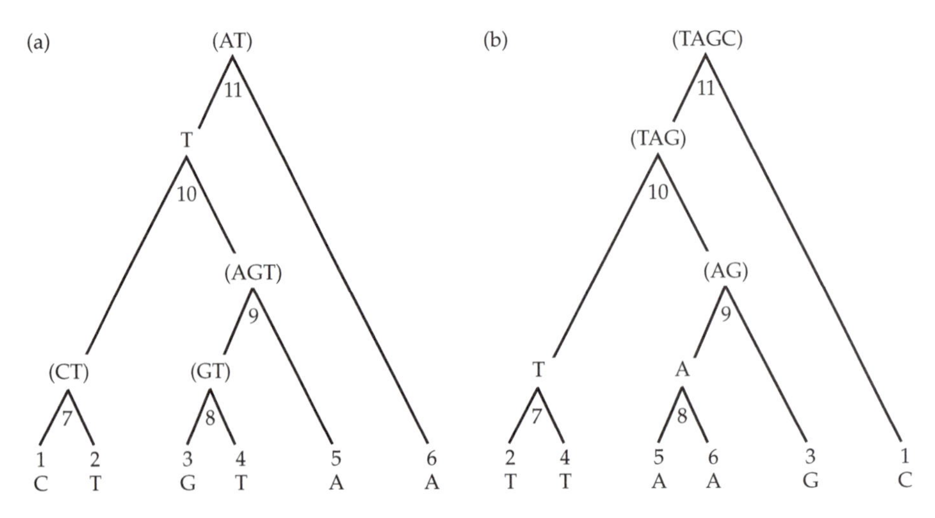

FIGURE 5.15 Nucleotides in six extant species (1–6) and inferred possible nucleotides in five ancestral species (7–11) according to the method of Fitch (1971). Unions are indicated by parentheses. Two different trees (a and b) are depicted. Note that the inference of an ancestral nucleotide at an internal node is dependent on the tree. Modified from Fitch (1971).

It is important to note that the informative sites that support the internal branches in the inferred tree are deemed to be synapomorphies, whereas all the other informative sites are deemed to be homoplasies.

When the number of OTUs under study is larger than four, the situation becomes more complicated because there are many more possible trees to consider (Table 5.1), and because inferring the number of substitutions for each alternative tree becomes more tedious. However, the basic goal remains simple: to identify the tree (or set of trees) requiring the minimum number of substitutions.

The inference of the number of substitutions for a given tree can be made by using Fitch's (1971) method. Let us consider a case of six OTUs (1–6) and assume that at a particular nucleotide site the nucleotides in the six sequences are C, T, G, T, A, and A (Figure 5.15a). We want to infer the nucleotides at the internal nodes 7, 8, 9, 10, and 11 from the nucleotides at the tips of the tree. The nucleotide at node 7 cannot be determined uniquely, but must be either C or T under the parsimony rule. (If we put A or G at node 7, two nucleotide substitutions rather than one must be assumed to have occurred at this site.) Therefore, the set of candidate nucleotides at node 7 consists of C and T. Similarly, the set of candidate nucleotides at node 8 consists of G and T, and that at node 9 consists of A, G, and T. However, at node 10, T is chosen because it is shared by the sets at the two descendant nodes, 7 and 9. Finally, the nucleotide at node 11 cannot be determined uniquely, but parsimony requires it be either A or T.

In mathematical terms, the rule used is as follows: The set at an internal node is the intersection (denoted by $\cap$) of the two sets at its immediate descendant nodes if the intersection is not empty (e.g., the nucleotide at node 10 is the intersection of the sets at nodes 7 and 9); otherwise, it is the union (denoted by $\cup$) of the sets at the descendant nodes (e.g., the set at node 9 is the union of the sets at nodes 8 and 5). When a union is required to form a nodal set, a nucleotide substitution at this position must have occurred at some point during the evolution of this position. Thus, the number of unions equals the minimum number of substitutions required to account for the descendant nucleotides from a common ancestor. In the example in Figure 5.15a, this number is 4. The alternative tree in Figure 5.15b requires only 3 unions, i.e., 3 nucleotide substitutions. By searching all the possible trees for the 6 OTUs, we find out that 3 is the minimum number of substitutions required to explain the differences among the nucleotides at this site. (In addition to the one shown in Figure 5.15b, there are many other equally parsimonious trees, each requiring 3 substitutions.)

Although inferring the minimum number of substitutions is straightforward, inferring the evolutionary path (i.e., the true sequence of nucleotide substitutions along each branch) is often difficult (Fitch 1971). Note further that the above procedure neglects all substitutions at uninformative sites. Such substitutions can be easily inferred because the number of substitutions at an uninformative site is equal to the number of different nucleotides present at that site minus one. For example, if the nucleotides at an uninformative site are A, T, T, C, T, and T in lineages 1, 2, 3, 4, 5, and 6, respectively, then the number of substitutions is $3 - 1 = 2$, because regardless of the tree topology the most parsimonious assumption is that a T $\leftrightarrow$ A substitution has occurred in the first lineage and a T $\leftrightarrow$ C substitution has occurred in the fourth lineage.

The total number of substitutions at both informative and uninformative sites in a particular tree is called the **tree length**.

Weighted and unweighted parsimony

As we have seen previously, the length of each tree was computed by adding up all the substitutions over all sites. In this computation, all the different nucleotide substitutions were given equal weight. This procedure is called **unweighted parsimony**. However, we may wish to give different weights to different types of substitution. For example, we may wish to give a greater weight to transversions, since they occur less frequently than transitions (Chapters 1 and 4). Maximum parsimony methods that assign different weights to the various character state changes are called **weighted parsimony**. If transitions are completely ignored and only transversions are used, the method is called **transversion parsimony**.

In the example used in the previous section (Figure 5.14), we see that the sites supporting tree I (5 and 7) involve transitions. On the other hand, site 9, which supports tree II, involves an A $\leftrightarrow$ T transversion. If we give a weight of 1 for each transition and a weight greater than 2 for each transversion, then tree II rather than tree I emerges as the maximum parsimony tree.

Searching for the maximum parsimony tree

When the number of sequences is small, it is possible to look at all the possible trees, determine their length, and choose from among them the shortest one (or ones). This type of search for the maximum parsimony tree(s) is called an **exhaustive search**. A simple algorithm can be used for the exhaustive search (Figure 5.16). In the first step we connect the first three taxa to form the only possible unrooted tree for three OTUs. In the next step, we add the fourth taxon to each of the three branches of the three-taxon tree, thereby generating all three possible unrooted trees for four OTUs. In the third step, we add the fifth taxon to each of the five branches of the three four-taxon trees, thereby generating $3 \times 5 = 15$ unrooted trees. We continue in a similar fashion, adding the next taxon in line to each of the branches in every tree obtained in the previous step.

However, as the number of possible trees increases rapidly with the number of OTUs (Table 5.1), it is virtually impossible to employ an exhaustive search when 12 or more OTUs are studied. Fortunately, there exist short-cut algorithms for identifying all maximum parsimony trees that do not require exhaustive enumeration. One such algorithm is the **branch-and-bound method** (Hendy and Penny 1982). We first consider an arbitrary tree or, better, a tree obtained from a fast method (e.g., the neighbor-joining method), and compute the minimum number of substitutions, L, for the tree. L is then considered as the **upper bound** to which the length of any other tree is compared. The rationale of the upper bound is that the maximum parsimony tree must be either equal in length to L or shorter. The branch-and-bound method works by searching for the maximum parsimony tree by using a similar procedure to that employed for the exhaustive search. In each step of the branch-and-bound algorithm, the length of each tree is compared with the previously determined L value (Figure 5.17). If the tree is longer than L, it is no longer used for addition of new taxa in the subsequent steps. The reason is that adding branches to a tree can only increase its length. For example, if a four-taxon tree is longer than L, then all the five-taxon trees descended from it will also be longer than L, and we can therefore ignore them. By dispensing with the evaluation of all the descendant trees from all the partial trees that are longer than L, we may greatly reduce the total number of trees to be considered. Depending on the efficiency of the implementation, the speed of the computer, and the type of data, the branch-and-bound method may be used for finding the maximum parsimony tree for up to 20 OTUs.

Above 20 OTUs, we need to use **heuristic searches**. In a heuristic search, only a manageable subset of all the possible trees is examined. Most heuristic searches are based on the same principle. An initial tree is constructed by using a certain procedure, say the neighbor-joining method, and we seek to

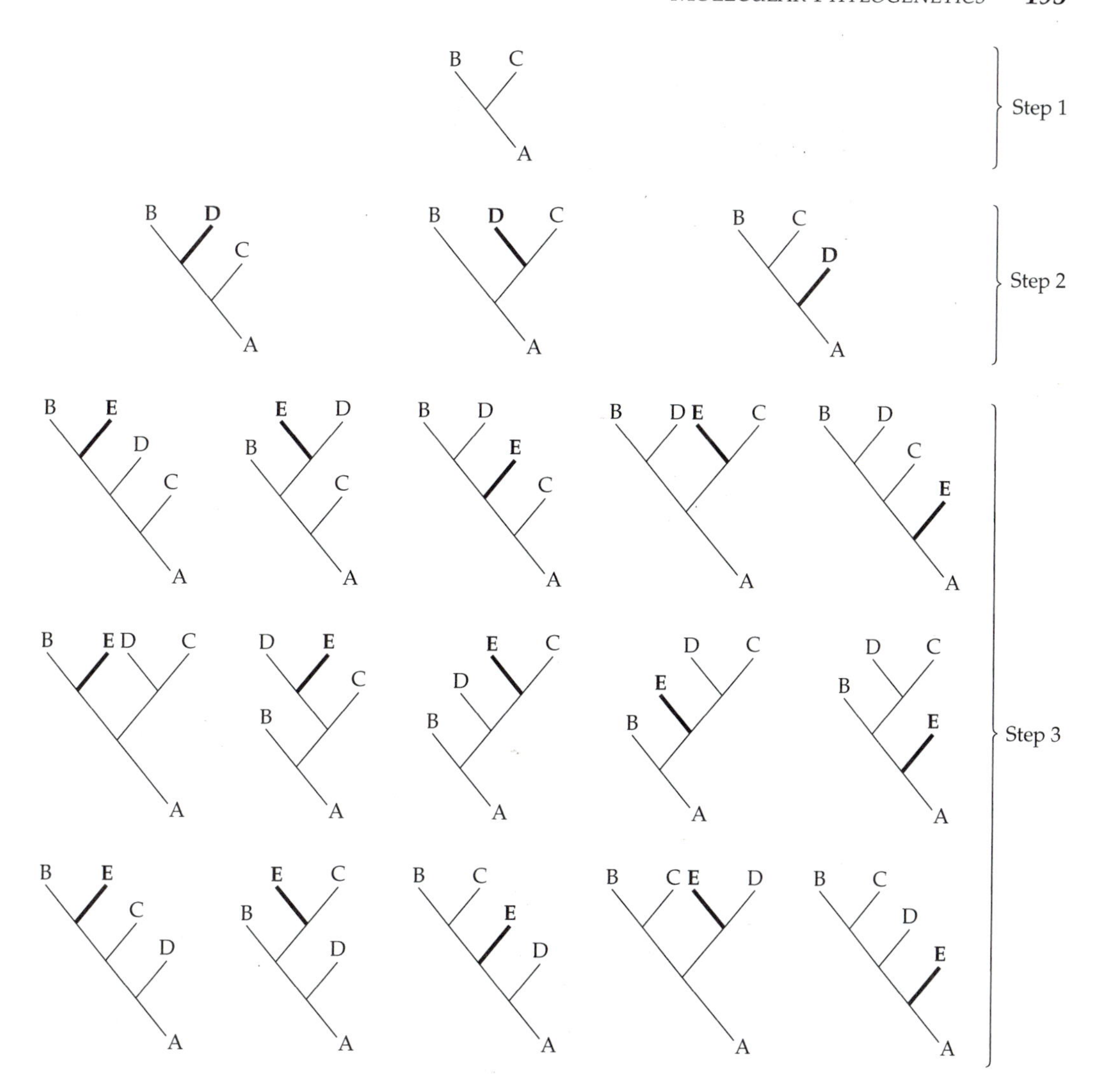

FIGURE 5.16 **Exhaustive stepwise construction of all 15 possible trees for five OTUs. In step 1, we form the only possible unrooted tree for the first three OTUs (A, B, and C). In step 2, we add OTU D to each of the three branches of the tree in step 1, thereby generating three unrooted trees for four OTUs. In step 3, we add OTU E to each of the five branches of the three trees in step 2, thereby generating 15 unrooted trees. Additions of OTUs are shown as heavier lines. Modifed from Swofford et al. (1996).**

find a shorter tree by examining trees that have a similar topology to the initial one. Of course, we must decide on a quantitative measure of similarity between two trees, below which a tree will be considered similar to the starting tree and above which it will not be (for measures of topological similarity or dissimilarity among trees, see page 206). If a shorter tree is found among the

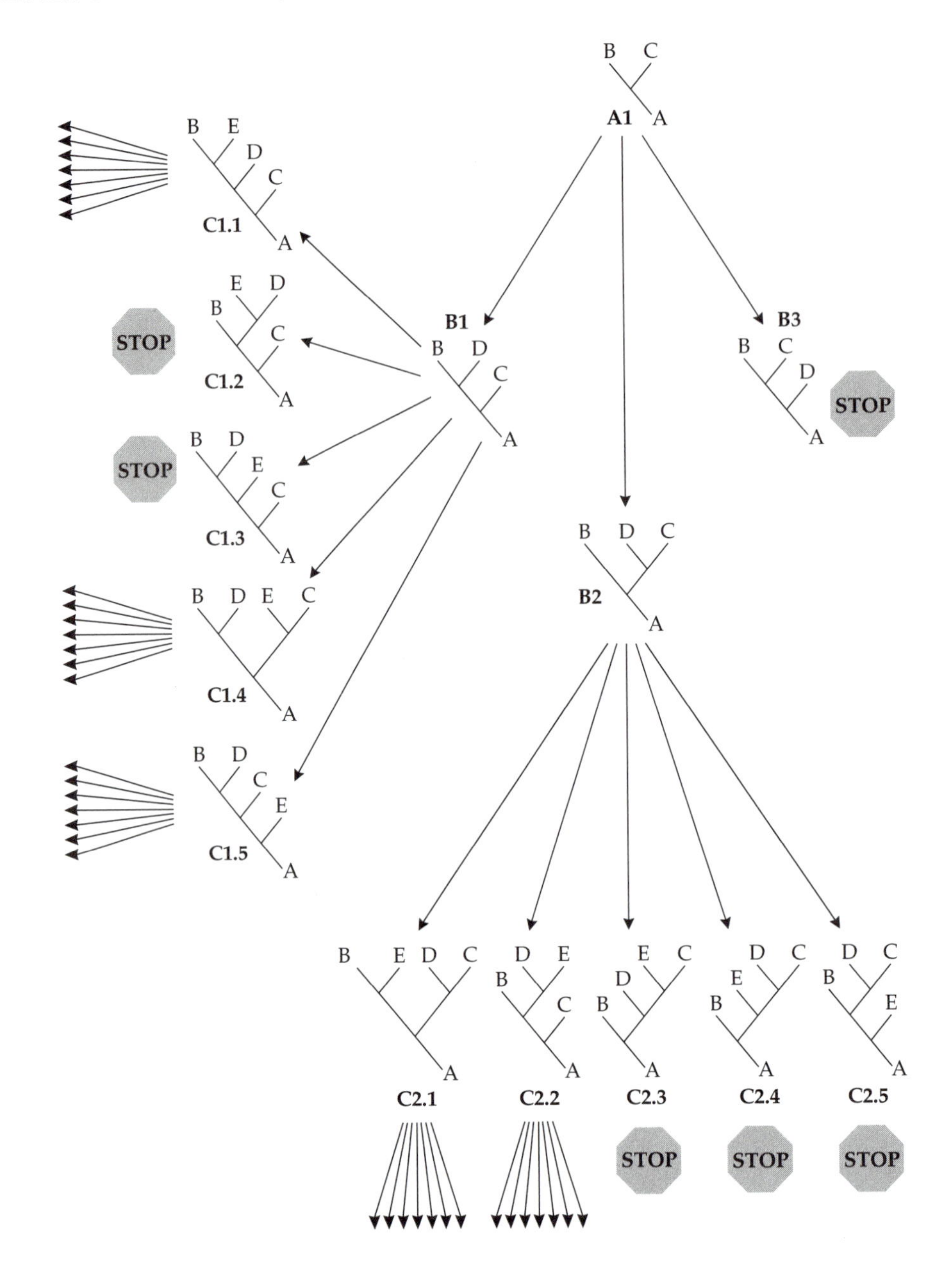

set of similar trees, a new round of exploration is initiated starting from this new tree. This iterative quest is terminated when at a certain round we fail to find a shorter tree within the set of similar ones.

With heuristic searches, there is no guarantee that the maximum parsimony tree will be found. The reason is that the most parsimonious tree may

◀FIGURE 5.17 An illustration of the branch-and-bound search algorithm for the maximum parsimony tree. We start with the only possible unrooted tree for OTUs A, B, and C (tree A1). Addition of OTU D to each of the three branches in tree A1 results in three unrooted trees (B1–B3). The length of each of these three trees is compared to the upper bound value of *L*. Tree B3 was found to be longer than *L*, and branch addition is no longer performed on it (stop sign). In the next step, OTU E is added to each of the five branches of the remaining two trees, B1 and B2, resulting in the formation of 10 trees (C1.1–C1.5 and C2.1–C2.5). Again, each of these trees is compared to the upper bound, and the process of branch addition is only continued for trees shorter than *L*. In the present case, we end up considering only 35 six-taxon trees (terminal arrows) instead of 105 possible ones. Modified from Swofford et al. (1996).

lack sufficient similarity to any of the intermediate trees identified in the iteration, i.e., identifying the maximum parsimony tree would require passing through trees that are longer than the ones already obtained. Nevertheless, it is possible to increase the probability of finding the maximum parsimony tree under certain conditions (Swofford et al. 1996).

There are several methods of **branch swapping** (or rearrangement) that can be used to generate toplogically similar trees from an initial one. In Figure 5.18 we show one such method, called **subtree pruning and regrafting**.

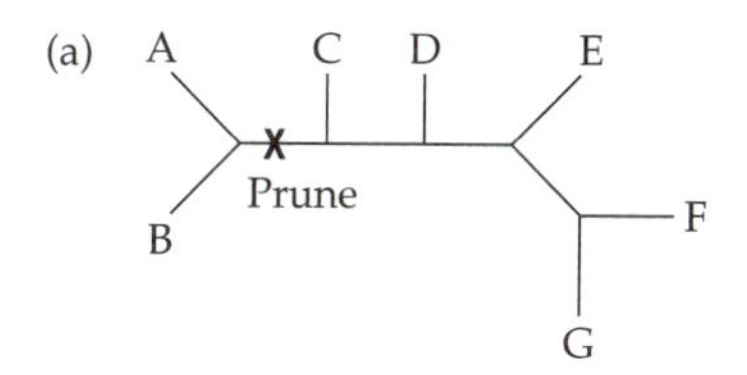

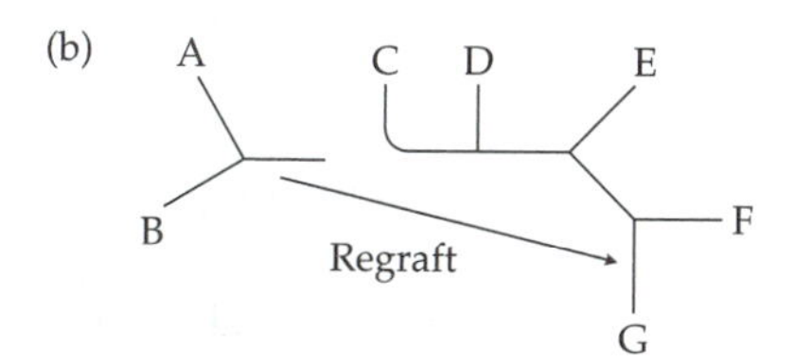

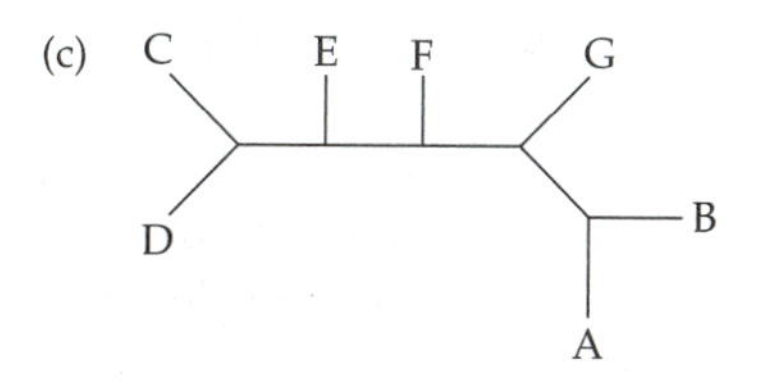

FIGURE 5.18 An example of branch swapping by subtree pruning and regrafting for an unrooted tree with 7 OTUs. (a) The initial tree is pruned. (b) The pruned part is regrafted on the branch leading to OTU G. (c) The resulting rearranged tree. Modified from Swofford et al. (1996).

MAXIMUM LIKELIHOOD METHODS

The first application of a maximum likelihood method to tree reconstruction was made by Cavalli-Sforza and Edwards (1967) for gene frequency data. Later, Felsenstein (1973, 1981) developed maximum likelihood algorithms for amino acid and nucleotide sequence data. Because this approach involves fairly sophisticated statistical theory, we present only some basic principles of the method without any mathematical details. For a more detailed presentation, see Swofford et al. (1996) and Li (1997).

The **likelihood**, L, of a phylogenetic tree is the probability of observing the data (e.g., the nucleotide sequences) under a given tree and a specified model of character state changes (e.g., the substitution pattern). This is usually written as $L = P(\text{data}|\text{tree})$. The aim of maximum likelihood methods is to find the tree (from among all the possible trees) with the highest L value.

The basic principles involved in calculating the likelihood of a tree are shown in Figure 5.19. Figure 5.19a shows a set of aligned nucleotide sequences from four taxa. Let us first evaluate the likelihood of the unrooted tree in Figure 5.19b; that is, What is the probability that this tree could have generated the data in Figure 5.19a under our chosen model of nucleotide substitution? Under the assumption that nucleotide sites evolve independently, we may calculate the likelihood for each site separately and combine the likelihoods at the end.

As an example, let us investigate the likelihood at site 5. At this site, OTUs 1, 2, 3, and 4 have C, C, A, and G, respectively. The unrooted tree in Figure 5.19b has two internal nodes, denoted as 5 and 6, each of which can have one of the four different nucleotides. Thus, we should consider $4 \times 4 = 16$ possibilities (Figure 5.19c). Obviously, some of these possibilities are less plausible than others, but each alternative has a non-zero probability of generating any pattern of observed nucleotides at the four tips of the tree. Therefore, the likelihood of observing the nucleotides that we do observe at site 5 is equal to the sum of 16 independent probabilities (Figure 5.19c). The same procedure is repeated for each site separately, and the likelihood for all the sites is computed as the product of the individual site likelihoods (Figure 5.19d).

For mathematical convenience, the likelihood is usually evaluated by the logarithmic transformation, which transforms multiplication into summation (Figure 5.19e). That is, we consider the **log likelihood** ($\ln L$) of the tree. We then proceed to compute the likelihood values for the other possible trees, and the tree with the highest likelihood value is chosen as the **maximum likelihood tree**.

A critical element missing from the above description is how the probabilities of the various changes are calculated. These probabilities depend on assumptions concerning the process of nucleotide substitution and the branch lengths, which in turn depend on the rate of substitution and the evolutionary time. (Thus, a long branch may either represent a long period of evolutionary time or a high rate of substitution.) We note that the branch lengths are usually unknown and must be estimated as part of the process of computing the likelihood. The methods for finding the branch lengths that maximize the likelihood value usually involve an iterative approach (e.g., Kishino et al. 1990).

Note that since the likelihoods depend on the model of nucleotide substitution, a tree with the largest likelihood value under one substitution model may not be the maximum likelihood tree under another model of nucleotide substitution.

The maximum likelihood method is computationally extremely time-consuming, and so was not used often in the past. With the development of fast computers, the method is now used fairly often, although in its exhaustive version it is still only applicable to a modest number of taxa.

(a)

	1	2	3	4	5	6	7	8	9	...*n*
OTU1	A	A	G	A	C	T	T	C	A	...N
OTU2	A	G	C	C	C	T	T	C	T	...N
OTU3	A	G	A	T	A	T	C	C	A	...N
OTU4	A	G	A	G	G	T	C	C	T	...N

(b)

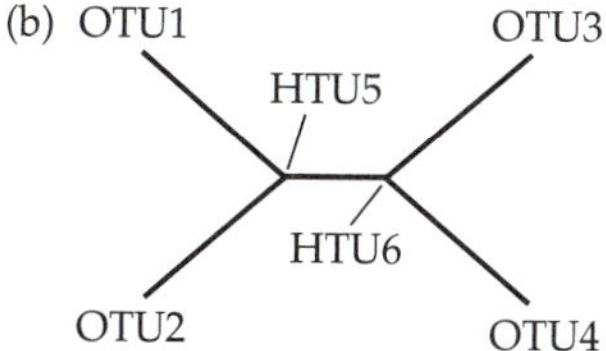

(c)

$L_{(5)}$ = Prob(C, C >A—A< A, G) + Prob(C, C >A—C< A, G) + Prob(C, C >A—T< A, G) + Prob(C, C >A—G< A, G)
+ Prob(C, C >C—A< A, G) + Prob(C, C >C—C< A, G) + Prob(C, C >C—T< A, G) + Prob(C, C >C—G< A, G)
+ Prob(C, C >T—A< A, G) + Prob(C, C >T—C< A, G) + Prob(C, C >T—T< A, G) + Prob(C, C >T—G< A, G)
+ Prob(C, C >G—A< A, G) + Prob(C, C >G—C< A, G) + Prob(C, C >G—T< A, G) + Prob(C, C >G—G< A, G)

(d) $L = L_{(1)} \times L_{(2)} \times L_{(3)} \times ... \times L_{(n)} = \prod_{i=1}^{n} L_{(i)}$

(e) $\ln L = \ln L_{(1)} + \ln L_{(2)} + \ln L_{(3)} + ... + L_{(n)} = \sum_{i=1}^{n} \ln L_{(i)}$

FIGURE 5.19 Schematic representation of the calculation of the likelihood of a tree. (a) Data in the form of sequence alignment of length *n*. (b) One of three possible trees for the four taxa whose sequences are shown in (a). (c) The likelihood of a particular site, in this case site 5, equals the sums of the 16 probabilities of every possible reconstruction of ancestral states at nodes 5 and 6 in (b). (d) The likelihood of the tree in (b) is the product of the individual likelihoods for all *n* sites. (e) The likelihood is usually evaluated by summing the logarithms of the likelihoods at each site, and reported as the log likelihood of the tree. Modified from Swofford et al. (1996).

ROOTING UNROOTED TREES

The majority of tree-making methods yield unrooted trees. To root an unrooted tree, we usually need an outgroup (an OTU for which external information, such as paleontological evidence, clearly indicates that it has branched off earlier than the taxa under study). The root is then placed between the outgroup and the node connecting it to the other OTUs, which are the ingroup.

While we must be certain that the outgroup did indeed diverge prior to the taxa under study, it is not advisable to choose an outgroup that is too distantly related to the ingroup, because in such cases it is difficult to obtain reliable estimates of the distances between the outgroup and the ingroup taxa. For example, in reconstructing the phylogenetic relationships among a group of placental mammals, we may use a marsupial as an outgroup. Birds may serve as reliable outgroups only if the DNA sequences used have been highly conserved in evolution. Plants or fungi would have clearly qualified as outgroups in this example; however, by being only very distantly related to the mammals, their use as outgroups may result in serious topological errors. On the other hand, the outgroup must not be phylogenetically too close to the other OTUs, because then we cannot be certain that it diverged from the ingroup OTUs prior to their divergence from one another.

The use of more than one outgroup generally improves the estimate of the tree topology, provided again that they are not too distant from the in-

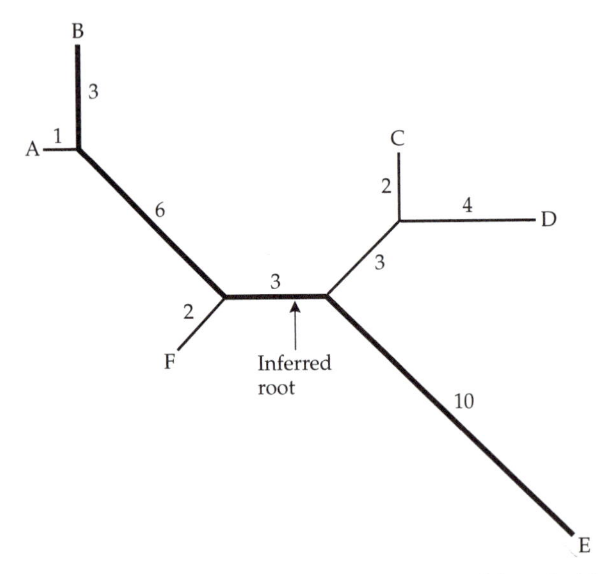

FIGURE 5.20 A hypothetical unrooted phylogenetic tree with scaled branches that has been rooted at the midpoint of the longest pathway (thick line) from among all possible pathways between two OTUs (i.e., B and E). The numbers of substitutions are marked on the branches.

group taxa. If the outgroups are very distant from the ingroup, the use of multiple outgroups may yield worse results than using a single outgroup because of the long branch attraction phenomenon (see page 215).

In the absence of an outgroup, we may position the root by assuming that the rate of evolution has been approximately uniform over all the branches. Under this assumption we put the root at the midpoint of the longest pathway between two OTUs. For example, in the hypothetical unrooted tree in Figure 5.20, the longest path is between OTUs B and E. The length of this path is 3 + 6 + 3 + 10 = 22, so we position the root at a distance of 22/2 = 11 from either B or E.

As mentioned previously, an unrooted tree cannot be said to represent the evolutionary history of divergence among a group of taxa. For example, the unrooted tree in Figure 5.21a can be converted into any of the five rooted trees in Figure 5.21b, depending on the placement of the root. For example, by placing a root in the unrooted tree at the position marked by R_1, OTU B becomes the representative of the most ancient divergence event among the four taxa, whereas by putting the root at R_5, OTU B comes out as representing a relatively recent divergence event. Note that the direction of the evolutionary time arrow changes with the position of the root. For example, by placing the

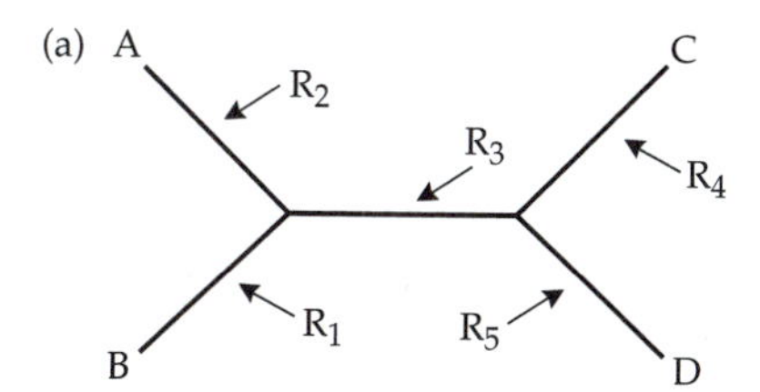

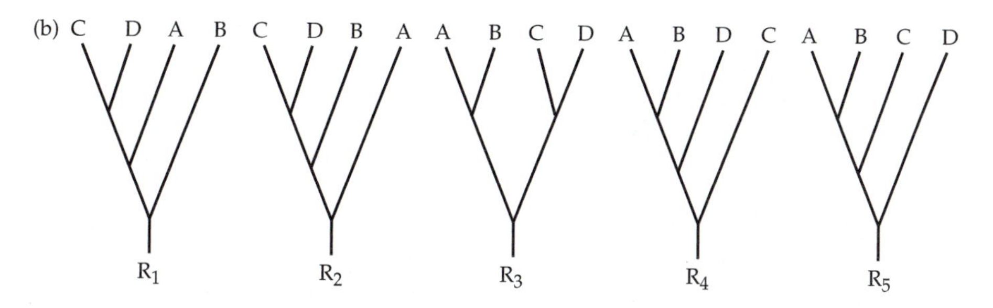

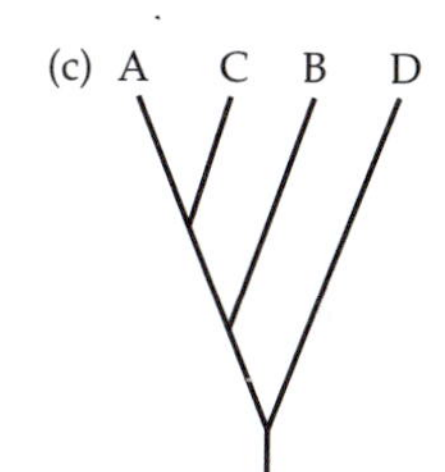

FIGURE 5.21 **An unrooted four-taxon tree (a) can be rooted on each of its branches (R_1–R_5) to obtain five compatible rooted trees (b). The rooted tree in (c) is incompatible with the unrooted tree in (a).**

root at R_1 in Figure 5.21a, we in fact claim that the common ancestor of all the four taxa gave rise to OTU B and the common ancestor of A, C, and D, and thus the existence of the common ancestor of A and B preceded that of the common ancestor of C and D. In comparison, by placing the root at R_5, we obtain the opposite situation, i.e., the existence of the common ancestor of C and D preceded that of the common ancestor of A and B.

We note, however, that unrooted trees are useful in (1) reducing the number of rooted phylogenetic trees that need to be considered in subsequent studies, and (2) answering specific phylogenetic questions concerning the monophyly or paraphyly of certain OTUs. For example, out of 15 possible rooted trees with four OTUs, only 5 rooted trees are compatible with the unrooted tree in Figure 5.21a. The tree in Figure 5.21c, for instance, is incompatible with the unrooted tree in Figure 5.21a. The general rule is that two OTUs cannot form a monophyletic clade in the rooted tree if the path leading from one to the other in the unrooted tree passes through two or more internal nodes. Consequently, even with an unrooted tree we can provide an answer to such questions as whether or not two OTUs constitute a natural clade.

ESTIMATING BRANCH LENGTHS

So far, we have mainly considered the topology of phylogenetic trees. However, branch lengths provide us with useful information about divergence times and rates of evolution.

For UPGMA and maximum likelihood, branch lengths are estimated together with the topology. For the maximum parsimony method, the procedure for inferring the minimum number of substitutions on each branch has been described (see page 192). This procedure tends to underestimate the branch lengths, for it is intended to minimize the number of substitutions. The degree of underestimation may not be severe if the tree contains only short branches; otherwise, the underestimation may be quite serious (see Saitou 1989; Tateno et al. 1994).

In the following, we describe Fitch and Margoliash's (1967) method for estimating branch lengths, assuming that the tree topology has already been inferred by a distance matrix procedure, such as Sattath an Tversky's neighbors-relation method.

First let us consider the simplest case, i.e., an unrooted tree with three OTUs (A, B, and C) and a single node (Figure 5.22a). Let x, y, and z be the lengths of the branches leading to A, B, and C, respectively. It is easy to see that the following equations hold:

$$d_{AB} = x + y \quad \textbf{(5.13a)}$$

$$d_{AC} = x + z \quad \textbf{(5.13b)}$$

$$d_{BC} = y + z \quad \textbf{(5.13c)}$$

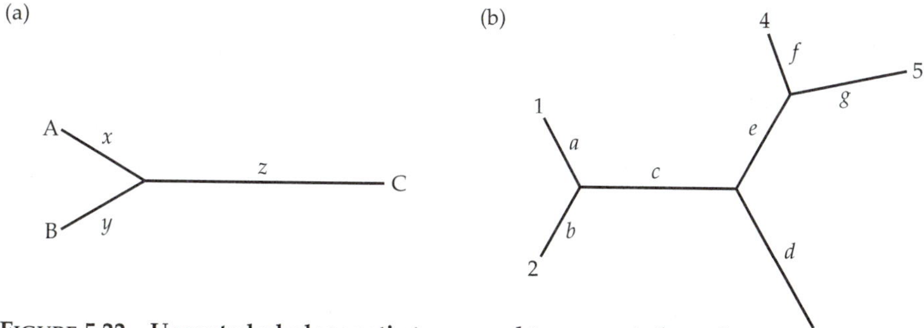

FIGURE 5.22 Unrooted phylogenetic trees used to compute branch lengths by the Fitch and Margoliash's (1967) method. (a) A tree with three OTUs. (b) A tree with five OTUs.

From these equations, we obtain the following solutions:

$$x = \frac{d_{AB} + d_{AC} - d_{BC}}{2} \quad \textbf{(5.14a)}$$

$$y = \frac{d_{AB} + d_{BC} - d_{AC}}{2} \quad \textbf{(5.14b)}$$

$$z = \frac{d_{AC} + d_{BC} - d_{AB}}{2} \quad \textbf{(5.14c)}$$

Let us now deal with the case of more than three OTUs. For simplicity, let us assume that there are five OTUs (1, 2, 3, 4, and 5) and that the topology and the branch lengths are as in Figure 5.22b. Suppose that OTUs 1 and 2 were the first OTUs to be clustered together in the tree reconstruction process. We then use A and B to denote OTUs 1 and 2, respectively, and put all the other OTUs (3, 4, and 5) into a composite OTU denoted as C. By this arrangement, we can apply Equations 5.14a–c to estimate the lengths of the branches leading to A, B, and C, except that now $d_{AC} = d_{1(345)} = (d_{13} + d_{14} + d_{15})/3$, and $d_{BC} = d_{2(345)} = (d_{23} + d_{24} + d_{25})/3$. Then we have $a = x$ and $b = y$. OTUs 1 and 2 are subsequently considered as a single composite OTU. In the next step, suppose that the composite OTU (12) and the simple OTU 3 were the next pair to be joined together. Then we denote OTUs (12) and 3 by A and B, respectively, and put the other OTUs (i.e., 4 and 5) into the new composite OTU C. In the same manner as above, we obtain x, y, and z. Note that $d = y$ and $c + (a + b)/2 = x$. From the values for a and b, which have been obtained previously, we can calculate c. The process is continued until all branch lengths are obtained.

Note that sometimes an estimated branch length can be negative. Since the true length can never be negative, it is better to replace such an estimate by 0.

As an example of using the above method, let us compute the branch lengths of the tree in Figure 5.11c. For convenience, we again present the distance matrix that was used to infer the topology of this tree. To avoid confusion with the notation in Equations 5.13a–c, we rename OTUs A, B, C, and D as OTUs 1, 2, 3, and 4, respectively.

OTU	1	2	3
2	8		
3	7	9	
4	12	14	11

Since OTUs 1 and 2 were clustered first, we first compute the lengths (a and b) of the branches leading to these two OTUs by putting OTUs 3 and 4 into a composite OTU C. We then have $d_{AB} = d_{12} = 8$, $d_{AC} = (d_{13} + d_{14})/2 = (7 + 12)/2 = 9.5$, and $d_{BC} = (d_{23} + d_{24})/2 = 11.5$.

From Equations 5.14a–c, we have $a = x = (8 + 9.5 - 11.5)/2 = 3$, and $b = y = (8 + 11.5 - 9.5)/2 = 5$. Next we treat OTUs 1 and 2 as a single OTU (12) and denote it by A. Since we are left with only three OTUs, we denote OTU 3 by B and OTU 4 by C. We then have $d_{AB} = d_{(12)3} = (d_{13} + d_{23})/2 = (7 + 9)/2 = 8$; $d_{AC} = d_{(12)4} = (d_{14} + d_{24})/2 = (12 + 14)/2 = 13$; and $d_{BC} = d_{34} = 11$. From Equations 5.14a–c we have $x = (8 + 13 - 11)/2 = 5$; $d = y = (8 + 11 - 13)/2 = 3$; and $e = z = (13 + 11 - 8)/2 = 8$. We note from Figure 5.11c that $(a + b)/2 + c = x$, and so $c = 1$. This completes the computation.

Note, however, that since we do not know the exact location of the root, we cannot estimate the length of the branch connecting the root and OTU D but can only estimate the length from the common ancestral node of OTUs A, B, and C through the root to OTU D, i.e., $e = 8$.

ESTIMATING SPECIES DIVERGENCE TIMES

Because the paleontological record is far from complete, we are often ignorant of the dates of divergence between taxa. DNA sequence data can be of great help in this respect. Let us assume that the rate of evolution for a DNA sequence is known from a previous study to be r substitutions per site per year. To obtain the divergence time, T, between species A and B, we compare the sequences from both species and compute the number of substitutions per site, K. As shown in Chapter 4 (Equation 4.1), the rate of substitution is $r = K/2T$. Therefore, T is estimated as

$$T = \frac{K}{2r} \qquad \textbf{(5.15)}$$

As noted in Chapter 4, the rate of nucleotide substitution obtained from one group of organisms may not be applicable to another group. To avoid this problem, we estimate the substitution rate by adding a third species, C, whose divergence time (T_1) from the species pair A and B is known (Figure

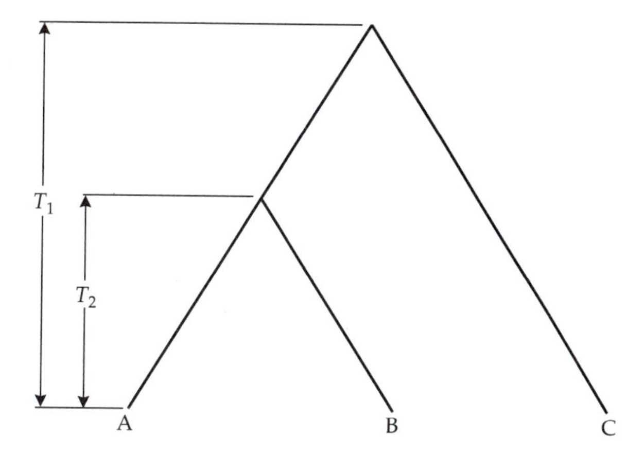

FIGURE 5.23 Model tree for estimating times of divergence. T_1 = divergence time between species C and the ancestor of species A and B. T_2 = divergence time between species A and B.

5.23). Let K_{ij} be the number of nucleotide substitutions per site between species i and j. Then, the rate of nucleotide substitution is estimated by

$$r = \frac{K_{AC} + K_{BC}}{2(2T_1)} \quad \textbf{(5.16)}$$

The unknown divergence time between species A and B (T_2) is estimated by

$$T_2 = \frac{K_{AB}}{2r} = \frac{K_{AB}T_1}{K_{AC} + K_{BC}} \quad \textbf{(5.17)}$$

Conversely, in the case that T_2 is known but T_1 is not, T_1 is given by

$$T_1 = \frac{(K_{AC} + K_{BC})T_2}{2K_{AB}} \quad \textbf{(5.18)}$$

The above formulations assume rate constancy. As discussed in the previous chapter, this assumption often does not hold, and so estimated divergence times should be treated with caution. Methods that can reduce the effects of unequal rates of substitution on divergence time estimates have been proposed (e.g., Li and Tanimura 1987b; Steel 1994; Lockhart et al. 1994; Lake 1994; Takezaki et al. 1995; Sanderson 1997). We also noted earlier that the divergence time between two sequences may pre-date the divergence between the species from which the sequences were obtained. However, this error is usually not very serious if we are concerned with very ancient divergence events. Note also that estimates of divergence time are usually subject to large stochastic errors. To reduce such errors, many sequences should be used in the estimation.

TOPOLOGICAL COMPARISONS

It is sometimes necessary to measure the similarity or dissimilarity among several tree topologies. Such a need may arise when dealing with trees that have been inferred from analyses of different sets of data or from different types of analysis of the same data set. Moreover, several methods of tree reconstruction (maximum parsimony, for example) may produce many trees rather than a unique phylogeny. In such cases, it may be advisable to draw a tree that summarizes the points of agreement among all the trees. When two trees derived from different data sets or different methodologies are identical, they are said to be **congruent**. Congruence can sometimes be partial, i.e., limited to some parts of the trees, other parts being incongruent.

Penny and Hendy's topological distance

A commonly used measure of dissimilarity between two tree topologies is Penny and Hendy's (1985) topological distance. The measure is based on tree partitioning, and is equal to twice the number of different ways of partitioning the OTUs between two trees.

$$d_{\mathrm{T}} = 2c \tag{5.19}$$

where d_{T} is the topological distance and c is the number of partitions resulting in different divisions of the OTUs in the two trees under consideration. (In comparisons between bifurcating trees, d_{T} is always an even integer.)

Consider, for instance, the trees in Figure 5.24. Tree (a) has six OTUs and three internal branches. If we partition this tree at branch 1, we obtain two groups of OTUs: A and B on the one hand, and C, D, E, and F on the other. Cutting tree (b) at branch 1 results in the same partitioning of the six OTUs. Cutting tree (a) at branch 2 results in the same partitioning of OTUs as the cutting of tree (b) at branch 3, i.e., A, B, E, and F on the one hand, and C and D on the other. Cutting tree (a) at branch 3 results in a partition of OTUs that cannot be obtained by cutting tree (b) at any of its three internal branches. Therefore, $d_{\mathrm{T}} = 2 \times 1 = 2$.

In comparing the trees in (a) and (c), we see that none of the partitions in (a) is mirrored in (c). Therefore, d_{T} achieves its maximal possible value, i.e., $d_{\mathrm{T}} = 2 \times 3 = 6$. Hence, we conclude that tree (a) is more similar to tree (b) than to tree (c).

Consensus trees

Trees inferred from the analysis of a particular data set are also called **fundamental trees**, i.e., they summarize the phylogenetic information in a data set. **Consensus trees** are trees that have been derived from a set of trees, i.e., they summarize the phylogenetic information in a set of trees. The purpose of a consensus tree is to summarize several trees as a single tree. For example, maximum parsimony may sometimes produce many equally parsimonious

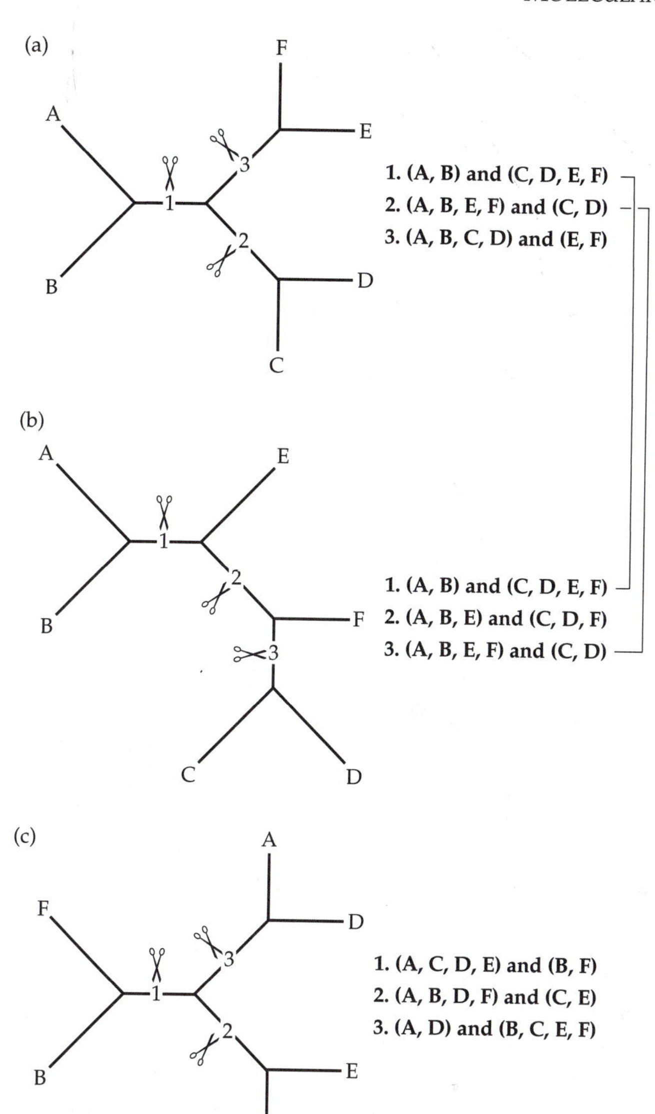

FIGURE 5.24 Measuring similarity between tree topologies by Penny and Hendy's (1985) method. Each tree can be partitioned three different ways by cutting the internal branches (1–3). The resulting partitions are shown on the right. Note that partitions 1 and 2 in tree (a) are identical, respectively, to partitions 1 and 3 in tree (b). There are no identical partitions between trees (a) and (c).

trees rather than a unique solution. In such cases, it is often difficult to present all the trees, and a consensus tree is usually shown.

In consensus trees the points of agreement among the fundamental trees are shown as bifurcations, whereas the points of disagreement are collapsed into

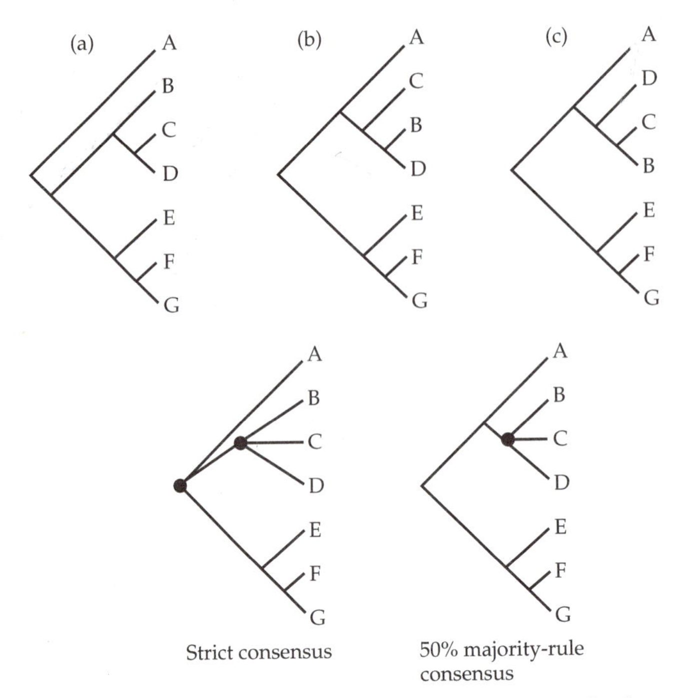

FIGURE 5.25 Three inferred trees (a, b, and c) can be summarized as a strict consensus tree (bottom left) or as a 50% majority-rule consensus tree (bottom right). Multifurcations are indicated by black circles.

polytomies. There are several different types of consensus trees, but the most commonly used are the **strict consensus** and **majority-rule consensus trees**.

Let us assume that we obtained three rooted trees for seven taxa (Figure 5.25). In a strict consensus tree, all conflicting branching patterns are collapsed into multifurcations. We therefore obtain a strict consensus tree that contains two multifurcations.

Among the majority-rule consensus trees, the most commonly used in the literature is the **50% majority-rule consensus tree**. In this tree, a branching pattern that occurs with a frequency of 50% or more is adopted. In the example in Figure 5.25, the position of taxon A relative to taxa B, C, and D is the same in two out of the three rival trees (Figures 5.25b and 5.25c), so this pattern is adopted. This tree, therefore, contains a single multifurcation. It is possible to change the majority-rule percentage to any value; at 100% the result will be identical with the strict consensus tree.

ASSESSING TREE RELIABILITY

Phylogenetic reconstruction is a problem of statistical inference (Edwards and Cavalli-Sforza 1964). Therefore, one should assess the reliability of the inferred phylogeny and its component parts. After inferring a phylogenetic tree, two questions may be asked: (1) How reliable is the tree? or, more particularly, Which parts of the tree are reliable? and (2) Is this tree significantly better than another tree? To answer the first question, we need to assess the reliability of the internal branches of the tree. This can be accomplished by several analytical or resampling methods. In phylogenetic studies, one resampling method, the bootstrap, has become very popular and is discussed in the next section. To answer the second question, we need statistical tests for the difference between two phylogenetic trees; in other words, Is tree A significantly better or worse than tree B, or are the differences within the expectation of random error?

The bootstrap

The **bootstrap** is a computational technique for estimating a statistic for which the underlying distribution is unknown or difficult to derive analytically (Efron 1982). Since its introduction into phylogenetic study by Felsenstein (1985), the bootstrap technique has been frequently used as a means to estimate the confidence level of phylogenetic hypotheses. The statistical properties of this technique in the context of phylogenetics are quite complex, but theoretical studies (e.g., Zharkikh and Li 1992a,b, 1995; Felsenstein and Kishino 1993; Hillis and Bull 1993) have led to a better understanding of the technique. The bootstrap belongs to a class of methods called **resampling techniques** because it estimates the sampling distribution by repeatedly resampling data from the original sample data set.

Figure 5.26a illustrates the bootstrapping procedure in phylogenetics. The data sample consists of five aligned sequences from five OTUs. From these data, a phylogenetic tree was constructed, in this case by the maximum parsimony method. The inferred tree is the null hypothesis to be tested by the bootstrap. Note that this particular null hypothesis consists of two subhypotheses: (1) OTUs 3 and 4 belong to one clade, and (2) OTUs 2 and 5 belong to another (Figure 5.26b).

To estimate the confidence levels of these subhypotheses, we generate a series of n pseudosamples (usually 500–1,000 pseudosamples) by resampling the sites in the sample data with replacement. Sampling with replacement means that a sampled site can be sampled again with the same probability as any other site. Consequently, each pseudosample may contain sites that are represented several times, and sites that are not represented at all. For example, in pseudosample 1 in Figure 5.26a, site 1 is represented four times, while sites 3 and 4 are not represented. Each pseudosample has the same aligned length as the original sample.

Each pseudosample is used to construct a tree by the same method used for the inferred tree. Subhypothesis (1) is given a score of 1 if OTUs 3 and 4 are sister taxa in a bootstrap tree, but a score of 0 otherwise. The score for subhypothesis (2) is similarly decided. The scores for each of the two subhypotheses are added up for all n trees, thus obtaining a **bootstrap value** for each subhypothesis. Bootstrap values are expressed as percentages, and are indicated on the internal branches defining the clades (Figure 5.26b). In our

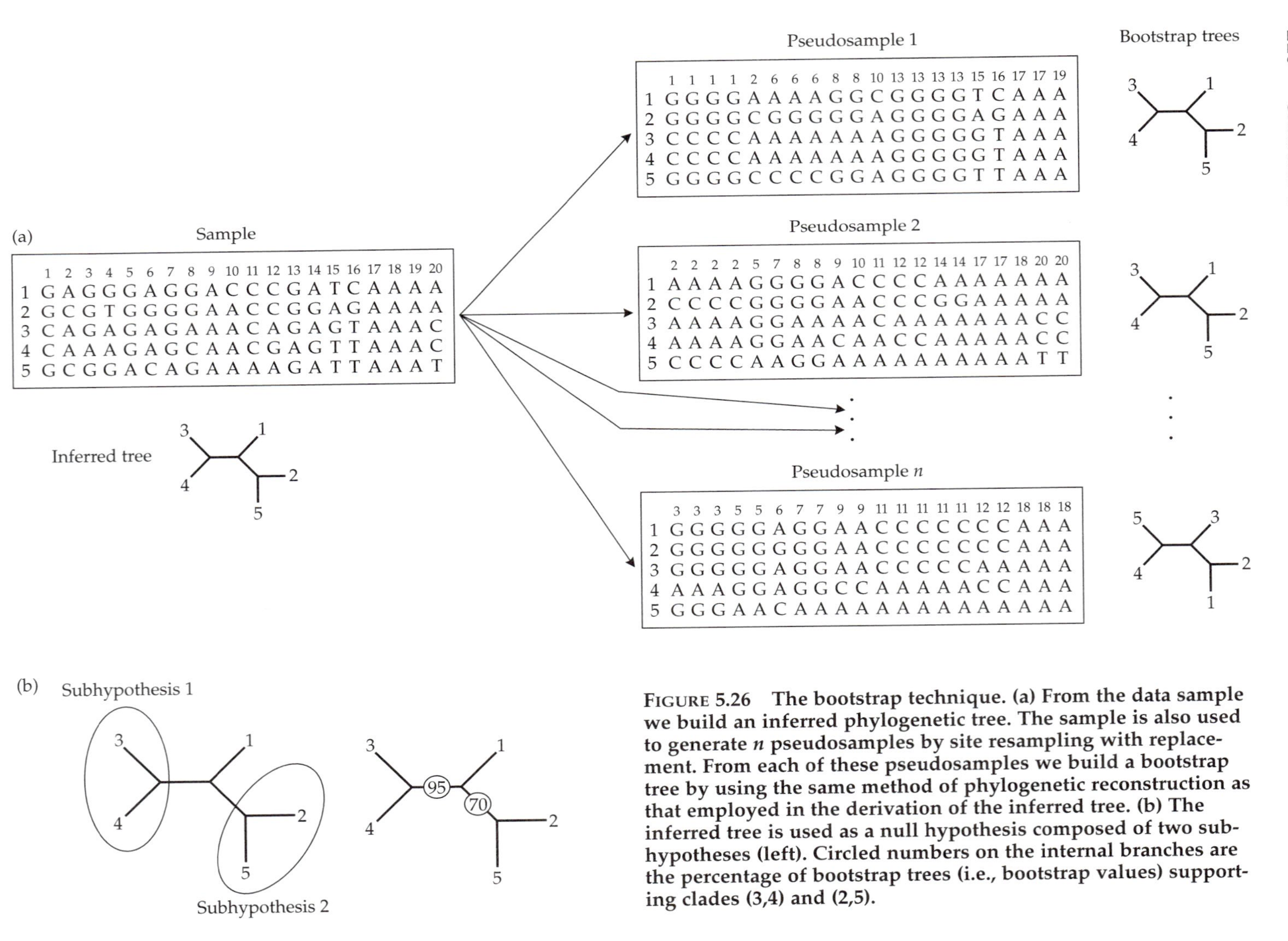

FIGURE 5.26 The bootstrap technique. (a) From the data sample we build an inferred phylogenetic tree. The sample is also used to generate *n* pseudosamples by site resampling with replacement. From each of these pseudosamples we build a bootstrap tree by using the same method of phylogenetic reconstruction as that employed in the derivation of the inferred tree. (b) The inferred tree is used as a null hypothesis composed of two subhypotheses (left). Circled numbers on the internal branches are the percentage of bootstrap trees (i.e., bootstrap values) supporting clades (3,4) and (2,5).

particular example, the clade consisting of OTUs 3 and 4 is supported by 95% of the bootstrap replicates, while the clade consisting of OTUs 2 and 5 is supported by only 70% of the bootstrap replicates.

Bootstrap values are usually interpreted as confidence levels for the clades, although this is not a rigorous practice. Zharkikh and Li (1992a,b) and Hillis and Bull (1993) have shown that bootstrapping tends to underestimate the confidence level at high bootstrap values and overestimate it at low values. Zharkikh and Li (1995) developed a method to correct bootstrap estimation biases.

There are additional difficulties in the interpretation of results obtained by the bootstrap approach (Felsenstein 1985). First, the bootstrap statements are not joint confidence statements: for two clades each supported by a bootstrap value of 95%, we might have lower confidence than $(0.95 \times 0.95) \times 100 = 90\%$ in the statement that both clades are present in the true tree. Second, there is the "multiple tests" problem: if there are 20 clades or more, then, on average, one may obtain statistical significance at the 5% level purely by chance. One way to overcome these difficulties is to rely only on very high bootstrap values—say 95% or higher. In the literature, the bootstrap resampling process is often repeated only 100 times, but this number is too low; at least several hundred pseudosamples should be used, particularly when many species are involved. This can be very time-consuming, especially if a computationally time-consuming method such as maximum parsimony is used.

A common practice in the literature is to "reduce" the inferred tree by collapsing branches that are associated with bootstrap values that are lower than a certain critical value (Figure 5.27). The resulting tree is, of course, multifurcated. By using topological comparisons (see page 206) between simulated "true" trees and inferred ones, it has been shown that collapsed trees are more similar to the true tree than the original inferred tree (Berry and Gascuel 1996).

Tests for two competing trees

Several tests have been devised for testing whether one phylogeny is significantly better than another. Such tests exist for each of the three types of tree reconstruction methods (distance matrix, maximum parsimony, and maximum likelihood). In the following we present a simple test for testing maximum parsimony trees against alternative phylogenies. For the other methods, readers should consult Rzhetsky and Nei (1992), Tateno et al. (1994), and Huelsenbeck and Crandall (1997).

Kishino and Hasegawa (1989) devised a parametric test for comparing two trees under the assumption that all nucleotide sites are independent and equivalent. The test uses the difference in the number of nucleotide substitutions at informative sites between the two trees, D, as a test statistic; where $D = \Sigma D_i$, and D_i is the difference in the minimum number of nucleotide substitutions between the two trees at the ith informative site. The sample variance of D is

$$V(D) = \frac{n}{n-1} \sum_{i=1}^{n} \left(D_i - \frac{1}{n} \sum_{k=1}^{n} D_k \right)^2 \qquad \textbf{(5.20)}$$

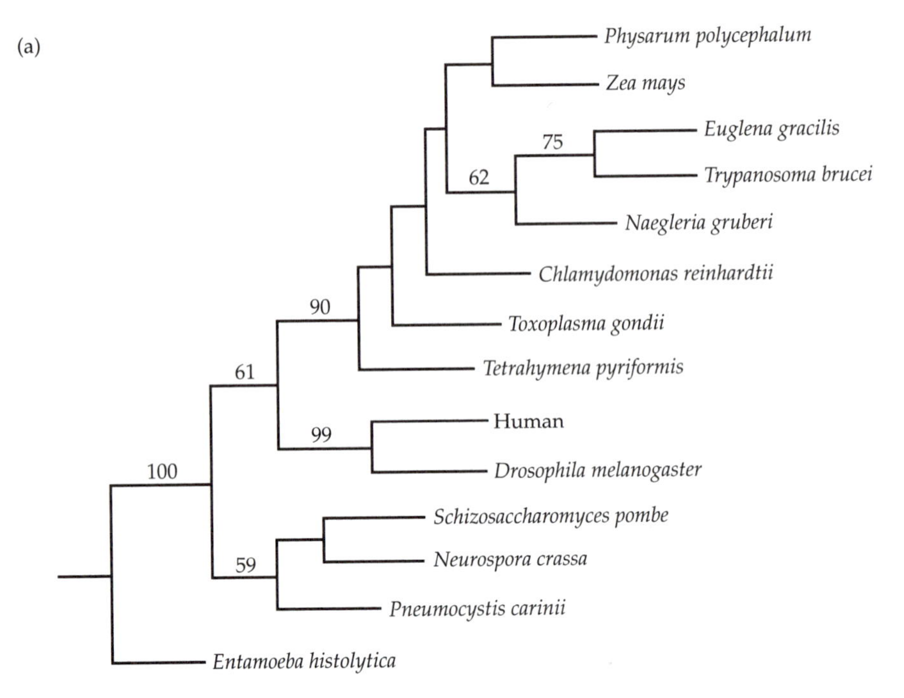

FIGURE 5.27 Reduction of a phylogenetic tree by the collapsing of internal branches associated with bootstrap values that are lower than a certain critical value. (a) Gene tree for α-tubulin sequences (430 amino acid residues) from eukaryotes. Bootstrap values greater than 50% are marked on the relevant internal branches. The tree was rooted with paralogous β-tubulin sequences. (b) Reduced tree in which all branches with bootstrap values lower than 50% were collapsed into polytomies (black circles). (c) Reduced tree in which all branches with bootstrap values lower than 90% were collapsed into polytomies. Data from Edlind et al. (1996).

where n is the number of informative sites. The null hypothesis that $D = 0$ can be tested with the paired t-test with $n - 1$ degrees of freedom, where

$$t = \frac{D / n}{\sqrt{V(D)}\sqrt{n}} \tag{5.21}$$

PROBLEMS ASSOCIATED WITH PHYLOGENETIC RECONSTRUCTION

No method of phylogenetic reconstruction can be claimed to be better than others under all conditions. Each of the methods of phylogenetic reconstruction has advantages and disadvantages, and each method can succeed or fail depending on the nature of the evolutionary process, which is by and large unknown. In the following we will review the strengths and weaknesses of different methods and outline several strategies for minimizing error in phylogenetic analysis.

Strengths and weaknesses of different methods

UPGMA works well only if the rate constancy holds at least approximately. Its main advantage is the high speed of computation. However, speedy algorithms are currently available for other distance matrix methods, and UPGMA is rarely used nowadays, except for pedagogic purposes.

The additive tree methods, including the transformed distance method, the neighbors-relation method, and the neighbor-joining method, are free of systematic error if the distance data satisfy the four-point condition. The performance of these methods, however, depends on the method used to transform the raw character state data into distances. To the extent that the methods used do not compensate adequately for multiple substitutions at a site, the performance of additive tree methods may be compromised. When the distances are small and the sequences used are long, fairly accurate estimates of distances can be obtained, and these methods may perform well even under nonconstant rates of evolution. Indeed, as noted by Saitou and Nei (1987), when the distances are small, one may even use the Hamming distances (the observed uncorrected number of differences between two sequences) and still get the correct tree. Note, however, that if the sequences are short, then the distance estimates are subject to large statistical errors. Moreover, if some distances are large or if the rate varies greatly among sites, then the accurate estimation of distances may be unattainable (Chapter 3). Under any of these situations, the performance of additive tree methods may not be good. An advantage of these methods is that computation time is usually very fast, and they may be used on enormous numbers of OTUs.

Maximum parsimony methods make no explicit assumptions except that a tree that requires fewer substitutions is better than one that requires more. Note that a tree that minimizes the number of substitutions also minimizes the number of homoplasies, i.e., parallel, convergent, and back substitutions (Chapter 3). When the degree of divergence between sequences is small so that homoplasies are rare, the parsimony criterion usually works well. However, when the degree of divergence is large so that homoplasies are common, maximum parsimony methods may yield faulty phylogenetic inferences. In particular, if some sequences have evolved much faster than others, homoplasies are likely to have occurred more often among the branches leading to these sequences than among others, and parsimony may result in erroneous trees. In other words, maximum parsimony methods may perform poorly whenever some branches of the tree are much longer than the other branches, because parsimony will tend to cluster the long branches together (Felsenstein 1978). This phenomenon is called **long-branch attraction** or the **Felsenstein zone** (Figure 5.28). Note also that the chance of homoplasy depends on the substitution pattern. For example, if transitions occur more often than transversions, then the chance of homoplasy will be higher than that for the case of equal substitution rates among the four nucleotides. Some of these effects can be remedied by using weighted parsimony (e.g., Swofford 1993), in which the transitional bias is taken into account.

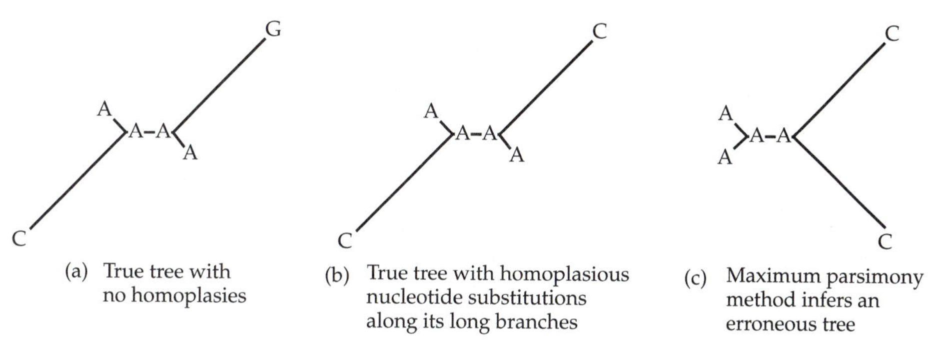

FIGURE 5.28 The long-branch attraction phenomenon. (a) The true unrooted tree has two long branches, each neighboring a short branch. The letters represent the nucleotides at the terminal and internal nodes. On the short branches, we assume that the probability of a nucleotide substitution is very small, so that the nucleotides at the tips of the short branches are likely to retain the same character state as that of the ancestral node. In contrast, on the long branches nucleotide substitutions are likely to occur with a high probability. If the nucleotide substitutions on the long branches are not homoplasious, then by using maximum parsimony we will obtain the correct tree. (b) By chance, however, a site may experience homoplasious nucleotide substitutions along the two long branches. As a consequence, the maximum parsimony method will yield an erroneous tree (c), in which the long branches are inferred to be neighbors. The reason for this error is that the correct tree (b) requires two nucleotide substitutions, whereas the erroneous tree (c) requires only a single nucleotide substitution.

It has been argued that character state methods are more powerful than distance methods because the raw data is a string of character states (e.g., the nucleotide sequence), and in transforming character state data into distance matrices some information is lost. We note, however, that while the maximum parsimony method indeed uses the raw data, it usually uses only a small fraction of the available data. For instance, in the example on page 190, only three sites are used while six sites are excluded from the analysis. For this reason, this method is often less efficient in identifying clades than some distance matrix methods. Of course, if the number of informative sites is large, the maximum parsimony method is generally very effective.

In maximum parsimony methods, we are required to compare all possible trees. This comparison is feasible only when the number of OTUs is small and the sequences under study are not too long. For example, for ten OTUs there are more than 2 million possible unrooted trees to be considered (Table 5.1), and the computer time required becomes very large if the sequences are long. Thus, when the number of OTUs is large, an exhaustive search may no longer be feasible and a heuristic search must be employed. Unfortunately, the heuristic approaches do not guarantee obtaining the maximum parsimony tree.

The maximum likelihood method uses the character state information at all sites, and thus can be said to use the "full" information; however, it requires explicit assumptions on the rate and pattern of nucleotide substitution.

It has been commonly believed that this method is relatively insensitive to violations of assumptions, but a simulation study (Tateno et al. 1994) suggests that the method may not be very robust. In other words, likelihood methods may perform poorly if the stochastic model used is unrealistic and if some sequences are highly divergent. The main disadvantage of the maximum likelihood method is that its computation is very tedious and time-consuming. As in maximum parsimony, the maximum likelihood method requires consideration of all the possible alternative trees, and for each tree it searches for the maximum likelihood value (Kuhner and Felsenstein 1994). Thus, when the number of OTUs is large, it uses heuristic approaches to reduce the number of trees to be considered (e.g., Felsenstein 1981; Saitou 1988). In such cases, we may not obtain the maximum likelihood tree.

Minimizing error in phylogenetic analysis

Several strategies are available to minimize random and systematic errors in phylogenetic analysis. However, it is not always possible to identify the potential sources of error or bias. In the following, we list several do's and don'ts that may increase our chances of recovering the true phylogenetic tree.

The best way to minimize random errors is to use large amounts of data. All other things being equal, a tree based on large amounts of molecular data is almost invariably more reliable than one based on a more limited amount of data. When the sequences do not provide sufficient phylogenetic information (e.g., because they are too short or lacking in variation), no phylogenetic method will produce a sensible result. This said, one should only include reliable data in the analysis. By that we mean that the analysis should be limited to sequences that have been reliably determined, and for which positional homology is certain. (We note, however, that elimination of data deemed "unreliable" may be subjective and arbitrary.) Moreover, we should only use sequences that evolve at an appropriate rate for the phylogenetic question under investigation. Fast-evolving sequences (or parts of sequences, such as third codon positions) should be used for questions regarding close phylogenetic relationships, and slowly evolving sequences should be used for distant phylogenetic relationships. Choosing incorrectly may result in lack of phylogenetic information in the case of slowly evolving sequences, or saturation effects in the case of fast-evolving sequences.

One way to reduce the chance of systematic error leading to inconsistency—i.e., resulting in a wrong inference even when the amount of data is large—is to use more realistic models or more suitable methods of analysis to better match the data. For example, base-composition biases are known to have a pronounced effect on phylogenetic reconstruction, and most methods will incorrectly group OTUs with similar base composition. Several additive distances, e.g., the log-determinant or paralinear distances (Steel 1994; Lockhart et al. 1994; Lake 1994; Galtier and Gouy 1995, 1998), are quite robust to base composition variability among the taxa under study.

Sometimes it is worth examining the assumption of independent evolution among sites. For instance, when dealing with DNA or RNA sequences that tend to form internal hairpin structures, nucleotide changes on one side of the stem are frequently matched by complementary changes on the other side of the stem. In these cases, we may count two such changes as one.

Phylogenetic studies often use sequence data from different DNA regions. If all the regions studied have similar rates of nucleotide substitution, then all the data can be combined into a single set. However, if considerable variation in rates exists, regions with different rates should be analyzed separately, particularly when the distance matrix approach is used. In this case, however, it may be difficult to combine the results from the different data sets and to assess the reliability of the clades within the consensus tree.

Phylogenetic errors are expected to be worse with larger distances among the OTUs than with smaller distances. Therefore, having many long distances will tend to compound the problems arising from the long-branch attraction phenomenon, and it is advisable to remove long branches from the analysis.

For a character to be useful in a phylogenetic context, it should be informative and reliable. That is, it should provide us with true evolutionary information. Some characters are both informative and reliable. Others are reliable, but they do not tell us anything useful about the phylogenetic relationships of interest and are thus uninformative. The third category, consisting of "misinformative" characters, is the most problematic. Identification of such unreliable characters is of crucial importance. For example, we know that maximum parsimony methods yield wrong phylogenies when there are many homoplasies in the data. Since rapidly evolving characters are likely to result in homoplasies more often than slowly evolving characters, it is advisable to give such characters a lower weight in the analysis. One extreme form of weighting is the elimination of such characters, for example the elimination of all transitions in transversion parsimony.

Lastly, we should realize that inferred trees often contain errors regardless of the precautions taken.

MOLECULAR PHYLOGENETIC EXAMPLES

The application of molecular biology techniques and advances in tree reconstruction methodology have led to tremendous progress in phylogenetic studies, resulting in a better understanding of the evolutionary history of almost every taxonomic group. In this section we present several examples where molecular studies have (1) resolved a longstanding issue, (2) led to a drastic revision of the traditional view, or (3) pointed to a new direction of research. The field of molecular phylogenetics is progressing rapidly, however, and some of the views presented here may eventually be revised.

Phylogeny of humans and apes

The issue of the closest living evolutionary relative of humans has always intrigued biologists. Darwin, for instance, claimed that the African apes, the chimpanzee (*Pan*) and the gorilla (*Gorilla*), are our closest relatives, and hence he suggested that the evolutionary origins of humans were to be found in Africa (Darwin 1871). Darwin's view fell into disfavor for various reasons, and for a long time taxonomists believed that the genus *Homo* was only distantly related to the extant apes and, thus, *Homo* was assigned to a family of its own, Hominidae. Chimpanzees, gorillas, and orangutans (*Pongo*), on the other hand, were usually placed in a separate family, the Pongidae (Figure 5.29a). The gibbons (*Hylobates*) were classified either separately (Hylobatidae) or with the Pongidae (Figure 5.29b; see Simpson 1961). Goodman (1963) correctly recognized that this systematic arrangement is anthropocentric in presupposing that humans represent "a new grade of phylogenetic development, one which is 'higher' than the pongids and all other preceding grades." Indeed, placing the various apes into one family and humans into another implies that the apes share a more recent common ancestry with one another than with humans. When *Homo* was put in the same clade with an extant ape, it was usually with the Asian orangutan (Figure 5.29c; Schultz 1963; Schwartz 1984).

By using a serological precipitation method, Goodman (1962) was able to demonstrate that humans, chimpanzees, and gorillas constitute a natural

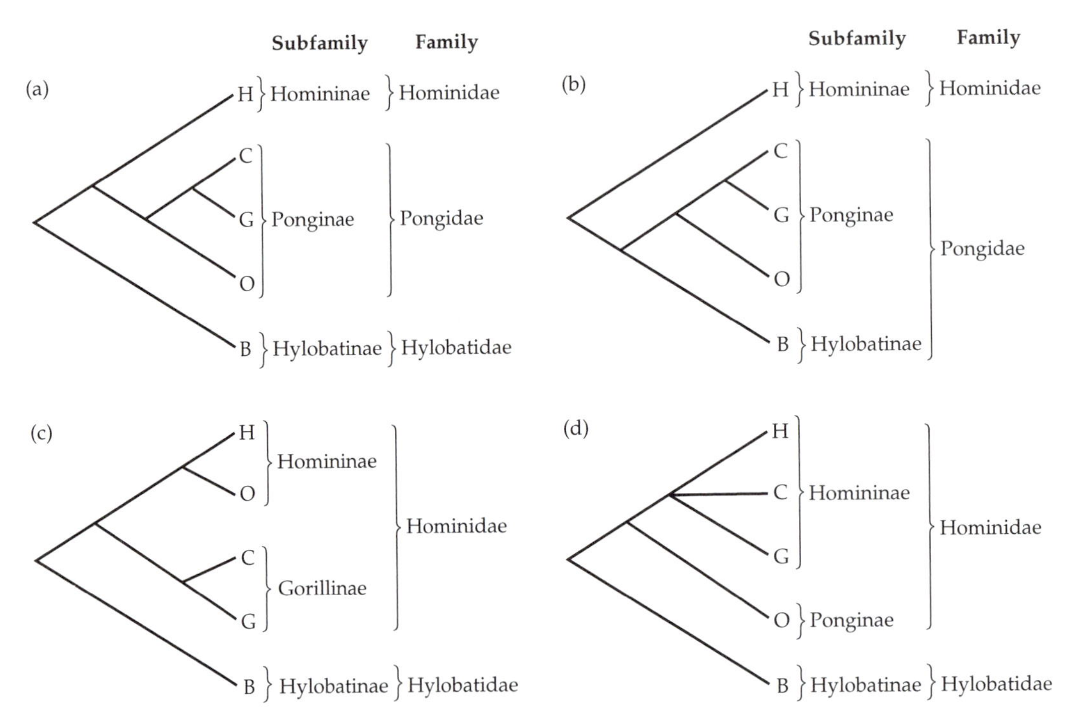

FIGURE 5.29 Four alternative phylogenies and classifications of extant apes and humans (Hominoidae). Traditional classifications setting humans apart are shown in (a) and (b). The clustering of humans with the orangutan is shown in (c). Cumulative molecular as well as morphological evidence favors the classification in (d). Species abbreviations: H, human (*Homo*); C, chimpanzee (*Pan*); G, gorilla (*Gorilla*); O, orangutan (*Pongo*); and B, gibbon (*Hylobates*).

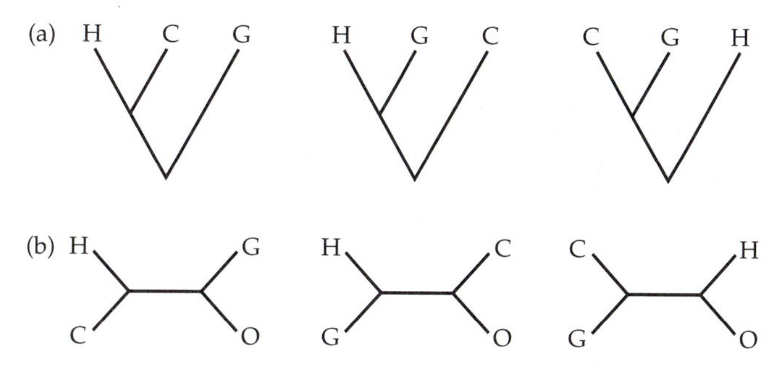

FIGURE 5.30 (a) Three possible rooted trees for humans, chimpanzees, and gorillas. (b) Comparable unrooted tree with the orangutan as an outgroup. Species abbreviations: H, human (*Homo sapiens*); C, chimpanzee (*Pan troglodytes*); G, gorilla (*Gorilla gorilla*); and O, orangutan (*Pongo pygmaeus*).

clade (Figure 5.29d), with orangutans and gibbons having diverged from the other apes at much earlier dates. From microcomplement fixation data, Sarich and Wilson (1967) estimated the divergence time between humans and chimpanzees or gorillas to be as recent as 5 million years ago, rather than a minimum date of 15 million years ago, as was commonly accepted by paleontologists at that time.

However, serological, electrophoretical, and DNA–DNA hybridization studies, as well as amino acid sequences, could not resolve the evolutionary relationships among humans and the African apes, and the so-called human–gorilla–chimpanzee trichotomy remained unsolved and continued to be an extremely controversial issue (Figure 5.30), with some data favoring a chimpanzee–gorilla clade (Ferris et al. 1981; Brown et al. 1982; Hixon and Brown 1986), and others supporting a human–chimpanzee clade (Sibley and Ahlquist 1984; Caccone and Powell 1989; Sibley et al. 1990).

In the following, we shall use the DNA sequence data from Miyamoto et al. (1987) and Maeda et al. (1988) to show that the molecular evidence supports the human–chimpanzee clade and, at the same time, to illustrate some of the tree-making methods discussed in the previous sections.

Table 5.2 shows the number of nucleotide substitutions per 100 sites between each pair of the following OTUs: humans (H), chimpanzees (C), gorillas (C), orangutans (O) and rhesus monkeys (R). Let us first apply UPGMA to these distances. The distance between humans and chimpanzees is the shortest ($d_{HC} = 1.45$). Therefore, we join these two OTUs first, and place the node at $1.45/2 = 0.73$ (Figure 5.31a). We then compute the distances between the composite OTU (HC) and each of the other species, and obtain a new distance matrix:

OTU	HC	G	O
G	1.54		
O	2.96	3.04	
R	7.53	7.39	7.10

TABLE 5.2 Mean (below diagonal) and standard error (above diagonal) of the number of nucleotide substitutions per 100 sites between OTUs[a]

	OTU				
OTU	**Human**	**Chimpanzee**	**Gorilla**	**Orangutan**	**Rhesus monkey**
Human		0.17	0.18	0.25	0.41
Chimpanzee	1.45		0.18	0.25	0.42
Gorilla	1.51	1.57		0.26	0.41
Orangutan	2.98	2.94	3.04		0.40
Rhesus monkey	7.51	7.55	7.39	7.10	

From Li et al. (1987b).

[a]The sequence data used are 5.3 Kb of noncoding DNA, which is made up of two separate regions: (1) the η-globin locus (2.2 Kb) described by Koop et al. (1986b) and (2) 3.1 Kb of the η-δ globin intergenic region sequenced by Maeda et al. (1983, 1988).

Since (HC) and G are now separated by the shortest distance, they are the next to be joined together, and the connecting node is placed at $1.54/2 = 0.77$. Continuing the process, we obtain the tree in Figure 5.31a. We note that the estimated branching node for H and C is very close to that for (HC) and G. In fact, the distance between the two nodes is smaller than all the standard errors for the estimates of the pairwise distances among H, C, and G (Table 5.2). Thus, although the data suggest that our closest living relatives are the chimpanzees, the data do not provide a conclusive resolution of the branching order. The position of the orangutan, however, as an outgroup to the human-chimpanzee-gorilla clade is unequivocal.

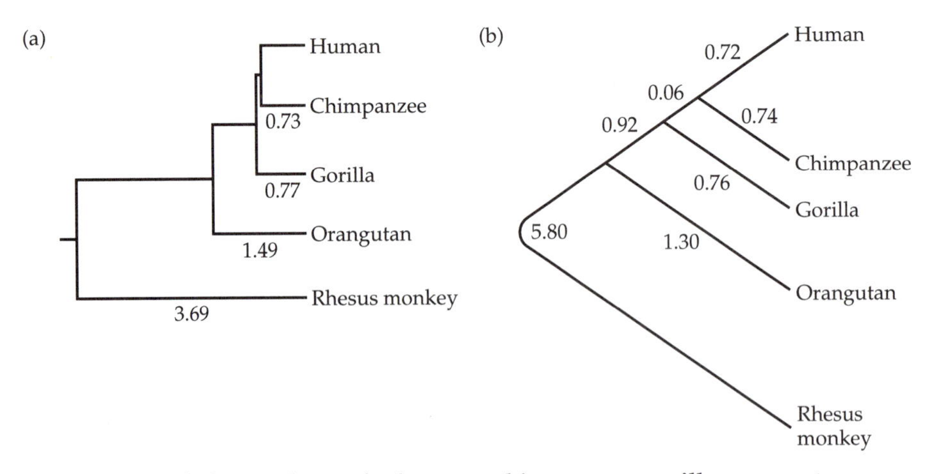

FIGURE 5.31 Phylogenetic tree for humans, chimpanzees, gorillas, orangutan, and rhesus monkeys inferred from UPGMA (a) and from Sattath and Tversky's neighbors-relation method (b).

TABLE 5.3 Neighbors-relation scores obtained from the distance matrix in Table 5.1

OTUs compared[a]	Sum of pairwise distances	Neighbor pairs chosen
H,C,G,O	$d_{HC} + d_{GO} = 4.49$	(HG), (CO)
	$d_{HG} + d_{CO} = 4.45$	
	$d_{HO} + d_{CG} = 4.55$	
H,C,G,R	$d_{HC} + d_{GR} = 8.84$	(HC), (GR)
	$d_{HG} + d_{CR} = 9.06$	
	$d_{HR} + d_{CG} = 9.08$	
H,C,O,R	$d_{HC} + d_{OR} = 8.55$	(HC), (OR)
	$d_{HO} + d_{CR} = 10.53$	
	$d_{HR} + d_{CO} = 10.45$	
H,G,O,R	$d_{HG} + d_{OR} = 8.61$	(HG), (OR)
	$d_{HO} + d_{GR} = 10.37$	
	$d_{HR} + d_{GO} = 10.55$	
C,G,O,R	$d_{CG} + d_{OR} = 8.67$	(CG), (OR)
	$d_{CO} + d_{GR} = 10.33$	
	$d_{CR} + d_{GO} = 11.59$	

Total scores: (HC) = 2, (HG) = 2, (HO) = 1, (HR) = 0, (CG) = 1, (CO) = 1, (CR) = 1, (GO) = 1, (GR) = 1, (OR) = 3

[a]H, human; C, chimpanzee; G, gorilla; O, orangutan; R, rhesus monkey.

Next, we use Sattath and Tversky's neighbors-relation method (see page 186). We consider four OTUs at one time. Since there are five OTUs, there are $5!/[4!(5-4)!] = 5$ possible quadruples. We start with OTUs H, C, G, and O and compute the following sums of distances (data from Table 5.2): $d_{HC} + d_{GO} = 1.45 + 3.04 = 4.49$, $d_{HG} + d_{CO} = 4.45$, and $d_{HO} + d_{CG} = 4.55$. Since the second sum is the smallest, we choose H and G as one pair of neighbors and C and O as the other. Similarly, we consider the four other possible quadruples; the results are shown in Table 5.3. Noting from the bottom of that table that (OR) has the highest neighbors-relation score among all neighbor pairs, we choose (OR) as the first pair of neighbors. Treating this pair as a single composite OTU, we obtain the following new distance matrix:

OTU	H	C	G
C	1.45		
G	1.51	1.57	
(OR)	5.25	5.25	5.22

As only four OTUs are left, it is easy to see that $d_{HC} + d_{G(OR)} = 6.67 < d_{HG} + d_{C(OR)} = 6.76 < d_{H(OR)} + d_{CG} = 6.82$. Therefore, we choose H and C as one pair of neighbors, and G and (OR) as the other. The final tree obtained by this method is shown in Figure 5.31b. The topology of this tree is identical to that in Figure 5.31a. Note, however, that in this method O and R rather than H and C were the first pair to be joined to each other. This is because, in an unrooted tree, O and R are in fact neighbors. The branch lengths in Figure 5.31b were obtained by the Fitch-Margoliash method (see page 202). By using the neighbor-joining method we obtain exactly the same tree as that obtained by the neighbors-relation method (Figure 5.31b).

Finally, let us consider the maximum parsimony method. For simplicity, let us consider only humans, chimpanzees, gorillas, and orangutans (Figure 5.30b). Table 5.4 shows the informative sites for the 10.2-Kb region including the η-globin pseudogene and its surrounding regions (Koop et al. 1986a; Miyamoto et al. 1987; Maeda et al. 1988; Bailey et al. 1991). For each site, the hypothesis supported is given in the last column. If we consider nucleotide substitutions only, there are 15 informative sites, of which eight support the human–chimpanzee clade (hypothesis I), four support the chimpanzee–gorilla clade (hypothesis II), and three support the human–gorilla clade (hypothesis III). Moreover, there are four informative sites involving a gap, and they all support the human–chimpanzee clade. Therefore, the human–chimpanzee clade is chosen as the best representation of the true phylogeny. In another analysis with more sequence data, Williams and Goodman (1989) showed that the support for the human–chimpanzee clade is statistically significant at the 1% level.

The clustering of humans and chimpanzees in one clade, however, is not supported by the involucrin gene, which favors instead the chimpanzee–gorilla clade (Djian and Green 1989), and by the Y-linked *RPS4Y* locus, which favors the human–gorilla clade (Samollow et al. 1996). However, given the well-known phenomenon of possible incongruence between gene trees and species trees (see page 173), an agreement among all gene trees is not expected. The overall molecular evidence is now strongly and significantly in favor of the human–chimpanzee clade. In addition to the 10.2-Kb sequence data discussed above, this clade is supported by extensive DNA–DNA hybridization data (Sibley and Ahlquist 1987; Caccone and Powell 1989), by two-dimensional protein electrophoresis data (Goldman et al. 1987), restriction site variation in the spacers of the genes specifying ribosomal RNA (Suzuki et al. 1994), and especially by extensive mitochondrial DNA (Ruvolo et al. 1991; Horai et al. 1992; Árnason et al. 1996) and nuclear DNA sequence data (Bailey et al. 1991; Ruvolo 1997).

Thus, the closest extant relatives of humans are the two chimpanzee species (*Pan troglodytes* and *P. paniscus*), followed in order of decreasing relatedness by gorillas, orangutans, a clade consisting of gibbons and siamangs, Old World monkeys (Catarrhini), and New World monkeys (Platyrrhini). In contrast to the molecular data, morphological and physiological comparisons among apes usually support the clustering of chimpanzee and gorilla into a monophyletic clade to the exclusion of humans (Tuttle 1967; Ciochon 1985; Andrews 1987). Recently, however, morphological, anatomical, and parasito-

TABLE 5.4 Informative sites among human, chimpanzee, gorilla, and orangutan sequences

	Sequence				
Site[a]	Human	Chimpanzee	Gorilla	Orangutan	Hypothesis supported[b]
Data from Miyamoto et al. (1987)					
34	A	G	A	G	III
560	C	C	A	A	I
1287	*[c]	*	T	T	I
1338	G	G	A	A	I
3057–3060	****	****	TAAT	TAAT	I
3272	T	T	*	*	I
4473	C	C	T	T	I
5153	A	C	C	A	II
5156	A	G	G	A	II
5480	G	G	T	T	I
6368	C	T	C	T	III
6808	C	T	T	C	II
6971	G	G	T	T	I
Data from Maeda et al. (1988)					
127–132	******	******	AATATA	AATATA	I
1472	G	G	A	A	I
2131	A	A	G	G	I
2224	A	G	A	G	III
2341	G	C	G	C	III
2635	G	G	A	A	I

Modified from Williams and Goodman (1989).

[a]Site numbers correspond to those given in the original sources. The total length of the sequence used is 10.2 Kb, about twice that used in Table 5.2.

[b]Hypotheses: I, human and chimpanzee are sister taxa; II, chimpanzee and gorilla are sister taxa; and III, human and gorilla are sister taxa.

[c]Each asterisk denotes a one-nucleotide indel at the site.

logical data are beginning to accumulate in favor of the so-called Troglodytian hypothesis, i.e., that *Pan* and *Homo* are sister taxa (Shoshani et al. 1996; Thirnanagama et al. 1991; Retana Salazar 1996).

Surprisingly, the phylogenetic and taxonomic picture that emerges from the molecular data (Goodman 1999) is quite close to the one suggested by the father of taxonomy, Carolus Linnaeus, who wrote: "As a naturalist, and following naturalistic methods, I have not been able to discover up to the present a single character which distinguishes Man from the anthropomorphs, since they comprise specimens … that resemble the human species … to such

TABLE 5.5 Estimates of times of divergence between humans and other Old World primates

	Divergence time estimates[a]	
Species pair	**Hasegawa et al. (1987) (η-globin pseudogene, 2,040 nucleotides)[b]**	**Árnason et al. (1996) (11 mitochondrial proteins, 3,033 amino acids)[c]**
Human–rhesus monkey	25.3 ± 2.4	Not available
Human–gibbon	Not available	36.1 ± 4.1
Human–orangutan	11.9 ± 1.7	24.4 ± 2.7
Human–gorilla	5.9 ± 1.2	18.0 ± 2.9
Human–chimpanzee	4.9 ± 1.2	13.7 ± 2.5

[a]In million years ± standard error

[b]Calibrated from the Catarrhini–Platyrrhini divergence = 38 million years ago.

[c]Calibrated from the bovine–cetacean divergence = 60 million years ago.

a degree that an inexpert traveler may consider them varieties of Men" (Linnaeus 1758, translated in Chiarelli 1973). In fact, Linnaeus assigned humans, chimpanzees, and orangutans to the same genus, and their original scientific names were *Homo sapiens*, *H. troglodytes*, and *H. sylvestris*, respectively. (The gorilla was only discovered in 1799 and described in 1847.) The orangutan and the chimpanzee were removed from the genus *Homo* by Lacepede in 1799 and by Oken in 1816, respectively. In light of the molecular decipherment of the phylogeny of humans and the great apes, isn't it time to abide by the rules of precedence in taxonomy and restore Linnaeus' terminology?

In the literature, there are many estimates of the divergence times between humans, on the one hand, and various species of apes and monkeys, on the other. These estimates differ from one another considerably. For example, by using sequence comparisons involving the η-globin pseudogene locus (2,040 nucleotides), and by setting the date of divergence between the Old World monkeys (Catarrhini) and the New World monkeys (Platyrrhini) at 38 million years ago, Hasegawa et al. (1987) estimated the divergence time between humans and chimpanzees to be about 5 million years ago (Table 5.5). In contrast, by using amino acid sequence comparisons of eleven mitochondrial proteins, and by using as reference the time of divergence between artiodactyls and cetaceans (60 million years ago), Árnason et al. (1996) estimated the time of divergence between humans and chimpanzees to be almost three times as large (Table 5.5). There are several possible reasons for the large difference between the two estimates. First the assumption of a constant rate (a molecular clock) may not hold. Second, the reference dates for calibration may not be accurate. Third, each estimate is subject to stochastic errors. This example shows that divergence date estimates should be taken with extreme caution.

Cetartiodactyla and SINE phylogeny

The more than 80 species of whales, dolphins, and porpoises, which form the order Cetacea, are among the most fascinating and spectacular of all placental mammals (eutherians). They possess an elaborate communication system indicative of an advanced social structure, and the physical bulk of some cetaceans far exceeds that of the largest dinosaurs. The origin of Cetacea has been an enduring evolutionary mystery since Aristotle, for the transition from terrestriality to an exclusive aquatic lifestyle required an unprecedented number of unique yet coordinated changes in many biological systems. For example, living cetaceans are unique among mammals in completely lacking external hindlimbs and in swimming by dorsoventral oscillations of a heavily muscled tail. In the context of phylogenetics, these unique morphological, anatomical, and behavioral traits constitute autapomorphies for the Cetacea, and cannot be used to determine the phylogenetic affinities of this order within the eutherian tree.

A link between cetaceans and ungulates (hoofed mammals) was suggested more than a century ago by Flower (1883) and Flower and Garson (1884) on the basis of comparative anatomical information. This view was accepted by Gregory (1910), but two of the most influential paleontologists of the century, Simpson (1945) and Romer (1966), suggested that the cetacean lineage goes back to the very root of the eutherian tree. Flower's view was later endorsed by Van Valen (1966) and Szalay (1969), who argued, mainly on the basis of dental characters, for a connection between cetaceans and mesonychid condylarths, a Tertiary assemblage of ungulates. The first paleontological evidence for a connection between cetaceans and artiodactyls (even-hoofed ungulates) was provided by the remains of a middle Eocene (~45 million-year-old) whale exhibiting an artiodactyl-like paraxonic arrangement of the digits on its vestigial hindlimbs (Gingerich et al. 1990; Wyss 1990). The discovery of the 50-million-year-old fossil cetacean, *Ambulocetus natans*, in Pakistan by Gingerich et al. (1994) and Thewissen et al. (1994), provided some insight into the terrestrial–aquatic transition (Novacek 1994).

The molecular evidence for a close relationship between Cetacea and Artiodactyla has been increasing since the 1980s. Goodman et al. (1982) analyzed seven protein sequences and concluded that Cetacea is a sister taxon of the Artiodactyla. This conclusion received further support from studies on mitochondrial DNA sequences (e.g., Irwin et al. 1991; Milinkovitch et al. 1993; Cao et al. 1994).

The order Artiodactyla is traditionally divided into three suborders: Suiformes (pigs and hippopotamuses), Tylopoda (camels and llamas), and Ruminantia (deer, elk, giraffes, pronghorn, cattle, goats, and sheep). Graur and Higgins (1994) inferred the phylogenetic position of Cetecea in relation to the three artiodactyl suborders by using protein and DNA sequence data from cow, camel, pig, several cetacean species, and an outgroup. Their phylogenetic analysis suggested that cetaceans are not only intimately related to the artiodactyls, but are deeply nested within the artiodactyl phylogenetic tree;

i.e., they are more closely related to some members of the order Artiodactyla (e.g., Ruminantia) than some artiodactyls are to one another. Thus, the artiodactyls do not constitute a monophyletic clade, unless the cetaceans are included within the order. The term Cetartiodactyla (Montgelard et al. 1997) is currently used for the clade consisting of artiodactyls and cetaceans.

An unambiguous resolution of cetacean evolutionary affinities has been obtained by Shimamura et al. (1997) and Nikaido and Okada (in press), who used the insertion patterns of short interspersed repeated sequences (SINEs; Chapters 7 and 8) to resolve the cetartiodactyl phylogenetic tree. Figure 5.32 illustrates the principles of phylogenetic inference by using SINEs. First, a SINE is identified in a certain species. Then the 5′ and 3′ primers around the SINE unit are used to identify uniquely its genomic location (Figure 5.32a). If the surroundings of the SINE are conserved during evolution, they may be used with the polymerase chain reaction (PCR) to amplify the homologous loci from the genomic DNA of the other species under study (Figure 5.32b). The PCR products are then subjected to electrophoresis, which separates them according to length. A long PCR product indicates the presence of a SINE unit; a short

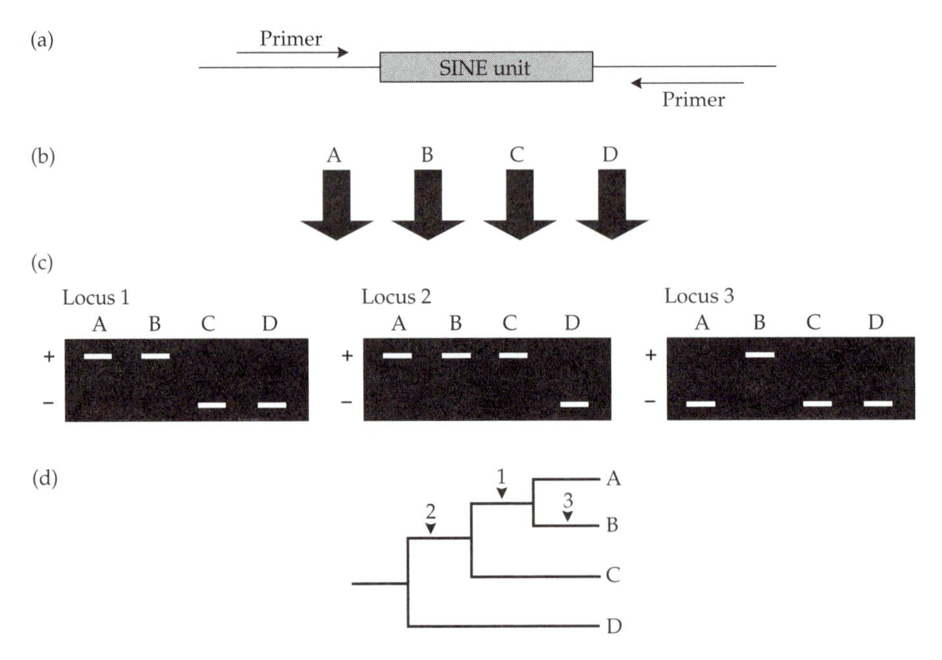

FIGURE 5.32 Inference of phylogeny from insertion patterns of SINEs. (a) The primers identify the genomic location (locus) of a SINE unit. (b) PCR is used to amplify the homologous loci from the genomic DNA of several species under study (A, B, C, and D). (c) The PCR products are subjected to electrophoresis and separation by length. A long PCR product indicates the presence (+) of a SINE unit; a short PCR product indicates absence (–). (d) Because SINE insertion is essentially an irreversible character state, the presence of a SINE at a certain locus may be treated as a synapomorphy defining monophyletic clades (arrowheads 1 and 2) or as an autapomorphy for a single taxon (arrowhead 3). Courtesy of Professor Norihiro Okada.

PCR product indicates absence (Figure 5.32c). To ensure that the insertions are indeed homologous (the same SINE at the exact same location), the PCR products may be subsequently sequenced and compared. Because a SINE insertion is essentially an irreversible character state, the presence of a SINE at a specific locus in several species may be treated as a synapomorphy defining a monophyletic clade (Figure 5.32d). For example, the pattern at locus 2 indicates that species A, B, and C belong to a monophyletic cluster.

The traditional phylogeny of cetaceans and artiodactyls is shown in Figure 5.33a. Shimamura et al. (1997) and Nikaido and Okada (in press) identi-

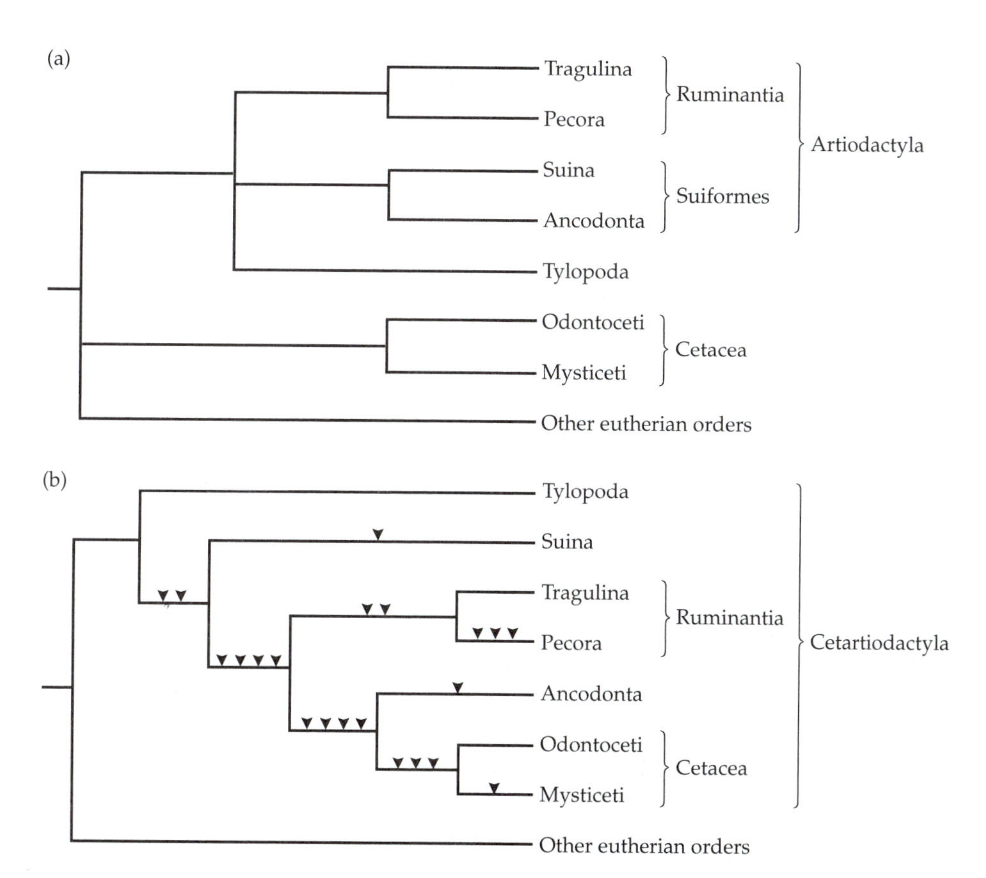

FIGURE 5.33 (a) Traditional phylogenetic tree and taxonomic nomenclature for cetaceans and artiodactyls. The order Artiodactyla is divided into three equidistant suborders: Tylopoda (camels and llamas), Suiformes, and Ruminantia. Suborder Suiformes is divided into two infraorders: Suina (pigs and peccaries) and Ancodonta (hippopotamuses). Suborder Ruminantia is divided into two infraorders: Tragulina (mouse deer or chevrotains) and Pecora (deer, elk, giraffes, pronghorn, cattle, goats, and sheep). The order Cetacea, composed of Odontoceti (toothed whales) and Mysticeti (baleen whales), may or may not be related to Artiodactyla. (b) Molecular phylogenetic tree and revised taxonomic nomenclature for cetartiodactyls. Arrowheads denote SINE insertions. Courtesy of Professor Norihiro Okada.

fied 21 synapomorphic and autapomorphic SINE insertions all over the genome of cetaceans and artiodactyls and used them to reconstruct the phylogeny of cetaceans and artiodactyls (Figure 5.33b). For example, they found two *CHR-1* SINEs in Pecora and Tragulina that were not found in any other organisms. One of these SINEs was found in the third intron of the gene for the α subunit of the pituitary glycoprotein hormone; the other was in the third intron of the gene for steroid 21-hydroxylase. These two SINEs demonstrate the monophyly of ruminants.

As shown in Figure 5.33b, four SINE synapomorphies unequivocally indicate that hippopotamuses (Ancodonta) are the closest extant relatives of whales. Incidentally, pigs and peccaries (Suina) were found to be unrelated to hippopotamuses, and the molecular evidence therefore invalidates the monophyly of the suborder Suiformes.

The origin of angiosperms

The origin of the angiosperms (flowering plants) was deemed "an abominable mystery" by Charles Darwin, and to this day remains a highly controversial issue. Paleontological evidence indicates that angiosperms, which are uniquely defined by their carpel-enclosed ovules and seeds, began to radiate rapidly in the middle Cretaceous (~115 million years ago), and became the dominant group of land plants about 90 million years ago (Lidgard and Crane 1988). About 275,000 extant angiosperm species are currently described (Appendix I), attesting to the immense success of this group. Angiosperms are generally thought to have descended from gymnosperm-like seed plants (Spermatopsida), and since the spermatopsid lineage extends back to at least 370–380 million years ago (Stewart 1983; Kenrick and Crane 1997), there is an enormous range of time during which angiosperms might have had their beginnings. Theories concerning the paucity of angiosperms in the fossil record prior to the Cretaceous fall into two basic types: either angiosperms did not exist until the early Cretaceous (e.g., Hickey and Doyle 1977; Doyle 1978; Thomas and Spicer 1987), or pre-Cretaceous angiosperms lived in habitats so refractory to fossilization that they left no record (Axelrod 1952, 1970; Takhtajan 1969).

One way to decide between these two views is to estimate the date of divergence between monocotyledons (monocots) and dicotyledons (dicots), the two major classes of angiosperms. This would provide us with a minimal estimate for the age of angiosperms. The first application of DNA sequence data to estimate this date was made by Martin et al. (1989), who used the sequences of the nuclear gene encoding cytostolic glyceraldehyde-3-phosphate dehydrogenase from plants, animals, and fungi. By using several divergence dates among animal taxa, and between the plant, animal and fungal kingdoms, they estimated the rate of evolution of this gene. From this rate, they inferred the monocot and dicot lineages to have diverged approximately 300–320 million years ago (Martin et al. 1993). This date appears to be too ancient, because the earliest land plant fossils are only about 420 million years old (Gensel and Andrews 1984), and so it would imply that all of the vascular plants (i.e., bryophytes, pteridophytes, gymnosperms, monocotydelons, and dicotydelons) appeared within less than 100 million years after the emergence of plants on land. Nevertheless, the data provided evidence for a pre-Cretaceous origin of angiosperms.

Wolfe et al. (1989b) obtained a different estimate by using three approaches. The first was based on a calibration of the rate of synonymous substitution in chloroplast genes with the maize–wheat divergence as a reference (50–70 million years ago). Using DNA sequence data, they first demonstrated that maize, wheat, and rice all originated at approximately the same time, i.e., that their phylogenetic relationships may be represented approximately as a trichotomy (Figure 5.34). From the average number of synonymous substitutions per site between maize and wheat chloroplast genes, they estimated the rate of synonymous substitution to be 1.73×10^{-9} or 1.24×10^{-9} substitutions per site per year, depending on whether the lower bound (50 million years) or the upper bound (70 million years) of the maize–wheat divergence event was used. The average number of synonymous substitutions per site between monocot (maize and wheat) and dicot (tobacco) genes was 0.577. Therefore, the date of the monocot–dicot divergence (Figure 5.34) was estimated to be 170–230 million years ago.

The second approach was based on a calibration of the rate of nonsynonymous substitution with the bryophyte–angiosperm divergence as reference

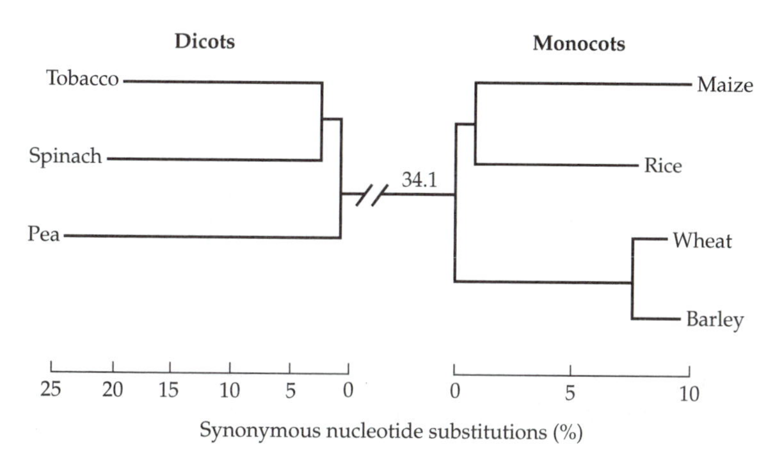

FIGURE 5.34 Phylogenetic tree for three dicot and four monocot species. The tree was inferred by the neighbor-joining method using the synonymous distances for three chloroplast genes: *rbcL*, *atpB*, and *atpE*. The length (0.7%) of the internal branch leading to the maize–rice pair is less than the standard error (~1.7%), and thus the maize, rice, and wheat/barley lineages are probably close to a trichotomy. Note that for purposes of clarity, the two scales are different from each other. From Wolfe et al. (1989b).

(350–450 million years). This approach gave an estimate of 150–260 million years for the monocot–dicot divergence event.

The third approach was based on a calibration of the substitution rate in nuclear ribosomal RNA genes with the plant–animal divergence as reference (1 billion years). The estimate obtained from 26S rRNA data was 200–250 million years, and that obtained from 18S rRNA was 200–210 million years.

From these estimates, Wolfe et al. (1989a) suggested that monocots and dicots diverged about 200 million years ago, with an uncertainty of about 40 million years. These results are supported by estimates from mitochondrial gene sequences and completely sequenced chloroplast genomes (Goremkyn et al. 1997; Laroche et al. 1995), and suggest that monocots and dicots may have diverged in the early Jurassic. Therefore, the molecular data strongly support the hypothesis that angiosperms existed long before they became prominent in the terrestrial paleoflora.

Two paleontological findings from late Jurassic rocks in the Liaoning province of northeast China corroborate this early estimate for the origin of flowering plants. Sun et al. (1998) reported the discovery of the oldest undisputed angiosperm, *Archaefructus liaoningensis*, a magnolia relative. Ren (1998) discovered several species of short-horned flies (Brachycera), and from functional morphology and comparisons with extant flies belonging to the same family, he concluded that these flies were pollinators of at least two floral types. Therefore, angiosperms not only existed in the Jurassic, but they had already speciated to a considerable extent.

MOLECULAR PHYLOGENETIC ARCHEOLOGY

DNA is an unstable molecule that decays spontaneously through hydrolysis and oxidation. The chances of unprotected DNA surviving over long periods of time are low, unless special conditions exist for its preservation. Theoretical calculations suggest that DNA should not survive for more than 10,000–100,000 years, and then only in a highly fragmented form (Lindahl 1993). All records of ancient DNA recovery from protected and unprotected sources, such as Miocene plant fossils (Golenberg et al. 1990), Cretaceous bones (Woodward et al. 1994), and amber-entombed organisms (Cano et al. 1993) have now been discredited (e.g., Austin et al. 1997; Waldan and Robertson 1997; Gutiérrez and Marín 1998). These disappointments notwithstanding, genetic information, albeit in minute quantities, may be preserved in biological material that is 100,000 years old or younger. This enables us to use molecular phylogenetic techniques on extinct species.

Ancient DNA can be detected by staining it on electrophoretic gels with ethidium bromide and by observing its activity as a template capable of directing the incorporation of radioactive nucleotides into newly synthesized DNA in the presence of DNA polymerase and a mixture of random primers. Hybridization with DNA from an extant species that is presumed to be closely related to the extinct species under study is usually used to determine whether the DNA found in the sample originated from the species from which the sample has been taken or from a contaminating source, such as bacterial DNA.

It is now possible to sequence segments of DNA from unpurified samples derived from micrograms of preserved tissues. The method employed is the polymerase chain reaction (PCR), which involves the amplification of unique sequences from a mixture of sequences via the use of two primers (Kocher et al. 1989; Mullis 1990). By using this procedure, it is possible to synthesize many copies of a chosen piece of DNA in the presence of vast excesses of other DNA sequences.

By employing PCR we can retrieve particular DNA sequences from museum specimens, such as preserved organic material (mainly skin and muscle), badly damaged archeological remains, and even bones (Table 5.6), and use this DNA to establish phylogenetic affiliations of extinct species and populations. Notable examples of such studies include the determination that Neandertals may have gone extinct without contributing to the gene pool of modern humans (Krings et al. 1997), and the recovery of DNA unique to *Mycobacterium tuberculosis* in a 1,000-year-old adult pre-Columbian female from southern Peru (Salo et al. 1994).

TABLE 5.6 Types of ancient biological material reported to contain DNA

Sample	Maximum age (years)
Dry remains	
Museum skins	140
Naturally preserved skins	13,000
Human mummies	7,500
Bones and teeth	30,000
Natural animal mummies	13,000
Feathers	130
Hair	1,300
Herbarium plant specimens	120
Charred seeds and cobs	4,500
Mummified seeds	45,000
Frozen remains	
Muscle tissue	53,000
Wet remains	
Pickled museum specimens	100
Human remains preserved in peat	8,000

Data from Pääbo et al. (1989), Hanni et al. (1990), Lawlor et al. (1991), Brown and Brown (1994), Lin et al. (1995), and Krings et al. (1997).

When dealing with ancient genetic material it is important to assess whether or not any postmortem changes in the DNA have occurred. In a concatenated 229-bp sequence from a 140-year-old skin sample, Higuchi et al. (1987) detected two postmortem modifications. Both modifications were transitions that could be attributed to the postmortem deamination of cytosine to uracil. Therefore, about 1% of all the nucleotides in this sample have changed following death and preservation.

Phylogeny of the marsupial wolf

The marsupial wolf *Thylacinus cynocephalus* is a carnivorous animal about the size of a large dog. It is often referred to as the Tasmanian tiger, in reference to the stripes on its back and rump. The marsupial wolf was already extinct on the mainland of Australia thousands of years ago. The last known wild marsupial wolf was captured in Tasmania in 1933, and died in Hobart Zoo in 1936. Sporadic sightings of live marsupial wolves are still reported (Douglas 1986). Despite its geographical distribution, which is restricted to Australia and adjoining islands, *T. cynocephalus* has been frequently classified on the basis of its morphology together with the South American marsupials but apart from the Australian ones.

By using the PCR method, Thomas et al. (1989) and Krajewski et al. (1992, 1997) (1989) sequenced three mitochondrial segments totaling 1,765 nucleotides from *T. cynocephalus* and compared them with homologous sequences from living Australian and South American marsupials, as well as with homologous sequences from placental mammals. On this basis, they were able to decide between two claims: (1) the marsupial wolf belongs to a South American group of marsupials called Didelphimorphia (opossums), or (2) the marsupial wolf is closely related to a diverse group of Australian marsupials called Dasyuromorphia (marsupial mice and cats). From these sequence comparisons, it was concluded that *Thylacinus* is closely related to two dasyurid Australian marsupials, the nearly extinct Tasmanian devil (*Sarcophilus harrisii*) and the Australian tiger cat (*Dasyurus maculatus*), but only distantly related to the South American marsupials such as the gray four-eyed opossum (*Philander opossum andersoni*) (Figure 5.35). Thus, the morphological similarity between *Thylacinus* and South American marsupials appears to represent an instance of convergent evolution at the morphological level that has no parallels in the mitochondrial DNA. Similar conclusions had been reached earlier by Lowenstein et al. (1981) on the basis of a radioimmunoassay comparison of albumins.

Is the quagga extinct?

When Boer settlers arrived in the Karoo regions of the Cape of Good Hope in the seventeenth century, they found grasslands teeming with herds of zebra-like animals, which the natives called quagga and taxonomists labeled accordingly *Equus quagga*. Due to uncontrolled hunting, the once-ubiquitous animal was driven to extinction within a little over 200 years (Hughes 1988). It is un-

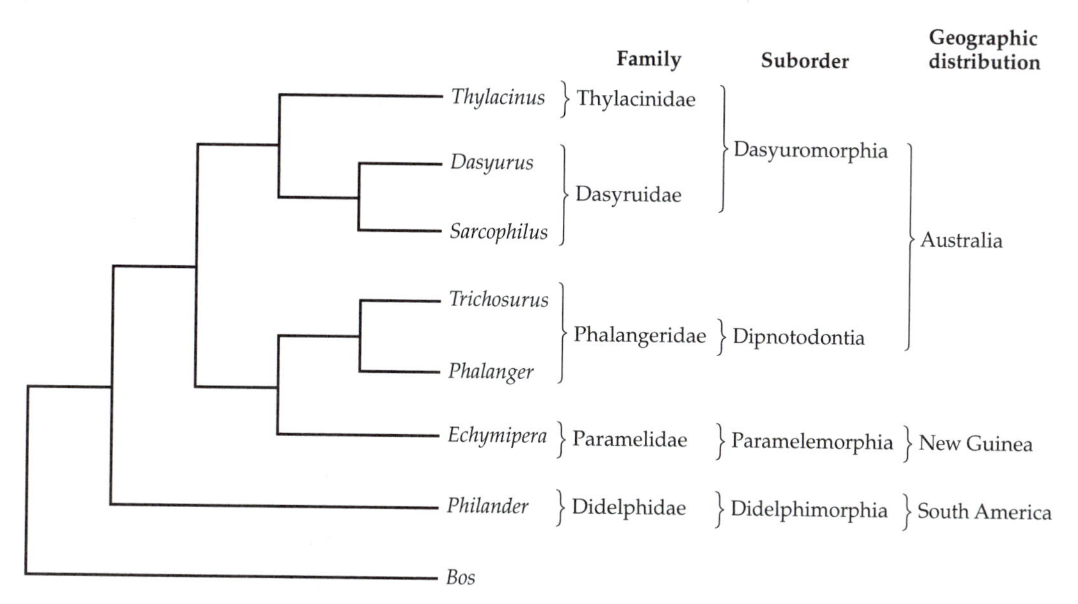

FIGURE 5.35 Phylogenetic tree, taxonomic nomenclature, and geographic distribution for the marsupial wolf (*Thylacinus*) and six other marsupials. The tree was built on the basis of mitochondrial 12S rRNA sequences, and was rooted with cow (*Bos*) as an outgroup. Modified from Thomas et al. (1989).

certain when the last quagga died; the last animal shot in the wild was reported in 1876, and a few specimens seem to have survived in European zoos into the 1880s. The female quagga that died in the Amsterdam Zoo on August 12, 1883 quite possibly marked the extinction of the species (Harley 1988).

The phylogenetic affinities of the quagga have always been controversial. Bennett (1980), for instance, placed the quagga in the same clade with the domestic horse (*E. caballus*), but apart from the plains zebra (*E. burchelli*), the mountain zebra (*E. zebra*), and Grevy's zebra (*E. grevyi*). In contrast, Eisenmann (1985) clustered the quagga with the zebras, apart from the horse, and Rau (1974) considered the quagga as merely a color variant of the plains zebra, a conclusion supported by radioimmunoassay comparisons (Lowenstein and Ryder 1985).

A comparison of a concatenated 229-nucleotide-long mitochondrial sequence revealed no differences between the quagga and the plains zebra (Higuchi et al. 1987), so the available molecular evidence to date strongly supports Rau's view that the quagga is at most a subspecies of *E. burchelli* (Figure 5.36). The suggestion that the quagga is a sister group of the horse (Bennett 1980) could be safely discarded on the basis of these molecular findings. The molecular taxonomic assessment requires that the morphological evidence used to cluster the horse and the quagga be reconsidered. For instance, the dental similarities between the quagga and the horse, which were thought to be shared derived characters (synapomorphies), must be reinterpreted as

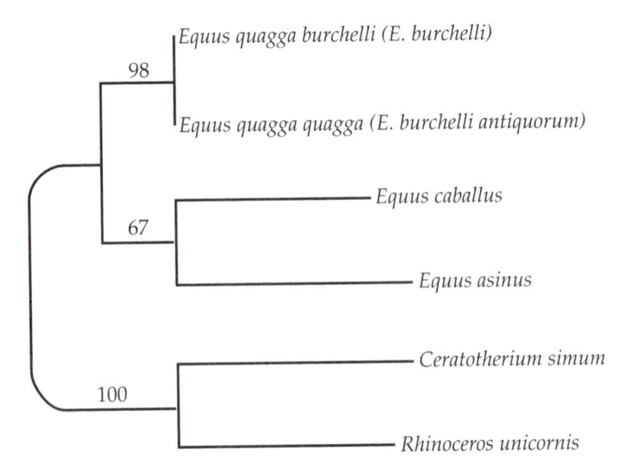

FIGURE 5.36 Scaled neighbor-joining phylogenetic tree for four *Equus* species (quagga, *E. quagga quagga*; plains zebra, *E. q. burchelli*; horse, *E. caballus*; and donkey, *E. asinus*). The tree has been rooted with two non-equid perissodactyl species (Asian one-horned rhinoceros, *Rhinoceros unicornis*; and white rhinoceros, *Ceratotherium simum*). The tree is based on a 229-bp concatenated mitochondrial sequence. The numbers on the internal branches represent bootstrap values based on 1,000 pseudosamples. Minor taxonomic synonyms are shown in parentheses.

primitive (symplesiomorphies). These dental characters have been retained in the horse and quagga lineages, but were lost in the zebras. In addition, the molecular findings require the renaming of the quagga and the plains zebra. According to the rules of taxonomic precedence, the name of the newly defined species that includes both the quagga and the plains zebra should be *Equus quagga*, this term having been coined earlier (1785) than *E. burchelli* (1824). If one wishes to distinguish between the two subspecies, the terms *E. quagga quagga* and *E. quagga burchelli* should be used for quagga and plains zebra, respectively. In the literature, however, the names *E. burchelli antiquorum* and *E. burchelli* are frequently used.

Most importantly, the newly established taxonomic status for the quagga provided a rationale for the South African Museum in Cape Town to establish the "Quagga Experimental Breeding Program" aimed at "reconstructing" the extinct subspecies. By breeding selected plains zebras, the program aims at obtaining the coat pattern that characterized the quagga, i.e., paucity or absence of stripes over the posterior part of the body and legs and a rich brown background color of the upper parts of the body. The result is expected to be a genuine reconstruction in terms of genetic constitution, rather than a quagga look-alike.

The dusky seaside sparrow: A lesson in conservation biology

The last dusky seaside sparrow died on June 16, 1987, in a zoo at Walt Disney World, near Orlando, Florida. Dusky seaside sparrows were discovered in

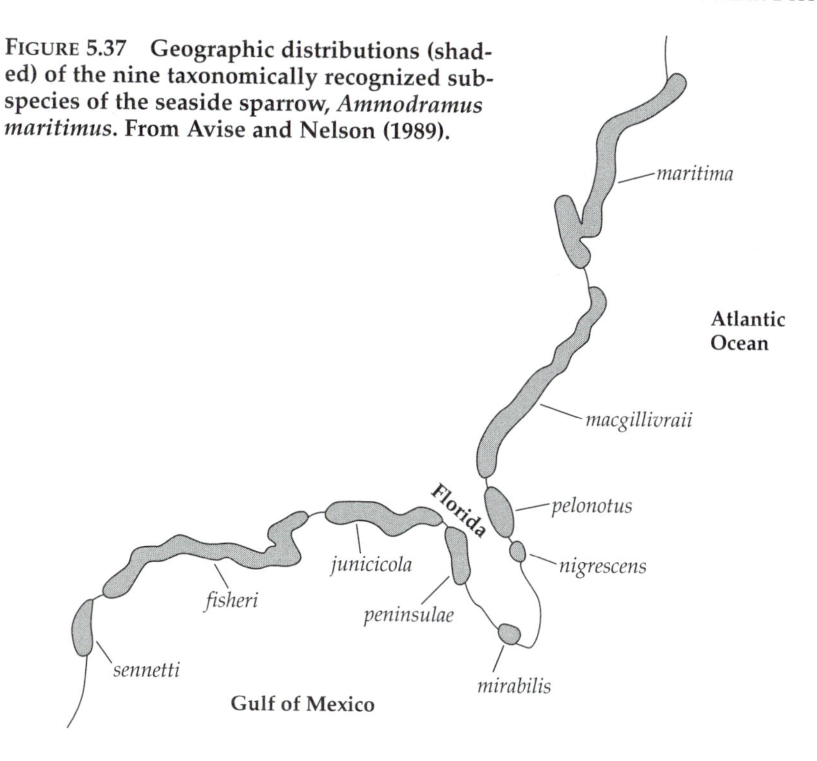

FIGURE 5.37 Geographic distributions (shaded) of the nine taxonomically recognized subspecies of the seaside sparrow, *Ammodramus maritimus*. From Avise and Nelson (1989).

1872, and their melanic spotted appearance led to their classification as a distinct subspecies (*Ammodramus maritimus nigrescens*). The geographic distribution of *A. m. nigrescens* was confined to the salt marshes of Brevard County, Florida (Figure 5.37). At the time of their discovery, the population of dusky seaside sparrows consisted of about 2,000 individuals. From 1900 onward, the bird was slowly edged out of its range as the salt marshes were flushed with fresh water to control mosquitoes. The subspecies was declared endangered in 1967. By 1980 only six individuals, all males, could be found in nature. Obviously, the population was doomed, and an artificial breeding program was launched as a last-ditch attempt to preserve the genes of this subspecies.

In such a case, the conservation program involves the mating of the males from the nearly extinct subspecies with females from the closest subspecies available. The female hybrids of the first generation are then backcrossed to the males, their offspring are again backcrossed to the original males, and the process is continued for as long as the original males live. The crux of such an experiment is to decide from which population to choose the females, i.e., which subspecies is phylogenetically closest to the endangered one.

In the case of *A. maritimus*, there were eight recognized subspecies from which to choose. The geographical ranges of these species are shown in Figure 5.37. On the basis of morphological and behavioral characters, as well as geographic proximity, it was decided that the closest subspecies to *A. m. nigrescens* is Scott's seaside sparrow (*A. m. peninsulae*), which inhabits Florida's Gulf shores. As a consequence of this decision, several *nigrescens* males were mated with *peninsulae* females. Two successful backcrosses were accomplished and the resulting population has since been kept inbred with the view of someday releasing the "reconstructed" subspecies into its original habitat.

In order to find out whether or not the choice of females was correct, Avise and Nelson (1989) compared the restriction enzyme pattern of mitochondrial DNA from the last pure *A. m. nigrescens* specimen with that of 39 individual birds belonging to five of the eight extant subspecies of *A. maritimus.* They chose mitochondrial DNA for several reasons. First, mitochondrial DNA in vertebrates is known to evolve very rapidly (Chapter 4), and hence it can provide a high resolution for distinguishing between closely related organisms. Second, mitochondria are maternally inherited, and thus complications

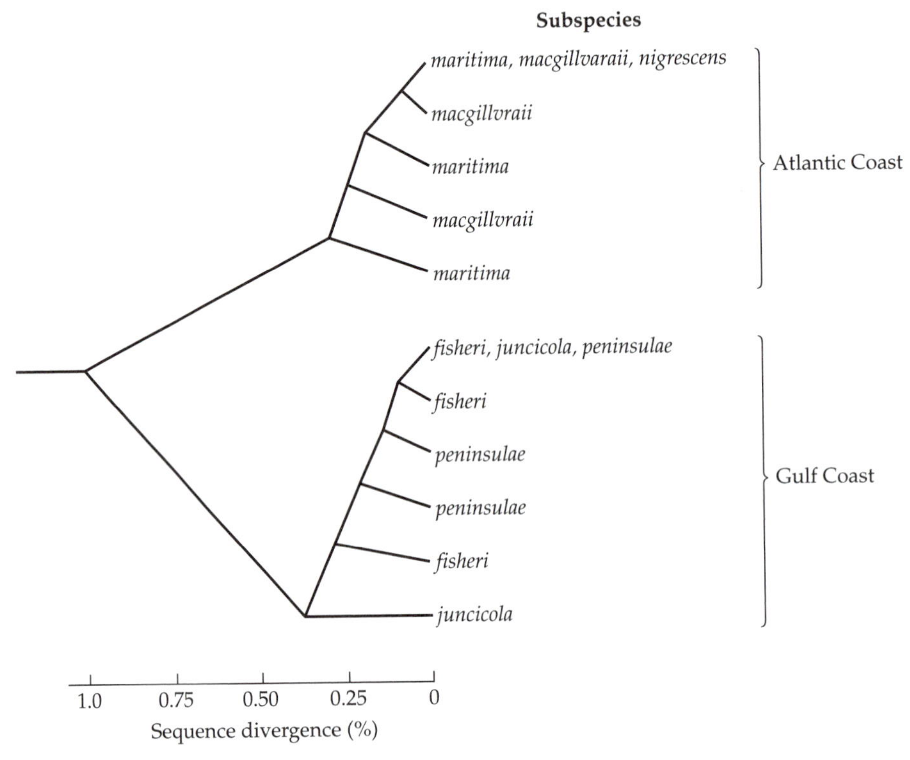

FIGURE 5.38 UPGMA dendrogram showing the distinction between mitochondrial DNA genotypes of the Atlantic Coast versus Gulf Coast populations of seaside sparrow. By using maximum parsimony methods many equally parsimonious trees were obtained, including one identical in topology to the tree shown. All alternative maximum parsimony trees involved minor branch rearrangements within either the Atlantic clade or the Gulf clade, while the distinction between the two groups remained unaltered. The multiple appearance of the same subspecies name in the dendrogram indicates that different individuals belonging to the same subspecies exhibit different restriction enzyme patterns. Conversely, the appearance of several subspecies names at the end of a single branch indicates that individuals classified as different subspecies on morphological and zoogeographical grounds exhibit identical patterns for the restriction enzymes used. From Avise and Nelson (1989).

due to allelic segregation do not arise. Finally, because of the maternal mode of transmission, mitochondrial DNA from the last male dusky seaside sparrow had not been transmitted to the hybrids in the restorative breeding program. Thus, unlike nuclear genes, some of which survive in the hybrids, the mitochondrial genes of the dusky seaside sparrow are truly extinct.

From the restriction enzyme patterns, Avise and Nelson (1989) reconstructed the evolutionary relationships among several subspecies of *A. maritimus* by UPGMA and maximum parsimony. As seen from Figure 5.38, the Atlantic Coast subspecies, including *A. m. nigrescens*, are nearly indistinguishable from one another. The same is true for the three Gulf Coast subspecies in this study. In comparison, the Atlantic Coast subspecies are quite distinct from the Gulf Coast subspecies. The number of nucleotide substitutions per site between the two groups has been estimated to be about 1%. If mitochondrial DNA in sparrows evolves at about the same rate as mitochondrial DNA in well-studied mammalian and avian lineages (2–4% sequence divergence per million years), then these two groups of subspecies have separated from each other some 250,000–500,000 years ago. While these estimates of the time of divergence must remain qualified due to uncertainties in calibration, they agree well with the dates obtained for the falling sea level that exposed the Florida peninsula, which serves as a reproductive barrier between the *A. maritimus* subspecies on the two sides of the peninsula.

Most importantly, Avise and Nelson's (1989) molecular study showed that, while the *A. m. nigrescens* subspecies is indistinguishable from the two other Atlantic subspecies (i.e., *A. m. maritima* and *A. m. macgillivraii*), it is quite different from the Gulf subspecies, such as *A. m. peninsulae*, whose females had been chosen for the breeding program.

In conclusion, the salvation program of the dusky seaside sparrow has rested on an erroneous phylogenetic premise and therefore, instead of reconstructing an extinct subspecies, the program created a new one. Indeed, the U.S. Department of the Interior ruled that the 1973 Endangered Species Act does not extend to the protection of "mongrels," and in 1990 the dusky seaside sparrow was officially declared extinct. Thus, knowledge of phylogenetic relationships is essential in making rational decisions for the conservation of biotic diversity. A faulty taxonomy may turn even the most well-intentioned effort into an irreparable fiasco.

THE UNIVERSAL PHYLOGENY

"All the organic beings which have ever lived on this earth have descended from one primordial form, into which life was first breathed." Thus Charles Darwin inaugurated in 1859 the monophyletic view of life. Everything biologists have learned since 1859 supports Darwin's conclusion: there is but one tree of life, one universal phylogeny that connects humans, onions, mushrooms, slime molds, and bacteria. Prior to the advent of molecular phylogenetic techniques, questions pertaining to the deepest branches of the universal phylogenetic tree could be answered only tentatively and involved a great deal of speculation. The reason was that at the morphological level—indeed, at the micromorphological level of the cell—there are almost no comparable (homologous) characters with which to resolve the evolutionary relationships among very distantly related organisms.

In the following, we shall survey three topics related to the universal phylogeny problem: (1) the rooting of the universal tree (i.e., the identification of the first branching events in the history of life); (2) the possibility of inferring some of the characteristics of the ancestor of all extant life forms; and (3) the origin of the multiple genomes (nuclear, mitochondrial, and chloroplast) in the eukaryotic cell.

The first divergence events

The living world has traditionally been divided dichotomously into **eukaryotes** and **prokaryotes**. Eukaryotes are organisms with a distinct nucleus and cytoplasm. Organisms that lack a well-defined, membrane-enclosed nucleus are called prokaryotes. In the traditional classification, the prokaryotes consist of a single kingdom, **Bacteria**, which also includes the cyanobacteria, formerly called blue-green algae. The eukaryotes were thought to consist of a single exclusively unicellular kingdom, **Protista**, which includes organisms such as ciliates, flagellates and amoebae; two kingdoms that consist of both unicellular and multicellular organisms, **Fungi** and **Plantae**; and an exclusively multicellular kingdom, **Animalia**. The entire living world was thus divided into five kingdoms (Margulis and Schwartz 1988).

Woese and coworkers (Woese and Fox 1977; Fox et al. 1980) have challenged the traditional view. Since the late 1960s, they have been studying bacterial relationships by comparing the ribosomal RNA (rRNA) sequences from different species. Woese and coworkers came across a totally unexpected finding when examining the rRNA of methanogenic bacteria. These unusual organisms are obligatory anaerobes, i.e., they live only in oxygen-free environments, such as sewage treatment plants and the intestinal tracts of animals. These bacteria generate methane (CH_4) by the reduction of carbon dioxide (CO_2). Methanogens are without a doubt bacteria because of their size, their lack of a nuclear membrane, and their low DNA content. Thus, they were expected to be more closely related to the other bacteria than to the eukaryotes. However, in terms of rRNA dissimilarity, methanogens turned out to be equally distant from both taxa. On the basis of this finding, and the fact that the methanogenic metabolism was thought to be suited to the kind of atmosphere believed to have existed on the primitive earth (rich in CO_2, but virtually devoid of oxygen), Woese and Fox (1977) proposed to include the methanogens and their relatives into a new taxon, **Archaebacteria**, a name which implied that this group of bacteria is evolutionarily at least as ancient as the "true" bacteria, which they renamed **Eubacteria**.

As it turned out, the archaebacterial group was found to include, in addition to methanogens, many bacteria that live in extremely harsh environments (extremophiles), such as the thermophiles and the hyperthermophiles, which live in hot springs at temperatures as high as 110°C, and the halophiles, which are highly salt-dependent and grow in such habitats as the Great Salt Lake and the Dead Sea. Currently, the archaebacteria are defined by a single biochemical synapomorphy: the absence of muramic acid from their cell wall.

Woese and Fox (1977) and Fox et al. (1980) proposed that archaebacteria, eubacteria, and eukaryotes were derived from a common ancestor and represent the three primary lines of descent in the tree of life, and are about equally distant from one another. A new taxonomic nomenclature for these clades was proposed by Woese et al. (1990). The most inclusive taxonomic units in this classification are three **urkingdoms** (literally, "primordial kingdoms") or **domains**, corresponding to the primary lines of descent in the tree of life: **Bacteria**, **Archaea**, and **Eucarya**. (The misnomer Eukarya is frequently used in the literature.)

An unrooted molecular phylogenetic tree of all living organisms is shown in Figure 5.39. Note that of the five traditional kingdoms, only Animalia re-

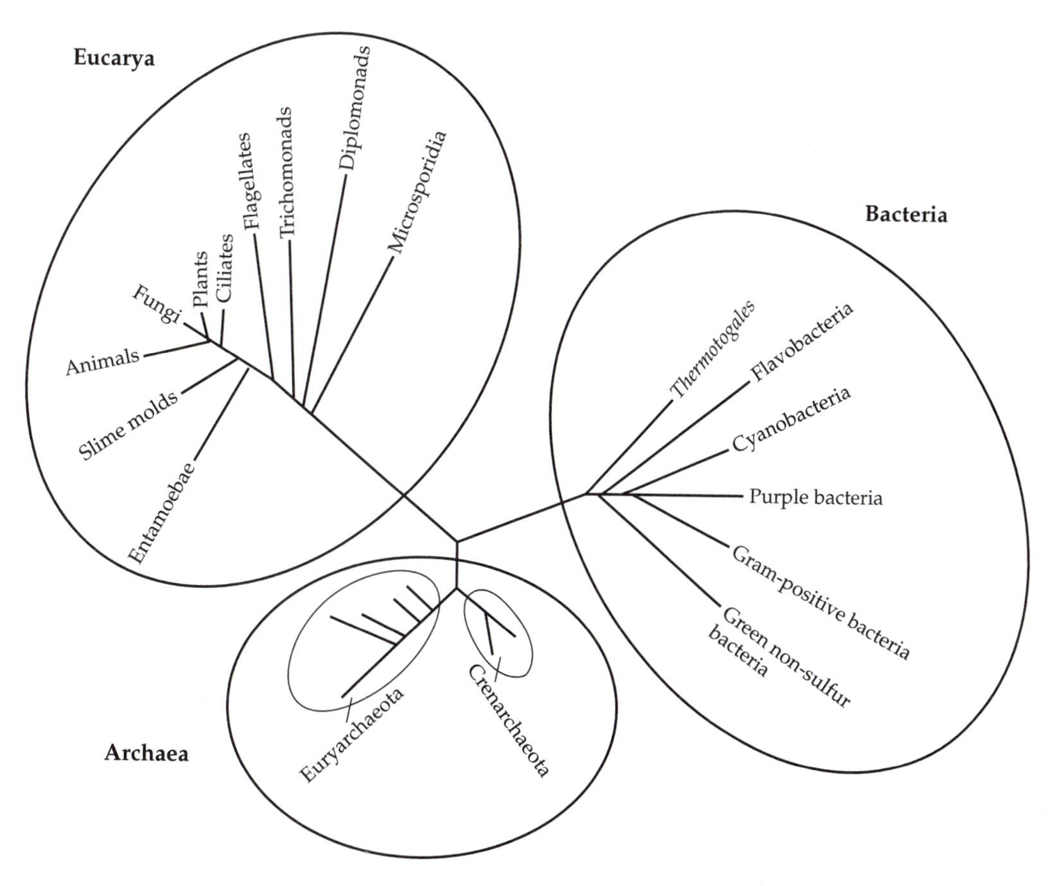

FIGURE 5.39 An unrooted tree of all living organisms. The three main lines of descent (domains) are Eucarya, Bacteria, and Archaea. A deep branching within the he Archaea divides it into two kingdoms, Crenarchaeota and Euryarchaeota. An additional kingdom within the Archaea, Korarchaeota, is only known from ribosomal RNA genes, and the organisms bearing these genes have not been identified. Data from Barns et al. (1996) and Woese (1996).

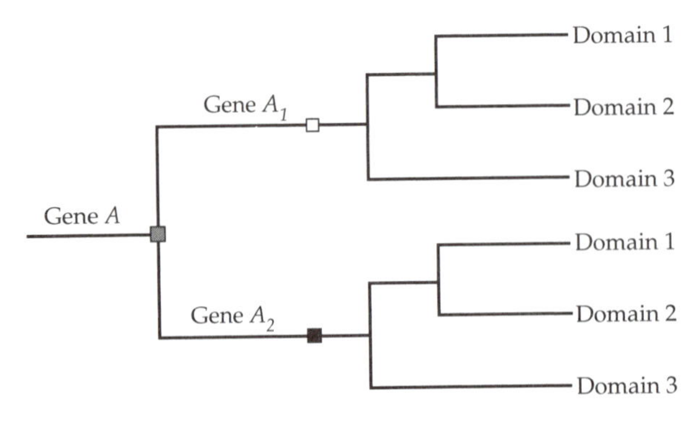

FIGURE 5.40 **Duplication of gene *A* (gray square) into *A1* (white) and *A2* (black) prior to the divergence of three domains, will result in two identical topologies for the two subtrees. Modified from Li (1997).**

mains unscathed by the molecular revision. Fungi must be redefined by the exclusion of such taxa as the slime molds, and Plantae by the exclusion of many groups of algae. The most extreme illustration of the departure from traditional taxonomic assessment is exemplified by Protista—a single kingdom in the traditional classification—which turns out to be paraphyletic and scattered all over the eucaryan tree. This universal phylogeny also indicates that kingdom Animalia along with the redefined kingdoms Plantae and Fungi may constitute a monophyletic clade. The term **Metakaryota** has been coined for this superkingdom. The other branches in the Eucarya are not monophyletic and were given the convenience name "**Archezoa**." Interestingly, animals, plants, and fungi (i.e., the kingdoms that have traditionally attracted most of the attention in biological studies) turn out to be mere "twigs" on the tip of one branch in the tree of life (Olsen and Woese 1996).

Identifying the first branching events in the history of life requires finding the root of the tree of life (i.e., the tree of all organisms). We note, however, that by definition, the evolutionary tree of all the organisms has no outgroup. In 1989, two research groups came up with an ingenious method to infer the root of the tree (Gogarten et al. 1989; Iwabe et al. 1989). The idea, first suggested by Schwartz and Dayhoff (1978), was to use a pair of genes that exist in all organisms and are therefore derived from a gene duplication event (Chapter 6) that occurred before the separation of the three domains. The idea is illustrated in Figure 5.40.

Suppose that gene *A* duplicated into *A1* and *A2* before the divergence of the three lineages. Subsequently, as the three lineages diverged, *A1* (and *A2*) should also diverge in the same order. Therefore, *A2* sequences may serve as outgroups to root the tree derived from the *A1* sequences. Similarly, *A1* sequences can be used to root the tree derived from the *A2*.

Iwabe et al. (1989) applied this concept to two homologous elongation factor genes, *EF-Tu* and *EF-G*, that are present in all prokaryotes and eukaryotes and must, therefore, have been derived from a duplication event that occurred before the divergence among the three domains. Thus, the *EF-Tu* sequences can be used as outgroups to infer the root of the tree for the *EF-G* sequences, and vice versa. The *EF-G* subtree in Figure 5.41 indicates that Eucarya (represented by a slime mold and a mammal) is a sister taxon of Archaea (represented by *Methanococcus*) to the exclusion of Bacteria (represented by *Micrococcus* and *Escherichia coli*). The *EF-Tu* sequences yield an identical topology.

From Figure 5.40, we note that in reconstructing the phylogenetic trees for duplicate genes, we must be sure that our identification of orthologous genes (genes whose homology is due to a speciation event) is correct. This is not always an easy task. An interesting solution for this problem was suggested by Lawson et al. (1996). In their study of the carbamoylphosphate synthetase, they took advantage of the fact that the gene for this enzyme contains an ancient internal gene duplication (Chapter 6) common to all three domains. Therefore, the duplicated sequences remain linked to one another in the same orientation, and the identification of the orthologous sequences is trivial.

The combined picture that emerges from the study of duplicated genes and gene regions, such as those for the α and β H^+-ATPase subunits (Gogarten et al. 1989), elongation factors (Baldauf et al. 1996), aminoacyl-tRNA synthetases (Brown and Doolittle 1995), and carbamylphosphate synthetase (Lawson et al. 1996), is that the first divergence event is the fundamental split between the Bacteria and the common ancestor of Eucarya and Archaea, which later diverged from each other (Figure 5.42a). At present, the monophylies of Bacteria and Eucarya are also undisputed. However, it is possible that the Archaea is not monophyletic, and that the Eucarya is nested within the archaean tree as a sister taxon of the Crenarchaeota (Figure 5.42b).

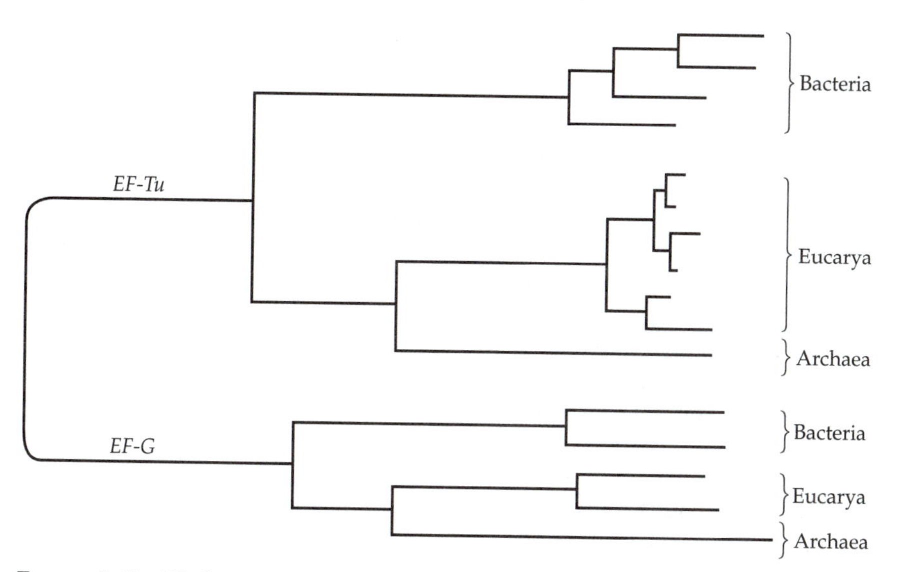

FIGURE 5.41 Phylogenetic tree inferred from a simultaneous comparison of the duplicated elongation factor genes, *EF-Tu* and *EF-G*, from Archaea, Bacteria, and Eucarya. Modified from Iwabe et al. (1989).

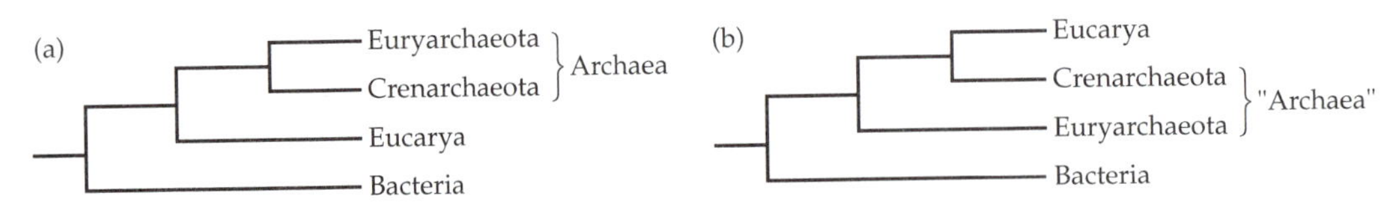

FIGURE 5.42 Two possible phylogenies for Eucarya, Bacteria, and the archaeal kingdoms Crenarchaeota and Euryarchaeota. (a) The Archaea is monophyletic. (b) The Eucarya arose from within the Archaea, which is therefore paraphyletic (indicated by the use of quotation marks). This tree is sometimes referred to as the Eocyta tree.

There is, however, the proposal that the eukaryotic genome is a chimera derived from the fusion of a Gram-negative bacterium and an archaebacterium (Zillig 1991; Gupta and Golding 1993; Golding and Gupta 1995; Koonin et al. 1997). In a maximum likelihood analysis of 273 protein sequences from eukaryotes, archaebacteria, and Gram-positive and Gram-negative eubacteria, Ribeiro and Golding (1998) found 76 topologies significant at the 5% level. Of these, 59 (78%) significantly supported the Archaea/Eucarya clade, 14 (18%) significantly supported the Gram-negative/Eucarya clade, and 3 (4%) supported the Gram-positive/Eucarya clade. They argued that such a large proportion of cases supporting the Gram-negative/Eucarya clade is unlikely to be due to convergent evolution or methodological errors. Ribeiro and Golding (1998) suggested two alternative explanations for the origin of the eukaryotic genome: either multiple horizontal gene transfer events (Chapter 7) from Gram-negative bacteria to the archaeal ancestor of the eukaryotes, or a chimerical fusion of an archaebacterial genome and a Gram-negative bacterial genome. These two alternatives are not easily distinguishable from one another because (1) the organelles of eukaryotes are of eubacterial origin (see page 245), (2) transfer of organelle genes to the nuclear genome is known to occur and (3) the genome of archaea may contain a considerable number of eubacterial genes (Figure 5.43).

Within the Eucarya, there are many candidates for the title the first eukaryotic lineage. In the last decades, some parasitic amitochondriate protists (eukaryotes devoid of mitochondria) such as *Giardia* were thought to represent the most ancient lineage. They were, however, "demoted" by the finding that some amitochondriate taxa are in fact related to Fungi, and that the lack of mitochondria most probably represents a derived state—a secondary loss of the organelle from an organismal lineage that once possessed mitochondria (Embley and Hirt 1998). Currently, the search for the first eukaryotic lineage centers around the Palebiontida, a group of free-living amitochondriates such as the giant amoeboid flagellate *Pelomyxa palustris* (Stiller et al. 1998). The focus of the search on amitochondriate organisms is driven by the idea that the eukaryotic state is independent of the mitochondriate state, i.e., that eukaryotes acquired mitochondria after they already possessed a membrane-enclosed nucleus. It is possible, however, that the mitochondrion is the defin-

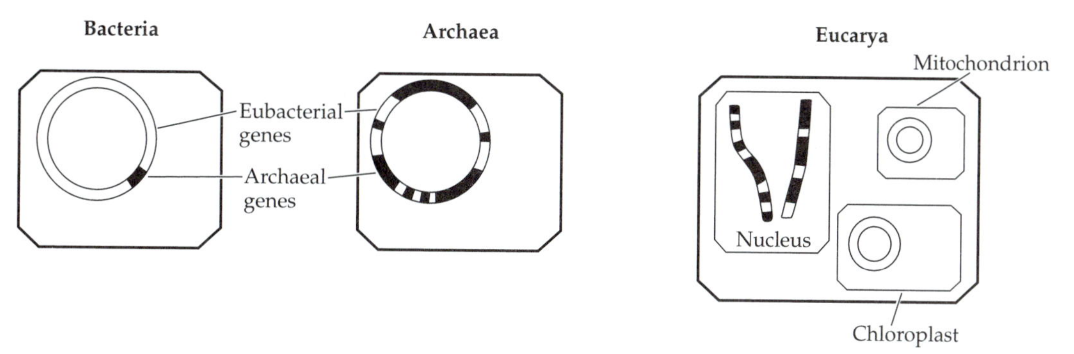

FIGURE 5.43 Origin and distribution of protein-coding genes in the three domains. Bacteria contain few archaeal genes (black segments), e.g., ATPase A in *Thermus* and *Enterococcus*. Archaea contain a large number of eubacterial genes (white segments), in particular genes involved in biosynthesis. The nuclear genome of Eucarya contains many archaeal genes, as well as several eubacterial genes derived from either the chimerical archaeal ancestor or from the organelles through horizontal gene transfer. The mitochondrial and chloroplast genomes are of exclusively eubacterial origin. Modified from Olendzenski et al. (1998).

ing trait of Eucarya, i.e., that the acquisition of mitochondria by a prokaryote is what turned it into an eukaryote.

The cenancestor

The putative ancestor of all extant organisms is referred to as the **cenancestor** (Fitch and Upper 1987). In attempting to infer some of the characteristics of the cenancestor, we note that the distribution of a particular binary character among the three domains (Bacteria, Eucarya, and Archaea) may come in seven patterns (Figure 5.44). The key to reconstructing the nature of the cenancestor lies in knowing the distribution of genetic traits across the three domains. We note, however, that we assume the acquisition and loss of a character state occur with equal probabilities. This assumption may not be correct. In the following we present some inferences on cenancestor characteristics based on the rationale presented in Figure 5.44.

The cenancestor had a circular genome with many genes grouped into operons. In some instances it is even possible to infer the particular operon structure in the cenancestor. For example, in *Escherichia coli* there are four different operons for ribosomal proteins, containing 3, 4, 8, and 11 genes, respectively. The same division is found in several archaeans, and in each instance the arrangement of the particular genes within the operon is the same. Because there is no known functional reason why the ribosomal protein genes need to be in a particular order, it is inferred that these operons also existed in the cenancestor (Doolittle and Brown 1994).

The cenancestor possessed at least one DNA polymerase, three genes encoding the subunits of a DNA-dependent RNA polymerase, and several DNA

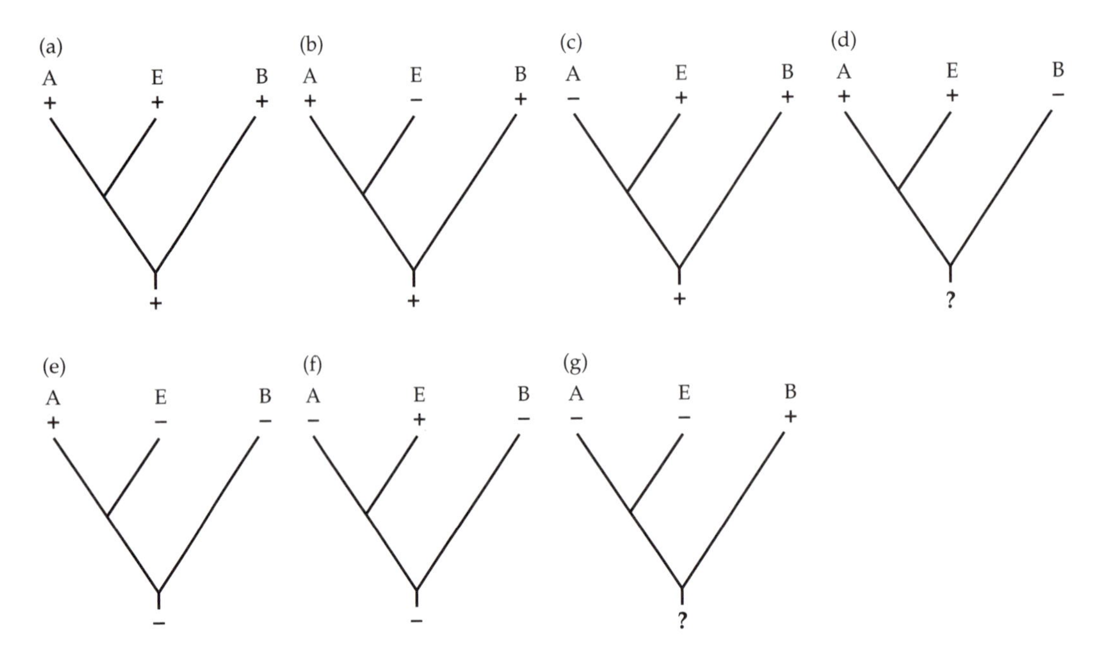

FIGURE 5.44 Inferring the characteristics of the cenancestor from the distribution of binary traits among Bacteria (B), Eucarya (E), and Archaea (A). Presence of a trait is denoted by +; absence by –. If the presence of the trait is universal (a), then the most parsimonious evolutionary scenario is that the trait existed in the cenancestor. If the trait is present in B and A but not in E (b), or if the trait occurs in B and E but not in A (c), then the most parsimonious scenarios are that the trait existed in the cenancestor but was lost along the lineages leading to E or A, respectively. If the trait occurs in E and A but not in B (d), then we cannot infer the character state in the cenancestor (?) because there are two equally parsimonious possibilities: either the cenancestor possessed the trait and it was subsequently lost in the lineage leading to B, or the cenancestor lacked the trait and it arose in the ancestor of the clade leading to E and A. If the presence of the trait is autapomorphous for A or E (e, f), then the cenancestor is assumed to lack the trait. If the autapomorphy is present in B, then the trait is not informative.

topoisomerases. It had an elaborate translation apparatus, performed by two-subunit ribosomes composed of RNA and proteins, and it used the universal genetic code. It is, however, impossible to infer the nature of the translation initiation factors, since these are similar in Archaea and Eucarya, but distinct in Bacteria.

Woese (1987) hypothesized a thermophylic (hot temperature) and auxotrophic (requiring complex nutrients) origin of life. In two unrelated studies, the two components of Woese's hypothesis have been tested. The G+C nucleotide composition of ribosomal RNA was found to be strongly correlated with the optimal growth temperature (Galtier and Lobry 1997). The reason is that G:C pairs are more stable than A:U pairs at high temperatures because of an additional hydrogen bond (Chapter 1). Thus, from knowledge of the G+C content of its rRNA sequences, it is possible to infer the environmental temperature in which the cenancestor lived. By using rRNA sequences from species representing all the major lineages of life and several models of sequence evolution, Galtier et al. (1999) estimated the cenancestral nucleotide composition to be 54%, as opposed to a minimum of 58% required by organisms living at high temperatures.

In a second study, Doolittle and Brown (1994) used the distribution of traits among Bacteria, Eucarya, and Archaea, and concluded that the cenancestor possessed a complex metabolism, including enzymes involved in amino acid, nucleotide, sugar, and fatty acid synthesis. Some of the enzymes involved in carbon fixation were found in Bacteria and Archaea, suggesting that the cenancestor may have been an autotroph, i.e., capable of metabolizing CO_2 and carbonates as the sole sources of carbon. These findings challenge Woese's hypothesis about the origin of life.

Endosymbiotic origin of mitochondria and chloroplasts

There are essentially two types of theories to explain the existence of separate nuclear, mitochondrial, and chloroplast genomes in eukaryotes. Theories in the first category (e.g., Cavalier-Smith 1975) stipulate that the genomes of organelles have autogenous origins and are descended from nuclear genes by **filial compartmentalization**, whereby part of the nuclear genome became incorporated into a membrane-enclosed organelle and subsequently assumed a quasi-independent existence. In contrast, the **endosymbiotic theory** (e.g., Margulis 1981) claims that the origin of extranuclear DNA is exogenous. According to this proposal, first made by Mereschkowsky (1905), the ancestor of eukaryotic organisms engulfed prokaryotes, which were subsequently retained because of a mutually beneficial or symbiotic relationship (Martin and Müller 1998). With time, the genomes of the endosymbionts were streamlined by loss of genes and became obligatory symbionts (i.e., incapable of an independent existence outside their host).

The molecular evidence is now overwhelmingly in favor of the endosymbiotic theory. A list of biochemical characters that distinguish the genomes of both chloroplasts and prokaryotes from the nuclear genome of eukaryotes is given in Table 5.7. The ultimate support, however, came from DNA sequence data, mainly rRNA sequences. Because of their low rates of substitution, rRNA sequences have proved to be very useful for addressing questions concerning very ancient evolutionary divergence events.

Schwarz and Kössel (1980) showed that the nucleotide sequence of the 16S rRNA gene from the chloroplast of maize (*Zea mays*) has regions with a strong similarity to those in the 16S rRNA from the bacterium *Escherichia coli*. The degree of similarity between nuclear and chloroplast rRNA sequences was much lower. A detailed analysis of 16S rRNA sequences from photosynthetic bacteria (Giovannovi et al. 1988) supports the proposal that green chloroplasts were derived from the cyanobacteria (Figure 5.45).

TABLE 5.7 Molecular characters that distinguish the genomes of chloroplasts and prokaryotes from the nuclear genomes of eukaryotes

1. Histoneless DNA
2. 120,000–150,000 base pairs in size
3. Circular genome
4. Sensitivity of transcription to rifampicin
5. Inhibition of ribosomes by streptomycin, chloramphenicol, spectromycin, and paromonycin
6. Insensitivity of translation to cycloheximide
7. Translation starts with formylmethionine
8. Polyadenylation of mRNA absent or very short
9. Prokaryotic promoter structure

Phylogenetic analyses of rRNA sequences suggested that mitochondria were derived from the α subdivision of the purple bacteria (Figure 5.45; Cedergren et al. 1988). There were initial indications that there were two independent endosymbiotic events giving rise to the mitochondria (Gray et al. 1989). Current data, however, strongly support a monophyletic origin of all mitochondrial genomes from a bacterial lineage that also gave rise to the rickettsias (Leblanc et al. 1997; Sogin 1997). Andersson et al. (1998) identified *Rickettsia prowazekii*, the causative agent of epidemic typhus, as the closest extant relative of the mitochondria.

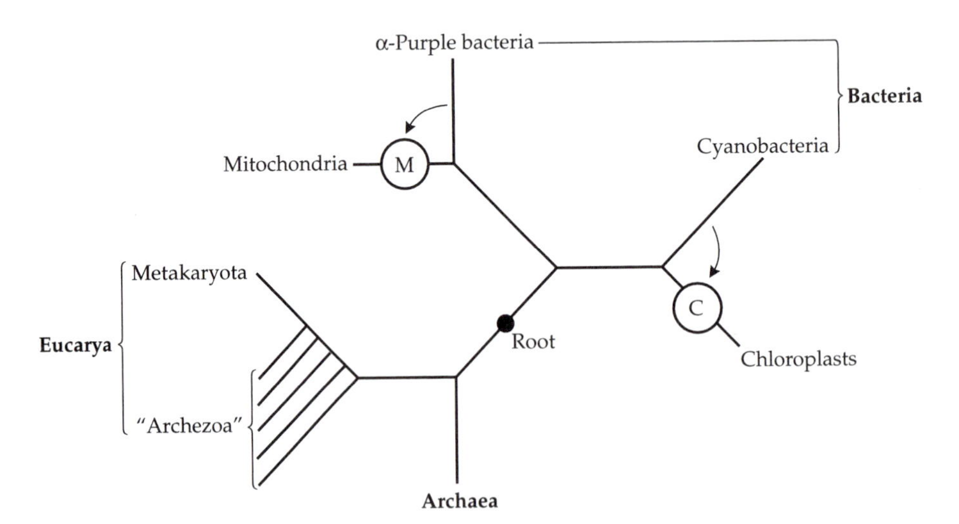

FIGURE 5.45 Schematic unrooted tree illustrating the phylogenetic affinities of chloroplasts and mitochondria. The mitochondria are derived from the α-purple bacteria via an endosymbiotic event (M). The chloroplasts are derived from the cyanobacteria via a second endosymbiotic event (C).

FURTHER READINGS

Brown, T. A. and K. A. Brown. 1994. Using molecular biology to explore the past. BioEssays 16: 719–726.

Crawford, D. J. 1990. *Plant Molecular Systematics: Macromolecular Approaches*. Wiley, San Francisco.

Durbin, R., S. Eddy, A. Krogh, and G. Mitchison. 1998. *Biological Sequence Analysis: Probabilistic Models of Proteins and Nucleic Acids*. Cambridge University Press, Cambridge.

Eldredge, N. and J. Cracraft. 1980. *Phylogenetic Patterns and the Evolutionary Process: Method and Theory in Comparative Biology*. Columbia University Press, New York.

Felsenstein, J. 1988. Phylogenies from molecular sequences: Inference and reliability. Annu. Rev. Genet. 22: 521–565.

Felsenstein, J. 1996. Inferring phylogenies from protein sequences by parsimony, distance, and likelihood methods. Methods Enzymol. 266: 418–427.

Harvey, P. H., A. J. Leigh Brown, J. Maynard Smith, and S. Nee. (eds.) 1996. *New Uses for New Phylogenies*. Oxford University Press, Oxford.

Hillis, D. M., C. Moritz, and B. K. Mable (eds.). 1996. *Molecular Systematics*, 2nd ed. Sinauer Associates, Sunderland, MA.

Huelsenbeck, J. P. and K. A. Crandall. 1997. Phylogeny estimation and hypothesis testing using maximum likelihood. Annu. Rev. Ecol. Syst. 28: 437–466.

Martin, R. D. 1990. *Primate Origins and Evolution: A Phylogenetic Reconstruction*. Princeton University Press, Princeton.

Nei, M. 1987. *Molecular Evolutionary Genetics*. Columbia University Press, New York.

Nei, M. 1996. Phylogenetic analysis in molecular evolutionary genetics. Annu. Rev. Genet. 30: 371–403.

Pagel, M. 1997. Inferring evolutionary processes from phylogenies. Zool. Scripta 26: 331–348.

Sneath, P. H. A. and R. R. Sokal. 1973. *Numerical Taxonomy: The Principles and Practice of Numerical Classification*. W.H. Freeman, San Francisco.

Wiley, E. O. 1981. *Phylogenetics: The Theory and Practice of Phylogenetic Systematics*. Wiley, New York.

Chapter 6

Gene Duplication, Exon Shuffling, and Concerted Evolution

The evolutionary significance of gene duplication was first recognized by Haldane (1932) and Muller (1935), who suggested that a redundant duplicate of a gene may acquire divergent mutations and eventually emerge as a new gene. A gene duplication was first observed by Bridges (1936) in the *Bar* locus in *Drosophila*. This discovery notwithstanding, few other examples of duplicate genes were found prior to the advent of biochemical and molecular biology techniques. The development of protein sequencing methods in the 1950s provided the first tool for the study of long-term evolutionary processes, and in the late 1950s the α and β chains of hemoglobin were recognized to have been derived from duplicate genes (Itano 1957; Rhinesmith et al. 1958; Braunitzer et al. 1961). Later, isozyme and cytological studies provided evidence for the frequent occurrence of gene duplication during evolution.

Using evidence from various types of studies, Ohno (1970) put forward a view according to which gene duplication is the only means by which a new gene can arise. Although other means of creating new functions are now known, Ohno's view remains largely valid. In this chapter, we discuss several evolutionary aspects of DNA duplication, including the curious phenomenon of concerted evolution, i.e., the coordinated manner that characterizes the evolution of repeated DNA sequences.

The discovery of split genes prompted Gilbert (1978) to suggest that recombination within introns provides a mechanism for the exchange of exon sequences between genes. Many such examples of exon exchange have been found, indicating that this mechanism has played a significant role in the evolution of eukaryotic genes with new functions. Several aspects of the evolution of eukaryotic genes and the origin of introns will also be discussed.

TYPES OF GENE DUPLICATION

An increase in the number of copies of a DNA segment can be brought about by several types of **gene duplication**. These are usually classified according to the extent of the genomic region involved. The following types of duplication are recognized: (1) **partial** or **internal gene duplication**, (2) **complete gene duplication**, (3) **partial chromosomal duplication**, (4) **complete chromosomal duplication**, and (5) **polyploidy**, or **genome duplication**. The first four categories are referred to as **regional duplications** because they do not affect the entire haploid set of chromosomes. Ohno (1970) has argued that genome duplication has generally been more important than regional duplication, because in the latter case only parts of the regulatory system of structural genes may be duplicated, and such an imbalance may disrupt the normal function of the duplicated genes. However, as discussed below, regional duplications have apparently played a very important role in evolution.

The principal molecular mechanism responsible for gene duplication is unequal crossing over (see Figure 1.15). Unequal crossing over between misaligned sequences gives rise to a tandemly duplicated region on one chromosome and a complementary deletion on the other, the length of which depends on the size of the misalignment. Unequal crossing over is greatly facilitated by the presence of repeated sequences around the sequence to be duplicated (Chapter 7). Once a DNA sequence is duplicated in tandem, the process of gene duplication can proceed in a progressively accelerated or cascading fashion because the chance of unequal crossing over increases with the number of duplicated copies.

DNA duplication has long been recognized as an important factor in the evolution of genome size. In particular, the duplication of the entire genome (or a major part of it, such as a chromosome) may result in a sudden substantial increase in genome size. Genome duplication events have been registered repeatedly during the evolution of different groups of organisms. Evolutionary pathways resulting in genome enlargement will be discussed in Chapter 8.

DOMAINS AND EXONS

A **domain** is a well-defined region within a protein that either performs a specific function, such as substrate binding, or constitutes a stable, independently folding, compact structural unit within the protein that can be distinguished from all the other parts. The former is referred to as a **functional domain**, and the latter as a **structural domain** or **module** (Gō and Nosaka 1987). Defining the boundaries of a functional domain is often difficult because in many cases functionality is conferred by amino acid residues that are scattered throughout the polypeptide. Structural modules, on the other hand, are co-linear with the amino acid sequence of a protein (i.e., a module consists of a continuous stretch of amino acids).

The above distinction is important when considering possible evolutionary mechanisms by which multidomain proteins have come into existence. If a functional domain coincides with a module, its duplication will increase the number of functional segments. In contrast, if functionality is conferred by amino acid residues scattered among different modules, the effects of duplicating a single module may not be functionally desirable. Indeed, the internal repeats found in many proteins often correspond to either structural modules or single modular functional domains (Barker et al. 1978).

The identification of modules in proteins is usually accomplished by means of a graphical method called the **Gō plot** (Gō 1981). In this method, the amino acid residues of a protein are listed consecutively on the two axes of a two-dimensional matrix. Given the known tertiary structure of a protein, a plus sign (+) is entered in the matrix if the distance between two corresponding residues is greater than a certain value (Figure 6.1). For globular proteins, the value used is usually the radius of the sphere containing the globular protein, denoted as R. In the ideal case, domains are identified unambiguously as empty, nonoverlapping, right-angle triangles whose hypotenuses are on the diagonal of the Gō plot, and whose sides are defined by distinct rectangles containing clusters of plus signs (Figure 6.1a). In less than ideal cases, a certain overlap may exist between adjacent empty triangles (Figure 6.1b). In real life, the situation may be much more complicated, and complex statistical methods are used to identify the most likely locations of the boundaries between modules. Figure 6.1c shows a Gō plot for the β subunit of human hemoglobin.

Theoretically, several possible relationships may be envisioned between the structural domains and the arrangements of the exons in the gene (Figure 6.2). Gō (1981) found that in many globular proteins for which the internal modular division has been determined, a more or less exact correspondence exists between the exons of the gene and the structural domains of the protein product (Figure 6.2a and b). In only a few cases, a single module was found to be encoded by more than one exon (Figure 6.2d). A complete discordance between the modular structure of a protein and the division of its gene into exons (Figure 6.2e) was not found in her study. In a considerable number of cases, however, several adjacent modules were found to be encoded by the same exon (Figure 6.2c).

The vertebrate hemoglobin α and β chains, for instance, consist of four domains, whereas their genes consist of only three exons, the second of which encodes two adjacent domains. Gō (1981) postulated that a merger occurred between two exons as a result of the loss of a central intron. Indeed, the

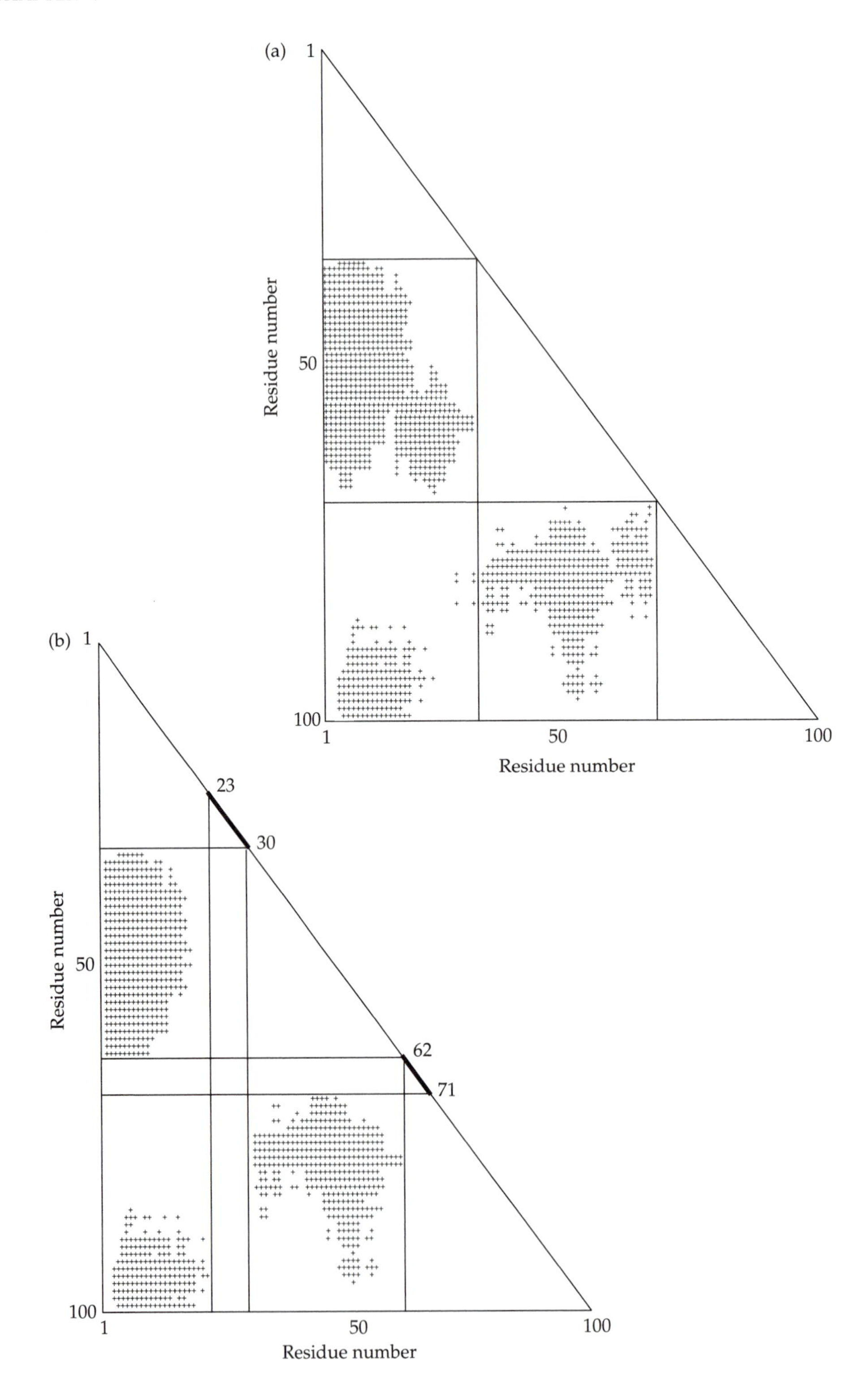
(a)
1
50
100
Residue number
1
50
100
Residue number
(b)
1
23
30
62
71
50
100
Residue number
1
50
100
Residue number

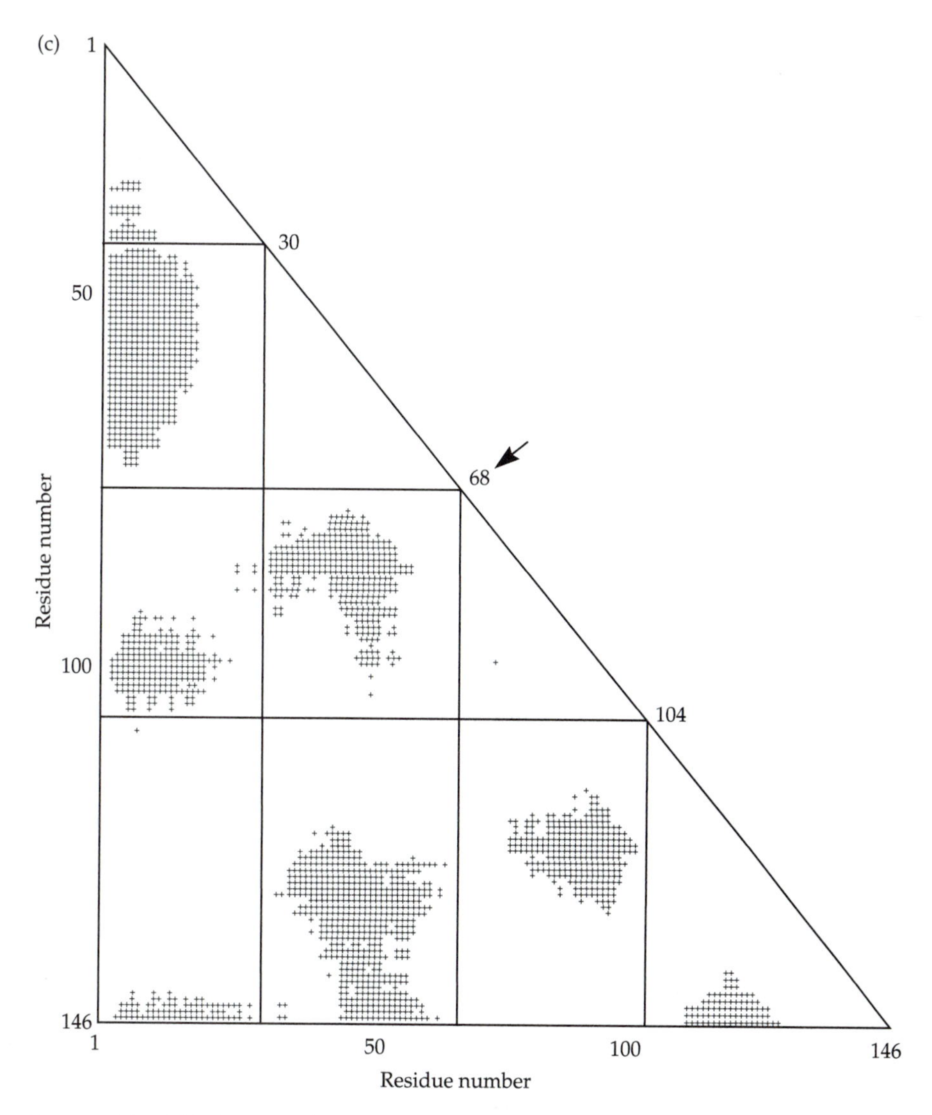

FIGURE 6.1 Gō plots. A plus sign is entered in the matrix if the distance between two corresponding amino acid residues is greater than the radius of the protein. (a) An ideal case in which the modules can be identified unambiguously, i.e., it is not possible to draw any other horizontal or vertical lines without cutting through blocks of plus signs. (b) A less ideal case, in which several alternative horizontal and vertical lines can be drawn, and the boundaries between adjacent modules are identified as ranges (thick lines). (c) An actual Gō plot for the β chain of hemoglobin from humans. The radius of the protein is 27Å. The predicted positions of the introns are shown on the hypotenuse. The human β chain gene consists of only three exons. Gō (1981) postulated that a merger occurred between two exons as a result of the loss of a central intron. An arrow marks the predicted position of the central intron. This prediction was borne out by the finding of just such an intron in plant leghemoglobin genes, as well as in several invertebrate globin genes. Modified from Gō (1981).

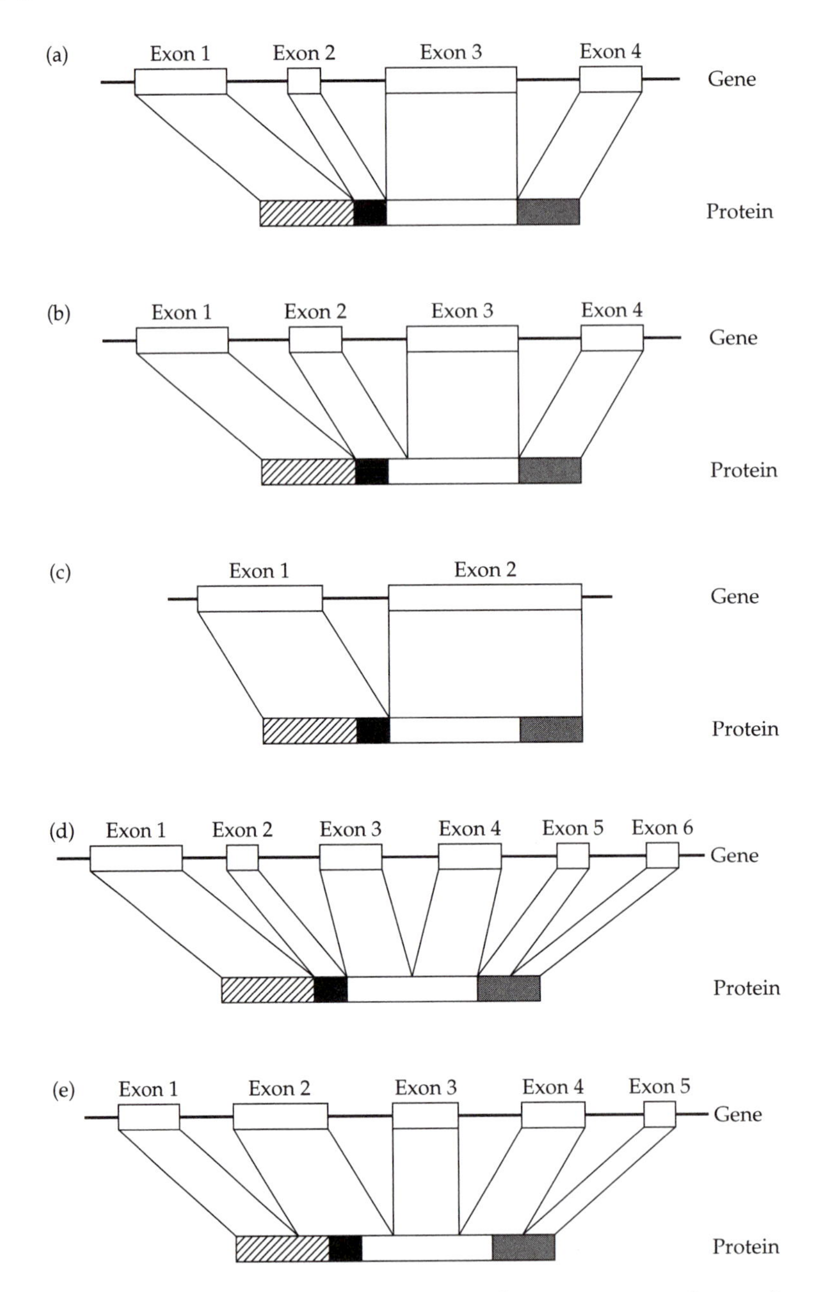

FIGURE 6.2 Five possible relationships between the arrangement of exons in a gene and the structural domains of its protein. (a) Each exon corresponds exactly to a structural domain. (b) The correspondence is only approximate. (c) An exon encodes two or more domains. (d) A single structural domain is encoded by two or more exons. (e) Lack of correspondence between exons and domains. The structural domains of the protein are designated by different boxes (hatched, black, white, and gray).

homologous globin genes in plants (leghemoglobins) were subsequently found to contain an additional intron at or very near the position predicted by the domain structure of vertebrate globins (Landsman et al. 1986). A similar intron was found in the globin genes of the nematode *Pseudoterranova decipiens* (Dixon et al. 1991). Interestingly, the globin-encoding gene of the free-living nematode *Caenorhabditis elegans* was found to contain a single intron corresponding in position to the central intron in leghemoglobins (Kloek et al. 1993). Thus, during the evolution of the globin gene family from a four-exon ancestral gene, several lineages lost some or all of their three introns, thereby generating a panoply of exon–intron permutations (Figure 6.3).

In the majority of cases, a domain duplication at the protein level indicates that an exon duplication has occurred at the DNA level. It has therefore been suggested that exon duplication is one of the most important types of internal gene duplication. Eukaryotic genes generally consist of many exons and introns (Chapter 1), and neighboring exons are often identical or very similar to one another. Moreover, many proteins of present-day organisms show internal repeats of amino acid sequences, and the repeats often correspond to functional or structural domains within the proteins (Barker et al. 1978). These observations suggest that the genes for many proteins were formed by internal gene duplication, and that the function of these proteins was improved by increasing their stability or the number of active sites. Internal duplications can also provide redundant DNA segments that may enable a gene to develop new functions. Many complex genes in present-day organisms might have evolved from small, simple primordial genes via internal duplication and subsequent modification (Li 1983).

DOMAIN DUPLICATION AND GENE ELONGATION

A survey of modern genes in eukaryotes shows that internal duplications have occurred frequently in evolution. This increase in gene size, or **gene elongation**, is one of the most important steps in the evolution of complex genes from simple ones. Theoretically, elongation of genes can also occur by other means. For example, a mutational change converting a stop codon into a sense codon can also elongate the gene (Chapter 1). Similarly, either insertion of a foreign DNA segment into an exon or the occurrence of a mutation obliterating a splicing site will achieve the same result. These types of molecular changes, however, would most probably disrupt the function of the elongated gene, because the added regions would consist of an almost random array of amino acids. Indeed, in the vast majority of cases, such molecular changes have been found to be associated with pathological manifestations. For instance, the α-hemoglobin abnormalities *Constant Spring* and *Icaria* resulted from mutations turning the stop codon into codons for glutamine and lysine, respectively, thus adding 30 additional residues to the α chains of these variants (Weatherall and Clegg 1979). By contrast, duplication of a structural domain is less likely to be problematic. Indeed, such a duplication can some-

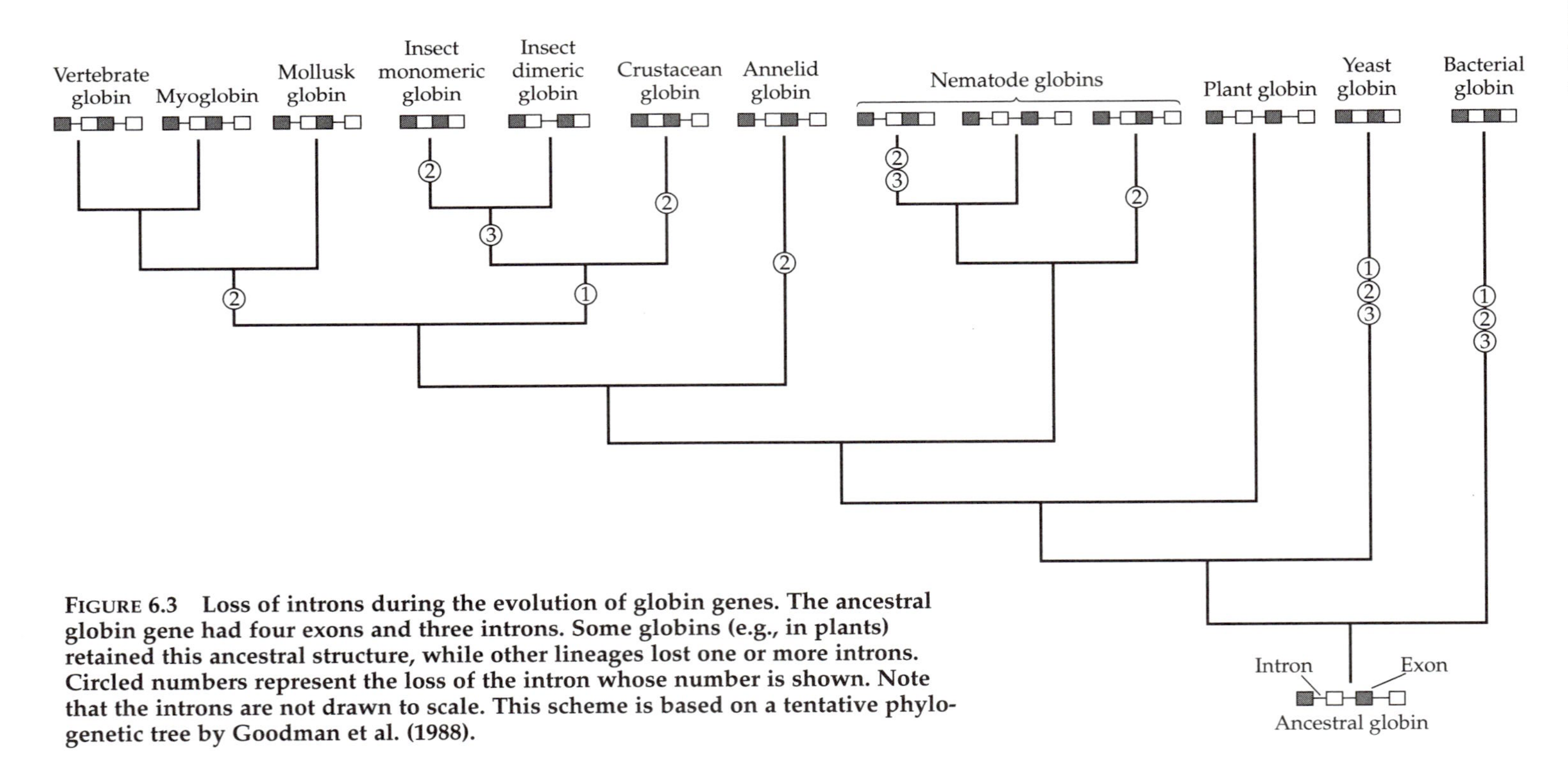

FIGURE 6.3 Loss of introns during the evolution of globin genes. The ancestral globin gene had four exons and three introns. Some globins (e.g., in plants) retained this ancestral structure, while other lineages lost one or more introns. Circled numbers represent the loss of the intron whose number is shown. Note that the introns are not drawn to scale. This scheme is based on a tentative phylogenetic tree by Goodman et al. (1988).

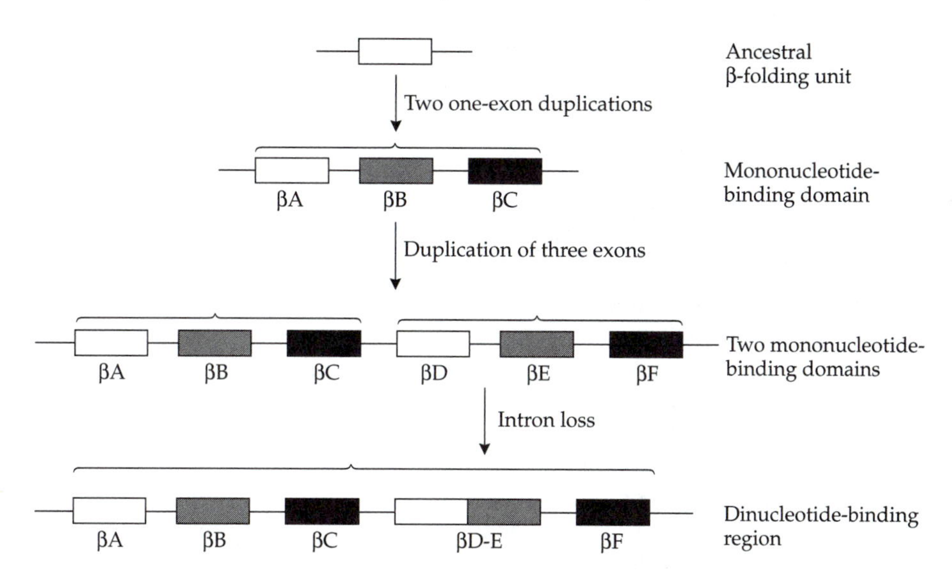

FIGURE 6.4 Hypothetical scenario for the evolution of a dinucleotide-binding region. Exons are shown as boxes connected by introns, shown as lines. An ancestral β-folding unit encoded by an exon underwent two duplication events to produce a triexonic mononucleotide-binding domain. The duplication of the three exons resulted in the creation of two mononucleotide-binding domains. Subsequent modifications of the primary sequence gave rise to a dinucleotide-binding region. The present pentaexonic arrangement is explained by the loss of an intron.

times even enhance the function of the protein produced, for example by increasing the number of active sites (a quantitative change), thus enabling the gene to perform its functions more rapidly and efficiently, or by having a synergistic effect yielding a new function (a qualitative change). One such example is the dinucleotide-binding regions of glyceraldehyde-3-phosphate dehydrogenase (GAPDH) and alcohol dehydrogenase (ADH). In many species, these binding regions consist of two domains (Figure 6.4). Each of these domains consists of three homologous β-folding units, denoted βA, βB, and βC in the N-terminal domain and βD, βE, and βF in the C-terminal domain. The entire dinucleotide-binding region is encoded by five exons, three for the N-terminal domain and two for the C-terminal domain. Each coenzyme-binding domain can only bind a mononucleotide. The entire duplicated region, however, can bind not only two mononucleotides, but also a dinucleotide. The duplicated domain, therefore, recognizes a different molecular entity than the unduplicated domain.

The second possibility for the emergence of a novel function following partial gene duplication is for the internal copies thus produced to diverge in sequence, ultimately resulting in each of them performing a different function. For instance, the variable and constant regions of immunoglobulin genes were probably derived from a common primordial domain, but have since acquired distinct properties (Leder 1982). Thus, despite common molecular ancestry, the variable region of immunoglobulins binds antigens, while the constant region mediates non-antigenic functions. Many complex genes might have arisen in this manner.

Even if an internal duplication does not involve any active sites, it may still be beneficial. For example, a duplication event involving a structural domain engaged in either conferring spatial stability to the protein or protecting its active parts may indirectly alter protein function or longevity. For example, so-called PEST polypeptides—proteins rich in proline (P), glutamic acid (E), serine (S), and threonine (T)—were found to be degraded rapidly inside eukaryotic cells (Rogers et al. 1986). In some cases, duplicated PEST domains were found in proteins, thus ensuring a very rapid degradation, which is evidently important in proteins, such as regulatory nuclear factors, that have a specific and transient function (Chevaillier 1993). It has also been suggested that, as long as they do not interfere with normal function, redundant duplicate domains may be maintained indefinitely within the genome, and may in time serve as raw materials for creating new functions.

The following sections present several examples of internal gene duplications to illustrate the consequences of gene elongation during evolution.

The ovomucoid gene

Ovomucoid is an inhibitor of trypsin, an enzyme that catalyzes the digestion of proteins. It is present in the albumen (egg white) of birds. The ovomucoid polypeptide can be divided into three functional domains (Figure 6.5). Each domain is capable of binding one molecule of either trypsin or another serine proteinase. The DNA regions coding for the three functional domains clearly share a common origin and are separated from each other by introns (Stein et al. 1980). Domains I and II, I and III, and II and III exhibit 46, 33, and 30% amino acid sequence identity, respectively, and 66, 42, and 50% nucleotide sequence identity. Each of the three regions consists of two exons interrupted by an intron, and the two exons exhibit no similarity between them. Thus, the ovomucoid gene appears to have been derived from a primordial single-domain gene by two internal duplications, each of which involved two neighboring exons. Since domains I and II are more similar to each other than either of them is to domain III, they were probably derived from the second duplication event, while domain III was the product of the first duplication.

Enhancement of function in the $\alpha2$ allele of haptoglobin

A well-known example of an enhancement in function as a consequence of an internal gene duplication is that of the haptoglobin $\alpha2$ allele in humans (Smithies et al. 1962). Haptoglobin is a tetrameric protein made out of two α and two β chains. Both chains are produced by the same gene as a single polypeptide, which is then cleaved at an arginine residue to generate the α and β subunits. Haptoglobin is found in the blood serum, where it functions

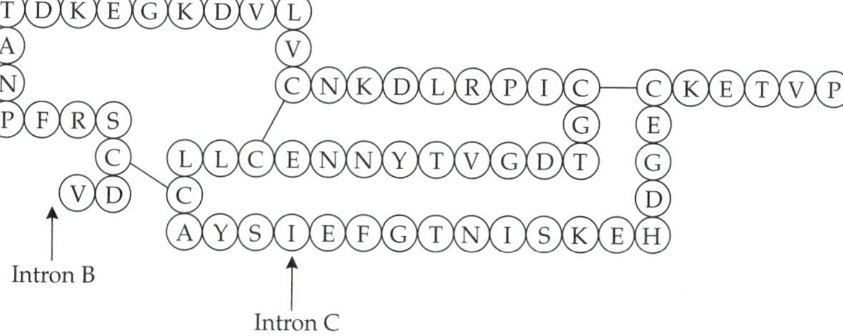

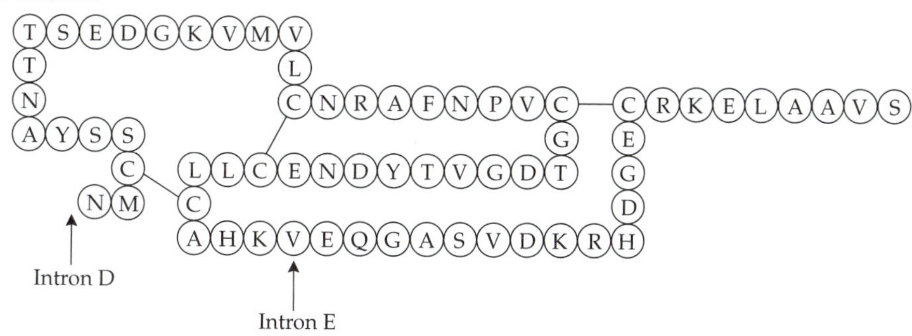

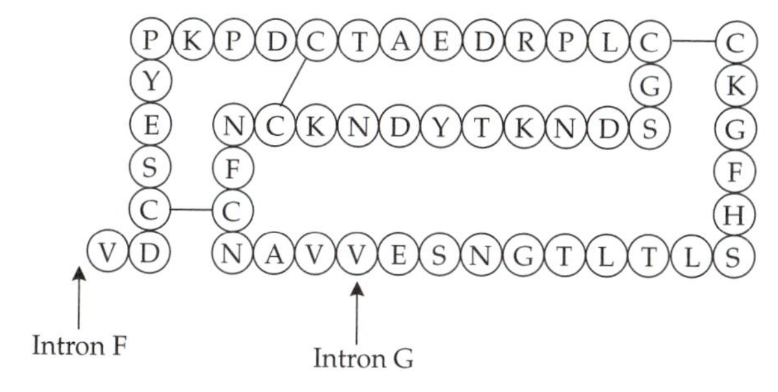

	Percent similarity	
Domains	Amino acids	Nucleic acids
I vs. II	46	66
II vs. III	30	42
I vs. III	33	50

FIGURE 6.5 The three functional domains of the secreted ovomucoid from chicken and the degree of sequence similarity between the domains at the amino acid and nucleotide levels. Introns B–G are indicated by arrows. Intron A interrupts the 5′ noncoding region and is not shown. Data from Stein et al. (1980) and O'Malley et al. (1982).

as a transport glycoprotein that removes free hemoglobin from the circulation of vertebrates. In human populations, haptoglobin α is polymorphic due to the existence of three common alleles: *slow α1* (*α1S*), *fast α1* (*α1F*), and *α2*. The *α2* allele was probably created by a nonhomologous crossing over within different introns of two *α1* alleles within a heterozygous individual carrying both the *α1S* and *α1F* electrophoretic variants. The internal duplication of about 1.7 kb, of which 177 bp are exonic, nearly doubled the length of the polypeptide (from 84 to 143 amino acids). As a consequence, the stability of the haptoglobin–hemoglobin complex and the efficiency of rendering the heme group of hemoglobin susceptible to degradation increased considerably (Black and Dixon 1968).

The *α2* allele is probably of recent origin, at least more recent than the human–chimpanzee split, but it has a fairly high frequency (30–70%) in Europe and in parts of Asia (Mourant et al. 1976). If indeed the individuals carrying the *α2* alleles enjoy a selective advantage over carriers of the *α1* alleles, it is likely that in the future the *α2* allele will become fixed in the human population at the expense of the *α1* variants.

Interestingly, an even longer allele, *α3* (or haptoglobin Johnson), was found in human populations. This allele contains a threefold tandem repeat of the same 1.7-kb segment implicated in the *α2* allele duplication (Oliviero et al. 1985).

Origin of an antifreeze glycoprotein gene: A "cool" tale of internal gene duplications

The body fluids of most teleosts (ray-finned fish) freeze at temperatures ranging from –1.0°C to –0.7°C. Therefore, most fish cannot survive the freezing temperatures (–1.9°C) of the Antarctic Ocean. Freezing resistance in Antarctic fish is due to the existence in the blood of a protein that lowers the freezing temperature by adsorbing to small ice crystals and inhibiting their growth, which might otherwise break the cell membranes. There are several such proteins, termed antifreeze protein types I, II, and III, and antifreeze glycoprotein. Where did these proteins come from? In the case of the antifreeze glycoprotein gene in the Antarctic toothfish (*Dissostichus mawsoni*), the answer provides an interesting example of acquisition of a novel function through internal gene duplication.

There are many different antifreeze glycoproteins in the cod, each of which is composed mostly of two simple tripeptide repeats: Thr-Ala-Ala and Thr-Pro-Ala. (Note that the proline codon family differs from the alanine codon family by a single nucleotide.) The antifreeze glycoproteins are encoded by a large gene family, in which each gene encodes a large polyprotein precursor that is cleaved posttranslationally to yield multiple antifreeze glycoprotein molecules. Chen et al. (1997) characterized one antifreeze glycoprotein gene from the Antarctic toothfish and discovered that it derives from a gene encoding pancreatic trypsinogen (Figure 6.6). The evolutionary history of the antifreeze glycoprotein gene could be reconstructed accurately, mainly

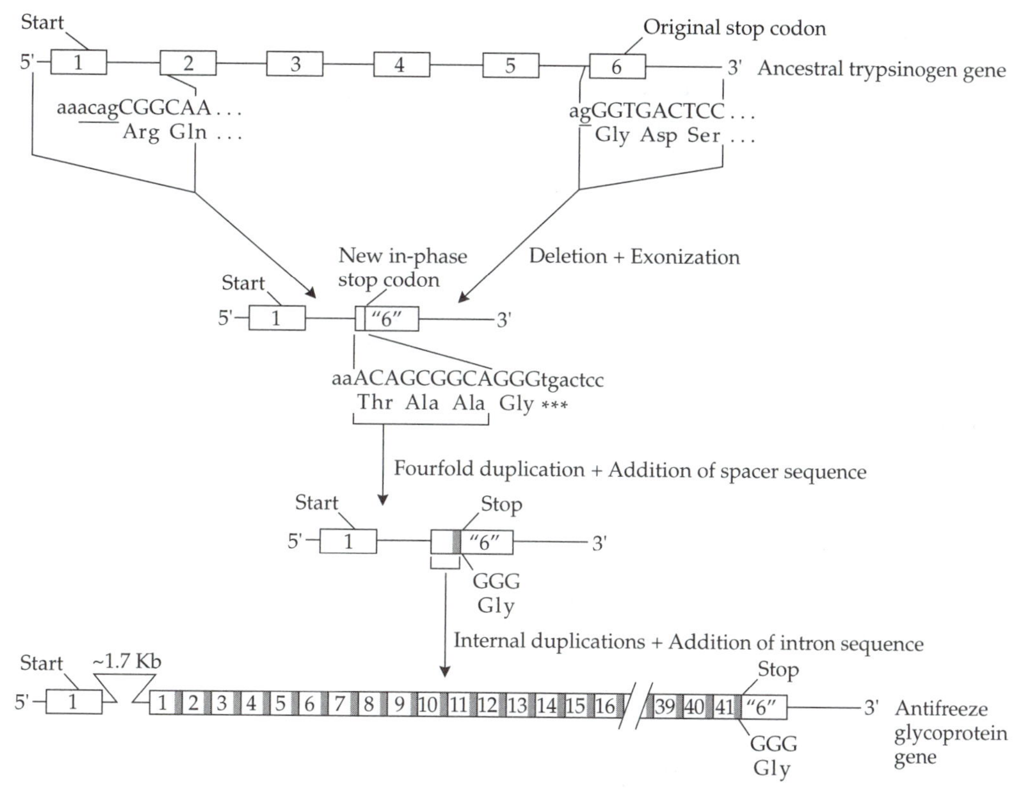

FIGURE 6.6 A likely evolutionary pathway by which an ancestral trypsinogen gene with six exons (numbered boxes) was transformed into an antifreeze glycoprotein gene in the Antarctic toothfish (*Dissostichus mawsoni*). Following a deletion and the exonization of five intronic nucleotides (underlined lowercase letters) in the ancestral trypsinogen gene, a new gene with two exons emerged. (The second exon is marked "6" to emphasize its ancestry, and the new stop codon that was brought into frame by the exonization is marked by asterisks.) The sequence encoding Thr-Ala-Ala was duplicated to create a fourfold repetition, and a short spacer sequence of unknown origin (shaded box) was added to the repeated unit. Multiple internal gene duplications resulted in 41 repeats. The addition of the ~1.7 Kb-long sequence to the intron is indicated as a triangular loop. This addition could have occurred in any one of the previous steps in the evolution of this gene, and was added at the end for graphical convenience only.

because the gene was inferred to have arisen only 5–14 million years ago, an estimate that agrees well with the presumed date of the freezing of the Antarctic Ocean (10–14 million years ago). Its relative newness essentially means that its evolutionary history has not yet been obscured by superimposed molecular changes.

The evolutionary history of the antifreeze glycoprotein gene can be summarized as shown in Figure 6.6. An initial deletion in the ancestral trypsinogen gene, starting at nucleotide 6 in the second exon and ending one nucleotide before the start of exon 6, created a new gene with two exons. Four nucleotides from the first intron and one from the fifth intron became part of this newly created exon, whose short frameshifted reading frame encoded a tetrapeptide (Thr-Ala-Ala-Gly) before reaching a new in-frame stop codon (TGA). The 12-bp sequence encoding the first three amino acids (Thr-Ala-Ala) was duplicated twice to create a fourfold repetition of this tripeptide. Following this initial round of duplication, a short spacer sequence of unknown origin was added to the 3′ end of the repeated unit. The resulting sequence was then duplicated multiple times to yield 41 repeats followed by the Gly codon from the original tetrapeptide and a stop codon. Some of the spacer sequences encode peptide motifs that serve as signals for the cleavage of the antifreeze glycoprotein polypeptide into the active proteins. Sometime along the line from the ancestral trypsinogen gene to the antifreeze glycoprotein gene, a sequence approximately 1.7 Kb long was added to the intron. Most probably, this insertion had no functional consequences.

We note that a completely new function has been created by multiple mutational events (mostly internal gene duplications) in a very short time span. Thus, the new gene must have been subjected to intense positive selection, most probably due to an abrupt shift in environmental conditions (Logsdon and Doolittle 1997).

Prevalence of domain duplication

Gene elongation during evolution has largely depended on the duplication of domains. Table 6.1 presents a list of genes for which there is evidence of internal duplication during their evolutionary histories. All involve one or more domain duplications, and some of the sequences (e.g., ferredoxin, serum albumin, and the tropomyosin α chain) were derived from multiplications of a primordial sequence, resulting in a repetitive structure that takes up the entire length of the protein. In each of these examples, the duplication event could easily be inferred from protein or DNA sequence similarity. There may be many other complex genes that have evolved by internal gene duplication, but their duplicated regions may have diverged from each other to such an extent that the sequence homology between them is no longer discernible. In some cases, such as the constant and variable regions of immunoglobulin genes, we can infer common ancestry by comparing the secondary structures of the domains, because the secondary structure has been preserved better than the amino acid sequence (Hood et al. 1975). Thus, internal duplications in proteins are probably much more ubiquitous than the empirical data indicate.

FORMATION OF GENE FAMILIES AND THE ACQUISITION OF NEW FUNCTIONS

A complete gene duplication produces two identical copies. How they will evolve varies from case to case. In principle three possibilities exist. The

TABLE 6.1 Examples of proteins with internal domain duplications taking up 50% or more of the total length of the protein

Sequence (organism)	Length of protein[a]	Length of repeat[a]	Number of repeats[b]	Percent repetition[c]
α1β-glycoprotein (human)	474	91	5	96
Angiotensin I-converting enzyme (human)	1,306	357	2	55
Calbindin (human, bovine)	260	43	6	99
Calcium-dependent regulator protein (human)	148	74	2	100
Ferredoxin (*Azobacter vinelandii*)	70	30	2	86
Ferredoxin (*Azobacter pasteurianum*)	55	28	2	100
Hemopexin (human)	439	207	2	94
Hexokinase (human)	917	447	2	97
Immunoglobulin γ chain C region (human)	329	108	3	98
Immunoglobulin ε chain C region (human)	423	108	4	100
Interleukin-2 receptor (human)	251	68	2	54
Interstitital retinol-binding protein (bovine)	1,263	302	4	96
Lactase-phlorizin hydrolase (human)	1,927	480	3	79
Lymphocyte activation gene-3 protein (human)	470	138	2	59
Multidrug resistance-1 P-glycoprotein (human)	1,280	609	2	95
Ovoinhibitor (chicken)	472	64	7	95
Parvalbumin (human)	108	39	2	72
Plasminogen (human)	790	79	5	50
Preproglucagon (rat)	180	36	3	60
Prepro-von Willebrand factor (human)[d]	3,817	586	3	
		30	2	
		117	2	
		354	5	85
Protease inhibitor, Bowman-Birk type (soybean)	71	28	2	79
Protease inhibitor, submandibular gland type (rat)	115	54	2	94
Ribonuclease/angiogenin inhibitor (human)	461	57	8	99
Serum albumin (human)	584	195	3	100
Tropomyosin α chain (human)	284	42	7	100
Twitchin (*Caenorhabditis elegans*)[d]	6,049	100	31	
		93	26	91
Villin (human)	826	360	2	87
Vitamin D-dependent calcium-binding protein (bovine)	260	54	3	63

[a]Number of amino acid residues.

[b]Some of the repeats may be truncated in comparison with the repeated consensus unit.

[c]Percent of the total length of the protein occupied by repeated sequences.

[d]In these proteins, several unrelated types of repeats are present.

copies may retain their original function, enabling the organism to produce a larger quantity of RNAs or proteins. Alternatively, one of the copies may be incapacitated by the occurrence of a deleterious mutation and become a functionless pseudogene (see page 274). More important, however, is the third possibility: that the gene duplication may result in the emergence of genetic novelties or new genes. This will happen if one of the duplicates retains its original function while the other accumulates molecular changes such that, in time, it can perform a different task.

Repeated genes can be divided into two types: variant and invariant repeats. **Invariant repeats** are identical or nearly identical in sequence to one another. In several cases, the repetition of identical sequences can be shown to be correlated with the synthesis of increased quantities of the gene product that is required for the normal function of the organism. Such repetitions are referred to as **dose repetitions**. Dose repetitions are quite common when there is a metabolic need to produce large quantities of specific RNAs or proteins (Ohno 1970). For example, a duplication of the acid monophosphatase locus in yeast enables the carrier to produce twice the amount of enzyme, thus exploiting available phosphate more efficiently when phosphate is a limiting factor to growth (Hensche 1975). Representative examples of gene duplication include the genes for rRNAs and tRNAs, which are required for translation, and the histone genes, which constitute the main protein component of chromosomes and therefore must be synthesized in large quantities especially during the S phase of the cell cycle, when DNA is replicated (Elgin and Weintraub 1975). However, the genome of eukaryotes is also known to contain invariant repeats that lack any function (Chapter 8).

Variant repeats are copies of a gene that, although similar to each other, differ in their sequences to a lesser or greater extent. Interestingly, variant repeats can sometimes perform markedly different functions. For example, thrombin, which cleaves fibrinogen during the process of blood clotting, and the digestive enzyme trypsin have been derived from a complete gene duplication in the past. Similarly, lactalbumin, a subunit of the enzyme that catalyzes the synthesis of the sugar lactose, and lysozyme, which dissolves certain bacteria by cleaving the polysaccharide component of their cell walls are related by descent to each other. Differentiation in function usually requires a large number of substitutions. However, in many cases, a novel function may be achieved through surprisingly few substitutions. For example, lactate dehydrogenase can be changed into malate dehydrogenase by replacing just one of its 317 amino acids (Wilks et al. 1988).

All the genes that belong to a certain group of repeated sequences in a genome are referred to as a **gene** or **multigene family**. Functional and nonfunctional members of a gene family may reside in close proximity to one another on the same chromosome, or they may be located on different chromosomes. A member of a gene family that is located alone at a different genomic location than the other members of the family is called an **orphon**.

When duplicate genes become too divergent from each other in either function or sequence, it may no longer be convenient to assign them to the same gene family. The term **superfamily** was coined by Dayhoff (1978) in order to distinguish closely related proteins from distantly related ones. Accordingly, proteins that exhibit at least 50% similarity to each other at the amino acid level are considered as members of a family, while homologous proteins exhibiting less than 50% similarity are considered as members of a superfamily. For example, the α- and β-globins are classified into two separate families, and together with myoglobin they form the globin superfamily (see page 278). However, the two terms cannot always be used strictly according to Dayhoff's criteria. For example, human and carp α-globin chains exhibit only a 46% sequence similarity, which is below the limit for assignment to the same gene family. For this reason, the classification of proteins into families and superfamilies is determined not only according to sequence resemblance, but also by considering auxiliary evidence pertaining to functional similarity, tissue specificity, or type of homology.

An important feature associated with gene duplication is that as long as two or more copies of a gene exist in proximity to each other, the process of gene duplication can be greatly accelerated in this region, and numerous copies may be produced. One practical outcome of gene duplication and the subsequent modification of the resulting copies is that many genes performing different functions have actually been derived from a common ancestral gene, and are thus homologous with each other. With the recent avalanche of DNA sequence data, a surprising number of unexpected similarities among proteins not previously known to be related to each other have been revealed (Table 6.2). One such example involves trypsin and chymotrypsin. Since their divergence from each other about 1.5 billion years ago, these two digestive enzymes have acquired distinct functions: trypsin cleaves polypeptide chains at arginine and lysine residues, whereas chymotrypsin cleaves polypeptide chains at phenylalanine, tryptophan, and tyrosine residues (Barker and Dayhoff 1980).

The number of genes within gene families varies widely. Some genes are repeated within the genome a few times; others may be repeated hundreds of times. In the following, rRNA and tRNA genes will be used to illustrate highly repetitive invariant genes. Lowly repetitive genes will be represented by the lactate dehydrogenase isozymes and the color-sensitive opsins.

RNA-specifying genes

Table 6.3 shows the numbers of rRNA and tRNA genes for a variety of organisms and organelles. The mitochondrial genome of vertebrates contains only one copy of both the 12S and the 16S rRNA genes. This is apparently sufficient for the mitochondrial translation system because the genome contains only 13 protein-coding genes. The mycoplasmas, which are the smallest self-replicating prokaryotes, contain two sets of rRNA genes. The genome of *Escherichia coli* is 4–5 times larger than that of *Mycoplasma capricolum*, and it contains seven sets of rRNA genes. The number of rRNA genes in yeast is approximately 140, and the numbers in fruitflies and humans are even larger.

TABLE 6.2 Similarity in amino acid sequence, regulation, chemical reactivity and specificity, aggregation properties, and place of expression between duplicate genes[a]

Gene pair (organism)	Amino acid similarity (%)	Time of duplication (million years)	Regulation[b]	Chemical attributes[c]	Aggregation properties[d]	Place of expression[e]
Trypsin and chymotrypsin (human)	36	1,500	- - -	- -	+	+
Hemoglobin and myoglobin (human)	23	800	- - -	- -	- - -	- -
Lactate dehydrogenase M and H chains (human)	74	600	- -	- -	+	-
Hemoglobin α and β chains (human)	41	500	- - -	-	-	+
Immunogobulin H and L chains (human)	25	400	- - -	- -	-	+
Lactalbumin and lysozyme (human)	37	350	- - -	- - -	- -	- -
Growth hormone and prolactin (human)	25	330	- -	- -	+	-
Chymotrypsins A and B (human)	79	270	+	+	+	+
Carbonic anhydrases B and C (human)	60	180	-	-	+	+
Insulins I and II (rat)	96	30	-	+	+	+
Growth hormone and lactogen (human)	85	23	- -	-	+	- -
Alcohol dehydrogenase A and S chains (horse)	98	10	?	-	+	+

Modified from Li (1983).

[a]+ = similar; - = slightly different; - - = moderately diferent; - - - = markedly diferent; ? = unknown.

[b]Regulation refers to differential expression over tissues or developmental stages, or to the rate of synthesis of the gene product if the two genes are expressed in the same tissue.

[c]Chemical attributes include catalytic properties and binding specificities to substrates, inhibitors, antigens, and so on.

[d]Aggregation refers to the number of subunits and the types of interactions between them.

[e]Place of expression refers to organs of the body or to types of differentiated cells.

TABLE 6.3 Numbers of rRNA and tRNA genes per haploid genome in various organisms

Genome source	Number of complete rRNA gene sets	Number of tRNA genes	Approximate genome size (bp)
Human mitochondrion	1	22	2×10^4
Nicotiana tabacum chloroplast	2	37	2×10^5
Mycoplasma capricolum	2	Not determined	1×10^6
Escherichia coli	7	~100	4×10^6
Neurospora crassa	~100	~2,600	2×10^7
Saccharomyces cerevisiae	~140	~360	5×10^7
Caenorhabditis elegans	~55	~300	8×10^7
Tetrahymena thermophila[a]	1	~800	2×10^8
Drosophila melanogaster	120–240	590–900	2×10^8
Gadus morhua (Atlantic cod)	~50	Not determined	3×10^8
Physarum polycephalum	80–280	~1,050	5×10^8
Euglena gracilis	800–1,000	~740	2×10^9
Homo sapiens	~300	~1,300	3×10^9
Rattus norvegicus	150–170	~6,500	3×10^9
Zea mays	3,000–9,000	Not determined	3×10^9
Xenopus laevis	500–760	6,500–7,800	8×10^9

Data from Long and Dawid (1980), Li (1983), and other sources.

[a]The tRNA data is taken from the congeneric species *Tetrahymena pyriformis*.

Xenopus laevis has a larger genome and more rRNA genes than humans. Thus, a rough positive correlation exists between the number of rRNA genes and genome size. This relationship with genome size also holds for the tRNA genes and other RNA-specifying genes (Table 6.3). There are, however, a few exceptions. For example, the maize (*Zea mays*) nuclear genome is about the same size as the human genome, but it contains about 45 times as many sets of rRNA genes as its human counterpart. One such exception to the rule is particularly interesting, since paradoxically it strengthens the case for a causal relationship in which the number of RNA-specifying genes (the dose) is dictated by the genome size. The ciliate *Tetrahymena* has a larger genome than the yeast *Saccharomyces cerevisiae*, but only one set of rRNA genes. This set, however, resides in the germinal nucleus, the micronucleus. In the derivation of the vegetative macronuclei from the micronucleus, the number of gene copies is amplified 200–600 times (Yao et al. 1974). It has been estimated that *Tetrahymena* possesses 600 extrachromosomal copies of rRNA genes and close to 1,500 copies of tRNA genes. Thus, a large number of rRNAs can be produced during vegetative growth, despite the small number of copies in the haploid set.

There may be two reasons for the general positive correlation between genome size and number of copies of RNA-specifying genes. Either large genomes require large quantities of RNA, or the number of RNA-specifying genes is simply a passive consequence of genome enlargement by duplication (Chapter 8).

Highly repetitive genes, such as the rRNA genes, are generally very similar to one other. One factor responsible for the homogeneity may be purifying selection, because these genes should abide by very specific functional and structural requirements. However, homogeneity often extends to regions devoid of any functional or structural significance, and thus the maintenance of homogeneity requires that other mechanisms be invoked (see page 304).

Isozymes

In addition to invariant repeats, the genomes of higher organisms contain numerous multigene families whose members have diverged to various extents. Good examples are families of genes coding for isozymes, such as lactate dehydrogenase, aldolase, creatine kinase, carbonic anhydrase, and pyruvate kinase. **Isozymes** are enzymes that catalyze the same biochemical reaction but may differ from one another in tissue specificity, developmental regulation, electrophoretic mobility, or biochemical properties. Note that isozymes are encoded by different loci, usually duplicated genes, as opposed to **allozymes**, which are distinct forms of the same enzyme encoded by different alleles at a single locus. The study of multilocus isozyme systems has greatly enhanced our understanding of how cells with identical genetic endowment can differentiate into hundreds of different specialized types of cells that constitute the complex body organization of vertebrates. Although all members of an isozyme family serve essentially the same catalytic function, different members may have evolved particular adaptations for different tissues or different developmental stages, thus increasing the physiological fine-tuning of the cell.

Let us consider the two genes encoding for the A and B subunits of lactate dehydrogenase (LDH) in mammals (Hiraoka et al. 1990). These two subunits form five tetrameric isozymes, A_4, A_3B, A_2B_2, AB_3, and B_4, all of which catalyze either the conversion of lactate into pyruvate in the presence of the oxidized coenzyme nicotinamide adenine dinucleotide (NAD^+) or the reverse reaction in the presence of the reduced coenzyme (NADH). It has been suggested that B_4 and the other isozymes rich in B subunits, which have a high affinity for NAD^+, function as true lactate dehydrogenase in aerobically metabolizing tissues such as the heart, whereas A_4 and the isozymes rich in A subunits, which have a high affinity for NADH, are especially geared to serve as pyruvate reductases in anaerobically metabolizing tissues such as skeletal muscle.

Figure 6.7 shows the developmental sequence of LDH production in the heart. We see that the more anaerobic the heart is (specifically, in the early stages of gestation), the higher the proportion of LDH isozymes rich in A subunits will be. Thus, the two duplicate genes have become specialized to differ-

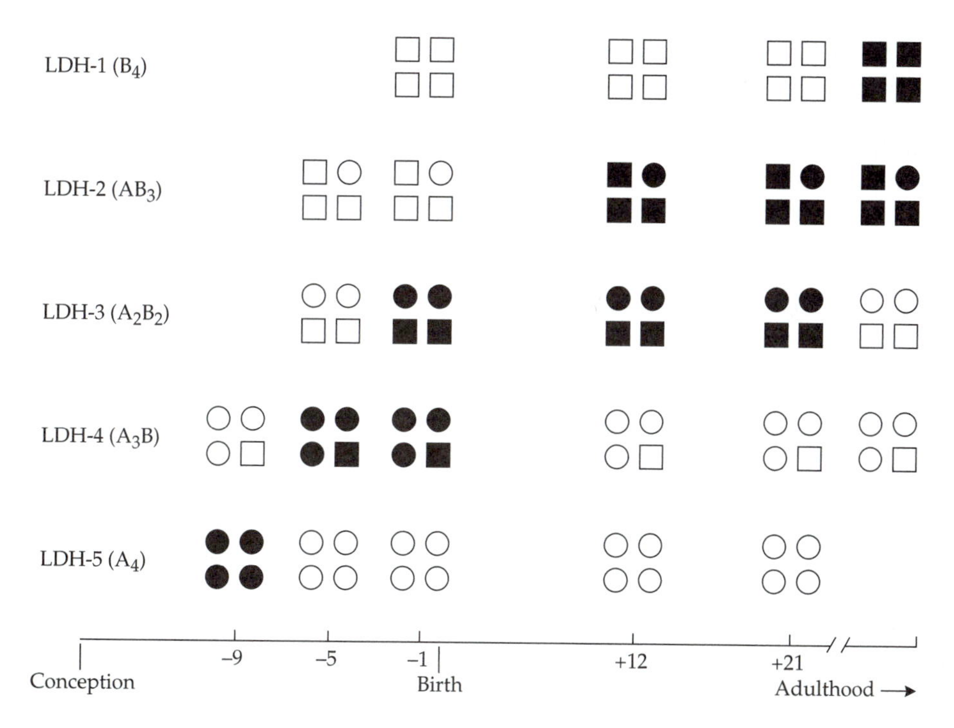

FIGURE 6.7 Developmental sequences of five lactate dehydrogenase (LDH) isozymes in the rat heart from conception to adulthood. Negative and positive numbers denote days before or after birth, respectively. Squares indicate B subunits, circles are A subunits. Solid symbols indicate quantitatively predominant forms. Notice the shift from A to B subunits during ontogenesis. Data from Markert and Ursprung (1971).

ent tissues and to different developmental stages. As the subunits are present in almost all the vertebrates studied to date, the duplication that produced the genes for LDH-A and LDH-B probably occurred either before or during the early stages of vertebrate evolution. An interesting feature of LDH is that the two subunits can form heteromultimers, thus further increasing the physiological versatility of the enzyme.

Many other examples of multimeric enzymes that are composed of polypeptides encoded by duplicated genes are known (Harris 1979, 1980/1981).

Opsins

Color vision in humans, apes, and Old World monkeys is mediated in the eye by three types of photoreceptor cells (cones), which transduce photic energy into electrical potentials. Each type of color-sensitive cone is maximally sensitive to a certain wavelength, depending on the kind of color-sensitive pigment (photopigment) present in the cone. In humans, the red, green, and blue cones are maximally sensitive at approximately 560, 530, and 430 nanometers, respectively. Each color stimulates one or more kinds of cones. For example, red light stimulates only red cones, blue light stimulates blue cones, yellow light stimulates red and green cones equally, and white light stimulates all three types of cones equally (Carlson 1991).

Each color-sensitive photopigment consists of two parts: a protein called opsin, and a lipid derivative of vitamin A1 called retinal. Color specificity is determined by the opsins, which are members of a superfamily of G-protein-coupled receptors. The blue opsin is encoded by an autosomal gene, while the red and green opsins are encoded by X-linked genes. Each X chromosome contains only one red opsin gene, but may contain more than one green opsin gene (Nathans et al. 1986). The amino acid sequences of the red and green opsins are 96% identical, but they share only 43% amino acid identity with the blue opsin. The blue opsin gene and the ancestor of the green and red opsin genes diverged about 500 million years ago (Yokoyama and Yokoyama 1989). In contrast, the close linkage and high similarity between the red and green opsin genes point to a very recent gene duplication. Because most New World monkeys have only one X-linked pigment gene (see below), whereas Old World monkeys (including apes and humans) have two or more, it is assumed that the duplication occurred about 25–35 million years ago in the ancestor of Old World monkeys after their divergence from the New World monkeys. As a consequence of this duplication, Old World monkeys are **trichromatic**; that is, any color perceived by these organisms can be reproduced by mixing various intensities of red, green and blue lights.

With the exception of the howler monkey (genus *Alouatta*), which has one autosomal and two X-linked genes (Jacobs et al. 1996), all other New World monkeys possess only one autosomal and one X-linked opsin genes. However, in many New World monkeys (e.g., squirrel monkeys and tamarins), the X-linked opsin locus is highly polymorphic (Jacobs et al. 1993; Boissinot et al. 1998). Two of these alleles have maximal-sensitivity peaks similar to those of human red and green opsin, respectively, while the third allele has an intermediate maximal-sensitivity peak. For this reason, a female that is heterozygous for two of these three alleles is trichromatic, while males and homozygous females are **dichromatic** (Figure 6.8). Dichromatic animals cannot distinguish between red and green, and in this sense they resemble people suffering from either protanopia (color blindness due to red photopigment deficiency) or deuteranopia (color blindness due to green photopigment deficiency).

Thus, in the case of humans, apes, and African monkeys, trichromatic vision is achieved by a mechanism akin to isozymes, (i.e., distinct proteins encoded by different loci). Heterozygous female squirrel monkeys, in contrast, achieve trichromacy through the use of two "allozymes," (i.e., distinct proteins encoded by different allelic forms at a single locus). If trichromacy confers a selective advantage on its carriers, then the long-term maintenance of several color-sensitive alleles at a locus in New World monkeys is achieved by maintaining a high level of polymorphism at the X-linked opsin locus. Such a

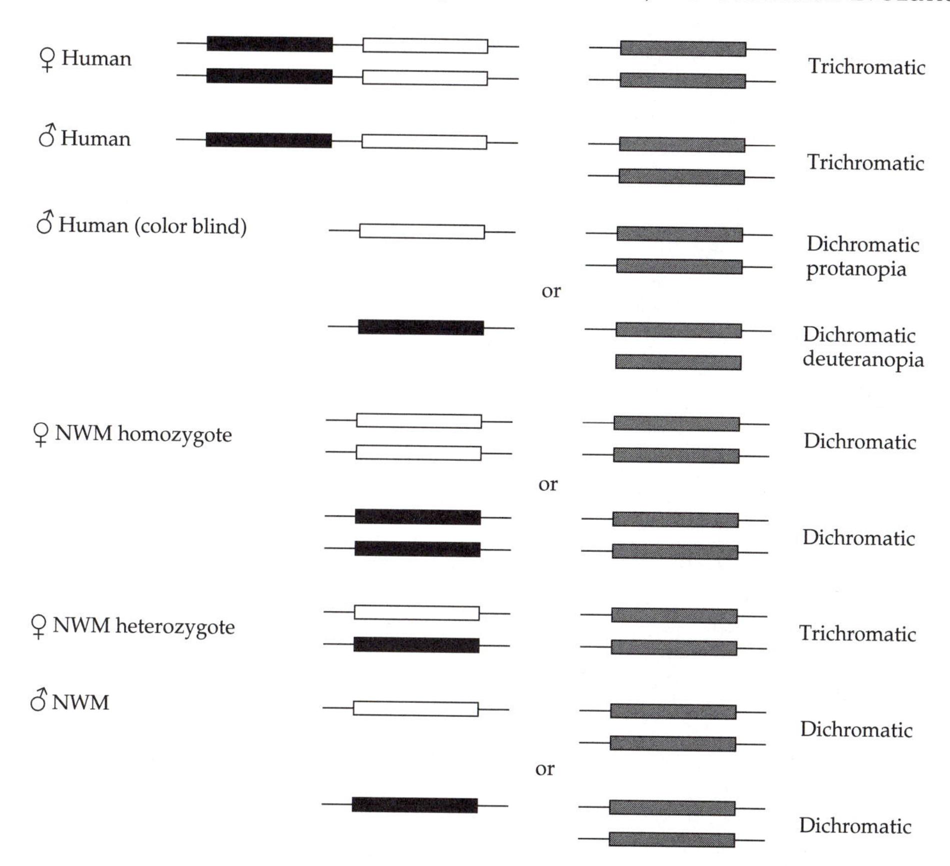

FIGURE 6.8 Molecular basis of dichromatic and trichromatic vision in males and females of humans and New World monkeys (NWM). Note that male New World monkeys cannot achieve trichromatic vision. The solid, empty, and shaded boxes denote the red, green, and blue pigment genes, respectively.

high level of polymorphism being maintained for such long periods of evolutionary time presumably requires a form of overdominant selection (Chapter 2). The selective advantage of trichromatic vision is thought to be the ability to detect ripe fruits against a background of dense green foliage.

DATING GENE DUPLICATIONS

Two genes are said to be **paralogous** if they are derived from a duplication event, but **orthologous** if they are derived from a speciation event. For example, in Figure 6.9, genes α and β were derived from the duplication of an ancestral gene and are therefore paralogous, while gene α from species 1 and gene α from species 2 are orthologous, as are genes β from species 1 and gene β from species 2.

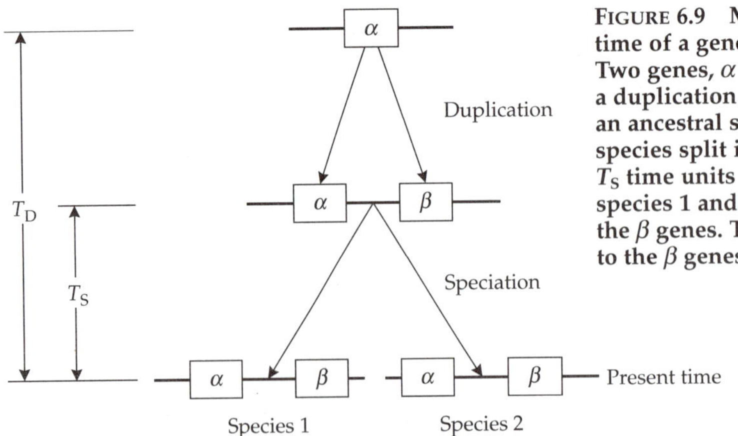

FIGURE 6.9 Model for estimating the time of a gene duplication event (T_D). Two genes, α and β, were derived from a duplication event T_D time units ago in an ancestral species. The ancestral species split into two species (1 and 2) T_S time units ago. The α genes in species 1 and 2 are orthologous, as are the β genes. The α genes are paralogous to the β genes.

We can estimate the date of duplication, T_D, from sequence data if we know the rate of substitution in genes α and β. The rate of substitution can be estimated from the number of substitutions between the orthologous genes in conjunction with knowledge of the time of divergence, T_S, between species 1 and 2 (Figure 6.9). We next show how an estimate of T_D can be obtained.

For gene α, let K_α be the number of substitutions per site between the two species. Then, the rate of substitution in gene α, r_α, is estimated by

$$r_\alpha = \frac{K_\alpha}{2T_S} \tag{6.1}$$

The rate of substitution in gene β, r_β, can be obtained in a similar manner. The average substitution rate for the two genes is given by

$$r = \frac{r_\alpha + r_\beta}{2} \tag{6.2}$$

To estimate T_D, we need to know the number of substitutions per site between genes α and β ($K_{\alpha\beta}$). This number can be estimated from four pairwise comparisons: (1) gene α from species 1 and gene β from species 2; (2) gene α from species 2 and gene β from species 1; (3) gene α and gene β from species 1; and (1) gene α and gene β from species 2. From these four estimates we can compute the average value for $K_{\alpha\beta}$ ($\overline{K}_{\alpha\beta}$), from which we can estimate T_D as

$$T_D = \frac{\overline{K}_{\alpha\beta}}{2r} \tag{6.3}$$

Note that in the case of protein-coding genes, by using the numbers of synonymous and nonsynonymous substitutions separately, we can obtain two independent estimates of T_D. The average of these two estimates may be used as the final estimate of T_D. However, if the number of substitutions per synonymous site between genes α and β is large, say larger than 1, then the number of synonymous substitutions cannot be estimated accurately, and so synonymous substitutions may not provide a reliable estimate of T_D. In such cases, only the number of nonsynonymous substitutions should be used. Conversely, if the number of substitutions per nonsynonymous site between the paralogous genes is small, then the estimate of the number of nonsynonymous substitutions is subject to a large sampling error, and in such cases, only the number of synonymous substitutions should be used.

In the above, we have assumed rate constancy. This assumption can be tested by the four pairwise comparisons mentioned above, and may not hold. In this case, the T_D estimate may be erroneous. As will be discussed later (see page 322), problems due to concerted evolutionary events may also arise and complicate the estimation of T_D.

Another method for dating gene duplication events is to consider the phylogenetic distribution of genes in conjunction with paleontological data pertinent to the divergence date of the species in question. For example, all vertebrates with the exception of jawless fish (hagfishes and lampreys) encode α- and β-globin chains. There are two possible explanations for this observation. One is that the duplication event producing the α- and β-globins occurred in the common ancestor of all vertebrates (Craniata), but the two jawless fish lineages (Myxini and Cephalospidomorphi) have lost one of the two duplicates. This is possible but not very likely, because such a scenario would require the losses to occur independently in at least two evolutionary lineages. The other explanation is that the duplication event occurred after the divergence of jawless fish from the ancestor of the jawed vertebrates (Gnathostomata), but before the radiation of the jawed vertebrates from each other. This latter explanation is thought to be more plausible, and the duplication date is commonly taken to be 450–500 million years ago.

Obviously, the above methods can only provide us with rough estimates of duplication dates, and all estimates should be taken with caution. Note that in estimating dates of divergence among species, one can use data from many genes belonging to many gene families. In comparison, in estimating dates of gene duplication, one must rely only on data from genes belonging to a single gene family. Because of the stringent limitations on the sequence data that can be used, estimates of gene duplication are often subject to very large standard errors.

GENE LOSS

The close to 7,000 genetic diseases that have been documented in the medical literature (McKusick 1998) attest to the fact that mutations can easily destroy the function of a protein-coding gene. The vast majority of such mutations are deleterious, and are either eliminated quickly from the population or maintained at very low frequencies due to overdominant selection or genetic drift (Chapter 2). However, as noted by Haldane (1932), as long as there are other copies of a gene that function normally, a duplicate gene can accumulate deleterious mutations and become nonfunctional without adversely affecting the fitness of the organism. Indeed, because deleterious mutations occur far more often than advantageous ones, a redundant duplicate gene is more likely to become nonfunctional than to evolve into a new gene (Ohno 1972).

Unprocessed pseudogenes

The **nonfunctionalization** or **silencing** of a gene due to deleterious mutations produces an **unprocessed pseudogene**, i.e., a pseudogene that has not gone through RNA processing (Chapter 7). The vast majority of unprocessed pseudogenes are derived via the nonfunctionalization of a duplicate functional gene. Some unprocessed pseudogenes, such as the $\psi\beta^X$ and $\psi\beta^Z$ in the goat β-globin multigene family, have been derived from a duplication of a preexisting pseudogene (Cleary et al. 1981). A very small number of unprocessed pseudogenes have been derived from a functional gene without a prior duplication.

Table 6.4 lists the structural defects found in several globin pseudogenes. Most of these unprocessed pseudogenes contain multiple defects such as frameshifts, premature stop codons, and obliteration of splicing sites or regulatory elements, so that it is difficult to identify the mutation that was the direct cause of gene silencing. In a few cases, identification of the "culprit" was possible. For example, human $\psi\xi 1$ contains only a single major defect—a nonsense mutation—that is probably the direct cause of its nonfunctionalization. In some cases, it is possible to identify the mutation responsible for the nonfunctionalization of a gene through a phylogenetic analysis. For example, human pseudogene $\psi\eta$ in the β-globin family contains numerous defects, each of which could have been sufficient to silence it. The β-globin clusters in chimpanzee and gorilla, our closest relatives, were found to contain the same number of genes and pseudogenes as in humans, indicating that the pseudogene was created and silenced before these three species diverged from one another. The three pseudogenes were found to have only three defects in common: a substitution in the initiation codon (ATG → GTA); a nonsense substitution in the tryptophan codon at position 15 (TGG → TGA); and a deletion in codon 20 resulting in a frameshift in the reading frame and a termination codon in the second exon (Chang and Slightom 1984). Thus, the "list of suspects" was reduced to three. Further studies showed that the same pseudogene exists in all primates and is therefore very ancient. A comparison of the defects among all primate sequences showed that the initial mutation responsible for the nonfunctionalization of $\psi\eta$ is the one in the initiation codon (Harris et al. 1984).

Interestingly, mutations that cause nonfunctionalization are only rarely missense mutations, most probably because such mutations result in the production of defective proteins that may be incorporated into final biological products, and thus may have deleterious effects. For example, there are dozens of chorion-coding genes in the genome of the silkworm *Bombyx mori*, yet if even one of them is rendered nonfunctional by a missense mutation, the entire eggshell becomes defective (e.g., Spoerel et al. 1989).

TABLE 6.4 Defects in unprocessed globin pseudogenes[a]

Pseudogene	TATA box	Initiation codon	Frameshift	Premature stop	Essential amino acid	Splice GT/AG rule	Altered stop codon	Polyadenylation signal
Human $\psi\alpha 1$		+	+	+	+	+	+	+
Human $\psi\zeta 1$				+				
Mouse $\psi\alpha 3$	+		+	+		+		
Mouse $\psi\alpha 4$			+		+			
Mouse $\beta h3$	?	+	+	+	+	+	?	?
Goat $\psi\beta^{x}$	+		+	+	+	+	+	+
Goat $\psi\beta^{z}$	+		+	+	+	+	+	+
Rabbit $\psi\beta 2$			+	+	+	+		

From Li (1983).

[a]A plus sign indicates the existence of a particular type of defect; a question mark indicates the possibility of a defect.

Because they are created by duplication, unprocessed pseudogenes are usually found in the neighborhood of the homologous functional genes from which they have been derived. There are, however, cases in which unprocessed pseudogenes become dispersed following genomic rearrangements (Chapter 8). For example, the α-globin cluster in the mouse is located on chromosome 11, and yet an unprocessed pseudogene was found on chromosome 17 (Tan and Whitney 1993).

Unitary pseudogenes

As stated previously, the loss of a single-copy gene is usually deleterious, and unlikely to be fixed in a population. This fact notwithstanding, a nonfunctional single-copy gene may become fixed in the population (most probably by random genetic drift) if the selection against the loss of the gene product no longer operates. A few such instances are known to have occurred during vertebrate evolution.

Unlike most vertebrates, guinea pigs, humans, and trout get the disease scurvy unless they consume L-ascorbic acid in their diet. For these organisms, ascorbic acid is a vitamin (vitamin C). The reason these animals cannot manufacture their own ascorbic acid is that they lack a protein called L-gulono-γ-lactone oxidase, an enzyme that catalyzes the terminal step in L-ascorbic acid synthesis. In animals that are not prone to scurvy, this protein is produced by a single-copy gene (Koshizaka et al. 1988). In humans, L-gulono-γ-lactone oxidase is a pseudogene, containing such molecular defects as the deletion of at least two exons (out of 12), deletions and insertions of nucleotides in the reading frame, and obliterations of intron–exon boundaries (Nishikimi et al. 1994).

Since this pseudogene has no functional counterpart in the human genome, it is called a **unitary pseudogene**. The unitary pseudogene for L-gulono-γ-lactone oxidase in the guinea pig contains different defects from those in the human pseudogene, indicating that the nonfunctionalization of this gene occurred independently in the two lineages (Nishikimi et al. 1992). It has been hypothesized that the guinea pig and human ancestors subsisted on a naturally ascorbic acid-rich diet, and therefore the loss of the enzyme did not constitute a disadvantage.

We would expect unitary pseudogenes to be rare because the loss of a biological function is usually deleterious. We note, however, that at least one other example of a unitary pseudogene is known: the α-1,3-galactosyltransferase in catarrhine (Old World) monkeys (Galili and Swanson 1991).

Nonfunctionalization time

The evolutionary history of an unprocessed pseudogene is assumed to consist of two distinct periods. The first period starts with the gene duplication event and ends when the duplicate copy is rendered nonfunctional. During this period, the would-be pseudogene presumably retains its original function, and the rate of substitution is expected to remain roughly the same as it was before the duplication event. After the loss of function, the pseudogene is freed from all functional constraints and its rate of nucleotide substitution is expected to increase considerably. From the evolutionary point of view, it is interesting to estimate how long a redundant copy of a functional gene may remain functional after the duplication event. To estimate this **nonfunctionalization time**, the following method has been suggested (Li et al. 1981; Miyata and Yasunaga 1981).

Consider the phylogenetic tree in Figure 6.10. T denotes the divergence time between species 1 and 2, i.e., the time since the separation between the orthologous functional genes A and B; T_D denotes the time since duplication, i.e., the time of divergence between the functional gene A and its paralogous pseudogene ψA; and T_N denotes the time since the nonfunctionalization of pseudogene ψA. The number of nucleotide substitution per site at the ith position of codons (i = 1, 2, or 3) between ψA and A, ψA and B, and A and B, are denoted as $d_{(\psi AA)i}$, $d_{(\psi AB)i}$, and $d_{(AB)i}$, respectively, and can be calculated directly from the sequence data (Chapter 3).

Let l_i, m_i, and n_i be the numbers of nucleotide substitutions per site at codon position i between points O and ψA, O and A, and O and B, respectively. We then have

$$d_{(\psi AA)i} = l_i + m_i \quad \textbf{(6.4a)}$$

$$d_{(\psi AB)i} = l_i + n_i \quad \textbf{(6.4b)}$$

$$d_{(AB)i} = m_i + n_i \quad \textbf{(6.4c)}$$

Therefore, l_i, m_i, and n_i can be estimated by

$$l_i = \frac{d_{(\psi AA)i} + d_{(\psi AB)i} - d_{(AB)i}}{2} \quad \textbf{(6.5a)}$$

$$m_i = \frac{d_{(\psi AA)i} - d_{(\psi AB)i} + d_{(AB)i}}{2} \quad \textbf{(6.5b)}$$

$$n_i = \frac{-d_{(\psi AA)i} + d_{(\psi AB)i} + d_{(AB)i}}{2} \quad \textbf{(6.5c)}$$

In the following, we assume that the rates of substitution at a given codon position are equal in the functional genes *A* and *B*. We denote these rates by a_i, where the subscript *i* stands for the codon position. We also assume that once ψA became nonfunctional, i.e., all functional constraints were obliterated, the rate of nucleotide substitution became the same for all three codon positions. We denote this rate by *b*. A reasonable expectation is that *b* would turn out to be much larger than a_1 and a_2, and possibly a little larger than a_3.

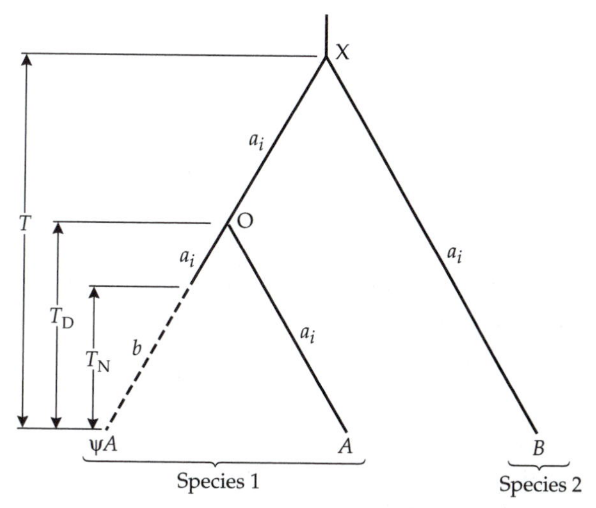

FIGURE 6.10 Schematic phylogenetic tree used to estimate the nonfunctionalization time of an unprocessed pseudogene. *T* denotes the divergence time between species 1 and 2, T_D denotes the time since duplication of gene *A*, and T_N denotes the time since nonfunctionalization of pseudogene ψA. a_i is the rate of nucleotide substitution per site per year at the *i*th codon position in the functional genes, and *b* is the rate of substitution for the pseudogene. The node connecting the orthologous genes is denoted by X, and the node connecting the paralogous genes is marked by O. Modified from Li et al. (1981).

From Figure 6.10, we obtain

$$d_{(\psi AA)i} = 2a_i T_D + (b - a_i)T_N \tag{6.6}$$

$$d_{(\psi AB)i} = 2a_i T + (b - a_i)T_N \tag{6.7}$$

$$d_{(AB)i} = 2a_i T \tag{6.8}$$

If we know T, a_i can be estimated from Equation 6.8 by

$$a_i = \frac{d_{(AB)i}}{2T} \tag{6.9}$$

Let us denote the difference $d_{(\psi AB)i} - d_{(AB)i}$ as y_i. Note that

$$y_i = d_{(\psi AB)i} - d_{(AB)i} = bT_N - a_i T_N \tag{6.10}$$

Therefore, T_D can be estimated from Equations 6.6 and 6.10 by

$$T_D = \frac{\Sigma d_{(\psi AA)i} - \Sigma y_i}{2\Sigma a_i} \tag{6.11}$$

where Σ is the summation over i.

Two simple formulae for estimating T_N and b have been suggested by Li et al. (1981).

$$T_N = \frac{y_{12} - y_3}{a_3 - a_{12}} \tag{6.12}$$

$$b = \frac{a_3}{y_{12} - y_3} \tag{6.13}$$

where $y_{12} = (y_1 + y_2)/2$, and $a_{12} = (a_1 + a_2)/2$.

By setting the time of divergence (T) between mouse, rabbit, and human at about 80 million years ago, Li et al. (1981) estimated that the mouse globin pseudogene $\psi\alpha 3$ was created by a gene duplication 27 ± 6 million years ago and became nonfunctional approximately 4 million years later. Similarly, the human globin pseudogene $\psi\alpha 1$ lost its function approximately 4 million years after it had been created by a gene duplication 49 ± 8 million years ago.

In general, it seems that those redundant duplicates that are ultimately destined to become pseudogenes retain their original function for only very short periods of time following the gene duplication event. Some unprocessed pseudogenes, such as the rabbit globin pseudogene $\psi\beta 2$, seem to have lost their function almost instantaneously following gene duplication.

THE GLOBIN SUPERFAMILY

The globin superfamily of genes has experienced all the possible evolutionary pathways that can occur in families of repeated sequences, i.e., (1) retention of original function, (2) acquisition of new function, and (3) loss of function. In

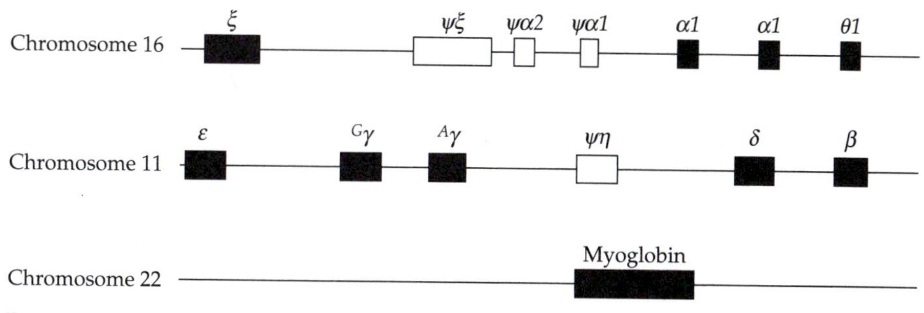

FIGURE 6.11 The chromosomal arrangement of the three gene families belonging to the globin superfamily of genes in humans: the α-globin family on chromosome 16, the β-globin family on chromosome 11, and myoglobin on chromosome 22. Solid black boxes denote functional genes; empty boxes denote pseudogenes.

humans, the globin superfamily consists of three families with at least one functional member: the myoglobin family, whose single member (an orphon) is located on chromosome 22; the α-globin family on chromosome 16; and the β-globin family on chromosome 11 (Figure 6.11). Together these three families produce two types of functional proteins: myoglobin and hemoglobin.

Judging by the fact that globin-like proteins exist in all life forms that have been studied in sufficient detail, the globins must be very ancient in origin. Myoglobin and hemoglobin diverged from one another before the emergence of annelid worms, i.e., more than 800 million years ago (Figure 6.12), and have become specialized in several respects. In terms of tissue specificity, myoglobin became the oxygen-storage protein in muscles, whereas hemoglobin became the oxygen carrier in blood. In terms of quaternary structure, myoglobin retained a monomeric structure, while hemoglobin became a tetramer. In terms of function, myoglobin evolved a higher affinity for oxygen than did hemoglobin, while the function of hemoglobin became much more refined and regulated. Mammalian hemoglobin, for instance, has acquired several capabilities that are absent in myoglobin. Among these are (1) binding of four oxygen molecules cooperatively, (2) responding to the acidity and carbon dioxide concentration inside red blood cells (the Bohr effect), and (3) regulating its own oxygen affinity through the level of organic phosphate in the blood. Apparently, the heteromeric structure of hemoglobin has facilitated these refinements of the function of hemoglobin.

The hemoglobin in humans and the vast majority of vertebrates is made up of two types of chains, one encoded by an α family member, the other by a member of the β family. The α and β families diverged following a gene duplication 450–500 million years ago (Figure 6.12). Since jawless fish contain only one type of monomeric hemoglobin, polymerization of hemoglobin in vertebrates must have occurred close to the time of divergence between α and β. The duplication that gave rise to the α- and β-globins was most probably a tandem duplication, resulting in two linked genes on the same chromosomes. The chromosomal linkage is still preserved in ray-finned fishes and amphib-

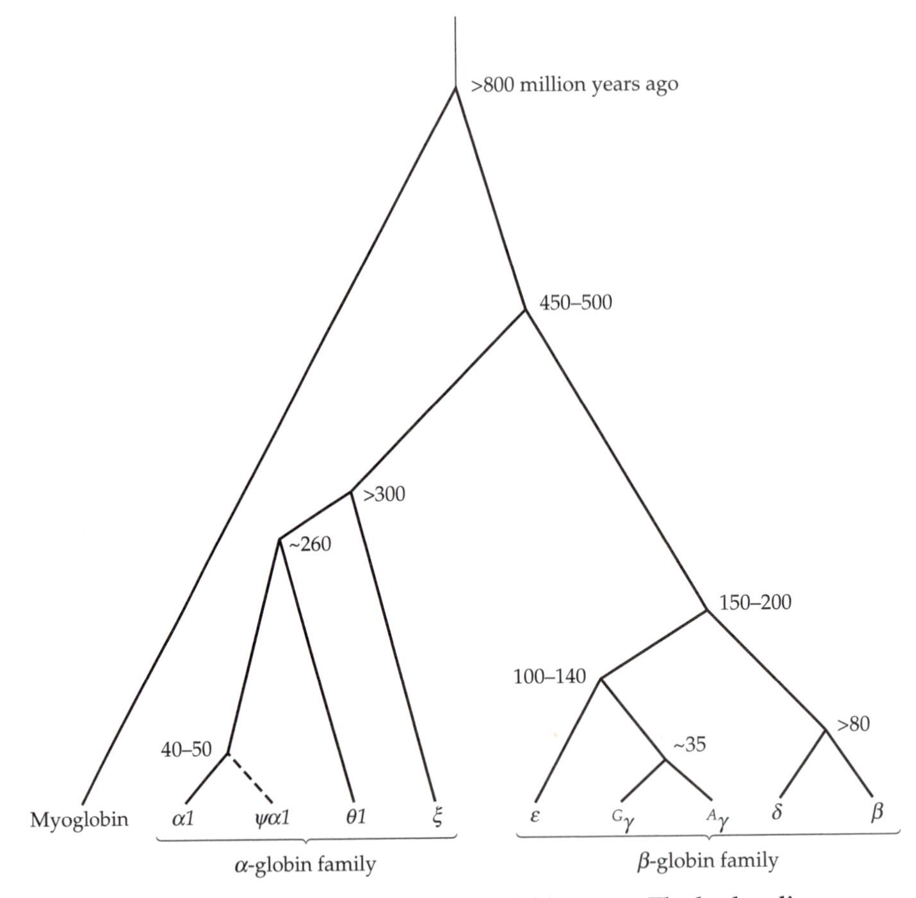

FIGURE 6.12 Evolutionary history of human globin genes. The broken line denotes a pseudogene lineage. Only one of the two α1-globin genes is shown because the date of their divergence from each other is uncertain.

ians. The date of the separation between the ancestors of the α and β gene families into different chromosomes must have occurred after the divergence of amphibians from amniotes but before the occurrence of the gene duplications that gave rise to the specific members of these two families. The date of the chromosomal separation is tentatively put at 300–350 million years ago.

In humans, the α family consists of four functional genes: the embryonic gene *ξ*; two adult genes, *α1* and *α2*; and the most recently discovered member of the family, *θ1* (Figure 6.11). It also contains three unprocessed pseudogenes, *ψε*, *ψα1*, and *ψα2*. The β family consists of five functional genes: the embryonic gene *ε*; two fetal genes, $^{G}\gamma$ and $^{A}\gamma$; and two adult genes, *β* and *δ*. The family also contains one unprocessed pseudogene, *ψη* (formerly *ψβ*). The two families have diverged in both physiological properties and ontological regulation. In fact, distinct hemoglobins appear at different developmental stages: $\zeta_2\varepsilon_2$ and

$\alpha_2\varepsilon_2$ in the embryo, $\alpha_2\gamma_2$ in the fetus, and $\alpha_2\beta_2$ and $\alpha_2\delta_2$ in adults. The *θ1* gene is mainly transcribed 5–8 weeks after conception, but at such low levels that the protein has not yet been detected *in vivo*. Differences in oxygen-binding affinity have evolved among these globins. For example, the embryonic and fetal hemoglobins ($\zeta_2\varepsilon_2$, $\alpha_2\varepsilon_2$, and $\alpha_2\gamma_2$) have a higher oxygen affinity than either adult hemoglobin ($\alpha_2\beta_2$ and $\alpha_2\delta_2$), mainly because they do not bind 2,3-diphosphoglycerate as strongly as the adult forms. Consequently, they function better in the relatively hypoxic (low oxygen) environment in which the embryo and the fetus reside. This example illustrates once again how gene duplication can result in evolutionary refinements of physiological systems.

Among members of the α family, the embryonic type, ζ, is the most divergent, having branched off more than 300 million years ago (Figure 6.12). The θ1 globin branched off about 260 million years ago. Because the divergence time between the two *α* genes is uncertain, only the *α1* gene is shown in the figure. The *α1* and *α2* genes have almost identical DNA sequences and produce identical polypeptides. This would seem to indicate a very recent divergence time. However, the similarity could also be the result of concerted evolution, a phenomenon that will be discussed later. The two genes are present in humans and all the apes (including gibbons), and so must have arisen more than 20 million years ago.

Among the β family members, the adult types (β and δ) diverged from the nonadult types (γ and ε) about 150–200 million years ago (Efstratiadis et al. 1980; Czelusniak et al. 1982). The ancestor of the two *γ* genes diverged from the *ε* gene less than 140 million years ago. The duplication that created $^G\gamma$ and $^A\gamma$ occurred after the separation of the simian lineage (Anthropoidea) from the prosimians about 55 million years ago (Hayasaka et al. 1992). Soon after that, the ancestor of the two *γ* genes, which was originally an embryonically expressed gene, became a fetal gene. The change in temporal ontogenetic expression was brought about by sequence changes within a 4-Kb region surrounding the *γ* gene (TomHon et al. 1997). The divergence between the *δ* and *β* genes is estimated to have occurred before the eutherian radiation, more than 80 million years ago (Hardison and Margot 1984; Goodman et al. 1984).

We note that in both the α and the β families, there is a good correlation between the time of divergence and the degree of functional or regulatory divergence between genes.

PREVALENCE OF GENE DUPLICATION, GENE LOSS, AND FUNCTIONAL DIVERGENCE

Gene duplications arise spontaneously at high rates in bacteria, bacteriophages, insects, and mammals, and are generally viable (Fryxell 1996). Thus, the creation of duplications by mutation is not the rate-limiting step in the process of gene duplication and subsequent functional divergence. However, only a small fraction of all duplicated genes are retained, and an even smaller fraction evolves new functions. The reason is the much higher probability of nonfunctionalization in comparison with that of evolving a new function. We note, however, that in large populations, the probability of evolving a new function may be considerable (Walsh 1995; Nadeau and Sankoff 1997).

As far as dose repetitions are concerned, we now have good evidence that an increase in gene number can occur quite rapidly under selection pressure for increased amounts of a gene product. For example, the genome of the wild type strain of the peach-potato aphid (*Myzus persicae*) contains two genes encoding esterases E4 and FE4. The two genes are very similar in sequence (98%), indicating that they have been duplicated recently (Field and Devonshire 1998). Following exposure to organophosphorous insecticides, which can be hydrolized and sequestered by esterase, resistant strains of *Myzus persicae* were found to contain multiple copies of E4 and FE4. The gene duplications and the subsequent increase in the frequency of the carriers of these duplications within the aphid population is likely to have occurred within the last 50 years, with the introduction of the selective agent. This is consistent with the finding that the individual copies of each duplicated gene, both within and between aphid clones, show no sequence divergence (Field and Devonshire 1998). Similar evolutionary responses involving dose repetitions were found in the mosquitoes *Culex pipens* and *C. quinquefasciatus*, and in *Drosophila melanogaster* (Mouchès et al. 1986; Maroni et al. 1987; Callaghan et al. 1998).

The evolution of functionally novel proteins following gene duplication is more problematic. A widely cited model of the evolution of functionally novel proteins (Ohno 1970; Kimura 1983) holds that after gene duplication, one copy is rendered redundant and is therefore free to accumulate substitutions at random. According to this view, by chance some of these substitutions may result in a new function. There are at least three difficulties with this proposal. First, unless the new function can be acquired through one or a few nucleotide substitutions, it is more than likely that the copy will become a pseudogene rather than a new functional member of a gene family. Second, evidence from tetraploid organisms, such as the South African frog (*Xenopus laevis*) and the common carp (*Cyprinus carpio*), indicate that after genes duplicate, both copies continue to be subjected to purifying selection (as inferred from the ratio of synonymous to nonsynonymous substitutions) as intense as before the duplication (Hughes and Hughes 1993; Larhammar and Risinger 1994). Third, for a number of divergent multigene families, there is evidence that functionally distinct proteins have arisen not as a result of chance fixation of neutral variants, but as a result of positive Darwinian selection (again inferred from the ratio of synonymous to nonsynonymous substitutions) (e.g., Baba et al. 1984; Zhang et al. 1998).

Hughes (1994) proposed a model for the evolution of a new function under which a single protein-coding gene first evolves into a gene that encodes a multifunctional protein (see page 302). Gene duplication then allows each copy to specialize for one of the functions of the ancestral gene. Alternatively, a gene may initially have a generalized function, and gene duplication may allow for the evolution of specializations, such as tissue or ontogenetic specificity. An example that seems to support Hughes' (1994) suggestion is seen in the evolution of carbamoylphosphate synthetase genes. Carbamoylphosphate synthetase catalyzes the formation of carbamoylphosphate from CO_2, ATP, and ammonia or glutamine. This reaction is needed in pyrimidine biosynthesis, arginine biosynthesis, and the urea cycle. Phylogenetic reconstructions by Lawson et al. (1996) indicate that the evolution of this enzyme was characterized by many cycles of gene duplication throughout the living world. Interestingly, in organisms that possess two or more carbamoylphosphate synthetases following gene duplication, the enzymes are always specific—arginine-specific, urea cycle-specific, or pyrimidine-specific—whereas in lineages in which gene duplication did not occur, the enzyme functions in a generalized or multifunctional manner.

EXON SHUFFLING

There are three types of **exon shuffling**: exon duplication, exon insertion, and exon deletion. **Exon duplication** refers to the duplication of one or more exons in a gene and so is a type of internal duplication, which was discussed in the context of gene elongation (see page 255). **Exon insertion** is the process by which structural or functional domains are exchanged between proteins or inserted into a protein. **Exon deletion** results in the removal of a segment of amino acids from the protein. All types of shuffling have occurred in the evolutionary process of creating new genes. In the following, we discuss the insertion of an exon from one gene into another, with the consequent production of mosaic proteins (Banyai et al. 1983; Doolittle 1985; Patthy 1985).

Mosaic proteins

A **mosaic** or **chimeric protein** is a protein encoded by a gene that contains regions that are also found in other genes. The existence of such proteins indicates that exon shuffling has occurred during the evolutionary history of their genes.

The first described mosaic protein was tissue plasminogen activator (Figure 6.13). Tissue plasminogen activator is activated by blood-clotting Factor XIIa. The active form of tissue plasminogen activator converts plasminogen into its active form, plasmin, which dissolves fibrin, a soluble fibrous protein in blood clots. The conversion of plasminogen into plasmin is greatly accelerated by the presence of fibrin, the substrate of plasmin. Fibrin polymers bind both plasminogen and tissue plasminogen activator, thus aligning them for catalysis. This mode of molecular alignment allows plasmin production only in the proximity of fibrin, thus conferring fibrin specificity to plasmin. The physiological significance of this molecular mechanism is that it ensures that plasminogen activation takes place predominantly on the surface of fibrin, thus restricting plasmin action to its proper substrate. By contrast, prourokinase, the precursor of the urinary plasminogen activator, lacks fibrin specificity.

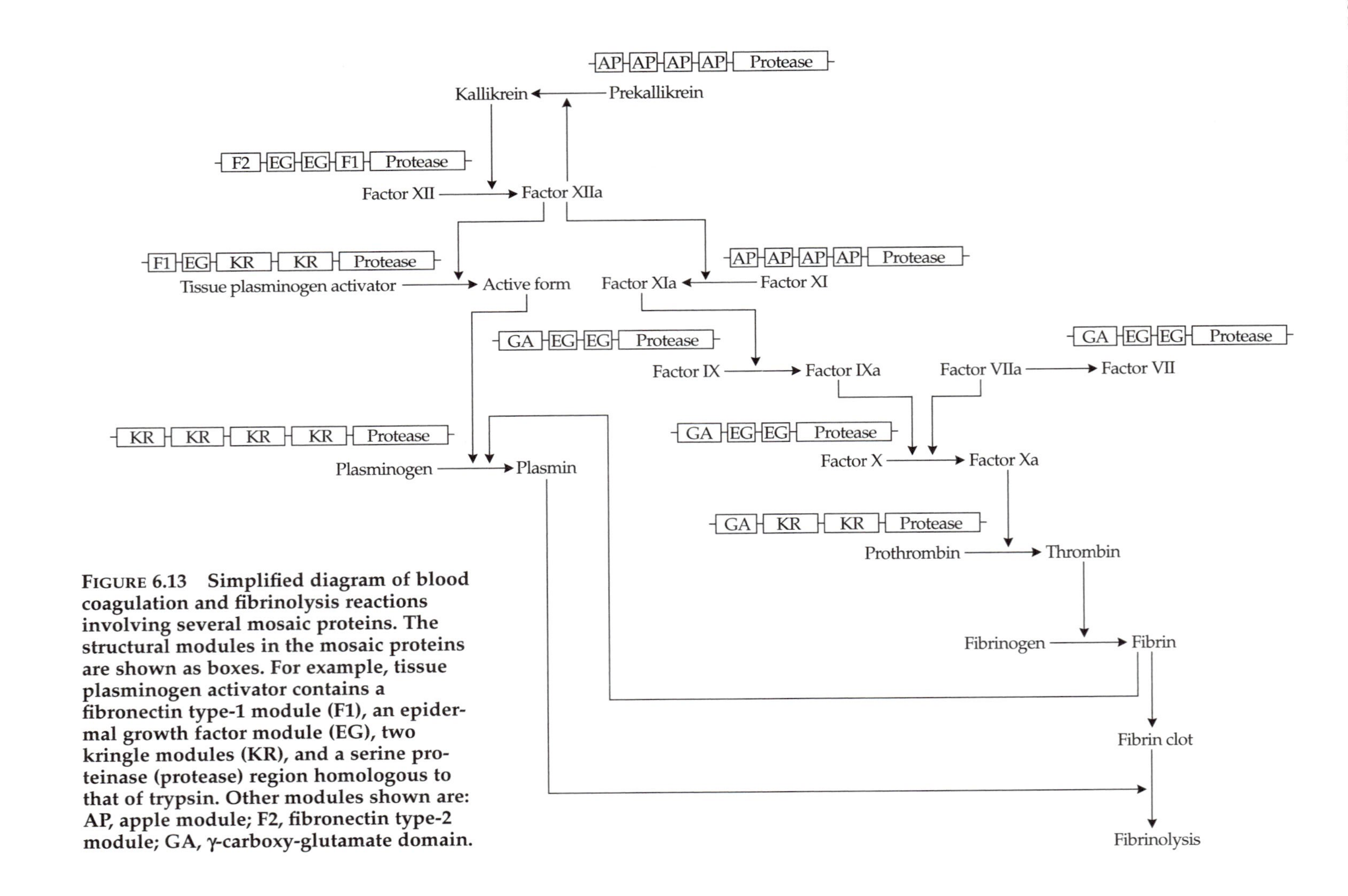

FIGURE 6.13 Simplified diagram of blood coagulation and fibrinolysis reactions involving several mosaic proteins. The structural modules in the mosaic proteins are shown as boxes. For example, tissue plasminogen activator contains a fibronectin type-1 module (F1), an epidermal growth factor module (EG), two kringle modules (KR), and a serine proteinase (protease) region homologous to that of trypsin. Other modules shown are: AP, apple module; F2, fibronectin type-2 module; GA, γ-carboxy-glutamate domain.

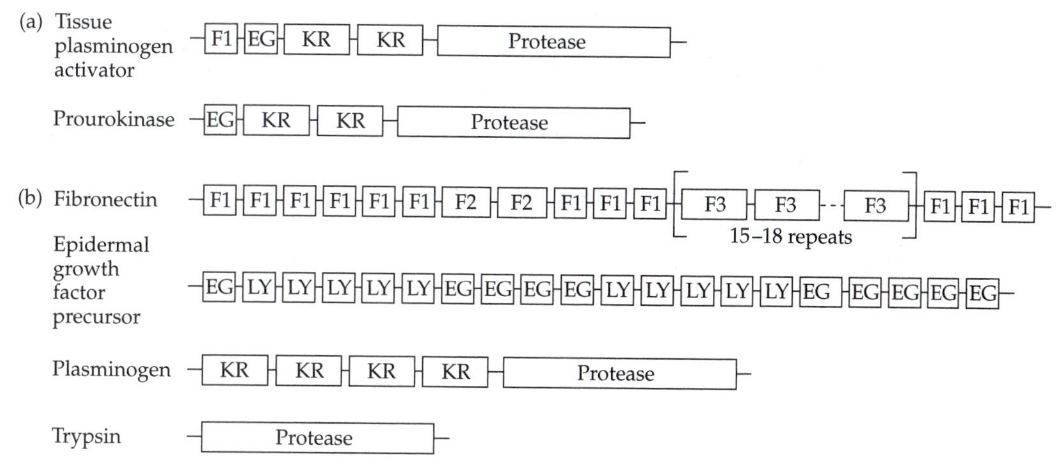

FIGURE 6.14 (a) Comparison of the modular structures of tissue plasminogen activator and prourokinase. (b) Possible origin of the modules acquired through exon insertion in the tissue plasminogen activator protein. Module abbreviations: EG, epidermal growth factor-like finger module; F1, fibronectin type-1 module; F2, fibronectin type-2 module; F3, fibronectin type-3 module; KR, kringle; LY, low-density lipoprotein receptor YWTD (tyrosine-tryptophan-threonine-aspartic acid) module; protease, serine proteinase module.

A comparison of the amino acid sequences of tissue plasminogen activator and prourokinase showed that the former contains a 43-residue sequence at its amino-terminal end that has no counterpart in prourokinase (Figure 6.14a). This segment can form a finger-like structure, and is homologous to one of the three finger domains responsible for the fibrin affinity of fibronectin—a large glycoprotein present in the plasma and on cell surfaces that promotes cellular adhesion (Figure 6.14b). Deletion of this segment leads to a loss of the fibrin affinity of tissue plasminogen activator. The homology of tissue plasminogen activator with fibronectin is restricted to this domain (currently denoted fibronectin type-1 domain). Thus, exon shuffling must have been responsible for the acquisition of this domain by tissue plasminogen activator from either fibronectin or a similar protein.

Tissue plasminogen activator also contains a segment homologous to portions of the epidermal growth factor precursor and the growth factor-like regions of other proteins, such as Factors VII, IX, X, and XII (Figures 6.13 and 6.14). In addition, the carboxy-terminal regions of tissue plasminogen activator are homologous to the protease parts of trypsin and other trypsin-like serine proteinases, such as prothrombin and plasminogen, which are enzymes that hydrolyze proteins into peptide fragments. Finally, the nonproteinase part of tissue plasminogen activator contains two structures similar to the kringles of plasminogen. (A "kringle" is a cysteine-rich sequence that contains three internal disulfide bridges and forms a pretzel-like structure resembling the Danish cake bearing this name.)

Thus, during its evolution, tissue plasminogen activator acquired at least five DNA segments from at least four other genes: plasminogen, epidermal growth factor, fibronectin, and trypsin (Figure 6.14b). Moreover, the junctions of these acquired units coincide precisely with the borders between exons and introns (Ny et al. 1984), lending further credibility to the idea that exons have been transferred from one gene to another.

For more examples of exon shuffling, mosaic proteins, and frequently shuffled domains, see Bork et al. (1996), Hegyi and Bork (1997), and Schultz et al. (1998).

Phase limitations on exon shuffling

For an exon to be inserted, deleted or duplicated without causing a frameshift in the reading frame, certain phase limitations of the exonic structure of the gene must be respected. In order to understand this phase limitation constraint, let us consider the different types of introns in terms of their possible positions

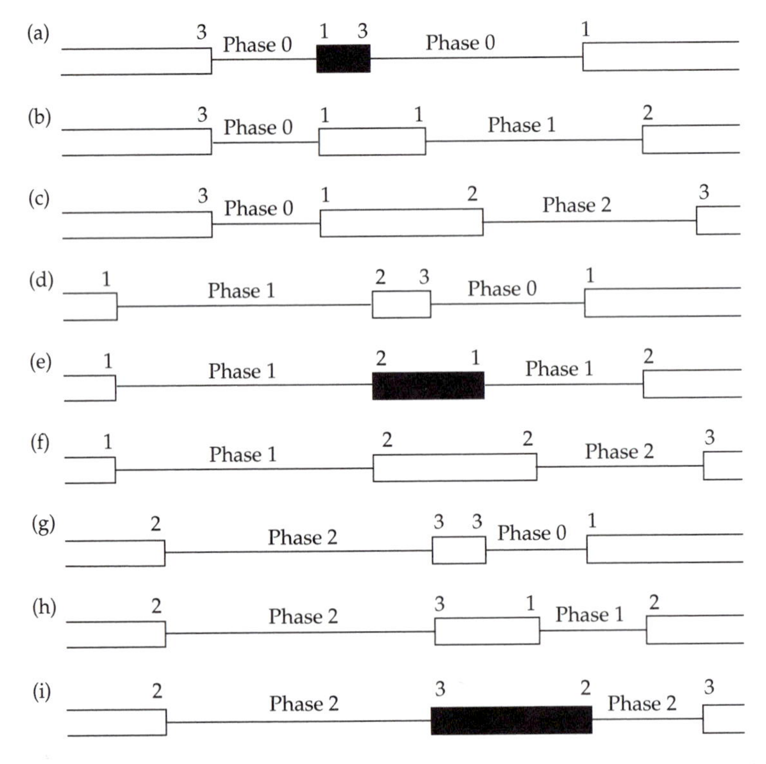

FIGURE 6.15 Phases of introns and classes of exons. Exons are represented by boxes. The numbers at the exon–intron junctions indicate the codon position of the last nucleotide of the exon, while the numbers at the intron–exon junctions indicate the codon position of the first nucleotide of the exon. Solid boxes represent symmetrical exons.

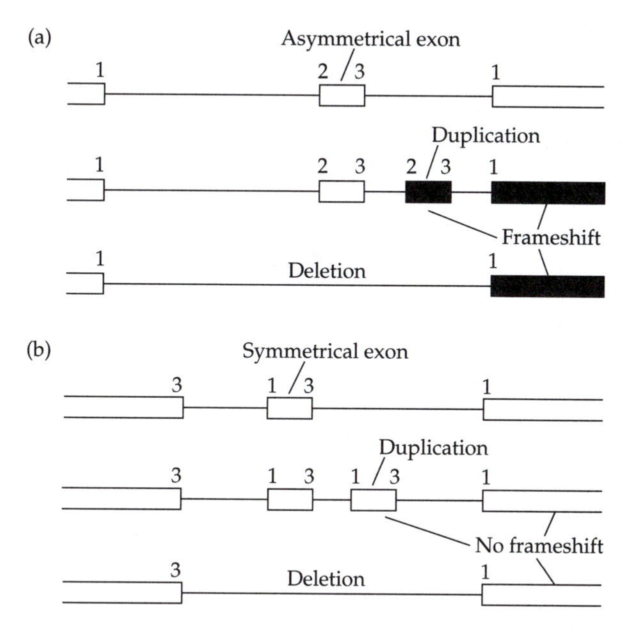

FIGURE 6.16 Consequences of exon duplication or deletion. Black boxes indicate frameshifts. (a) A gene containing a 1-2 asymmetrical exon. The duplication or deletion of this exon causes frameshifts in the downstream exons. In addition, the duplication results in frameshifts in the second copy of the duplicated exon itself. (b) A gene containing a 0-0 symmetrical exon. The duplication or deletion of this exon causes no frameshifts in the downstream exons.

relative to the coding regions. Introns residing between coding regions are classified into three types according to the way in which the coding region is interrupted. An intron is of **phase 0** if it lies between two codons, of **phase 1** if it lies between the first and second nucleotides of a codon, and of **phase 2** if it lies between the second and third nucleotides of a codon (Figure 6.15).

Exons are grouped into classes according to the phases of their flanking introns. For example, the middle exon in Figure 6.15b is flanked by a phase-0 intron at its 5′ end and by a phase-1 intron at its 3′ end; it is said to be a class 0-1 exon. An exon that is flanked by introns of the same phase at both ends is called a **symmetrical exon**, otherwise it is **asymmetrical**. For example, the middle exon in Figure 6.15a is symmetrical. Of the nine possible classes of exons, three are symmetrical (0-0, 1-1, and 2-2), and six are asymmetrical (0-1, 0-2, 1-0, 1-2, 2-0, and 2-1). The length of a symmetrical exon is always a multiple of three nucleotides.

Only symmetrical exons can be duplicated in tandem or deleted without affecting the reading frame (Figure 6.16). Duplication or deletion of asymmetrical exons would disrupt the reading frame downstream. Similarly, only symmetrical exons can be inserted into introns. For example, in Figure 6.17a the insertion of a 0-1 exon into a phase-0 intron causes a frame shift in all the

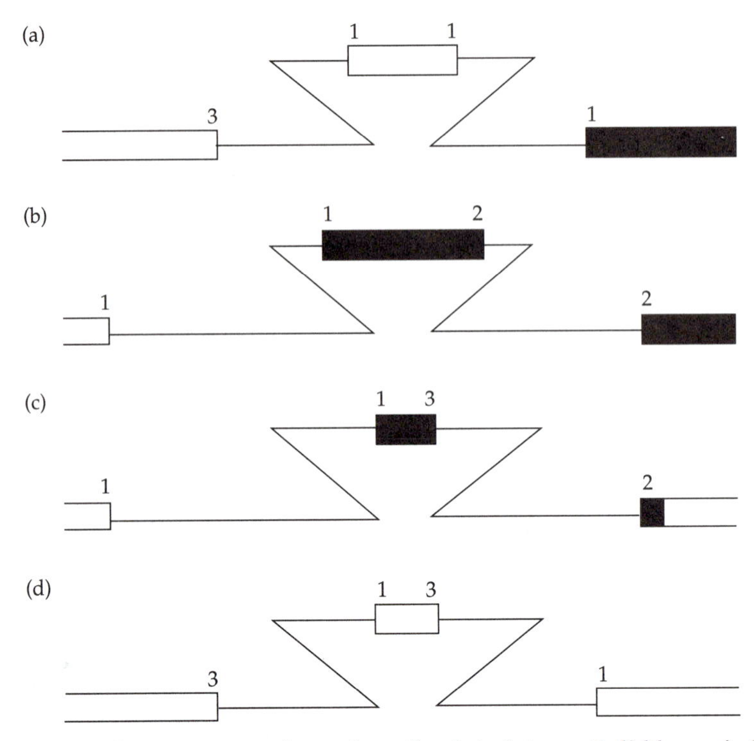

FIGURE 6.17 Consequences of exon insertion into introns. Solid boxes indicate frameshifts. (a) Insertion of a 0-1 asymmetrical exon into a phase-0 intron. (b) Insertion of a 0-2 asymmetrical exon into a phase-1 intron. (c) Insertion of a 0-0 symmetrical exon into a phase-1 intron. (d) Insertion of a 0-0 symmetrical exon into a phase-0 intron. The insertions in (a) and (b) cause frameshifts in the exons downstream of the insertion. The inserted exons in (b) and (c) are frameshifted. Only the insertion in (d) does not cause frameshifts in either the inserted exons or the downstream exons.

subsequent exons. Insertion of a 0-2 asymmetrical exon into a phase-0 intron not only causes a frameshift in all downstream exons but also in the inserted exon itself (Figure 6.17b).

Insertion of symmetrical exons is also restricted; a 0-0 exon can only be inserted in phase-0 introns, a 1-1 exon can only be inserted into phase-1 introns, and a 2-2 exon can only be inserted into phase-2 introns. For example, Figure 6.17c shows that the insertion of a 0-0 exon into an intron of phase 1 causes a frameshift in the inserted exon and at the 5′ end of the nearest downstream exon, while Figure 6.17d shows that the insertion of a 0-0 exon into a phase-0 intron causes no frameshift.

Not surprisingly, all the exons coding for the modules of mosaic proteins are symmetrical (Patthy 1987). For example, all the exons that encode the noncatalytic regions of the proteases taking part in the fibrinolytic, blood coagulation, and complement cascades are class 1-1 exons. The reason for this particular choice in the evolution of these proteases is not clear. A plausible explanation is that we are dealing with a "frozen accident," in which an ancestral phase-1 intron could only accept class 1-1 exons. Consequently, all subsequent insertions and duplications in this region could only involve symmetrical 1-1 exons. Thus, an initial bias led to a predominance of phase-1 introns.

Since nonrandom intron phase usage is a necessary consequence of exon duplication or insertion, this property may be used as a diagnostic feature of gene assembly through exon shuffling. For example, the genes coding for type-III collagen, β-casein, and the precursor of growth hormone have predominantly class 0-0 exons, consistent with the suggestion that these proteins have evolved by exon shuffling. On the other hand, phosphoglycerate kinase, glyceraldehyde-3-phosphate dehydrogenase, and triosephosphate isomerase genes contain a mixture of intron types, and consequently exon shuffling could not have played an important role in the formation of these genes.

In terms of splicing, introns are classified into two categories, self-splicing and spliceosomal, as described in Chapter 1. The vast majority of introns in eukaryotic nuclear genes are spliceosomal. Since a self-splicing intron plays a vital role in its own removal, some regions of the intron are involved in self-complementary interactions important for forming the three-dimensional structure possessing splicing activity. The need to preserve self-splicing activity obviously places severe restrictions on intronic recombination and the insertion of foreign exons. Archaic introns existing before the time of the prokaryote–eukaryote divergence are suggested to have been of the self-splicing type. Consequently, exon shuffling probably did not play a role in the formation of genes in the early stages of evolution. Exon shuffling came to full bloom with the evolution of spliceosomal introns, which do not play a role in their own excision. These introns contain mainly nonessential parts and therefore could accommodate quantities of "foreign" DNA.

Exonization and pseudoexonization

Because donor and acceptor splicing sites can be obliterated or created *de novo* quite easily by mutation, exons may appear or disappear by processes other than exon shuffling. **Exonization** is the process through which an intronic sequence becomes an exon. An exon created by exonization must abide by the same rules of exon insertion. Exonization has only rarely survived purifying selection during evolution. One such example involves the creation of exons of seemingly random sequence in the collagen IV gene, which may have been created by the inactivation of splicing sites (Butticé et al. 1990).

The opposite process is called **pseudoexonization**. It occurs when nonfunctionalization affects a single exon rather than the entire gene. The result is the creation of a **pseudoexon**, and the most obvious consequence of such a process is **gene abridgment** (as opposed to gene elongation). Pseudoexons are often created by the nonfunctionalization of internal gene duplications. For example, the aggrecan gene in rat contains 18 repeated exons and one pseudoexon (Doege et al. 1994). For pseudoexonization to occur without disruption of the reading frame, the rules pertaining to exon deletion must be respected.

TABLE 6.5 Principal biochemical reactions in the synthesis of fatty acids from malonyl CoA in eukaryotes and eubacteria

Reaction	Enzyme
1. Acetyl CoA + condensing enzyme domain ↔ acetyl-condensing enzyme	Acetyl transferase
2. Malonyl CoA + acyl-carrier peptide ↔ malonylacyl-carrier peptide	Malonyl transferase
3. Acetyl-condensing enzyme + malonylacyl-carrier peptide ↔ β-ketoacyl-carrier peptide	β-ketoacyl synthase
4. β-ketoacyl-carrier peptide + NADPH + H^+ ↔ β-hydroxyacyl-carrier peptide + $NADP^+$	β-ketoacyl reductase
5. β-hydroxyacyl-carrier peptide ↔ 2-butenoylacyl-carrier peptide + H_2O	β-hydroxyacyl dehydratase
6. 2-butenoylacyl-carrier peptide + NADPH + H^+ ↔ butyrylacyl-carrier peptide + $NADP^+$	Enoyl reductase
7. Butyrylacyl-carrier peptide + condensing enzyme domain ↔ butyryl-condensing enzyme + acyl-carrier peptide	Thioesterase

Different strategies of multidomain gene assembly

A corollary of exon shuffling is that some complex biological functions that require several enzymes may be specified by genes encoding different combinations of protein modules. In some species we may find single-module proteins, while in others we may find different combinations of multimodular proteins. One such instance is seen in the genes involved in the synthesis of fatty acids from acetyl-CoA. This multistep process requires seven enzymatic activities and an acyl-carrier protein (Table 6.5). In most bacteria, these functions are carried on by discrete monofunctional proteins. However, in fungi, these activities are distributed between two nonidentical polypeptides encoded by two unlinked intronless genes, *FAS1* and *FAS2*. *FAS1* encodes two of the seven enzymatic activities (β-ketoacyl synthase and β-ketoacyl reductase) as well as the acyl-carrier protein. *FAS2* encodes the rest of the five enzymatic activities (Chirala et al. 1987; Mohamed et al. 1988).

In animals, all the functions are integrated into a single polypeptide chain called fatty-acid synthase. Characterization of the fatty-acid synthase gene in the rat (Amy et al. 1992) revealed that the gene product contains eight modules, including one module that performs a dual function (acetyl transferase and malonyl transferase), and another whose function is most probably unrelated to fatty-acid synthesis but may have a role in determining the tertiary structure of this multimodular protein (Figure 6.18). The gene is composed of 43 exons separated by 42 introns. With only one exception, boundaries between adjacent modules coincide with the location of introns. Thus, the fatty-acid synthase genes in fungi and mammals are most probably mosaic proteins that have assembled from single-domain proteins like the ones found in bacteria. The fact that the arrangement of domains is different in fungi from that

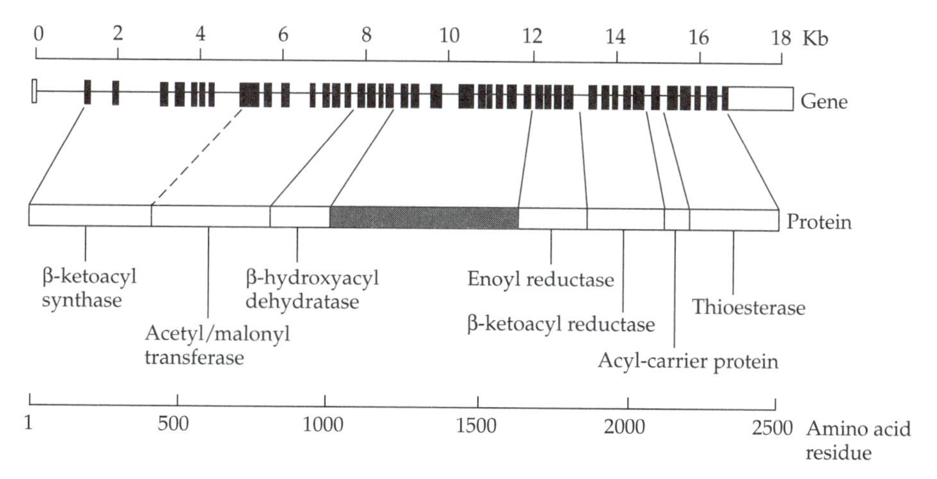

FIGURE 6.18 The rat fatty-acid synthase gene encodes a multifunctional protein. Exons are shown as boxes separated by introns (depicted as lines). Solid boxes indicate protein-coding regions. Seven domains are marked on the amino acid sequence. The function of the interdomain region (gray box) is unknown at this time, but its purpose may be to hold two fatty-acid synthase subunits together and generate a dimer with two centers of fatty-acid synthesis. With the exception of the boundary between the β-ketoacyl synthase and the acetyl/malonyl transferase domains (dashed line), all other boundaries between adjacent modules coincide with the location of introns.

in mammals indicates not only that the two lineages evolved multimodularity independently, but also that different strategies may be employed in the assembly of genes encoding multimodular proteins.

THE "INTRONS-EARLY" VERSUS "INTRONS-LATE" HYPOTHESES

Following the unexpected discovery of introns in eukaryotic genes, Gilbert (1978) suggested that this type of gene organization may be of evolutionary importance by facilitating the creation of novel proteins through the shuffling of exons. Doolittle (1978) and Darnell (1978) hypothesized that introns are a primitive feature of genes, while Blake (1978) proposed that exons originally corresponded to structural units of proteins. This view has been known as the **introns-early hypothesis**, later modified to become the **exon theory of genes** (Gilbert 1987). According to this theory, genes are descendants of ancient monoexonic minigenes, and introns are descendants of the spacers between them. According to the introns-early hypothesis, ancient genes possessed self-splicing introns, but most of these introns were lost in Bacteria and Archaea, whereas in Eucarya they evolved into spliceosomal introns.

The opposing **introns-late hypothesis** assumes that early genes had no introns, and that the addition of introns occurred after the emergence of the eukaryotic cell or the endosymbiotic process that gave rise to the mitochondria (e.g., Cavalier-Smith 1985a, 1991). According to the current version of the theory, classical nuclear spliceosomal introns were derived from group II self-splicing introns.

One line of evidence supporting the introns-early hypothesis was the correspondence between exons and protein modules (see page 251). For example, Gō (1981) identified four structural modules in α- and β-globins, but only three exons and two introns. She therefore predicted the existence of an additional intron between modules 2 and 3. The finding of such an intron in plants, nematodes, and insects (see Figure 6.3) was interpreted as strong evidence for the exon theory of genes. A similar positional prediction for the existence of an intron in the triosephosphate isomerase gene (Gilbert et al. 1986) was found to be true in a mosquito gene (Tittiger et al. 1993). However, a huge number of introns were found in positions not corresponding with the borders of modules (including about a dozen such introns in the globin superfamily alone), and these seemed to contradict the introns-early hypothesis. We note that molecular mechanisms for both intron gain and intron loss are known.

The 20-year-old debate between proponents of the "introns-early" and "introns late" hypotheses has been lively and scientifically very productive, with the weight of the evidence favoring sometimes one view and sometimes the other. The "introns-late" view has gained some support in recent years (e.g., Rzhetsky et al. 1997; Hurst and McVean 1996), but it is too early to say whether or not a final resolution has been achieved.

Intron sliding

Can the "introns-early" versus the "introns-late" controversy be solved by studying the correspondence between exons and structural modules of proteins? The answer is most likely negative for many reasons, chiefly among which is the process of **junctional** or **intron sliding**, i.e., a shift in the position of introns.

A shift in the position of an intron can occur through several mechanisms, such as a mutation resulting in the obliteration of a splicing site, or a mutation in either an intron or an exon giving rise to a new splicing site (see page 300). The intron sliding hypothesis has been raised to explain the increasing number of reports in the literature on intron positions that coincide neither with the boundaries between adjacent modules in proteins, nor with the position of introns in homologous genes. These intron positions are called **discordant intron positions**, as opposed to intron positions that conform to the expectations derived from the introns-early hypothesis, which are called **concordant intron positions**. The intron sliding mechanism was supposed to give an answer to the very large excess in the number of intron positions in several genes that poses a problem for the exon theory of genes. For instance, if all 205 different intron positions documented in published compilations of gene data for actins, gyceraldehyde-3-phosphate dehydrogenases, small G-proteins, triosephosphate isomerases, and tubulins are packed into a hypothetical ancestral gene (with a combined length of about 1,600 codons), they would

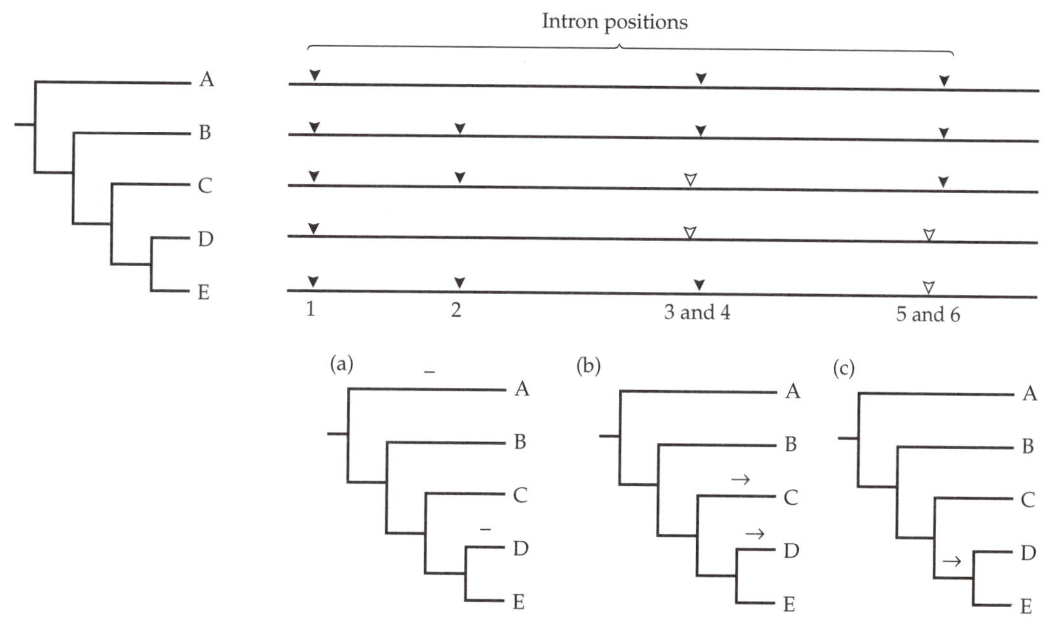

FIGURE 6.19 Hypothetical examples of phylogenetic distributions of intron positions for testing the introns-early hypothesis. The phylogenetic tree for five OTUs is in the upper left corner; the map of intron positions (arrowheads) is to its right. Concordant intron positions are marked by solid arrowheads; discordant codon positions by empty ones. Intron 1 shows a uniform conserved distribution consistent with both the introns-early and introns-late hypotheses. Intron 2 shows a conserved distribution, which would require two intron losses according to the introns-early hypothesis (a). The position of discordant intron 3 is not nested within that of concordant intron 4. Therefore, the distribution of the introns can be explained by the introns-early hypothesis only by invoking two independent intron-sliding events (b). The position of discordant intron 5 is nested within that of concordant intron 6, and can be explained by a single intron-sliding event (c). Absence or loss of an intron is marked by –, sliding by →.

break up this gene into exons with a mean length of only 23 bp, and a median length of only 14 bp (Stolzfus et al. 1997). Invoking intron sliding in this case would, in effect, reduce the number of ancestral intron positions from 205 to a much smaller (and explainable) number.

If intron sliding is indeed responsible for most of the excess in the number of intron positions and the discordance between the intron and the boundaries between protein structural domains, we would expect a certain spatial and phylogenetic distribution of intron positions. Specifically, we would expect a nested phylogenetic distribution of closely spaced introns, in which discordant intron positions are nested within the distribution of nearby concordant intron positions (Figure 6.19). In other words, the expectation of the introns-early hypothesis is for homologous discordant intron positions to be monophyletic. Using the methodology outlined in Figure 6.19, Stolzfus et al. (1997) studied the 205 intron positions mentioned above. Their conclusion was that the spatial distribution of intron positions can be explained by the introns-early hypothesis only if an unreasonably large number of parallel intron sliding events are assumed to have occurred independently in many lineages. Therefore, the observed intron position diversity is due primarily to the "late" addition of introns during eukaryotic evolution.

The relative fraction of "early" and "late" introns

In a series of studies, de Souza et al. (1997, 1998) investigated the problem of intron position by using a set of 44 "ancient genes," i.e., genes whose products are conserved in function between eukaryotes and prokaryotes. de Souza et al.'s (1998) dataset included 988 intron positions. Excluding the possibility of intron sliding, they reasoned that ancient introns should be of phase 0, and should be associated with compact modules, which were previously inferred to have diameters around 21, 27, and 33Å (de Souza et al. 1997). Their conclusion was that 35% of the introns present in their dataset were "early" introns, whereas about 65% of the introns have been added subsequently. These "late" introns exhibited neither a phase preference nor an association with modular structures.

ALTERNATIVE PATHWAYS FOR PRODUCING NEW FUNCTIONS

In addition to gene duplication and exon shuffling, there are many other mechanisms for producing new genes or polypeptides. Several such mechanisms are considered below.

Overlapping genes

A DNA segment can code for more than one gene product by using different reading frames or different initiation codons. This phenomenon of **overlapping genes** is widespread in DNA and RNA viruses, as well as in organelles and bacteria, but it is also known in nuclear eukaryotic genomes. Figure 6.20a shows the genetic map of ΦX174, which is a single-stranded DNA bacteriophage. Several overlapping genes are observed. For example, gene *B* is completely contained inside gene *A*, while gene *K* overlaps gene *A* on the 5′ end and gene *C* on the 3′ end. A more detailed illustration of the latter case is given in Figure 6.20b.

Overlapping genes can also arise by the use of the complementary strand of a gene. For example, the genes specifying $tRNA^{Ile}$ and $tRNA^{Gln}$ in the human mitochondrial genome are located on different strands and there is a three-nucleotide overlap between them that reads 5′—CTA—3′ in the former and 5′—TAG—3′ in the latter (Anderson et al. 1981).

The question arises as to how overlapping genes may have come into existence during evolution. To answer this question, we note that open reading

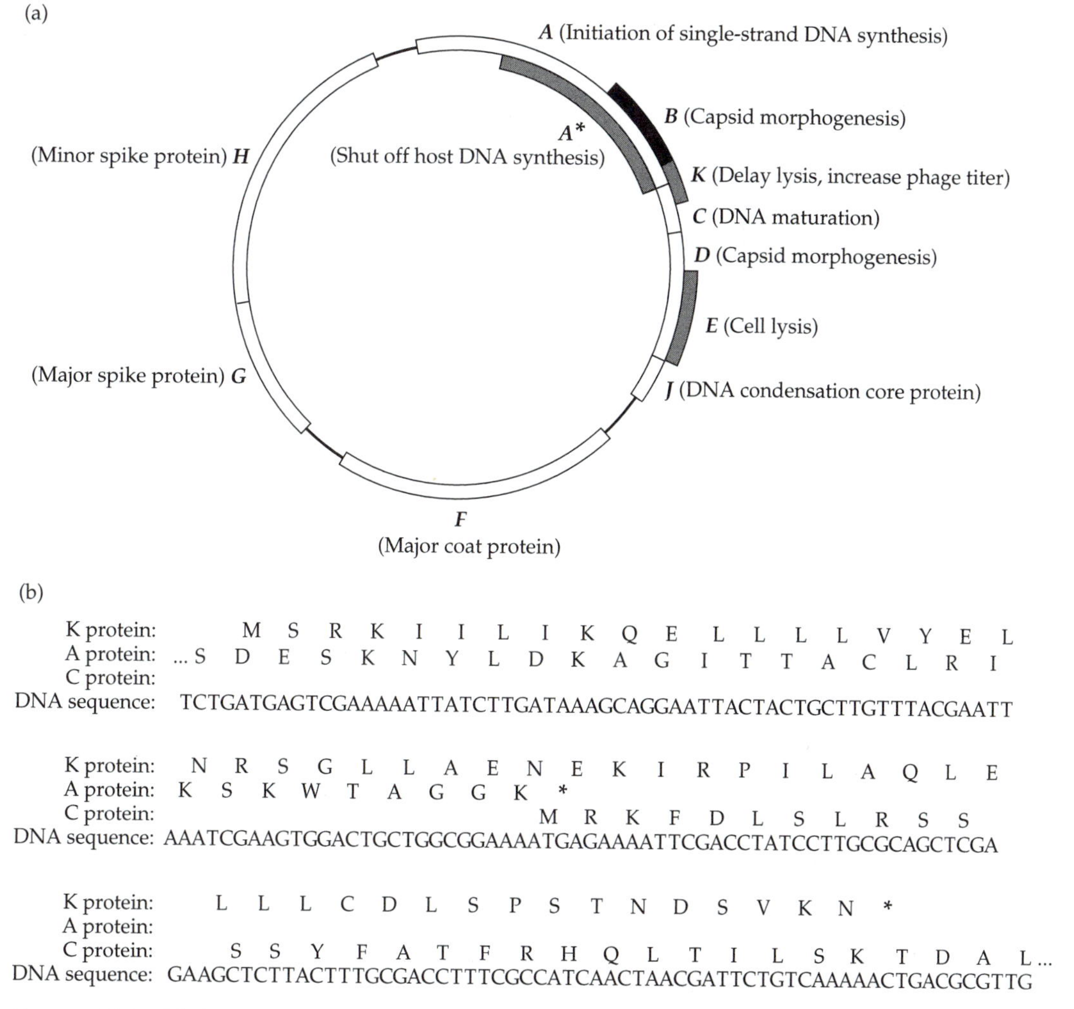

FIGURE 6.20 (a) Genetic map of the circular genome of the single-stranded DNA bacteriophage ϕX174. Note that the *B* protein-coding gene (black) is completely contained within the *A* protein-coding gene, and that gene *K* overlaps two genes, *A* and *C*. Gene *A is an abbreviated form of *A* that uses a different initiation codon to yield a shorter product. Modified from Kornberg (1982). (b) Sequence of the *K* gene, showing overlap with the 5′ part of the *A* gene and the 3′ of the *C* gene. Asterisks indicate stop codons. (See Table 1.2 for the one-letter amino acid abbreviations.)**

frames abound throughout the genome. Therefore, it is possible that potential coding regions of considerable length exist in either a different reading frame of an existing gene or on the complementary strand. Because only 3 of 64 possible codons are termination codons, even a random DNA sequence might contain open reading frames hundreds of nucleotides long. If by chance such a reading frame contains an initiation codon and a transcription initiation site, or if such sites are created by mutation, an additional mRNA will be transcribed and subsequently translated into a new protein. Whether the new product has a beneficial function or not is another matter, but if it does, the trait may become fixed in the population. Finally, we note that because of the structure of all genetic codes, codon-usage biases (Chapter 4) are conducive to the creation of long, nonstop reading frames on the complementary strands of protein-coding genes (Silke 1997).

The rate of evolution is expected to be slower in stretches of DNA encoding overlapping genes than in similar DNA sequences that only use one reading frame. The reason is that the proportion of nondegenerate sites is higher in overlapping genes than in nonoverlapping genes, thus vastly reducing the proportion of synonymous mutations out of the total number of mutations (Miyata and Yasunaga 1978). We note that since gene duplication is a widespread phenomenon, the maintenance of overlapping genes, as opposed to two nonoverlapping copies, would require quite strong selective pressure (say, against increasing the genome size).

Studies on aminoacyl-tRNA synthetases indicate that overlapping genes may have played a momentous role in the evolution of life. A crucial step in the process of translation is performed by 20 aminoacyl-tRNA synthetases, each of which activates a specific amino acid and attaches it to a specific tRNA. The aminoacyl-tRNA synthetases exist as two unrelated gene families. The class I family includes the aminoacyl-tRNA synthetases for valine, isoleucine, leucine, methionine, cysteine, arginine, tyrosine, tryptophan, glutamine, and glutamic acid; the class II family consists of aminoacyl-tRNA synthetases specific for the other ten amino acids. The lack of even marginal similarity between the two families have brought scientists to suggest that the two families evolved from two independent primordial synthetases that already existed in the cenancestor (e.g., Nagel and Doolittle 1995). Interestingly, Rodin and Ohno (1995) found that the two families do exhibit significant sequence similarity, but only when their coding DNA sequences are compared in the opposite direction. This finding prompted Rodin and Ohno (1995) to suggest that the two synthetase families originated as two protein-coding genes located on the complementary strands of the same primordial double-stranded nucleic acid.

Alternative splicing

Alternative splicing of a primary RNA transcript results in the production of different mRNAs from the same DNA segment, which in turn may be translated into different polypeptides. Because of alternative splicing, the distinction between exons and introns is no longer absolute but depends on the mRNA of reference. There are two types of exons: **constitutive** (i.e., exons that are included within all the mRNAs transcribed from a gene) and **facultative** (i.e., exons that are sometimes spliced in and sometimes spliced out). Many cases of alternative RNA processing have been found in multicellular organisms, as well as in eukaryotic transposable elements (Chapter 7) and animal viruses.

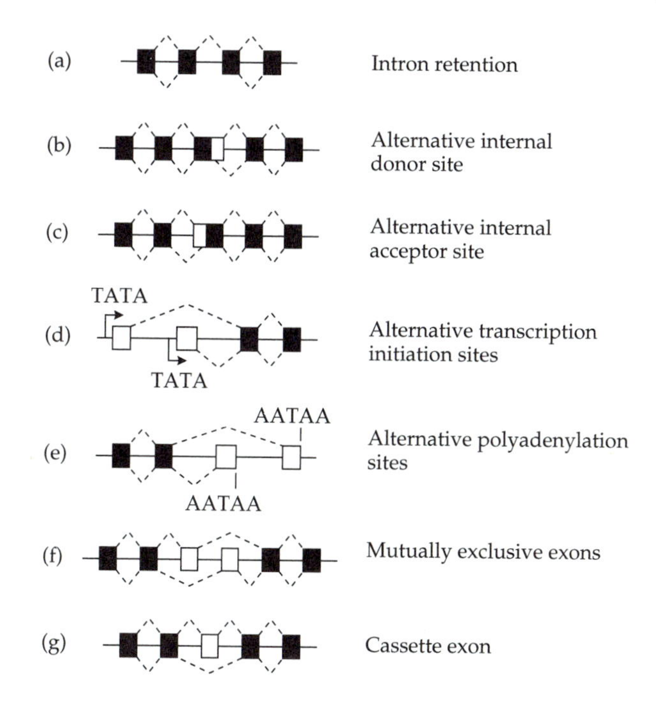

FIGURE 6.21 Types of alternative splicing. Constitutive and facultative exons are shown as black and empty boxes, respectively. Introns are depicted as lines. Alternative splicing pathways are shown by the diagonal dashed lines above and below the gene. Transcription initiation and polyadenylation sites are denoted by TATA and AATAA, respectively. Modified from Smith et al. (1989).

There are different types of alternative splicing. Perhaps the most trivial form of alternative splicing is **intron retention** (Figure 6.21a). An unspliced intron can result in the addition of a peptide segment if the reading frame is maintained. More commonly, however, intron retention results in the premature termination of translation due to frameshifts. One such example is the periaxin gene in mouse (Dytrich et al. 1998). The gene contains seven exons and encodes two proteins involved in the initiation of myelin deposition in peripheral nerves. The mRNA of the shorter protein, S-periaxin, retains the intron in between exons 6 and 7 and, as a consequence, translation stops 21 amino acids after exon 6. Since exons 1–6 of periaxin are very short and two of them precede the initiation codon, whereas exon 7 is very long, the larger mRNA (5.2 Kb) encodes a short protein (16 kilodaltons), and the shorter mRNA (4.6 Kb), encodes a much larger protein (147 kilodaltons).

Alternative splicing sometimes involves the use of **alternative internal donor** or **acceptor sites**, i.e., excisions of introns of different lengths with complementary variation in the size of neighboring exons (Figure 6.21b and c). Such use of competing splice sites was found in several transcription units of adenoviruses, as well as in eukaryotic cells such as the *transformer* gene in *Drosophila melanogaster* (see page 298).

In some instances, different mRNAs that are produced from the same gene differ from one another only at their 5′ or 3′ ends. This is usually the result of **alternative transcription initiation** or **alternative transcription termination** (due mostly to alternative polyadenylation sites), respectively (Figure

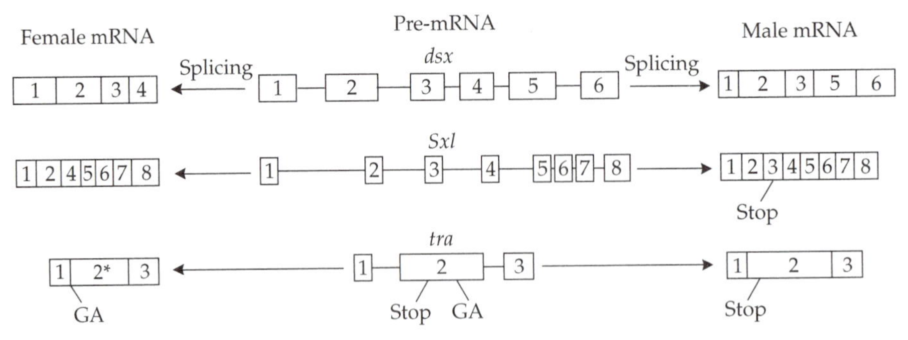

FIGURE 6.22 The patterns of splicing of the *doublesex* (*dsx*), *Sexlethal* (*Sxl*), and *transformer* (*tra*) genes in *Drosophila melanogaster* females (left) and males (right). "Stop" indicates a termination codon that truncates the coding region of the mature mRNA and renders the product nonfunctional. The internal acceptor site in *tra* is denoted as AG, and 2* denotes the transcript downstream. Modified from Baker (1989).

6.21d and e). An example of alternative transcription initiation can be seen in the gene encoding myosin light chains 1 and 3. Alternative polyadenylation sites are quite common in eukaryotic nuclear genes.

Some cases of alternative splicing involve the use of **mutually exclusive exons**, i.e., two exons are never spliced out together, nor are both retained in the same mRNA (Figure 6.21f). One such instance is the M1 and M2 forms of pyruvate kinase, which are produced from a single gene by mutually exclusive use of exons 9 and 10. A special case of mutual exclusivity is the **cassette exon** (Figure 6.21g). A cassette is either spliced in or spliced out in the alternative mRNA molecules. Usually the reading frame is maintained whether such an exon is in or out, although cases are known in which inclusion of a cassette can result in frameshifts.

Within a single gene, we may find different types of alternative splicing. For example, in the gene for troponin-T, the presence of five cassette exons at the 5′ end of the gene, in conjunction with a pair of mutually exclusive exons, allows the production of 64 different proteins.

Alternative splicing has often been used as a means of developmental regulation. A very intriguing situation is seen in several genes involved in the process of sex determination in *Drosophila melanogaster*. At least three genes, *doublesex* (*dsx*), *Sexlethal* (*Sxl*), and *transformer* (*tra*), are spliced differently in males and females (Figure 6.22). In the case of *dsx*, the gene has six exons; exons 1, 2, 3, and 4 are used in the female, and exons 1, 2, 3, 5, and 6 are used in the male. In *Sxl* and *tra*, the products of the alternative splicing in males contain premature termination codons and are therefore nonfunctional. For instance, exon 3 in *Sxl*, which is not included in the female transcript, contains an in-frame stop codon that prematurely terminates its translation in the male.

The evolution of alternative splicing requires that an alternative splice junction site be created *de novo*. Since splicing signals are usually 5–10 nucleotides long, it is possible that such sites are created with an appreciable frequency by mutation. Indeed, many such examples are known in the literature. Figure 6.23 illustrates a case in which a synonymous substitution in a glycine codon turned a coding region into a splice junction. In cases of pathological manifestations such as the β^+-thalassemia illustrated in Figure 6.23, the new splice site is usually stronger than the old splice site (i.e., most of the mRNA synthesized after such a mutation occurs is of the altered type). Such a mutation will obviously have deleterious effects and is not expected ever to become fixed in the population. However, if the newly created splice site is much weaker, then most mRNA will be of the original type, and only small quantities of the new mRNA will be made. Such a change will not obliterate the old function, and yet will create an opportunity to produce a new protein, possibly one with a new useful function.

Intron-encoded proteins and nested genes

An intron may sometimes contain an open reading frame that encodes a protein or part of a protein that is completely different in function from the one encoded by the flanking exons. In many cases, **intron-encoded protein genes** (Perlman and Butow 1989) are located within type-I self-splicing introns. For example, at least three genes in bacteriophage *T4 (td, nrdB,* and *sunY*) are known to have introns that contain protein-coding genes. Intron *a14α* in the yeast mitochondrial gene *cox I* is intriguing in that it encodes an enzyme called maturase that is required for the proper self-splicing of this intron from the pre-mRNA. This maturase also functions as an endonuclease in DNA recombination

From a mechanistic point of view, an intron-encoded protein gene that is transcribed from the same strand as the neighboring exons may be regarded as a special instance of alternative splicing. When an intron-encoded protein gene is transcribed from the opposite strand of the other gene, it is referred to as a **nested gene**. A case of nested genes was found in *Drosophila*, where a pupal cuticle protein gene is encoded on the opposite strand of an intron within the gene encoding the purine pathway enzyme glycinamide ribotide transformylase. Interestingly, the intronic gene is itself interrupted by an intron (Henikoff et al. 1986; Moriyama and Gojobori 1989).

Functional convergence

Given that the function of a protein is frequently determined by only a few of its amino acids, a protein performing one function may sometimes arise from a gene encoding a protein performing a markedly different function. If the new function is performed in other species by proteins of unrelated structure and descent, **functional convergence** may occur. One such case is the myoglobin in the red-muscle tissue of the buccal mass in the abalone *Sulculus diversicolor* and related prosobranchian mollusks (Suzuki et al. 1996). The myoglobin of *Sulculus* consists of 377 amino acids, which means it is 2.5 times

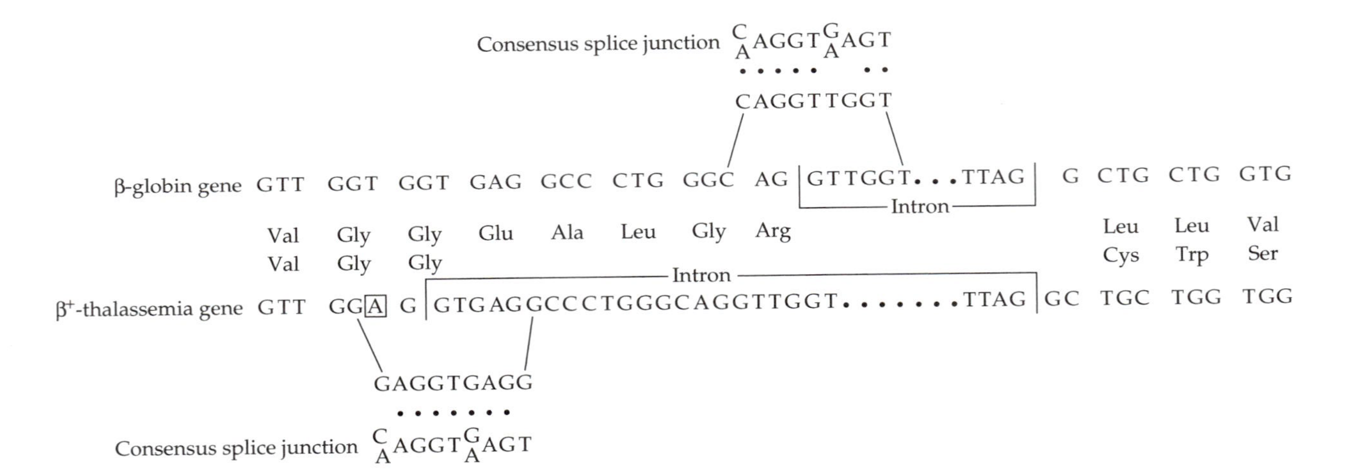

FIGURE 6.23 The nucleotide sequences at the borders between exon 1 and intron I and intron I and exon 2 in the β-globin gene from a normal individual and a patient with β^+-thalassemia. The mutated nucleotide is boxed. Splicing junctions are indicated. Each of the splice junctions is compared with the sequence of the consensus splice junction, and dots denote identity of nucleotides between the splice junction and the consensus sequence. Note that the nucleotide substitution in the β^+-thalassemia gene is synonymous, because both GGT and GGA code for the amino acid glycine. It is not silent, however, because the activation of the new splicing site in the β^+-thalassemia gene results in the production of a frameshifted protein. Data from Goldsmith et al. (1983).

larger than myoglobins belonging to the globin superfamily. The protein carries a heme group and can bind oxygen reversibly, but its oxygen affinity is somewhat lower than those of other vertebrate and invertebrate myoglobins.

Intriguingly, the amino acid sequence exhibits no similarity to any other myoglobin or hemoglobin, but was shown to be homologous with the enzyme indoleamine 2,3-dioxygenase, which degrades tryptophan and other indole derivatives into kinurenines. In mammals, indoleamine dioxygenase performs a very important function during pregnancy by preventing the immunological rejection of the fetus by the mother (Munn et al. 1998).

The taxonomic distribution of indoleamine dioxygenase-derived myoglobins in conjunction with a molecular phylogeny inferred from 18S rRNA sequences (Winnepenninckx et al. 1998) indicates that the recruitment of indoleamine dioxygenase as a myoglobin occurred once, about 270 million years ago, in the ancestor of *Sulculus* and its relatives *Nordotis, Battilus, Omphalius*, and *Chlorostoma*.

Evidence for convergent evolution of similar function among proteins (or domains) encoded by different gene families is accumulating steadily (Kuriyan et al. 1991; Bork et al. 1993; Alber and Ferry 1994; Hewett-Emmett and Tashian 1996).

RNA editing

RNA editing is the posttranscriptional modification of an RNA molecule that, in the case of protein-coding gene, changes its message (Chapter 1). One of the most common types of RNA editing is C-to-U conversion. This conversion may occur partially or completely in some tissues but not in others, leading to differential gene expression. Occasionally, it can produce a new protein with a different function from that of the unedited transcript. Such is the case with the apolipoprotein B gene, one of the lipid carriers in the blood. There are two types of apolipoprotein B, apoB-100, and apoB-48. In humans, apoB-100 is primarily synthesized by the liver and is the major protein constituent in low- and very low-density lipoproteins, which are recognized and bound by the low-density lipoprotein receptor. ApoB-48, in contrast, is synthesized by the intestine and does not bind the low-density lipoprotein receptor.

Despite differences in length, when one aligns the amino acid sequence of the gigantic protein apoB-100 (4,536 amino acids) with that of apoB-48 (2,152 amino acids), the result for the alignable part is 100% identity. Such a situation could have been achieved by either alternative mRNA splicing or two genes, one of which is a partial duplicate of the other. Unexpectedly, it was found that apoB-48 is translated from a very long mRNA that is identical to that of apoB-100 with the exception of an in-frame stop codon resulting from the RNA editing of codon 2153 from CAA (Gln) to UAA (Figure 6.24; Chen et al. 1987; Powell et al. 1987). Thus, by using RNA editing, two quite different proteins are produced from the same gene.

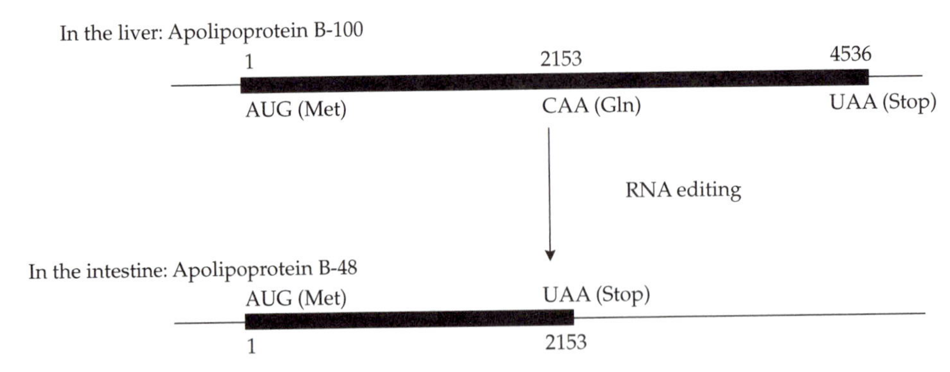

FIGURE 6.24 Two different proteins produced from the same gene. RNA editing of human apolipoprotein-B mRNA at codon 2153 produces a termination codon. The reading frames of the unedited and edited mRNAs are shown as boxes.

Gene sharing

An extremely intriguing situation arises when a gene product is recruited to serve an additional function. This phenomenon has been termed **gene sharing** (Piatigorsky et al. 1988). Gene sharing means that a gene acquires and maintains a second function without divergent duplication and without loss of the primary function. Gene sharing may, however, require a change in the regulation system of tissue specificity or developmental timing. (In the literature, the term "multifunctional protein" is frequently used instead of "gene sharing." We prefer to use the term "multifunctional protein" to refer exclusively to the multiple functions of multidomain mosaic proteins.)

Gene sharing was first discovered in crystallins, which are the major water-soluble proteins in the eye lens, and whose function is to maintain lens transparency and proper light diffraction. The first recognition of the phenomenon of gene sharing was for crystallin ε from birds and crocodiles, which was found to be identical in its amino acid sequence with lactate dehydrogenase B (see page 268) and to possess an identical enzymatic activity (Wistow et al. 1987). Subsequent work has shown that these "two" proteins are in fact one and the same, encoded by the same single-copy gene (Hendriks et al. 1988). Similarly, vertebrate crystallin τ has been shown to be identical to and encoded by a single gene, α-enolase—a glycolytic enzyme converting 2-phosphoglycerate into phosphoenolpyruvate (Piatigorsky and Wistow 1989). Thus, crystallins ε and τ illustrate instances of gene sharing, whereby a gene acquires additional roles without being duplicated. Since the eye is a recent evolutionary invention, it is assumed that the enzymatic function came first and the optical function later. For unknown reasons, carbonyl-metabolizing enzymes take part in gene sharing more often than other enzymes (Lee et al. 1993).

Crystallin δ, which exists throughout class Reptilia (including birds), has also been shown to be identical in sequence with an enzyme, argininosuccinate lyase, which catalyzes the conversion of argininosuccinate into the amino acid arginine. There are, however, two almost identical crystallin δ/argininosuccinate lyase genes, most probably derived from a very recent gene duplication (Piatigorsky 1998a, b). A similar situation was found for crystallin αB, which is identical in sequence to the small heat-shock protein, but this gene also underwent a duplication. Thus, crystallins δ and αB exhibit a type of gene sharing with gene duplication, but with no divergence of the duplicate genes.

Other crystallins, such as β and γ in vertebrates, are classic examples of proteins that evolved by means of gene duplication and subsequent sequence divergence from ancestral genes specifying different proteins (e.g., heat-shock genes, which encode proteins expressed following exposure to excessive heat). Similarly, the S-crystallins of cephalopods were derived from duplication and subsequent divergence of the gluthatione S-transferase gene, but they do not possess enzymatic activity (Tomarev and Zinovieva 1988).

Gene sharing might be a fairly common phenomenon. In fact, in the above examples, the enzymes/crystallins may have more than two functions. For instance, crystallin τ/α-enolase also serves as a heat-shock protein, and crystallin αB also functions as a molecular chaperone preventing the aggregation of denatured proteins. Gene sharing is also suspected for several proteins in the cornea and other tissues (Piatigorsky 1998a,b). For example, aldehyde dehydrogenase and transketolase are abundant in the cornea, and it has been suggested that they may also have a structural function by directly absorbing ultraviolet light and reduce scattering in the corneal epithelium.

Gene sharing clearly adds to the compactness of the genome, even though compactness does not seem to have a high priority in eukaryotes (Chapter 8). Also note that, in the case of crystallin gene sharing, the same polypeptide serves both as an enzyme and as a structural protein, thus blurring the traditional distinction between enzymes and nonenzymatic structural proteins.

There seems to be a pleasant historical subtext to the story of gene sharing and crystallin evolution, for it may contribute to the unraveling of a puzzle that has distressed students of evolution for almost two centuries. It concerns the evolution of the eye and the possibility of evolving a highly complex, totally novel character by natural means. In *The Origin of Species*, in a chapter entitled "Organs of extreme perfection and complication," Charles Darwin wrote: "To suppose that the eye, with all its inimitable contrivances ... could have been formed by natural selection, seems, I freely confess, absurd to the highest possible degree." At least as far as the lens and the cornea of the eye are concerned, it seems that their "inimitability" is not all that unique, and that some of their constituent proteins can be traced back in evolution to ubiquitous housekeeping enzymes that are found throughout all life forms on earth.

MOLECULAR TINKERING

The more we learn about the evolution of genes, the more we recognize that true innovations are only rarely produced during evolution. Many proteins that were originally considered to be relatively recent evolutionary additions turned out to be derived from ancient proteins. Collagen, for example, was thought to be a "recent vintage modern protein" (Doolittle et al. 1986; Doolittle 1987)—i.e., a protein found only in animals, with no counterparts in fungi, plants, or prokaryotes. As it turns out, collagens are abundant in fungi (Celerin et al. 1996), and thus they must have existed in the common ancestor of Fungi and Animalia. Carbonic anhydrases and histones were supposed to be "middle-aged" proteins, i.e., eukaryotic proteins with no counterparts in Bacteria. In fact, homologous open-reading frames were found in the *E. coli* genome.

True novelty is almost unheard of during evolution; rather, preexisting genes and parts of genes are transformed to produce new functions, and molecular systems are combined to give rise to new, often more complex systems. Molecular biologist François Jacob called this process of disassembly and reassembly "molecular tinkering," or in French, *bricolage moleculaire* (Jacob 1977, 1983; Duboule and Wilkins 1998). Doolittle (1988) referred to this feature of the evolutionary process as "molecular opportunism."

Examples of molecular tinkering abound in nature. Lactalbumin, one of the main proteins in mammalian milk, has been for a very long time something of an evolutionary puzzle, having been considered to have arisen *de novo* in the ancestor of mammals. Molecular studies, however, have shown that lactalbumin is not a mammalian innovation, but rather a modification of lysozyme, a ubiquitous enzyme found in a wide range of organisms from bacteria to plants, and a very ancient protein.

Lysozyme is a hydrolytic enzyme that destroys the mucopolysaccharide component of bacterial cell walls. Lactalbumin, on the other hand, has no known enzymatic activity. Therefore, a few amino acid replacements were sufficient for lysozyme to become a new molecular entity, and another classical tale of gene duplication and creation of a new function without the loss of the original one was written in the chronicles of evolution.

The tinkering does not end here, however, for lactalbumin also serves as the B chain of lactose synthetase, a heterodimeric enzyme present in the lactating mammary gland. Therefore, lactalbumin also represents a case of gene sharing. The other polypeptide chain of lactose synthetase, the A chain, also serves an enzymatic function as *N*-acetyllactosamine synthetase, representing an additional case of gene sharing.

We have listed many mechanisms that facilitate tinkering at the molecular level. Other mechanisms, such as gene conversion and transposition, will be discussed later. We may therefore deduce that molecular tinkering is most probably the paradigm of molecular evolution, and it is reasonable to assume that tinkering also characterizes the evolution of morphological, anatomical, and physiological traits as well.

CONCERTED EVOLUTION

From the mid-1960s to the mid-1970s, a large number of DNA reannealing and hybridization studies were conducted to explore the structure and organization of eukaryotic genomes. These studies revealed that the genome of multicellular organisms is composed of highly and moderately repeated sequences as well as single-copy sequences (Chapter 8). They also revealed an unexpected evolutionary phenomenon, namely that the members of a repeated-sequence family are generally very similar to each other within one species, although members of the family from even fairly closely related species may differ greatly from each other (Figure 6.25). A fine illustration of

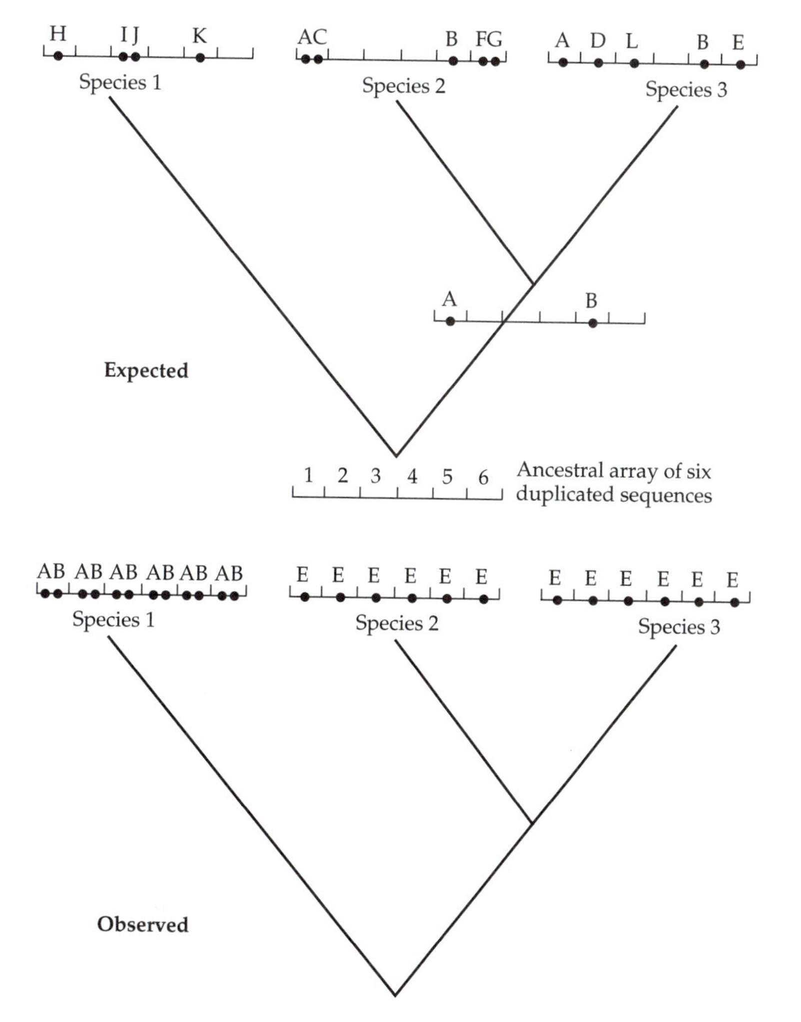

FIGURE 6.25 Schematic representation of the expected and observed patterns of variation in an array of six duplicated sequences shared by three species. If each duplicate sequence evolves independently, then the similarity between any two randomly chosen sequences within a species is expected to be the same as that between two sequences chosen between the species. Observed patterns, on the other hand, reveal a high degree of within-species homogeneity among duplicated sequences. Differences from the ancestral sequence are shown as letters.

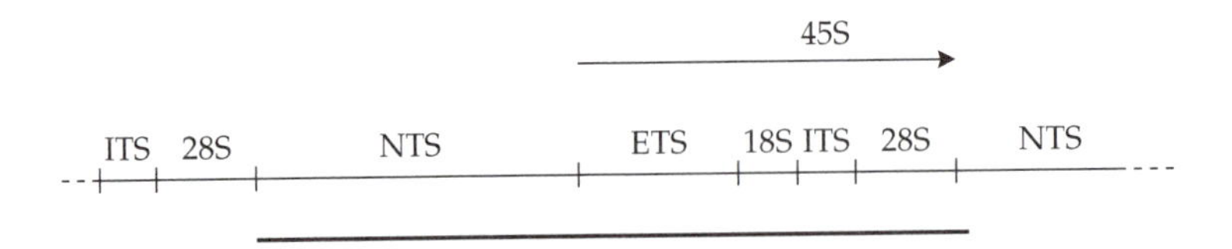

FIGURE 6.26 Diagrammatic representation of a typical repeated unit of rRNA genes in vertebrates. The thick bar designates the repeat unit, and the arrow indicates the transcribed unit. ETS, external transcribed spacer; ITS, internal transcribed spacer; NTS, nontranscribed spacer. Modified from Arnheim (1983).

this phenomenon was provided by Brown et al. (1972) in a comparison of the ribosomal RNA genes from the African frogs *Xenopus laevis* and *X. borealis* (the latter being misidentified at the time as *X. mulleri*).

In *Xenopus* and most other vertebrates, the genes specifying the 18S and 28S ribosomal RNA are present in hundreds of copies and are arranged in one or a few tandem arrays (Pardue 1974; Long and Dawid 1980). Each repeated unit consists of a transcribed and a nontranscribed segment (Figure 6.26). The transcribed segment produces a 45S RNA precursor from which the functional 18S and 28S ribosomal RNAs are produced by means of enzymatic cleavage. The transcribed repeats are separated from each other by a nontranscribed spacer (NTS). In a comparison of the ribosomal RNA genes of *X. laevis* and *X. borealis*, Brown et al. (1972) found that, while the 18S and 285 genes of the two species were virtually identical, the NTS regions differed greatly between the two species. In contrast, the NTS regions were very similar within each individual and among individuals within a species. Thus, it appears that the NTS regions in each species have evolved together, although they have diverged rapidly between species.

One simple explanation for the intraspecific homogeneity is that the function of the repeats depends strongly upon their specific nucleotide sequence, so that most mutations have been eliminated by purifying selection or fixed by positive selection (Figure 6.27a). However, the NTS regions have no known function and do not appear to be subject to stringent selective constraints. Another simple explanation is that the family has arisen from a recent amplification of a single unit (Figure 6.27b). In this case, the homogeneity would simply reflect the fact that there has not been enough time for the members of the multigene family to diverge from each other. If this is the case, it is expected that the homogeneity of the family would gradually decrease, because over evolutionary time mutations would accumulate in the family members through genetic drift, particularly in regions that are not subject to stringent structural constraints. Under this model of independent evolution, the degree of intraspecific variation among the repeated elements is expected to be approximately equal to the degree of interspecific variation.

The empirical data from *Xenopus* supports neither model. Rather, the intraspecific homogeneity among the repeated units seems to be maintained by

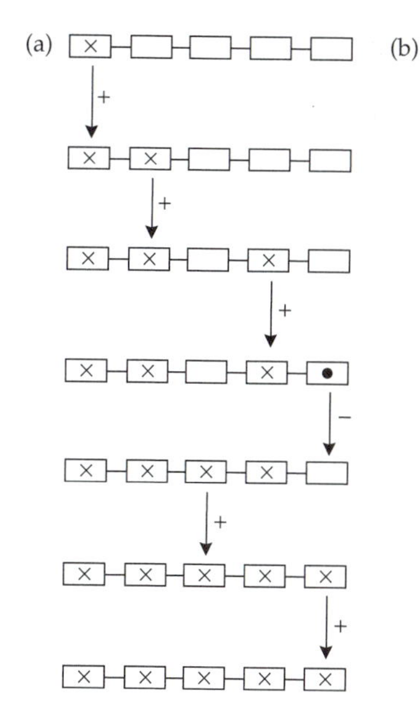

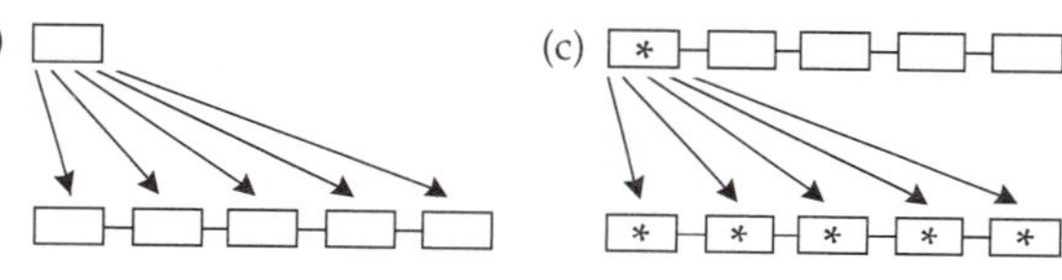

FIGURE 6.27 Three possible evolutionary scenarios for obtaining a homogenized tandemly repeated array. (a) The function of the repeats (boxes) depends strongly upon their specific nucleotide sequence, so that beneficial mutations (×) are fixed by positive selection (+) and deleterious mutations (•) are eliminated by purifying selection (–). (b) The repeated family arises through the amplification of a single unit. (c) A mutation (*) occurring in one repeat spreads to all other repeats through gene conversion or unequal crossing over.

a mechanism through which mutations can spread horizontally to all members in a multigene family. Brown et al. (1972) concluded that a "correction" mechanism must have operated to spread a mutation from one spacer sequence to the neighboring spacers faster than new changes can arise in these sequences (Figure 6.27c). They called this phenomenon, which manifests itself within a single individual, **horizontal evolution**, in contrast to **vertical evolution**, which refers to the spread of a mutation in a breeding population from one generation to the next. The terms "sequence coevolution" (Edelman and Gally 1970) and "coincidental evolution" (Hood et al. 1975) were also suggested. The term "concerted evolution" (Zimmer et al. 1980) is now most commonly used in the literature.

Concerted evolution essentially means that an individual member of a gene family does not evolve independently of the other members of the family. Rather, repeats in a family exchange sequence information with each other, either reciprocally or nonreciprocally, so that a high degree of intrafamilial sequence homogeneity is maintained. Through genetic interactions among its members, a multigene family evolves as a unit in a concerted fashion. The result of concerted evolution is a homogenized array of nonallelic homologous sequences. It is very important to note that concerted evolution requires not only the horizontal transfer of mutations among the members of the family (homogenization), but also the spread of mutations to all individuals in the population (fixation).

With the advent of restriction enzyme analysis and DNA sequencing techniques, a large body of data has attested to the generality of concerted evolution in multigene families (see reviews by Ohta 1980; Dover 1993; Arnheim 1983; Schimenti 1994; Elder and Turner 1995).

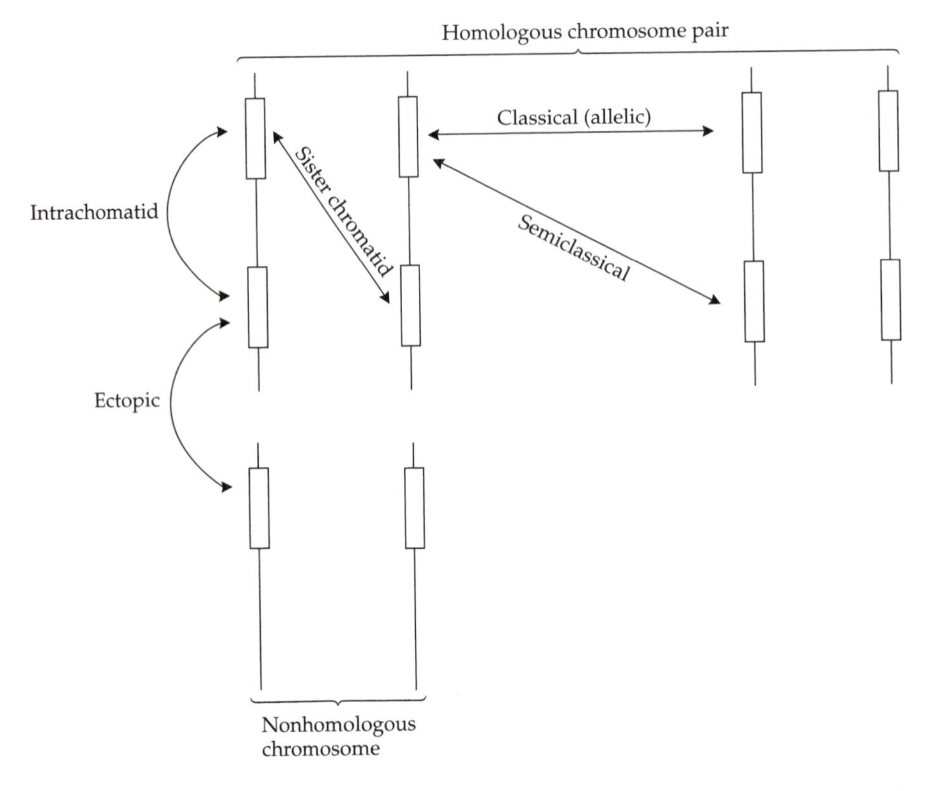

FIGURE 6.28 Types of gene conversion (double-headed arrows) between repeated sequences (boxes). The two chromatids of a chromosome are shown as neighboring pairs. Homologous double-stranded chromosomes are shown at the top; a nonhomologous chromosome is shown at the bottom.

MECHANISMS OF CONCERTED EVOLUTION

Gene conversion and unequal crossing over are currently considered to be the two most important mechanisms responsible for the occurrence of concerted evolution. They are also the two mechanisms that have received the most extensive quantitative coverage in the literature.

In addition to unequal crossing over and gene conversion, there are other mechanisms, such as slipped-strand mispairing and duplicative transposition (Chapters 1, 7, and 8), that can result in the creation of homogeneous families of repeated sequences.

Gene conversion

Gene conversion is a nonreciprocal recombination process in which two sequences interact in such a way that one is converted by the other (Chapter 1).

According to the chromatids involved in the process, gene conversion can be divided into several different types (Figure 6.28). When the exchange occurs between two paralogous sequences on the same chromatid, the process is called **intrachromatid conversion**. An exchange between two paralogous sequences from complementary chromatids is called **sister chromatid conversion**. **Classical conversion** involves exchanges between two alleles at the same locus. **Semiclassical conversion** involves an exchange between two paralogous genes from two homologous chromosomes. If the exchange occurs between paralogous sequences located on two nonhomologous chromosomes, the process is called **ectopic conversion**. From the point of view of the concerted evolutionary process, the most important types of gene conversion are the **nonallelic conversions** (i.e., conversions between genes located at different loci and not between allelic forms).

Gene conversion may be biased or unbiased. **Unbiased gene conversion** means that sequence A has as much chance of converting sequence B as sequence B has of converting sequence A. **Biased gene conversion** means that the probabilities of gene conversion between two sequences in the two possible directions occur with unequal probabilities. If deviation from parity occurs, we may speak of **conversional advantage** or **disadvantage** of one sequence over the other. If the conversional advantage of one sequence over the other is absolute (i.e., conversion is directional), the former sequence is referred to as the **master**, the latter as the **slave**.

Gene conversion between duplicated genes has been found in every species and at every locus that has been examined in detail. Preliminary data indicate that biased gene conversion is more common than the unbiased type. The amount of DNA involved in a gene conversion event varies from a few base pairs to a few thousands base pairs. Finally, the rate and the probability of occurrence of gene conversion varies with location; some locations being more prone to conversions than others.

Theoretical studies have shown that gene conversion can produce concerted evolution (Figure 6.29; Birky and Skavaril 1976; Nagylaki and Petes 1982; Nagylaki 1984a,b; Ohta 1990). Gene conversion has been suggested as a mechanism of homogenization in the human α- and γ-globin gene families (Jeffreys 1979; Slightom et al. 1980; Liebhaber et al. 1981; Scott et al. 1984), as well as of the heat-shock protein genes in *Drosophila* (Brown and Ish-Horowicz 1981). Moreover, it has been suggested as an important mechanism for the generation of polymorphism in the major histocompatibility complex (MHC) genes (Weiss et al. 1983; Ohta 1998).

Unequal crossing over

Unequal crossing over may occur either between the two sister chromatids of a chromosome during mitosis in a germline cell, or between two homologous chromosomes at meiosis. It is a reciprocal recombination process that creates a sequence duplication in one chromatid or chromosome and a corresponding deletion in the other (Chapter 1). A hypothetical example, in which an unequal

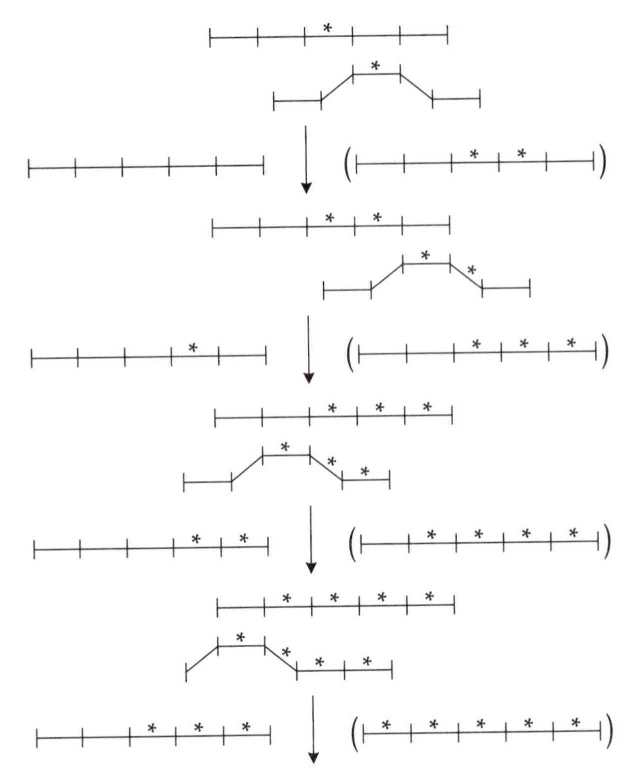

FIGURE 6.29 Concerted evolution of repeated sequences by gene conversion. There are two types of repeats; one type is marked with an asterisk. Repeated cycles of gene conversion events (shown as abutting segments) cause the duplicated sequences on each chromosome to become progressively more homogenized. On both sides of an arrow, the two possible resulting sequences from an unbiased gene conversion event are shown. The sequences in parentheses on the right are the ones selected for the next round of gene conversion. Note that the process of gene conversion does not affect the number of repeated sequences on each chromosome. Modified from Arnheim (1983).

crossing over has led to the duplication of three repeats in one daughter chromosome and the deletion of three repeats in the other, is shown in Figure 6.30. As a result of this exchange, both daughter chromosomes have become more homogeneous than the parental chromosomes. If this process is repeated, the numbers of each variant repeat on a chromosome will fluctuate with time, and eventually one type will become dominant in the family. Figure 6.31 illustrates how one type of repeat may spread throughout a gene family due to repeated rounds of unequal crossing over. Unequal crossing over has been investigated mathematically in detail and has received considerable experimental support (Ohta 1984; Li et al. 1985a). For example, unequal crossing over has been suggested to have played an important role in the concerted evolution of the immunoglobulin V_H gene family in mouse (Gojobori and Nei 1984).

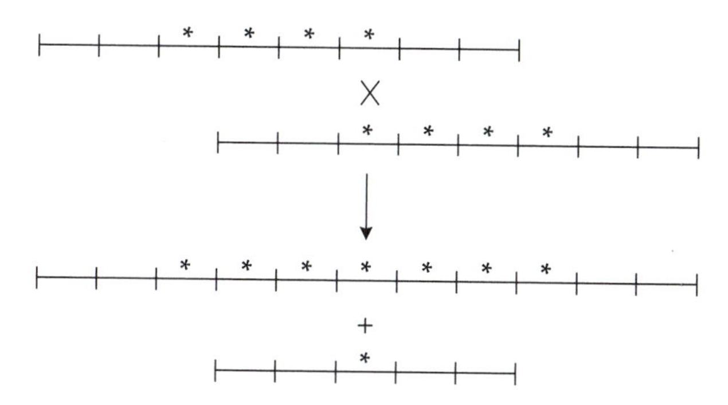

FIGURE 6.30 Model of unequal crossing over (×). As a result of unequal crossing over, both daughter chromosomes have an altered number of repeats and an altered frequency of the two repeat types (one of which is marked by an asterisk) when compared to the parental frequencies (50%). Moreover, both daughter chromosomes have a more homogenous repeat makeup than the parental chromosome. (The frequencies of the most common repeats are 64% and 80% in the first and second chromosomes, respectively.) Modified from Arnheim (1983).

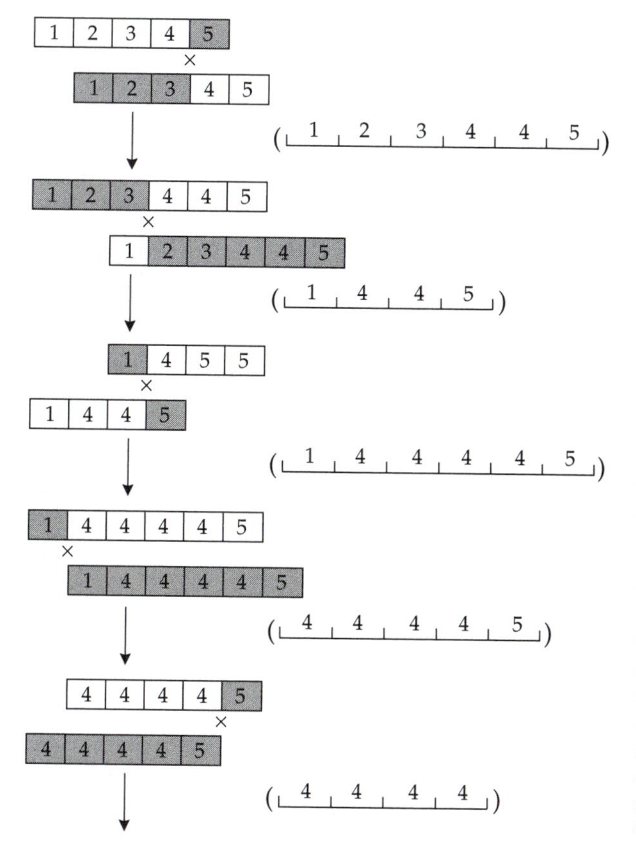

FIGURE 6.31 Concerted evolution by unequal crossing over (×). Repeated cycles of unequal crossover events cause the duplicated sequences on each chromosome to become progressively more homogenized. Different sequences are marked by different numbers. The sequences in parentheses on the right are the ones selected for the next round of unequal crossing over. The reciprocal sequence is shown as shaded boxes. The end result is a takeover by the type-4 repeat. Note that unequal crossing over affects the number of repeated sequences on each chromosome. From Ohta (1980).

Like unequal crossing over, slipped-strand mispairing is an expansion–contraction process that leads to the homogenization of the members of a tandem repeat family. However, as will be noted in Chapter 8, while unequal crossing over usually affects large tracts of DNA, slipped-strand mispairing is involved in the generation of tandem arrays of short repeats.

The relative roles of gene conversion and unequal crossing over

As a mechanism for concerted evolution, gene conversion appears to have several advantages over unequal crossing over. First, unequal crossing over generates changes in the number of repeated genes within a family, which may sometimes cause a significant dosage imbalance. For example, the deletion of one of the two α-globin genes following an unequal crossover gives rise to a mild form of α-thalassemia in homozygotes. Gene conversion, on the other hand, causes no change in gene number.

Second, gene conversion can act as a correction mechanism not only on tandem repeats but also on dispersed repeats within a chromosome (Jackson and Fink 1981; Klein and Petes 1981), between homologous chromosomes (Fogel et al. 1978), or between nonhomologous chromosomes (Scherer and Davis 1980; Ernst et al. 1982). In contrast, unequal crossing over is severely restricted when repeats dispersed on nonhomologous chromosomes are involved. It can probably act effectively on nonhomologous chromosomes only if the repeated genes are located on the telomeric parts of the chromosome (the ends of the chromosome arms), as in the case of rRNA genes in humans and apes, but will be greatly restricted if the dispersed repeats are located in the middle of chromosomes, as in the case of rRNA genes in mice, lizards, and *Drosophila melanogaster*. If the repeats are dispersed on a chromosome, unequal crossing over can result in the deletion or duplication of the genes that are located between the repeats. For example, Figure 6.32 shows a hypothetical case of unequal crossing over between two repeated clusters, resulting in the deletion of a unique gene in one chromosome and a corresponding duplication in the other. Either one or both chromosomes could have a deleterious effect on their carriers.

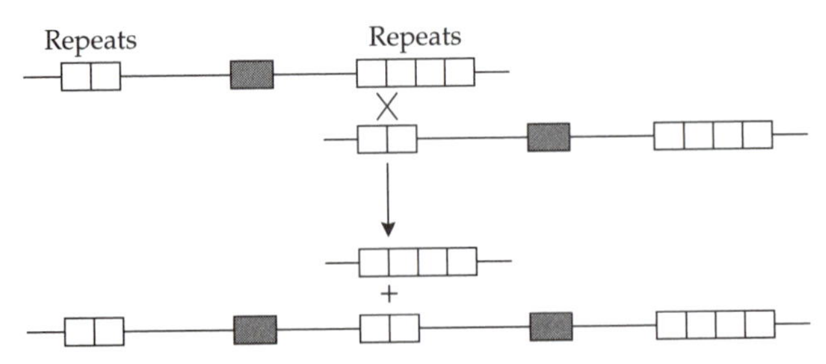

FIGURE 6.32 Crossing over involving dispersed repeats (empty boxes). The shaded box denotes a unique gene. In the crossing over event, the gene is deleted in one chromosome and duplicated in the other.

Third, gene conversion can be biased, i.e., have a preferred direction. Experimental data from fungi have shown that bias in the direction of gene conversion is common and often strong (Lamb and Helmi 1982), and theoretical studies have shown that even a small bias can have a large effect on the probability of fixation of repeated mutants (Nagylaki and Petes 1982; Walsh 1985).

For the above reasons, some authors (Baltimore 1981; Nagylaki and Petes 1982; Dover 1982) have proposed that gene conversion plays a more important role in concerted evolution than unequal crossing over. This is probably true for dispersed repeats, because in this case gene conversion can act more effectively than unequal crossing over. It is also probably true for small size multigene families (e.g., the duplicated α-globin genes in humans), because in such families unequal crossing over may cause severe adverse effects. In large families of tandemly repeated sequences, however, unequal crossing over may be as acceptable a process as gene conversion. Indeed, in such cases, unequal crossing over may be faster and more efficient than gene conversion in bringing about concerted evolution, for several reasons.

First, in such families, the number of repeats apparently can fluctuate greatly without causing significant adverse effects. This is suggested by the observations that the number of RNA-specifying genes in *Drosophila* varies widely among individuals of the same species and among species (Ritossa et al. 1966; Brown and Sugimoto 1973). Moreover, in humans, several families of tandem repeats that exhibit extraordinary degrees of variation in copy number have been found (Nakamura et al. 1987). Second, in a gene conversion event, usually only a small region (the heteroduplex region) is involved, whereas in unequal crossing over the number of repeats that are exchanged between the chromosomes can be very large. For example, in yeast, a single unequal crossover event was shown to involve on average seven repeats of rRNA genes, i.e., ~20,000 bp (Szostak and Wu 1980), whereas a gene conversion track may not exceed 1,500 bp (Curtis and Bender 1991). Obviously, the larger the number of repeats exchanged, the higher the rate of concerted evolution will be (Ohta 1983). In some cases, this advantage of unequal crossing over may be large enough to offset those of gene conversion. Finally, the empirical data shows that in some organisms (e.g., yeast), unequal crossing over occurs more frequently than nonallelic gene conversion. Of course, the observed lower rate of gene conversion might have been due to a detection bias, for it is generally much easier to detect unequal crossing over than gene conversion.

Detection and Examples of Concerted Evolution

From the point of view of evolutionary studies, an unfortunate feature of concerted evolution is that it erases the record of molecular divergence during the evolution of paralogous sequences. Thus, when dealing with very similar paralogous sequences from a species, it is usually impossible to distinguish between two possible alternatives: (1) the sequences have only recently diverged from one another by duplication, or (2) the sequences have evolved in concert. One way to distinguish between the alternatives is to use a phylogenetic approach. For example, the two α-globin genes in humans are almost identical to one another. Initially they were thought to have duplicated quite recently, so that there had not been sufficient time for them to have diverged in sequence. However, duplicated α-globin genes were also discovered in distantly related species, and so one had to assume either that multiple gene duplication events occurred independently in many evolutionary lineages, or that the two genes are quite ancient, having been duplicated once in the common ancestor of these organisms, but their antiquity was subsequently obscured by concerted evolution. Ultimately, the most parsimonious solution was to choose the latter alternative.

There is, however, a more direct and unambiguous method to detect instances of concerted evolution. We note that concerted evolution, be it due to gene conversion or unequal crossing over, only affects DNA segments of limited length, while neighboring segments remain unaffected. Thus, it is possible to detect concerted evolution at a genomic location by reference to neighboring sequences that did not evolve in concert. In short, we take advantage of the fact that the sequences involved in concerted evolution will be inordinately more similar to each other than their neighboring sequences. In the following sections we shall discuss several instances that illuminate important aspects of concerted evolution.

The $^{A}\gamma$- and $^{G}\gamma$-globin genes in the great apes

An interesting case of concerted evolution involves the $^{G}\gamma$- and $^{A}\gamma$-globin genes, which were created by a duplication that occurred approximately 55 million years ago, after the divergence between prosimians and simians. Since the African apes (humans, chimpanzees, and gorillas) diverged from each other at a much later date, we would expect the $^{G}\gamma$ orthologous genes from apes to be much more similar to each other than to any of the $^{A}\gamma$ paralogs. However, as shown in Figure 6.33a, this is only true for 3′ part of the gene, which contains exon 3. The 5′ part, which contains exons 1 and 2 (Figure 6.33b), exhibits a different phylogenetic pattern, i.e., paralogous exons within each species resemble each other more than they resemble their orthologous counterparts in other apes (Slightom et al. 1985). This discrepancy is obvious when counting the nucleotide differences between two paralogous genes from the same species. In humans, for example, the 5′ parts of $^{G}\gamma$ and $^{A}\gamma$ differ from one another at only 7 out of 1,550 nucleotide positions (0.5%). In contrast, a comparison of the 3′ part shows a difference that is 20 times larger, 145 out of 1,550 nucleotides (9.4%). Assuming that the 5′ and 3′ parts are subject to similar functional constraints, we may conclude that the 5′ end of the gene underwent gene conversion. This conclusion is strengthened by the fact that the second intron in both genes in all apes contains a stretch of the simple repeated DNA sequence $(TG)_n$ that can serve as a hotspot for the recombination events involved in the process of gene conversion.

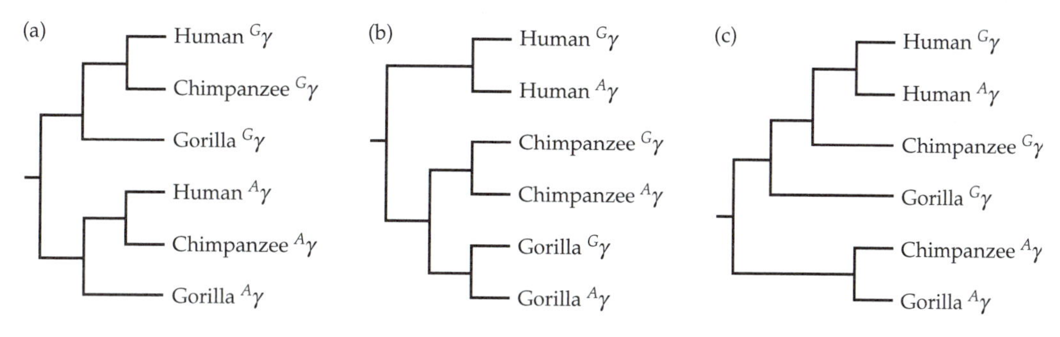

FIGURE 6.33 Phylogenetic trees for (a) exon 3 and (b) exons 1 and 2 of the $^{G}\gamma$- and $^{A}\gamma$-globin genes from human, chimpanzee, and gorilla. (c) Expected phylogenetic tree for exons 1 and 2 if gene conversion had only occurred in the human lineage.

Indeed, the data indicate that each of the three lineages has experienced multiple independent gene conversion events. There are two reasons for making such an inference. First, the tree in Figure 6.33b contains an additional phylogenetic anomaly besides the clustering of the paralogous genes before the orthologous ones: it also clusters chimpanzee and gorilla as a clade (Chapter 5). Second, if gene conversion had only occurred in one of the lineages, say the human lineage, we would expect to obtain the phylogenetic tree in Figure 6.33c.

Assuming that both the converted and unconverted parts of the genes evolve at equal rates, it is possible to date the last gene conversion event by using the degrees of similarity between the two sequences in conjunction with the date for the gene duplication event. The last conversion event in the human lineage has thus been calculated to have occurred about 1–2 million years ago, i.e., after the divergence between human and chimpanzee, which strengthens the previous inference that the conversions in the chimpanzee and gorilla lineages occurred independently.

The concerted evolution of genes and pseudogenes: When death is not final, life is precarious, and distinguishing between the two is difficult

Pancreatic ribonuclease is a ubiquitous protein secreted by the pancreas of all vertebrates. In mammals, this protein is usually encoded by a single-copy gene. In the ancestor of true ruminants (suborder Pecora; see Chapter 5), the gene underwent two rounds of duplication, from which emerged three paralogous genes encoding pancreatic, seminal, and cerebral ribonucleases. Interestingly, a functional gene for seminal ribonuclease was only found in the closely related bovine species *Bos taurus* (cattle), *Bubalus bubalis* (Asian water buffalo), and *Syncerus caffer* (Cape buffalo), whereas in all other pecorans, such as giraffe and deer, the orthologous sequence was found to be a pseudogene. Even in *Tragelapus imberbis* (the lesser kudu), which belongs to the same subfamily (Bovinae) as *Bos*, *Bubalus*, and *Syncerus*, the orthologous sequence is a pseudogene (Confalone et al. 1995; Breukelman et al. 1998).

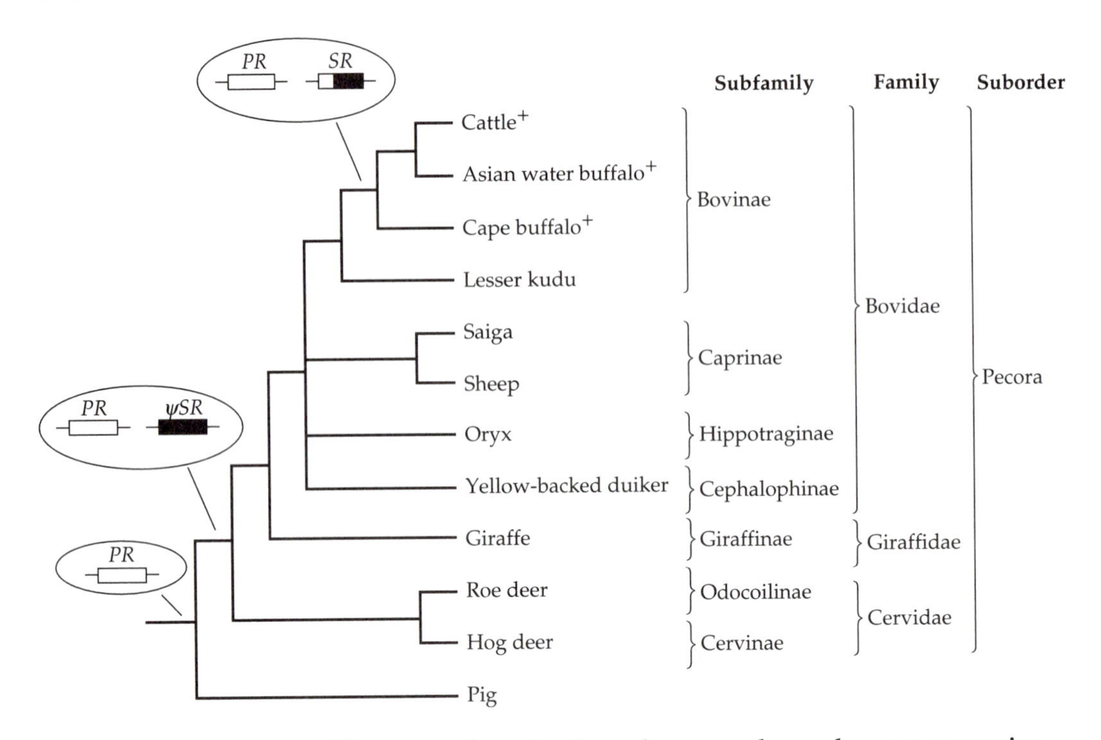

FIGURE 6.34 The resurrection of a ribonuclease pseudogene by gene conversion in a bovine lineage. Species in which the seminal ribonuclease protein is functional are marked with a plus sign (+). In the ancestor of the pecorans, the pancreatic ribonuclease (*PR*) gene (empty box) was triplicated to yield a nonfunctional seminal ribonuclease (*ψSR*) pseudogene (solid box) and a functional cerebral ribonuclease gene (not shown). In the ancestor of cattle and buffalos, the 5′ end of the pseudogene was converted by the functional gene, and it became a functional gene encoding seminal ribonuclease (*SR*). Species: cattle, *Bos taurus*; Asian water buffalo, *Bubalus bubalis*; Cape buffalo, *Syncerus caffer*; lesser kudu, *Tragelaphus imberbis*; saiga, *Saiga tatarica*; sheep, *Ovis aries*; yellow-backed duiker, *Cephalolaphus sylvicultor*; giraffe, *Giraffa camelopardis*; roe deer, *Capreolus capreolus*; hog deer, *Axis porcinus*; pig, *Sus scrofa*.

The most parsimonious explanation for these data is that the original seminal ribonuclease gene in the ancestor of the true ruminants was a pseudogene (Figure 6.34). It was subsequently "resurrected" in one lineage and became expressed in the seminal fluid. In the other lineages, it stayed "dead." Because of the taxonomic distribution of the seminal ribonuclease genes and pseudogenes, it is possible to date this resurrection at between 5 and 10 million years ago, after the divergence of the lesser kudu, but before the divergence of the Asian water buffalo. A detailed analysis of the ribonuclease sequences indicated that the resurrection may have involved a gene conversion event, i.e., a transfer of information from the gene for the pancreatic enzyme to that for the seminal ribonuclease (Trabesinger-Ruef et al. 1996).

The gene conversion event involved only a small region at the 5′ end of the gene, including about 70 nucleotides in the untranslated region. Interestingly, this conversion not only removed a deletion that caused a frameshift in the reading frame, but also restored two amino acid residues that are vital for the proper functioning of ribonuclease: histidine at position 12 and cysteine at position 31.

Thus, thanks to concerted evolution, death (as far as genes are concerned) does not by necessity connote finality. Pseudogene sequences may represent reservoirs of genetic information that participate in the evolution of new genes rather than relics of inactivated genes whose inevitable fate is genomic extinction (Fotaki and Iatrou 1993).

However, the proximity of a gene to a pseudogene may not only spell rebirth for the pseudogene, but can also spell death for the gene. One such example of gene death by concerted evolution concerns the 21-hydroxylase (cytochrome P21) gene. In humans, this ten-exon gene is located on chromosome 6, in a region in which many major histocompatibility and complement genes are interspersed with one other. The gene has a paralogous unprocessed pseudogene in the vicinity. Interestingly, in many organisms one of the genes became nonfunctional; however, the nonfunctionalization event occurred independently in many lineages. Thus, for instance, the ortholog of the human functional gene is a pseudogene in mouse, and the ortholog of the human pseudogene is a functional gene in mouse.

Hundreds of mutations in the 21-hydroxylase gene have been characterized in the clinical literature, and about 75% of them were found to be due to gene conversion (Mornet et al. 1991). In most cases, it is difficult to distinguish between unequal crossing over and gene conversion. In at least one case, however, because of the discovery of a *de novo* mutation in an individual, it was possible to identify directly the molecular mechanism responsible for the nonfunctionalization of the gene. It was due to an intrachromatid conversion event involving 390 nucleotides in the maternal chromosome (Collier et al. 1993).

Finally, we have the case of the ζ-globin locus in the horse. The equine ζ-globin locus consists of a gene and a pseudogene. The duplication of the ζ-globin genes is quite ancient; it most probably pre-dated the eutherian radiation. Because of repeated gene conversion events in their alignable parts, the gene and the pseudogene are almost identical (Flint et al. 1988). Were it not for the fact that the pseudogene is truncated, we would be hard pressed to say which is the gene and which is the pseudogene.

FACTORS AFFECTING THE RATE OF CONCERTED EVOLUTION

How cohesively the members of a repeated sequence family evolve together depends on several factors, including the number of repeats (i.e., the size of the gene family), the arrangement of the repeats, the structure of the repeated unit, the functional constraints imposed on the repeated unit, the mechanisms of concerted evolution, and the selective and nonselective processes at the population level.

Number of repeats

It is quite easy to see that the rate of concerted evolution is dependent on the number of repeats. For example, if there are only two repeats on a chromosome, a single intrachromosomal gene conversion will lead to homogeneity of the repeats on the chromosome. On the other hand, when there are more than two repeats on the chromosome, more than a single conversion may be required to homogenize the sequences.

Smith (1974) seems to be the first author to conduct a quantitative study of the effect of family size on the rate of homogenization in a multigene family. His simulation study indicated that the number of unequal crossing over events required for the fixation of a variant repeat in a single chromosomal lineage increases roughly with n^2, where n is the number of repeats on the chromosome.

Arrangement of repeats

There are, roughly speaking, two types of arrangement of repeated units. In some gene families, the members are highly dispersed all over the genome. One example is the human *Alu* family, whose approximately one million members are interspersed with single-copy sequences throughout the genome (Chapter 8). This type of arrangement is the least favorable for concerted evolution because it greatly reduces the chance of unequal crossing over and gene conversion, and because unequal crossing over often leads to disastrous genetic consequences. The high similarities among *Alu* sequences are most probably due to relatively recent amplification events of source sequences (Chapter 7) rather than to concerted evolution.

In the second type of arrangement, all members of a family are clustered either in a single tandem array or in a small number of tandem arrays located on different chromosomes. This arrangement is the most favorable for unequal crossing over and gene conversion to operate. If the repeats are located on more than one chromosome, the rate of unequal crossing over is greatly reduced, unless the clusters occur at the ends of chromosome arms (as in the case of the rDNA family in humans). Moreover, the rate of gene conversion would also be reduced. However, Ohta and Dover (1983) have shown that such a reduction in gene conversion rate has only a minor effect on the extent of identity between genes, unless the conversion rate between genes on nonhomologous chromosomes becomes very low, or unless the number of nonhomologous chromosomes on which gene family members reside is large.

Structure of the repeat unit

The structure of the repeat unit refers to the numbers and sizes of coding (i.e., exons) and noncoding regions (i.e., introns and spacers) within the repeat unit. As noncoding regions generally evolve rapidly, it is difficult to maintain a high degree of similarity among the repeats if each repeat contains large or numerous noncoding regions. We note that homogeneity and concerted evolution go hand in hand, since both unequal crossing over and gene conversion depend on sequence similarity for misalignment of repeats. Thus, the higher the homogeneity among the repeats in a family, the higher the rates of unequal crossing over and gene conversion.

Zimmer et al. (1980) estimated that in the great apes, the rate of concerted evolution in the α-globin gene region is 50 times higher than that in the β-globin gene region. They suggested that the rate in the β region has been greatly reduced because the introns and flanking sequences are highly divergent between the two β genes. It is interesting to note that the β genes have introns that are several times longer than those of the α genes, and that the intergenic region between the two β genes is 2,400 bases longer than that between the two α genes. Indeed, Zimmer et al. (1980) suggested that the larger introns and intergenic region in the β genes arose as a response to selection against unequal crossing over, which may produce a single gene out of the β- and δ-globin genes (hemoglobin Lepore), whose expression is under the control of the δ promoter, and is deleterious in the homozygous state.

We note, however, that qualitative (as opposed to quantitative) arguments concerning putative advantages associated with protection against mutational events (e.g., avoidance of pretermination codons, prevention of crossing over events) are usually highly exaggerated, because the selective advantage for a reduction in the rate of a mutational event would at most be as large as the rate itself. Assuming that mutational events occur at rates of 10^{-5} to 10^{-9}, the selective advantage would be insignificant. Thus, the larger introns and intergenic region might have arisen by chance rather than by selection. It is possible that the introns and the intergenic region were already large before the divergence of the apes, and this has promoted the divergence between the two β genes, rather than vice versa.

Functional requirement

Here again, we shall consider two extreme situations. One is that the function has an extremely stringent structural requirement, often requiring large amounts of the same gene product (dose repetitions). The rRNA genes and the histone genes are well-known examples. The other extreme is that the function requires a large amount of diversity. The immunoglobulin and histocompatibility genes belong to this category.

In general, the rate of concerted evolution is expected to be higher in the former type than in the latter type of families. Indeed, according to Gojobori and Nei's (1984) estimate, the rate of concerted evolution is 100 times higher in the rDNA family than in the immunoglobulin V_H family. In rRNA genes, purifying selection will tend to eliminate new variants and promote homogeneity, which in turn will facilitate unequal crossing over and gene conversion among members of multigene families, thus accelerating the process of concerted evolution. In V_H genes, on the other hand, an individual who had many identical copies owing to concerted evolution would be at a severe disadvantage, as its arsenal of immunoglobulins against pathological antigens would be limited. As mentioned previously, the rates of unequal crossing over and gene conversion are expected to decrease with decreasing intraspecific homogeneity. Thus, the process of concerted evolution is expected to be slower for V_H genes than for rRNA genes. Clearly, functional constraints play an important role in concerted evolution.

Populational processes

Population size affects the rate of concerted evolution because concerted evolution requires not only the horizontal spread of genetic variation among members of a gene family, but also the fixation of such homogeneous variants within the population. Obviously, the time required for a variant to be eliminated from a population or to become fixed in a population is dependent on the population size (Chapter 2).

Positive natural selection will accelerate the process of concerted evolution because the rate and probability of fixation for a variant favored by natural selection will be larger than those for selectively neutral variants. The effect of biased gene conversion on the evolution of multigene families would be similar to that of positive selection, albeit somewhat weaker. In addition, biased gene conversion will be more effective when the number of repeats is large (Walsh 1985). Both natural selection and biased gene conversion work more effectively in large populations than in small ones, because the effect of random genetic drift decreases with population size.

Finally, we note that unequal crossing over will create a large variation in the number of repeats among individuals in a population. Purifying selection against too many or too few repeats (**centripetal selection**) may thus be an important force shaping the genetic makeup of the population.

EVOLUTIONARY IMPLICATIONS OF CONCERTED EVOLUTION

Concerted evolution allows the spreading of a variant repeat to all gene family members. This capability has profound evolutionary consequences. In the following we discuss the effects of concerted evolution on the spread of advantageous mutations, the rate of divergence between duplicate genes, and the generation of genic variation.

Spread of advantageous mutations

Through concerted evolution, an advantageous mutant can spread rapidly and replace all other repeats within a gene family. We note that the selective advantage that a single variant can confer on an organism is usually very small. The advantage would, however, be greatly amplified if the mutation were to spread within the genome. Thus, through concerted evolution, a small selective advantage can become a great advantage. In this respect, concerted evolution surpasses independent evolution of individual gene family members (Arnheim 1983; Walsh 1985).

Arnheim (1983) compared the evolution of RNA polymerase I transcriptional control signals with that of RNA polymerase II transcriptional control signals. RNA polymerase I transcribes rRNA genes, whereas RNA polymerase II transcribes protein-coding genes (Chapter 1). RNA polymerase I transcriptional control signals appear to have evolved much faster than the signals for RNA polymerase II. For example, in cell-free transcription systems, a mouse rDNA clone does not work in a human cell extract, but clones of protein-coding genes from astonishingly diverse species can be transcribed in heterologous systems (e.g., silkworm genes in human cell extracts, and mammalian genes in yeast). Arnheim (1983) argues that in the case of transcription units for RNA polymerase I, mutations that favorably affect transcription initiation were propagated throughout the rDNA multigene family as a consequence of concerted evolution, and could become species-specific. On the other hand, in the case of transcription units for RNA polymerase II, advantageous mutations affecting transcription initiation that occur in any one gene would not be expected to be propagated throughout all genes, for they belong to many different families.

Retardation of paralogous gene divergence

The traditional view concerning the creation of a new function is that a gene duplication event occurs, and one of the two resultant genes gradually diverges and becomes a new gene. It is now clear that the process may not be as simple as previously assumed. As long as the degree of divergence between the two genes is not large (as is the case immediately after the duplication event), the divergent copies may be deleted by unequal crossing over or converted to the original form by gene conversion. In the former case, an additional duplication would be required to create a new redundant copy, while in the latter case divergence must start again from scratch. Thus, divergence of duplicate genes may proceed much more slowly than traditionally thought, or very strong positive selection should be invoked. On the other hand, gene conversion may prevent a redundant copy from becoming nonfunctional for long periods of time or, alternatively, may enable a "dead" gene to be "resurrected" (see page 315).

Generation of genic variation

From an evolutionary point of view, there is an analogy between the evolution of multigene families and the evolution of subdivided populations. We may regard each repeat in a multigene family as a deme in a subdivided population. The transfer of information between repeats is then equivalent to the migration of genes or individuals between demes. It is well known that migration reduces the amount of genetic difference between demes but increases the amount of genic variation (i.e., the number of alleles) in a deme. Similarly, transfer of information between repeats will reduce the genetic difference between repeats, but will increase the amount of genic variation at a locus (Ohta

1983, 1984; Nagylaki 1984). Indeed, some loci in the mouse major histocompatibility complex are highly polymorphic, with as many as 50 alleles being observed at a locus, and it has been suggested that the high polymorphism is due to concerted evolution (Weiss et al. 1983). An alternative explanation is that the alleles have persisted in the population for very long periods of time (Figueroa et al. 1988), probably being maintained by overdominant selection (Hughes and Nei 1989). Note that the two mechanisms are not mutually exclusive, and they may both operate at these loci.

METHODOLOGICAL PITFALLS DUE TO CONCERTED EVOLUTION

It has been customary to assume that, following a gene duplication, the two resultant genes will diverge monotonically with time. Under this assumption, we have previously shown that it is rather simple to infer the time of the duplication event (see page 271). Unfortunately, concerted evolution tends to erase the divergent history of duplicated genes; hence, gene duplications frequently appear younger than they really are. Phylogenetic reconstructions based on sequence comparisons can only go back to the last erasure of the evolutionary history. We must therefore use taxonomic information concerning the distribution of duplicated genes versus unduplicated ones to infer the time of gene duplication. In large multigene families, gene correction events are expected to occur frequently, and in such cases it will be even more difficult to trace the evolutionary relationships among the family members.

Thus, concerted evolution should be taken into account when attempting to reconstruct the evolutionary history of paralogous genes. Failure to consider this possibility may result in faulty phylogenetic reconstructions.

FURTHER READINGS

Bork, P., A. K. Downing, B. Kieffer, and I. D. Campbell. 1996. Structure and distribution of modules in extracellular proteins. Q. Rev. Biophys. 29: 119–167.

Cold Spring Harbor Symposium on Quantitative Biology. 1987. *Evolution of Catalytic Function*. Vol. 52. Cold Spring Harbor Laboratory, Cold Spring Harbor, NY.

Dean, A. M. 1998. The molecular anatomy of an ancient adaptive event. Am. Sci. 86: 26–37.

Graw, J. 1997. The crystallins: Genes, proteins, and diseases. Biol. Chem. 378: 1331–1348.

Henikoff, S., E. A. Greene, S. Pietrokovski, P. Bork, T. K. Attwood, and L. Hood. 1997. Gene families: The taxonomy of protein paralogs and chimeras. Science 278: 609–614.

Nei, M. and R. K. Koehn (eds.). 1983. *Evolution of Genes and Proteins*. Sinauer Associates, Sunderland, MA.

Ohno, S. 1970. *Evolution by Gene Duplication*. Springer, Berlin.

Ohta, T. 1980. *Evolution and Variation of Multigene Families*. Springer, Berlin.

Petes, T. D. and C. W. Hill. 1988. Recombination between repeated genes in microorganisms. Annu. Rev. Genet. 22: 147–168.

Chapter 7

Evolution by Transposition

Genomes used to be thought of as rather static entities in which genes could be assigned to well-defined loci. Accordingly, genes were supposed to retain their precise location over long periods of evolution. The static picture of the genome started crumbling in the 1940s, when Barbara McClintock discovered that certain genetic elements in maize can "jump" from one genomic location to another, sometimes altering the expression of structural genes. Her studies showed that genes associated with the development of color pigments in maize kernels could be turned on or off at abnormal times by the action of certain "controlling elements" that apparently had the ability to move from one chromosome to another. The rigid static picture, however, was so ingrained in scientific thought that it took nearly 40 years for the significance of McClintock's seminal discovery to be appreciated. Today we recognize that the structural organization of genomes is much more fluid than previously thought.

In this chapter we describe the myriad of transposable genetic elements that facilitate the movement of genetic material from one genomic location to another, from one genome to another, and from one organism to another, and we discuss the possible impacts such elements may have had on the evolutionary process.

Transposition and Retroposition

Transposition is defined as the movement of genetic material from one chromosomal location, the **donor site**, to another, the **target site**. DNA sequences that possess an intrinsic capability to change their genomic location are called **mobile elements** or **transposable elements**. There are two types of transposi-

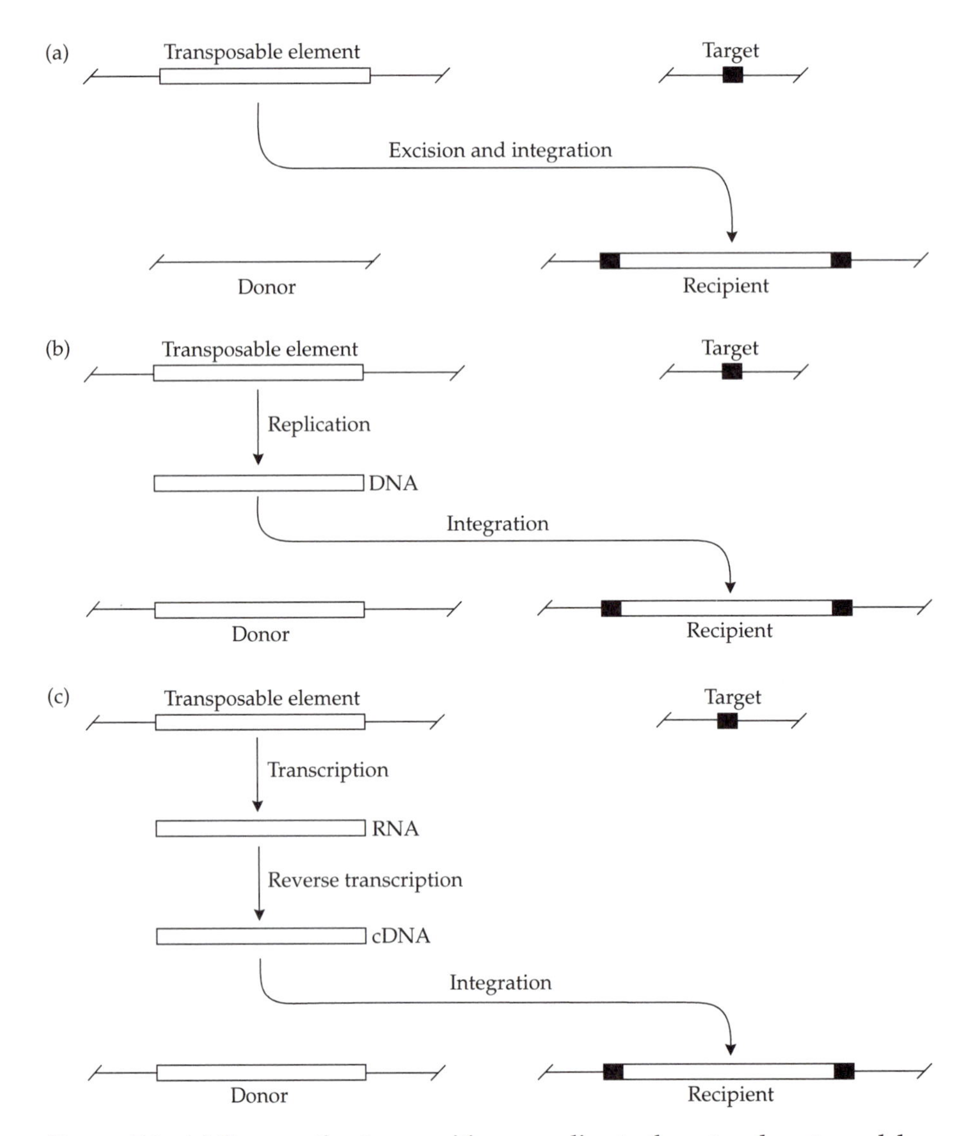

FIGURE 7.1 (a) Conservative transposition according to the cut-and-paste model. Following excision, the donor site is ligated by the host repair system. (b) Replicative transposition. The element is replicated, and one copy is inserted at a target site while the other copy remains at the donor site. (c) Retroposition. The element is transcribed into RNA, which is then reverse-transcribed into cDNA. The cDNA copy is inserted into the host genome. (For an example of retroposition, see Figure 7.4.) Note that both transposition and retroposition create a short repeat (black box) at each end of the newly inserted element.

tion, distinguished by whether the transposable element is replicated or not. In **conservative transposition**, the element itself moves from one site to another (Figure 7.1a). What happens to the donor site is unclear. The **donor suicide model** proposes that the strand ends of the donor DNA are not joined to one another and that the remnant molecule is destroyed. However, loss of the donor DNA from the cell lineage is avoided if the cell contains a duplicate of the donor sequence. In this case, although one copy is consumed, the other survives, and the resulting lineage will contain one element at the original site and a second at the new site (Berg et al. 1984; Weiner et al. 1986). If conservative transposition occurs predominantly according to this model, then it will not differ from replicative transposition (see below). The alternative **cut-and-paste model** proposes that the double-stranded break is ligated by the host repair system.

In **replicative transposition**, the transposable element is copied, and one copy remains at the original site while the other inserts itself at a new site (Figure 7.1b). Thus, replicative transposition is characterized by an increase in the number of copies of the transposable element. Some transposable elements use only one type of transposition; others use both the conservative and the replicative pathways. A single nucleotide substitution is sometimes sufficient to switch the mode of transposition from conservative to replicative (e.g., May and Craig 1996).

In the above types of transposition, the genetic information is carried by DNA. However, genetic information can also be transposed through RNA. In this mode of transposition, the DNA is transcribed into RNA, which is then reverse-transcribed into cDNA (Figure 7.1c). In order to distinguish between the DNA- and the RNA-mediated modes, the latter has been termed **retroposition**. Both transposition and retroposition are found in eukaryotic and prokaryotic organisms (see Weiner et al. 1986; Temin 1989). In contrast with DNA-mediated transposition, retroposition is always of the replicative type, because it is a reverse-transcribed copy of the element, rather than the element itself, that is transposed.

When a transposable element is inserted into a host genome, a small segment of the host DNA (usually 4–12 bp) is duplicated at the insertion site (Figure 7.1). This occurs because the double-stranded target site is cleaved in a staggered manner prior to the insertion of the transposable element. The single-stranded flanks are then repaired, and two **direct repeats** in the same orientation are created on both sides of the integrated transposable element. These repeats are the hallmarks of transposition and retroposition.

TRANSPOSABLE ELEMENTS

According to their mode of transposition and the number and kinds of genes they contain, transposable elements can be classified into three types: insertion sequences, transposons, and retroelements.

Insertion sequences

Insertion sequences are the simplest transposable elements. They carry no genetic information except that which is necessary for transposition. Insertion sequences are usually 700–2,500 bp in length and have been found in bacteria, bacteriophages, plasmids, and plants. (Barbara McClintock's original "controlling elements" are, in fact, insertion sequences.) Bacterial insertion sequences are denoted by the prefix *IS* followed by the type number.

The structure of one such insertion sequence, *IS1* from the intestinal bacteria *Escherichia coli* and *Shigella dysinteria*, is shown schematically in Figure 7.2a. *IS1* is approximately 770 nucleotides in length, including two inverted, nonidentical terminal repeats of 23 bp each. It contains two out-of-phase reading frames, *InsA* and *InsB*, from which a single protein is produced by translational frameshifting at a run of adenines. The N-terminal part of the protein encoded by *insA* is an inhibitor of transposition; the C-terminal part is a transposase, an enzyme that catalyzes the insertion of transposable elements into insertion sites (Sekino et al. 1995). A more specialized case is illustrated by the 1,500-bp long *IS50* from Enterobacteria, which produces two overlapping transcripts that are translated in the same reading frame. One mRNA mole-

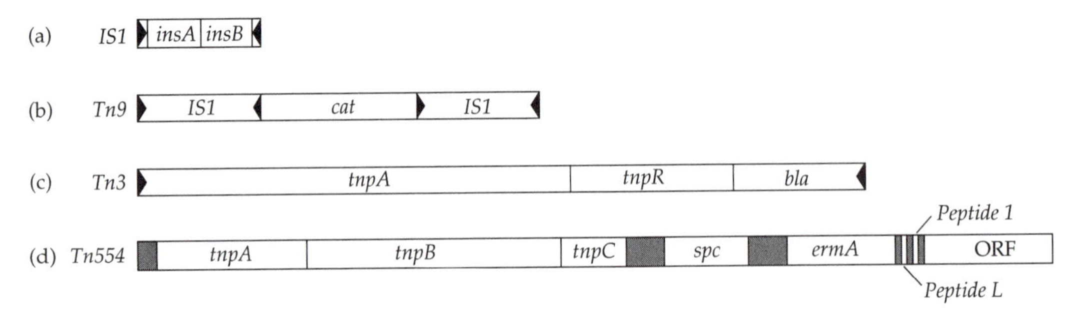

FIGURE 7.2 Schematic representation of four transposable elements in bacteria. Black triangles denote inverted repeats. (a) Insertion sequence *IS1* from *Escherichia coli* contains two out-of-phase reading frames flanked by two imperfect inverted terminal repeats. (b) Complex transposon *Tn9* from *E. coli* contains two copies of *IS1* flanking the *cat* gene, which encodes a protein conferring chloramphenicol resistance. (c) Transposon *Tn3* from *E. coli*, which confers streptomycin resistance, contains three genes, two of which (*tnpR* and *bla*) are transcribed on one strand, and the third (*tnpA*) on the other. *Tn3* is flanked by two perfect inverted repeats, 38 bp long. (d) Transposon *Tn554* from *Staphylococcus aureus* lacks terminal repeats and contains seven identified protein-coding genes and an open reading frame (ORF). Three of the genes encode transposases (*tnpA*, *tnpB*, and *tnpO*), and are transcribed as a unit. The *spc* and *ermA* genes confer spectinomycin and erythromycin resistance, respectively. The *spc* gene, which encodes an S-adenosyl methionine-dependent methylase, as well as the *peptide L* and *peptide 1* genes, are transcribed on a different strand from the other genes. The ORF is abundantly transcribed and translated, but its function is not known. The gray boxes contain no open reading frames.

cule is translated into transposase, the other into a repressor protein that regulates the rate of transposition. There are numerous types of insertion sequences in *E. coli*, and the genomes of most strains isolated from the wild contain varying numbers of each (Sawyer et al. 1987).

Transposons

Transposons are mobile elements, usually about 2,500–7,000 bp long, that exist mostly as families of dispersed repetitive sequences in the genome. Transposons are distinguished from insertion sequences by also carrying so-called **exogenous genes**, i.e., genes that encode functions other than those related to transposition. (Note that the nomenclature is muddled in the literature, and the term "transposon" is sometimes used to denote any transposable element, including insertion sequences, retrotransposons, etc.) In bacteria, transposons are denoted by the prefix *Tn* followed by the type number.

Transposons in bacteria often carry genes that confer antibiotic resistance (e.g., *Tn554*), heavy-metal resistance (e.g., *Tn21*), or heat resistance (e.g., *Tn1681*) on their carriers. Plasmids, which are autonomously replicating extrachromosomal molecules distinct from the bacterial genome, can carry such transposons from cell to cell, and as a consequence, resistance can quickly spread throughout the populations of bacteria exposed to such environmental factors.

Several bacteriophages (viruses of bacteria) are in fact transposons or **transposing bacteriophages**. For example, bacteriophage *Mu* (short for *Mutator*) is a very large transposon (~38,000 bp) that encodes not only the enzymes that regulate its transposition, but also a large number of structural proteins necessary to construct the virion packaging.

Some bacterial transposons are **complex** (or **composite**) **transposons**, so named because two complete, independently transposable insertion sequences in either orientation flank one or more exogenous genes (Figure 7.2b). Interestingly, in complex transposons, not only can the entire transposon move as a unit, but one or both of the flanking sequences can transpose independently. Since the transposition functions are encoded by the insertion sequences, complex transposons do not usually contain additional transposase genes. Complex transposons can be either symmetrical or asymmetrical. **Symmetrical complex transposons** contain the same insertion sequence on both sides (Figure 7.2b). **Asymmetrical complex transposons** contain different insertion sequences. For instance, *Tn1547* from *Enterococcus faecalis*, which confers vancomycin resistance, is flanked by an *IS16* element on one side and an *IS256* element on the other.

Some bacterial transposons do not contain insertion sequences, and are sometimes referred to as **simple transposons**. Simple transposons also come in two varieties, symmetrical and asymmetrical. Those flanked by short repeated sequences are **symmetrical** (Figure 7.2c); those that do not are **asymmetrical** (Figure 7.2d).

Hypercomposite transposons are made up of two or more transposons. For instance, *Tn5253* in *Streptococcus pneumoniae* conveys resistance to both tetracyclines and chloramphenicol because it is made of two transposons (*Tn5251* and *Tn5252*), each carrying a single set of resistance genes.

The coding regions of some transposons in animals (e.g., *P* elements in *Drosophila*) are interrupted by spliceosomal introns. In bacterial transposons (e.g., *Tn5397* in the bacterium *Clostridium difficile*), self-splicing introns have been found (Mullany et al. 1996). Alternative splicing (Chapter 6) has also been detected in transposable elements. In fact, in the case of *P* elements in *Drosophila*, the mechanism of alternative splicing was found to regulate transposition (see page 354).

Transposition of many types is widespread in the genomes of animals, plants, and fungi. *D. melanogaster*, for instance, contains multiple copies of 50–100 different kinds of transposons (Rubin 1983).

Taxonomic, developmental, and target specificity of transposition

Some mobile elements can transpose themselves in all cells; others are highly specific. For example, *Tc1* elements in the nematode *Caenorhabditis elegans* and *P* elements in *D. melanogaster* are usually mobile only in germ cells. In maize, the transposition of many elements was found to be regulated by the host developmental stage. From an evolutionary point of view, the developmental timing of transposition is particularly important, because it affects the propagation of the transposable element to future generations. One pertinent example involves the *LINE-1* (*L1*) transposable elements in mammals. Branciforte and Martin (1994) found that these elements are particularly active during two stages of meiosis (leptotene and zygotene), during which DNA strand breakages occur. This offers an opportunity for transposable elements to insert themselves into new sites.

The genomic locations of the target sites for transposition are extremely variable among different transposable elements. Some elements show an exclusive preference for a specific genomic location. For example, *IS4* incorporates itself exactly and always at the same point in the galactosidase operon of *E. coli*, and thus each bacterium can contain only one copy of *IS4* (Klaer et al. 1981). Others, such as bacteriophage *Mu*, can transpose themselves at random to almost any genomic location. Many transposable elements show intermediate degrees of genomic preference. For example, 40% of all *Tn10* transposons in *E. coli* are found in the *lac*Z gene, which constitutes a minute fraction of the host genome. Some transposable elements exhibit higher affinities for a particular type of nucleotide composition. For example, *IS1* favors AT-rich insertion sites (Devos et al. 1979), and *IS630* shows a special affinity for 5′—CTAG—3′ sequences (Tenzen and Ohtsubo 1991). Chromosomal preference has also been found. For example, the *TRIM* element in *Drosophila miranda* exhibits a preference for the Y chromosome, while the *D. melanogaster P* element prefers target sites on the homologous chromosome of that in which the donor site is located (Golic 1994; Tower and Kurapati 1994). One of the most peculiar biases in transposition was found in the *DIRS-1* transposable element in the slime mold *Dictyostelium discoideum*, which preferentially inserts itself into other *DIRS-1* sequences (Cappello et al. 1984). *D. discoideum* contains, on the average, about 40 intact copies of *DIRS-1* and 200–300 *DIRS-1* fragments. The self-affinity of *DIRS-1* may be the reason for the existence of so many defective *DIRS-1* fragments within the slime mold genome.

Finally, some transposable elements are known to be species-specific, while others are relatively independent of host specificity. In the case of the *mariner* transposable element, it can maintain efficient transposition in many different species, even if the species belong to different taxonomic kingdoms (Gueiros-Filho and Beverley 1997). It moves easily from species to species through horizontal gene transfer (see page 359).

Autonomy of transposition

Insertion sequences and transposons are **autonomous** mobile elements in the sense that they contain the gene for transposase that is needed in the process of transposition. Some transposable elements are **non-autonomous**, i.e., they do not encode transposase, and are mobile by virtue of their ability to use transposases from autonomous mobile elements. For example, the autonomous *Ac* (*Activator*) elements in maize can transpose regardless of genetic background. In contrast, the non-autonomous *Ds* (*Dissociation*) elements cannot transpose unless one or more copies of *Ac* are also present in the maize genome. (Non-autonomous mobile elements that use the process of retroposition will be discussed later in this chapter.)

Some non-autonomous elements are derived from autonomous elements through deletion of internal segments. Such elements have lost their capability to produce transposase, but retained the ability to respond to the enzyme. Alternatively, a non-autonomous element may be created when, through mutations, two short genomic sequences at a close distance from one another become sufficiently similar to the receptors of transposase (usually short inverted repeats), thereby conferring mobility on the sequence in between them. *Ds* type-2 elements exhibit extensive sequence similarity to *Ac* elements, and hence may be considered as derived (or defective) *Ac* elements. *Ds* type-1 elements, on the other hand, have no discernable sequence similarity to *Ac*, and have probably been derived from random genomic sequences that happened to reside between two appropriate inverted repeats.

RETROELEMENTS

Retroelements are DNA or RNA sequences that contain a gene for the enzyme **reverse transcriptase**, which catalyzes the synthesis of a DNA molecule from an RNA template. The resulting DNA molecule is called **complementary DNA (cDNA)**.

Not all retroelements possess the intrinsic capability to transpose. Therefore, not all retroelements are transposable elements. Those retroelements that do transpose do so by the process of retroposition. There are different classes

TABLE 7.1 Nomenclature and classification of retroelements and retrosequences

Element	Reverse transcriptase	Transposition	LTRs[a]	Virion particles
Retron	Yes	No	No	No
Retroposon (Non-LTR retrotransposon)	Yes	Yes	No	No
Retrotransposon (LTR retrotransposon)	Yes	Yes	Yes	No
Retrovirus	Yes	Yes	Yes	Yes
Pararetrovirus	Yes	No	Yes	Yes
Retrosequence	No	No	No	No

Modified from Temin (1989).

[a]LTR = long terminal repeat.

of retroelements, and in Table 7.1 we adopt a classification based on Temin (1989).

Retroviruses

Retroviruses are RNA viruses that are similar in structure to transposons. Although they are the most complex of all retroelements, we discuss them first because the concept of retroposition originated with the discovery of the life cycle of retroviruses (Figure 7.3).

After the retroviral particle, called the **virion**, invades a host cell, its genomic RNA is reverse-transcribed into viral cDNA. This DNA can integrate into the host genome and become a **provirus**. Next, the proviral DNA is transcribed into RNAs, which can serve both as mRNAs for synthesizing viral proteins and as viral genomes that can be packaged into infectious virion particles. Once a virion is formed, the cycle can start again.

Retroviruses possess at least three protein-coding genes: *gag*, *pol*, and *env* (Figure 7.4a). These genes encode several internal proteins, several enzymes (including a reverse transcriptase), and an envelope protein, respectively. Many retroviruses possess additional genes. HIV (the AIDS virus), for example, possesses at least six additional genes. The coding region of a retrovirus is flanked by **long terminal repeats (LTRs)**. The LTRs contain promoters for transcription (in the proviral stage), for reverse transcription (in the viral stage), as well as an enhancer and a polyadenylation signal.

Retroposons and retrotransposons

Retroposons and **retrotransposons** are transposable elements that do not construct virion particles: they lack the *env* gene, and so, unlike retroviruses, cannot independently transport themselves across cells. They are distinguished

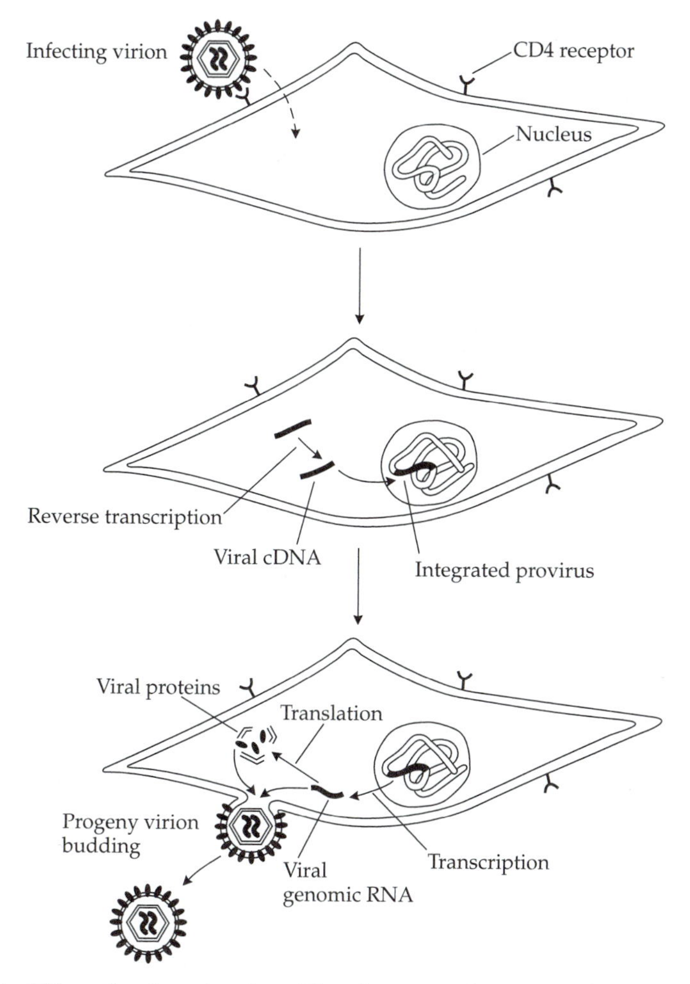

FIGURE 7.3 Life cycle of a retrovirus. The virion attaches to a receptor on the surface of the cell. The two copies of genomic RNA are injected into the cytoplasm, where they are reverse-transcribed by the enzyme reverse transcriptase. The cDNA penetrates the nucleus and may become integrated within the genome of the host cell. The integrated provirus is transcribed into mRNA that serves both a genomic RNA and as mRNA for the synthesis of viral proteins. The genomic RNA and the structural and enzymatic viral proteins assemble into a new infectious virion particle that buds off the cell membrane.

from each other by the absence or presence of terminal LTRs, respectively (Table 7.1). (Note that some authors use the terms **non-LTR retrotransposons** and **LTR retrotransposons** for retroposons and retrotransposons, respectively). The *Ty* elements in yeast and the *copia* elements in *Drosophila* represent typical retrotransposons; they contain LTRs at both ends and a single long open-reading frame with regions of similarity to the *pol* gene of retroviruses (Figure 7.4b).

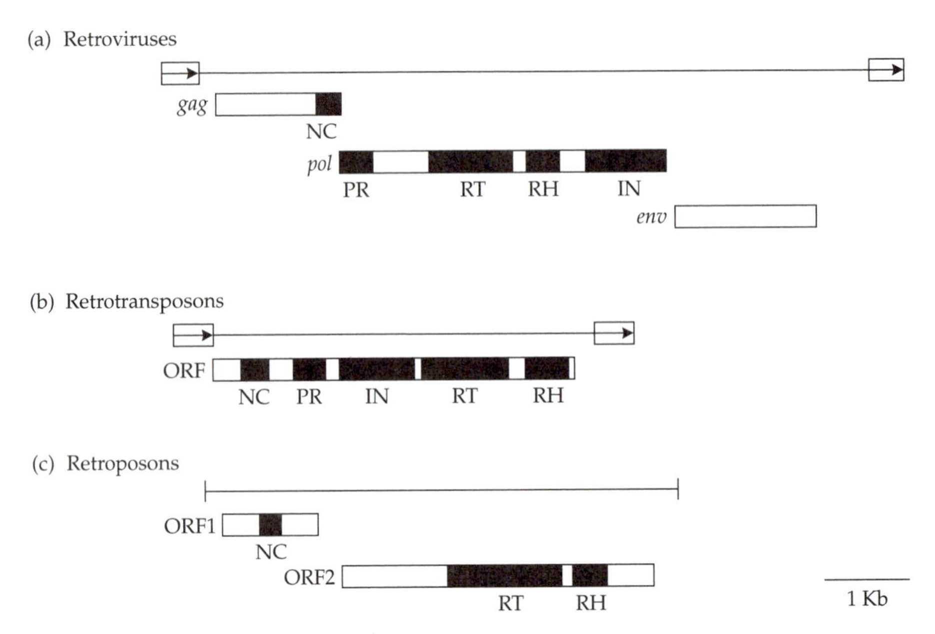

FIGURE 7.4 Structural comparison of transposable retroelements. Open-reading frames separated by termination codons are shown as boxes at different levels below the DNA diagram. Long terminal repeats are represented by arrows inside boxes. (a) A consensus retrovirus containing only the *gag*, *pol*, and *env* genes. (b) A *copia* retrotransposon from *Drosophila melanogaster*. (c) An *I*-factor retroposon from *D. melanogaster*. Identified domains are indicated by solid boxes. NC, nucleocapsid protein; PR, aspartate protease; RT, reverse transcriptase; RH, RNase H; IN, integrase. Modified from Eickbush (1994).

The *G3A* element and the *I* factor in *D. melanogaster* are typical retroposons (Figure 7.4c). These retroposons contain two open-reading frames (ORFs). ORF-2 in *G3A* consists of seven exons separated by very short intervening sequences (introns). Some families of long interspersed repeated sequences, e.g., the *L1* elements that appear thousands of times in the human genome (Chapter 8), are also retroposons. The consensus *L1* structure has a poly(A) tail at one end and contains two large open-reading frames, ORF-1 and ORF-2, with about 375 and 1300 codons, respectively. ORF-2 contains amino acid sequence motifs characteristic of reverse transcriptases.

Some group II self-splicing introns encode proteins belonging to the reverse transcriptase family. Because of their reverse transcriptase activity, this group II intron subset possesses the ability to change genomic location. Thus, according to the phenetic classification scheme adopted in this chapter, they should be considered retroposons. However, in phylogenetic reconstructions, these group II introns cluster with the bacterial and organelle retrons (see page 335).

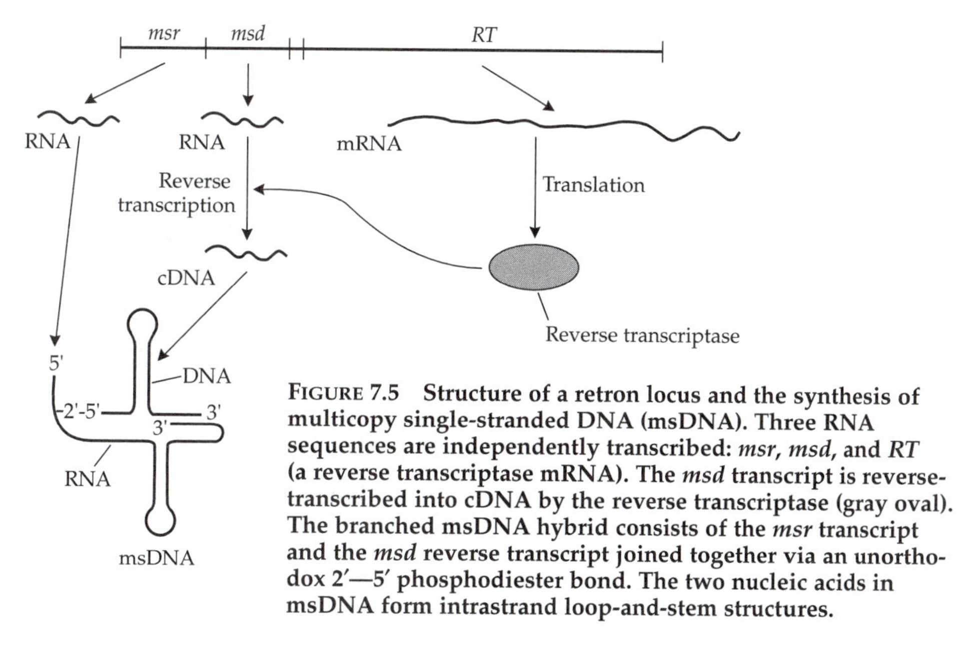

FIGURE 7.5 **Structure of a retron locus and the synthesis of multicopy single-stranded DNA (msDNA). Three RNA sequences are independently transcribed: *msr, msd,* and *RT* (a reverse transcriptase mRNA). The *msd* transcript is reverse-transcribed into cDNA by the reverse transcriptase (gray oval). The branched msDNA hybrid consists of the *msr* transcript and the *msd* reverse transcript joined together via an unorthodox 2′—5′ phosphodiester bond. The two nucleic acids in msDNA form intrastrand loop-and-stem structures.**

Retrons

Retrons are the simplest retroelements (Figure 7.5). They do not possess the capability to transpose. Retrons have been found in some bacterial genomes (Inouye et al. 1989; Lampson et al. 1989; Inouye and Inouye 1991), as well as in mitochondrial and plastid genomes. Retrons appear as single-copy genes in the genome of many bacteria (Rice et al. 1993). Their open-reading frame has sequence similarity to other genes for reverse transcriptase. However, retrons do not excise, and therefore are integral parts of the genome. In essence, retrons are endogenous genomic sequences that encode the enzyme reverse transcriptase. Retrons have no LTRs, and do not construct virion particles. Interestingly, with few exceptions, all retrons take part in the production of a mysterious DNA–RNA hybrid molecule of unknown function called multicopy single-stranded DNA (msDNA).

Pararetroviruses

Pararetroviruses, such as the hepatitis-B virus (a hepaDNAvirus) and the cauliflower mosaic virus (a caulimovirus), are double-stranded or partially double-stranded circular DNA viruses. The structure of the hepatitis-B virus is shown in Figure 7.6. The uniqueness of pararetroviruses lies in their mode of replication. Rather than replicating the two strands as in other double-stranded DNA viruses, pararetroviruses use one strand for synthesizing both daughter strands. Thus, one of the strands is replicated once in the usual manner to yield the complementary strand, and once transcribed into RNA that is subsequently reverse-transcribed to yield the original strand.

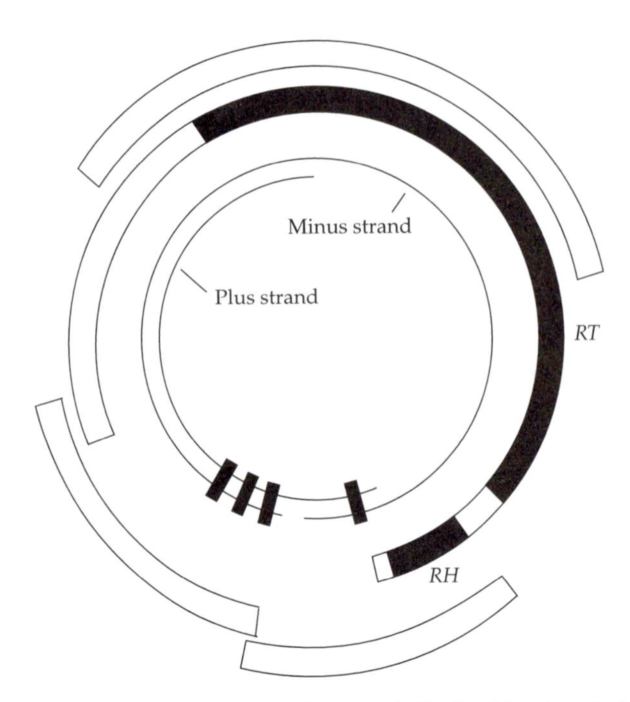

FIGURE 7.6 **Schematic representation of the partially double-stranded hepatitis-B pararetrovirus genome. The lengths of the minus and plus strands are 3.5 and 2.1 Kb, respectively. Bars spanning the two strands indicate sequence motifs involved in interstrand interactions. Four overlapping open-reading frames are shown as arcs. Identified domains are indicated by solid arcs: *RT*, reverse transcriptase; *RH*, RNase H.**

The amino acid sequence encoded by the longest reading frame in pararetroviral genomes was found to be homologous to the N-terminal region of the *pol* gene product of the retroviruses (Toh et al. 1983). Therefore, pararetroviruses are similar to retroviruses in their endogenous capability to produce reverse transcriptase, but they seem to have lost the ability to insert themselves into the host genome. This disqualifies them as transposable elements, although they clearly share a common evolutionary origin with the retroviruses.

Evolutionary origin of retroelements

The fact that the reverse transcriptases of all retroelements have some amino acid identity with one another suggests a common evolutionary origin. Because of the simplicity of retrons as opposed to the complexity of retroviruses, and because of the antiquity of bacteria, Temin (1986, 1989) suggested that the path of evolution went from retrons to retroposons to retrotransposons to retroviruses to pararetroviruses (Figure 7.7). One must note that this evolutionary scheme is mostly characterized by a progressive increase in the struc-

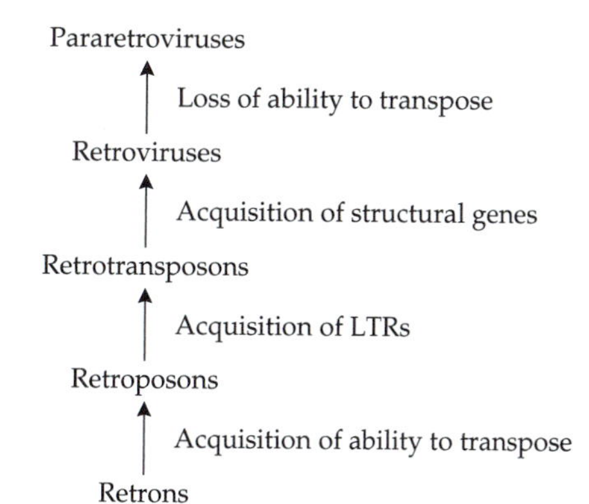

FIGURE 7.7 Schematic representation of possible evolutionary relationships among retroelements.

tural complexity of the retroelements. Of course, it is possible that some of the present-day retrotransposons and retroposons have been derived from retroviruses, rather than the other way around.

Many authors have studied the evolutionary relationships among retroelements (e.g., Miyata et al. 1985; Doolittle et al. 1989; Xiong and Eickbush 1988, 1990). Figure 7.8 shows a phylogeny of retroelements derived from reverse transcriptase sequences (Eickbush 1994). The tree can be roughly divided into two major clades. One clade contains the retrons, the reverse transcriptase-containing group II introns, and a paraphyletic group of retroposons. The second branch contains the retroviruses, the caulimoviruses and the hepaDNAviruses (as two unrelated lineages), as well as five paraphyletic lineages of retrotransposons. The various paraphyletic lineages of retroposons

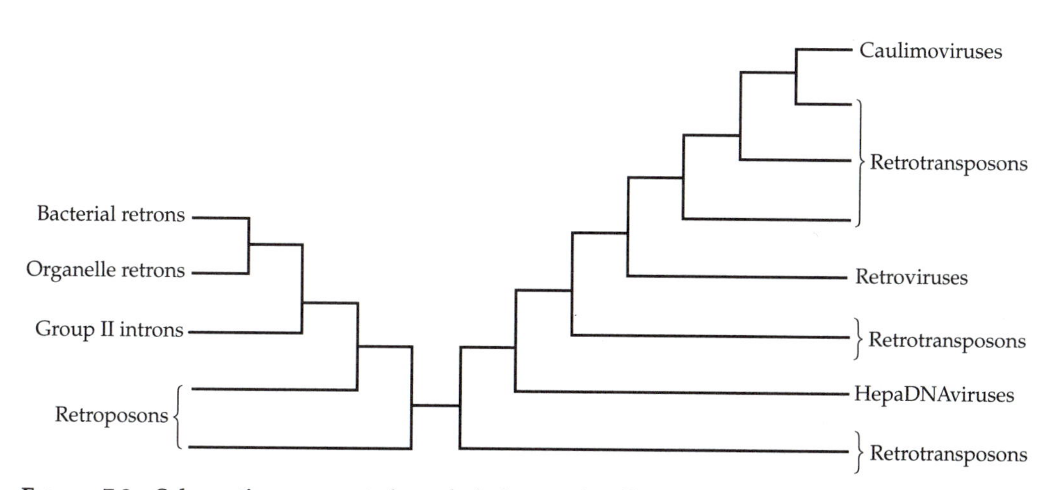

FIGURE 7.8 Schematic representation of phylogenetic affinities among retroelements. The unrooted unscaled tree was derived from the amino acid sequences of the reverse transcriptase domain (178 aligned amino acids) in 79 retroelements. Paraphyletic groups are indicated by brackets. Data from Eikbush (1994).

and retrotransposons do not seem to confirm with any well-established taxonomic divisions for the host organisms of these elements. For example, one retroposon lineage contains elements from such diverse organisms as amphibians, fruitflies, and slime molds. We note, however, that the tree is based on only 178 amino acids, so the branching order should therefore be treated with caution.

RETROSEQUENCES

Restrosequences (or **retrotranscripts**) are genomic sequences that have been derived through the reverse transcription of RNA and subsequent integration of the resulting cDNA into the genome. However, retrosequences lack the ability to produce reverse transcriptase (Table 7.1), and have been produced through the use of a reverse transcriptase from a retroelement. The template from which the retrosequence has been derived is usually the RNA transcript of a gene. A process of producing retrosequences is shown in Figure 7.9. If a gene is not transcribed within a germline cell, but only in specialized somatic cells, the creation of a retrosequence requires the RNA to cross cell barriers from the somatic cell to the germline cell. This can happen when an RNA molecule becomes encapsulated within the virion particle of a retrovirus and is then transported to the germline cell where it is reverse-transcribed (Linial 1987). This process is referred to as **retrofection**. Retrofection appears to be very common in some taxa, such as mammals, but not in others.

Since retrosequences originate from RNA sequences, they bear marks of RNA processing and are hence also referred to as **processed sequences**. The diagnostic features of retrosequences include (1) the lack of introns, (2) precise boundaries coinciding with the transcribed regions of the gene from which they were derived, (3) stretches of poly(A) at the 3′ end, (4) short direct repeats at both ends (indicating that transposition may have been involved), (5) various posttranscriptional modifications, such as the addition or removal of short stretches of nucleotides, and (6) chromosomal positions different from the locus of the original gene from which the RNA was transcribed.

Retrogenes

A **retrogene** or **processed gene** is a functional retrosequence that produces an identical or nearly identical protein to the one produced by the gene from which the retrogene has been derived. There are three main reasons why it is highly unlikely that a reverse-transcribed sequence will retain its functionality. First, the process of reverse transcription is very inaccurate, and many differences (mutations) between the RNA template and the cDNA may occur. Second, unless a processed gene has been derived from a gene transcribed by RNA polymerase III, it usually does not contain the necessary regulatory sequences that reside in the untranscribed regions, so it will likely be inactive

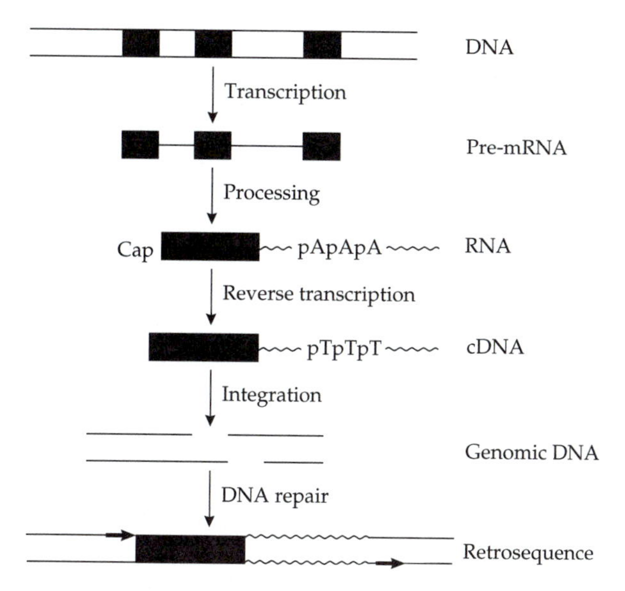

FIGURE 7.9 Creation of a processed retrosequence. The black boxes represent exons. The wavy lines indicate a poly(A) tail in mRNA and the complementary poly(T) in cDNA. The DNA is transcribed into pre-mRNA, then processed into mRNA. The mRNA is reverse-transcribed into cDNA, which becomes integrated into the genomic DNA. The gaps are repaired, so two direct short repeats flanking the inserted retrosequence are created (black horizontal arrows).

even if its coding region remains intact after retroposition. Two processed metallothionein-I pseudogenes, one in human and one in rat, appear to represent such a case (Karin and Richards 1984; Andersen et al. 1986). Their coding regions are intact, and in the case of the rat pseudogene even the 5′ regulatory sequences are intact. Nevertheless, these sequences are not transcribed, presumably because of the lack of some regulatory sequences or the formation of unstable transcripts. Third, a processed gene may be inserted at a genomic location that may not be adequate for its proper expression. Indeed, in the vast majority of cases, a processed sequence is "dead on arrival."

Surprisingly, functional processed genes have been found, although they seem to be very rare. The human phosphoglycerate kinase (PGK) multifamily consists of an active X-linked gene, a processed X-linked pseudogene, and an additional autosomal gene. The X-linked gene contains 11 exons and 10 introns. Its autosomal homolog, on the other hand, is unusual in that it has no introns and is flanked at its 3′ end by remnants of a poly(A) tail, i.e., two diagnostic features of a processed sequence. Interestingly, the autosomal *PGK* gene has not only maintained an intact reading frame and the ability to transcribe and produce a functional polypeptide, but has also acquired a novel tissue specificity: it is produced only in the male testis (McCarrey and Thomas 1987).

The muscle-specific calmodulin gene in chicken is also intronless and was apparently produced by a reverse transcriptase-mediated event (Gruskin et al. 1987). This might also be the case for the intronless globin gene in gnats belonging to the genus *Chironomus* (Antoine et al. 1987), and all but one of the intronless actin genes in the slime mold *Dictyostelium* (Romans and Firtel 1985).

Semiprocessed retrogenes

The vast majority of mammals contain a single preproinsulin gene. In contrast, myomorph (mouse-like) rodents contain two. The rat preproinsulin I gene (and its ortholog in mouse) may represent an instance of a **semiprocessed retrogene**. The gene contains a single 119-bp intron in the 5′ untranslated region. In comparison, its paralog, preproinsulin II, contains an additional 499-bp intron within the region coding for the C peptide (Lomedico et al. 1979). All preproinsulin genes from other mammals, including other rodents, contain two introns as well. Moreover, the preproinsulin I gene is flanked by short repeats, and the polyadenylation signal is followed by short poly(A) tract (Soares et al. 1985). These features suggest that the preproinsulin I gene has been derived from a partially processed prepronsulin II premRNA, from which only one intron has been excised.

Preproinsulin I appears to have been derived from an aberrant premRNA transcript that initiated 500 bp upstream of the normal cap site. This retrogene might have maintained its function following integration into a new genomic location precisely because the aberrant transcript contained 5′ regulatory sequences not normally transcribed.

Retropseudogenes

A **retropseudogene** or **processed pseudogene** is a retrosequence that has lost its function. It bears all the hallmarks of a functional retrosequence but has molecular defects that prevent it from being expressed. A comparison between a functional gene and a processed pseudogene is shown in Figure 7.10. Many processed pseudogenes are truncated during reverse transcription and retrofection; both 5′ and 3′ truncations are very common, and some retropseudogenes are truncated on both sides (Ophir and Graur 1997). Truncation can occur during (1) transcription initiation or termination (upstream or downstream of the normal sites), (2) RNA processing (e.g., faulty splicing), or (3) reverse transcription (e.g., failure of the enzyme to start properly or to complete the reverse transcription of the RNA molecule).

Most retropseudogenes are derived from processed RNA transcripts, though in rare cases retropseudogenes have been found that have been derived from unprocessed or semiprocessed RNA transcripts (Weiner et al. 1986). Retropseudogenes derived from all types of RNA transcripts are known. However, the vast majority of retropseudogenes are derived from RNA polymerases II and III transcripts (i.e., mRNA, snRNA, tRNA, 7SL RNA, and 5S rRNA). Only very few retropseudogenes derived from RNA poly-

```
         M   A   T   K   A   V   C   V   L   K   G   D   G   P   V
SOD-1   ATG GCG ACG AAG GCC GTG TGC GTG CTG AAG GGC GAC GGC CCA GTG
          • ••   •       •   •    •  •   •           • •  •       •
Ψ69.1   ATA ATG ATG AAG GTC ATG TAC ATG TTG AAG GGC CAG AGC CCG GTG
         I   M   M   K   V   M   Y   M   L   K   G   Q   S   P   V

         Q   G   I   I   N   F   E   Q   K            E   S   N   G
SOD-1   CAG GGC ATC ATC AAT TTC GAG CAG AAG  Intron  GAA AGT AAT GGA
             ••   –       •   –   •                        •  ---  •
Ψ69.1   CAG GCG A C ATC CAT TT  GAG CAG AAG          GAA AAT     GAA
         Q   A   T   S   I   *  **

         P   V   K   V   W   G   S   I   K   G   L   T   E   G   L
SOD-1   CCA GTG AAG GTG TGG GGA AGC ATT AAA GGA CTG ACT GAA GGC CTG
          •  •   •        –  •            •       •           •   •
Ψ69.1   CCA TTT ATG GTG T C AGA AGC ATT ACA GGA TTG ACT GAA CGC CAG

         H   G   F   H   V   H   E   F   G   D   N   T   A
SOD-1   CAT GGA TTC CAT GTT CAT GAG TTT GGA GAT AAT ACA GCA  G Intron
          •  •                                –   –   •   •
Ψ69.1   CAC AGA TTC CAT GTT CAT CAG TTT GGA G T A T AAC ACA  G
```

FIGURE 7.10 Comparison between the first two exons of the human Cu/Zn superoxide dismutase gene (*SOD-1*) and the homologous parts of a processed pseudogene (*ψ69.1*). Dots denote substitutions; minus symbols denote deletions. Note the absence of introns and the premature termination codon (indicated by asterisks). See Table 1.2 for the one-letter amino acid abbreviations. Data from Danciger et al. (1986).

merase I transcripts (i.e., 18S, 28S, and 5.8S rRNAs) are known (e.g., Wang et al. 1997). It is not clear whether this rarity reflects the fact that RNA polymerase I transcripts are only seldom retrotransposed, or that the sequences of processed rRNA pseudogenes are difficult to distinguish from those of their functional paralogs.

Transfer RNA retropseudogenes are particularly interesting because they provide one of the most compelling pieces of evidence that processed pseudogenes are indeed derived through the reverse transcription of RNA. All nuclear tRNA molecules possess a CCA sequence at their 3′ terminal (Figure 7.11). This sequence is not encoded by the tRNA-specifying gene, but is added enzymatically after transcription. In contrast, genomic tRNA retropseudogenes often contain the CCA sequence at their 3′ end (Reilly et al. 1982).

Processed pseudogenes have been found in animals, plants, and even bacteria and viruses. However, although processed pseudogenes are abundant in mammals, they are relatively rare in other animals, including chicken, amphibians, and *Drosophila*. For example, mammals have 20–30 tubulin retropseudogenes, while chicken and *Drosophila* have none, although, like mammals, chicken and *Drosophila* have many α- and β-tubulin genes. It is not clear how this conspicuous difference occurred. One hypothesis is that it is due to differences in gametogenesis between mammals and other animals (Weiner et

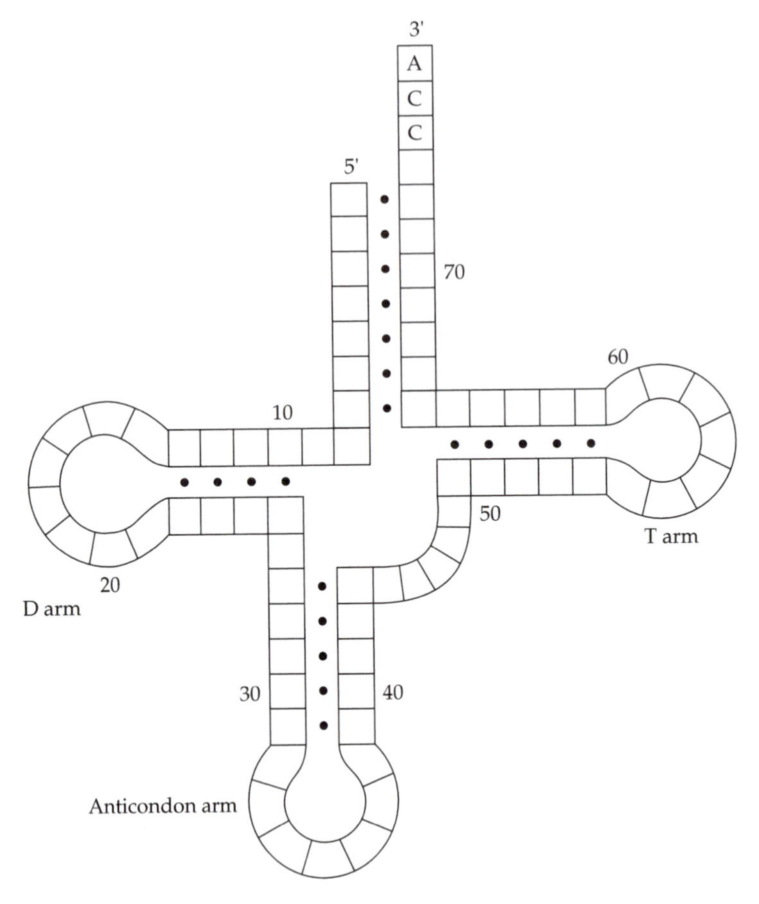

FIGURE 7.11 Cloverleaf structure of a tRNA molecule. The sequence CCA at the 3′ end is added posttranscriptionally in the functional molecule, but it often appears in the genomic sequence of tRNA retropseudogenes.

al. 1986). While spermatogenesis is very similar among mammals, chicken, amphibians, and *Drosophila*, mammalian oogenesis differs from that in the other animals by a prolonged lambrush stage that lasts from birth to ovulation. This state of relatively suspended animation may last up to 40 years in humans, but for only a few months in amphibians, for less than three weeks in birds, and is virtually absent in *Drosophila*. If this hypothesis is correct, and retroposition in mammals does occur predominantly in the female germline, then it is expected that mammals should possess a greater number of retrosequences than other animals.

An additional prediction derived from this hypothesis is that retrosequences should be found in high numbers on the X chromosome, in intermediate numbers on autosomes, and should be rare on the Y chromosome, which is not carried by females. At present, there are no quantitative empirical data with which to test this hypothesis. Admittedly, there are a few reports of retropseudogenes on the Y chromosome, but these are located within the pseudoautosomal region which pairs with the X chromosome during meiosis, and may have been transferred from the X chromosome to the Y chromosome through recombination.

Table 7.2 shows a list of processed pseudogenes from humans and rodents for which both the number of functional protein-coding genes and the number of processed pseudogenes are known or have been estimated. As far as polymerase II retropseudogenes are concerned, mammalian species seem, on average, to possess more processed pseudogenes than functional genes. In some cases, the number of processed pseudogenes exceeds the number of functional counterparts by orders of magnitude. For example, in the genome of the mouse, about 200 processed pseudogenes have been produced from a single gene for glyceraldehyde-3-phosphate dehydrogenase. In fact, the number of processed pseudogenes may even be underestimated in many cases, because old pseudogenes may have diverged in sequence from their parental functional genes to such an extent that they are no longer detectable by molecular probes derived from their functional homologs.

Sequence evolution of retropseudogenes

Due to the ubiquity of reverse transcription, the genomes of mammals are literally bombarded with copies of reverse-transcribed sequences. The vast majority of these copies have been nonfunctional from the moment they were integrated into the genome. Moreover, such sequences cannot be easily rescued by gene conversion, since they are mostly located at great chromosomal distances from the parental functional gene (Chapter 6). The phenomenon of a functional locus pumping out defective copies of itself and dispersing them all over the genome has been likened to a volcano generating lava, and the process has been termed the **Vesuvian mode of evolution** (P. Leder, cited in Lewin 1981).

By using Li et al.'s (1981) formula for calculating the nonfunctionalization time of pseudogenes (Chapter 6), Itoh et al. (unpublished) found out that the vast majority of retropseudogenes became nonfunctional, and hence free from all selective constraints, as soon as they were integrated as chromosomal sequences within the genome.

Because of their lack of function, pseudogenes are affected by two evolutionary processes (Graur et al. 1989b). The first involves the rapid accumulation of point mutations. This accumulation eventually obliterates the sequence similarity between the pseudogene and its functional homolog, which evolves much more slowly. The nucleotide composition of the pseudogene will come to resemble its nonfunctional surroundings more and more, eventually "blending" into it. This process has been called **compositional assimilation**.

The second evolutionary process is characterized by pseudogenes becoming increasingly shorter compared to the functional gene. This **length abridg-**

TABLE 7.2 The number of retropseudogenes and the number of parental functional genes

Gene	Number of genes	Number of retropseudogenes
Human		
Argininosuccinate synthetase	1	14
β-Actin	1	~20
β-Tubulin	2	15–20
Cu/Zn superoxide dismutase	1	4
Cytochrome *c*	2	20–30
Dihydrofolate reductase	1	~5
Glyceraldehyde-3-phosphate dehydrogenase	1	~25
Lactate dehydrogenase A	1	10
Lactate dehydrogenase B	1	3
Lactate dehydrogenase C	1	6
Laminin	1–9	20
Nonmuscle tropomyosin	1	3
Nucleophosmin B23	1	7–9
Phosphoglycerate kinase	2[a]	1
Prohibitin	1	4
Prothymosin α	1	5
Ribosomal protein L32	1	~20
Triosephosphate isomerase	1	5–6
Mouse		
α-Globin	2	1
Cytokeratin endo A	1	1
Glyceraldehyde-3-phosphate dehydrogenase	1	~200
Heat shock protein 86 kDa	1	~4
High mobility group protein 14	1	14
Lactate dehydrogenase A	1	10
Myosin light chain	1	1
Phosphoglycerate kinase	2[a]	1
Proopiomelanocortin	1	1
Ribosomal protein L7	1–2	20
Ribosomal protein L19	1	12
Ribosomal protein L30	1	15
Ribosomal protein L32	1	16–20
Tumor antigen p53	1	1

[a]One of which is a retrogene.

TABLE 7.2 Continued

Gene	Number of genes	Number of retropseudogenes
Rat		
Aldose reductase	1	3
α-Tubulin	2	10–20
Cytochrome c	1	20–30
High mobility group protein 14	1	25
Ornithine aminotransferase	1	3
Ribosomal protein L19	1	5

Data from Weiner et al. (1986) and later sources.

ment is caused by the excess of deletions over insertions. This process is very slow, however. It has been estimated that it would take about 400 million years for a mammalian processed pseudogene to lose half of its length. Of course, mammals are only about 200 million years old, so the mammalian genome is expected to contain major chunks of very ancient pseudogenic DNA. Obviously, these ancient pseudogenes have by now lost almost all similarity to the functional genes from which they have been derived. In other words, mammalian processed pseudogenes are created at a much faster rate than the rate by which they are obliterated by deletion. It has therefore been concluded that abridgment is too slow a process to offset the steady Vesuvian increase in genome size (Graur et al. 1989b).

The rate of DNA loss in *Drosophila* is approximately 75 times faster than that in mammalian pseudogenes (Petrov and Hartl 1997). This high rate of DNA loss leads to the rapid elimination of nonessential DNA, and may explain the dearth of pseudogenes in *Drosophila* species.

LINEs AND SINEs

The genomes of all multicellular eukaryotes contain several types of highly repetitive interspersed sequences (Chapter 8). These sequences, originally detected as rapidly reannealing components of genomic DNA, have been divided into two major classes, referred to as **short interspersed repetitive elements (SINEs)** and **long interspersed repetitive elements (LINEs)**. Over a third of the human genome is estimated to consist of interspersed repetitive sequences. LINEs were originally described as DNA sequences longer than 5 Kb and present in 10^4 or more copies per genome, while SINEs were defined as sequences shorter than 500 bp occurring in 10^5 copies or more in the haploid genome (Singer 1982). Following sequencing studies, however, the definitions of SINEs and LINEs were changed by Hutchison et al. (1989). That is, instead of using length and copy number as diagnostic features, LINEs and SINEs were redefined according to their ability or lack of ability to mediate their own transposition.

LINEs typically range in length from 3 to 7 Kb. DNA sequencing studies have indicated that LINEs are retroposons or degenerate copies of retroposons. Each functional LINE contains an open-reading frame that encodes domains corresponding to an endonuclease and a reverse transcriptase. The reverse transcriptase domain exhibits LINE specificity, i.e., a reverse transcriptase from one LINE will only recognize the 3′ end of that LINE, or a similar sequence, and will be much less efficient at recognizing and subsequently reverse transcribing other LINEs. Inactive LINEs are, in fact, partial LINE sequences or pseudogenes derived from active LINEs.

SINEs typically range in length from 75 to 500 bp. They do not code for proteins required for retroposition; in fact, they do not possess any open-reading frame of significant length. Thus, SINEs are non-autonomous elements that must be aided in the process of retroposition by other genetic elements. There are two known types of SINEs: those derived from 7SL RNA and those derived from tRNAs. With the exception of the primate *Alu* and rodent *B1* families, which are derived from 7SL RNA, all SINEs reported to date, from sources as diverse as humans, fish, turtles, plants, midges, and fungi, are derived from tRNAs.

SINEs derived from 7SL RNA

Alu sequences, so named because they contain a characteristic restriction site for the *Alu*I endonuclease, are approximately 300 bp long. They form a family of repeated sequences that appear more than a million times in the human genome, constituting a remarkable 10% of the genome.

Ullu and Tschudi (1984) found that *Alu* sequences are in fact highly derivative processed pseudogenes of the gene specifying 7SL RNA, an abundant cytoplasmic component of the signal recognition particle that is essential in the process of removal of signal peptides from secreted proteins (Walter and Blobel 1982). The active gene is highly constrained, and its sequence is conserved among such divergent organisms as humans, *Xenopus*, and *Drosophila*.

Human *Alu* sequences have been derived from a functional 7SL sequence by a series of steps involving a duplication, two deletions, and many nucleotide substitutions and small indels (Figure 7.12a). Most human *Alu* sequences have a dimeric structure. The human genome also contains a number of tetrameric *Alu* sequences, but only a few monomeric *Alu* elements have been found. In contrast, the rodent *Alu* equivalent, *B1*, is almost exclusively monomeric (Figure 7.12b). Britten et al. (1988) dated the emergence of the first monomer to a time before the mammalian radiation, and the duplication that produced the dimer to a time after the primate lineage had been established.

If a retropseudogene retains its capability to be transcribed, a cascade process may ensue, whereby new retropseudogenes are created out of the

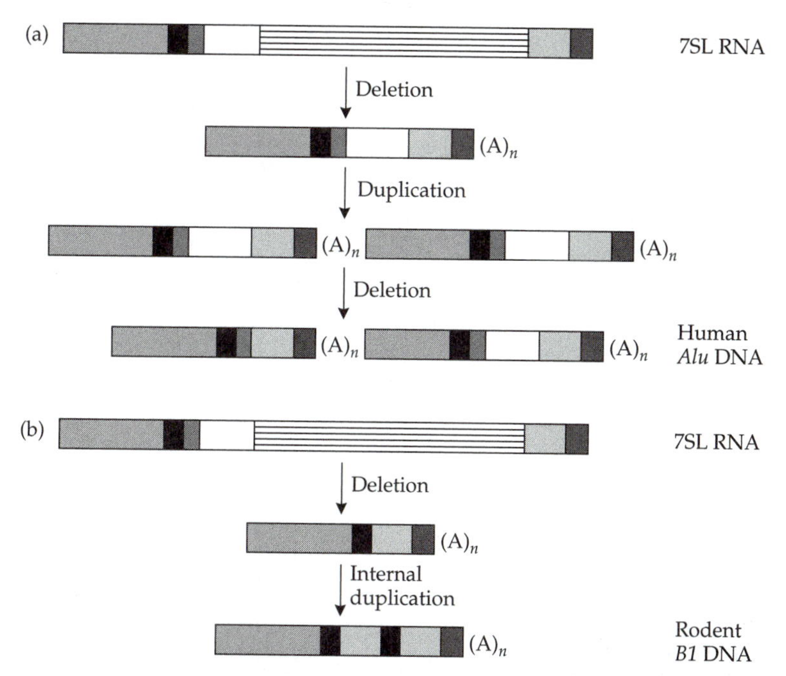

FIGURE 7.12 Origin of *Alu* sequences in humans and other primates (a), and *B1* sequences in rodents (b). Different regions in the 7SL RNA genes are shaded differently to emphasize the deletions and rearrangements in the *Alu* sequences. $(A)_n$ means that A is repeated *n* times. Note the dimeric structure in (a) and the monomeric structure in (b).

RNA transcripts of existing pseudogenes. This was suggested to be the case in the *Alu* family (Bains 1986). As the 7SL-RNA gene is transcribed by RNA polymerase III, which does not require promoters outside of the transcribed region, it is conceivable that some *Alu* sequences may have retained intact promoters and can be transcribed. However, *Alu* sequences are not regular retropseudogenes directly derived from the 7SL-RNA-specifying gene. Something else must have occurred to make them so effective in multiplying themselves to produce hundreds of thousands of copies. What this "something else" is, is at present a mystery as far as *Alu* elements are concerned; however, we do know why other SINEs are so efficient in the "be fruitful and multiply" business, despite lacking autonomous transposable capabilities (see the next section).

Willard et al. (1987) and Britten et al. (1988) recognized several subfamilies of *Alu* sequences. They proposed that these subfamilies have been sequentially derived from only three or four source sequences (Figure 7.13), rather than the cascade process suggested by Bains (1986). The more recently derived subfamilies have sequences that are more divergent from the progenitor 7SL sequence than the older subfamilies. Thus, most *Alu* sequences have been derived not directly from the 7SL functional gene, but from a small number of source sequences, which were originally derived from the 7SL-RNA

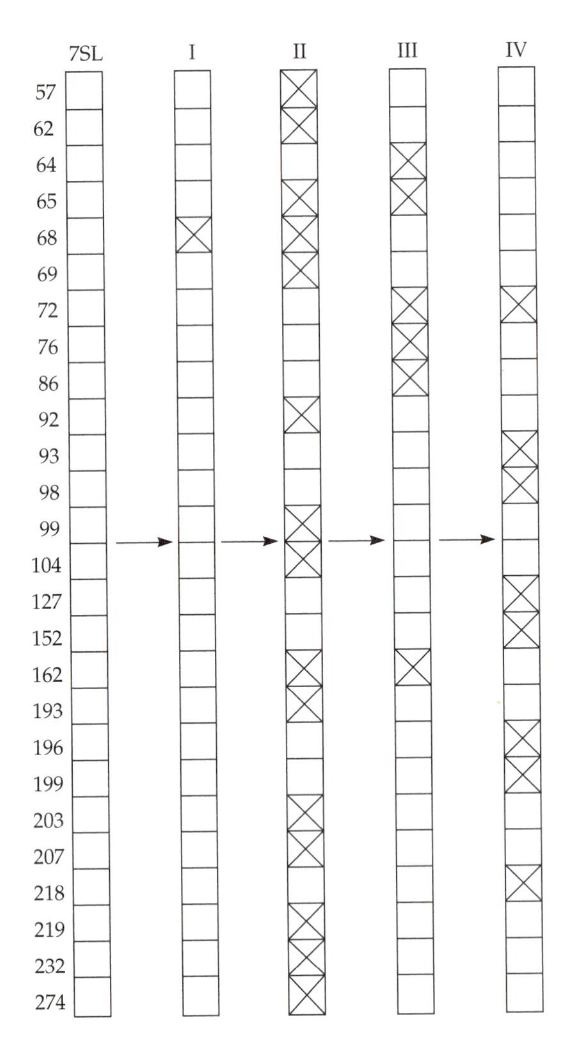

FIGURE 7.13 **Sequence of mutations at diagnostic positions (Arabic numerals) between different subfamilies of *Alu* sequences. Roman numerals indicate successive source sequences that served at various periods as the predominant source of *Alu* sequences. Substitutions distinguishing each subfamily from the preceding one are denoted by ×. For example, *Alu* subfamily I differs from 7SL RNA by a single substitution at position 68; subfamily II differs from subfamily I by 15 substitutions, including one at position 68 that was already substituted once in the previous step. Data from Britten at al. (1988).**

gene through many steps of changes. Each of these source sequences served at one time or another as the predominant source of *Alu* sequences and was superseded by a descendant line. The successive waves of fixation did not occur in sudden discrete bursts, but consecutive subfamilies continued to coexist within the genome for long periods of time (see also Quentin 1988).

SINEs derived from tRNAs

As mentioned previously, the vast majority of SINEs are tRNA-derived. A typical tRNA-derived SINE is a chimerical molecule consisting of two parts, one of which is a tRNA-related region and the other a tRNA-unrelated region (Figure 7.14). The latter in turn consists of two distinct regions: a 5′ region whose origin remains to be elucidated, and a LINE-derived region (Ohshima et al. 1996).

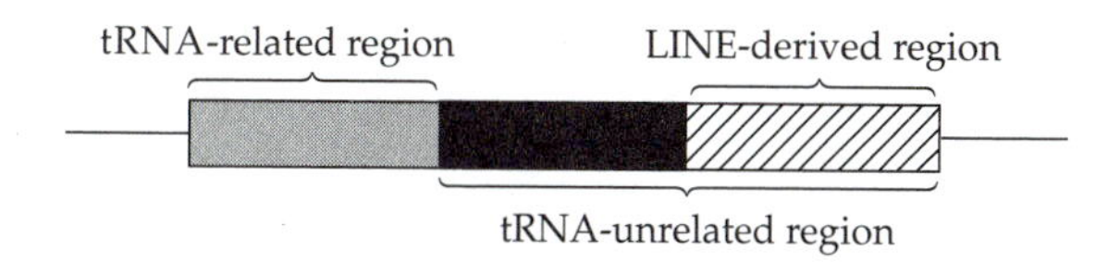

FIGURE 7.14 Structure of a tRNA-derived SINE. The element can be roughly divided into two parts: a tRNA-related region (gray box) and a tRNA-unrelated region. The latter region contains a sequence of unknown origin (solid box) and a LINE-derived region (hatched box). Courtesy of Professor Norihiro Okada.

Because of the rapid rate of SINE evolution, the identification of the tRNA species from which the tRNA-related region of a SINE had been derived is extremely difficult in the majority of cases, and errors in identification abound in the literature. To date, only four tRNA species have been identified unambiguously as origins of SINEs. These are: $tRNA^{Lys}$ (the most commonly used tRNA species), $tRNA^{Ala}$, $tRNA^{Arg}$, and $tRNA^{Glu}$ (Ohshima et al. 1993). At present, we do not know whether or not other types of tRNA can form SINEs.

Where there's a SINE, there's a LINE

An enduring mystery in molecular evolution has been the amazing efficiency with which SINEs, which are in themselves devoid of self-replicating mechanisms, could have produced hundreds of thousands of genomic copies. The discovery that the 3′ end of each tRNA-derived SINE exhibits sequence similarity with the 3′ end of a LINE (Ohshima et al. 1996) provided the key for solving this mystery.

Figure 7.15 shows one such example, the alignment of the 3′ end of a SINE and a LINE from the silkworm *Bombyx mori*. About 80 nucleotides at the

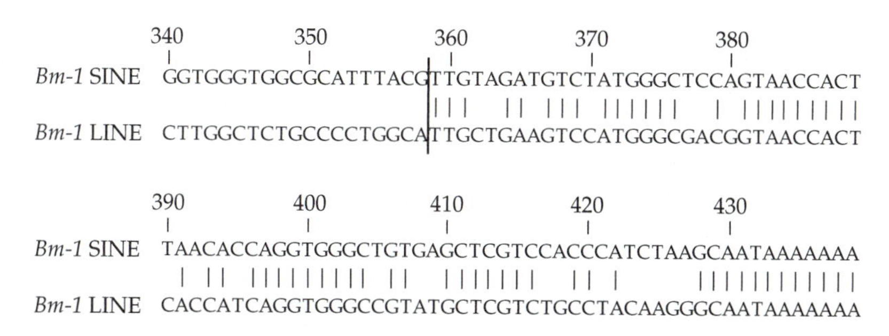

FIGURE 7.15 Alignment of the 3′ ends of the consensus sequences of the *Bm-1* SINEs and *Bm-1* LINEs from the silkworm *Bombyx mori*. Short vertical lines indicate identical nucleotides in the two sequences. No gaps were needed to align the two sequences. The long vertical line indicates the abrupt end in similarity between the SINE and LINE sequences. Position numbers are as in the original alignment (Okada et al. 1997).

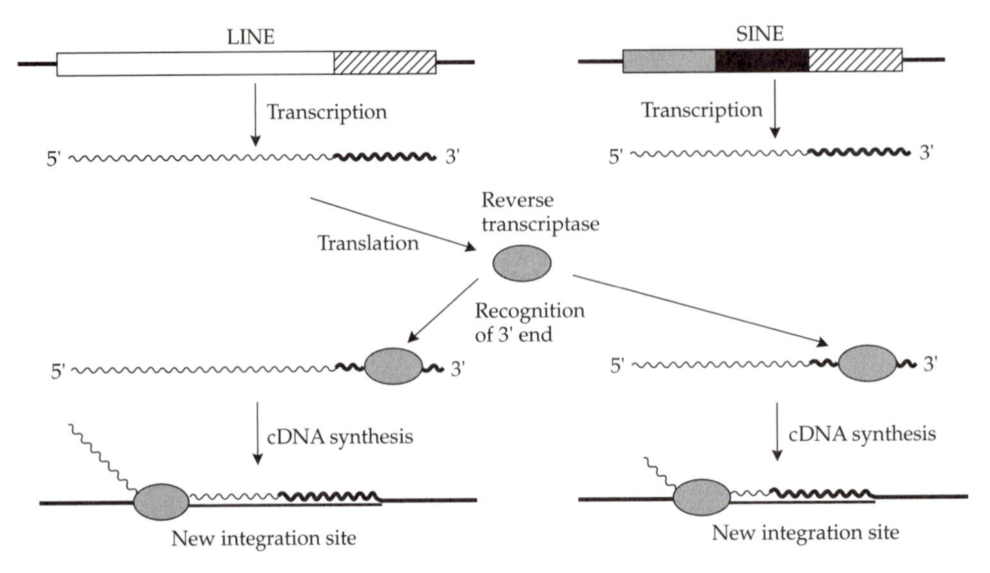

FIGURE 7.16 The reverse transcriptase (shaded oval) required for the retroposition of a tRNA-derived SINE is provided by the partner LINE. SINE transcripts (wavy lines) are recognized at their 3′ end (heavy lines), and are subsequently reverse-transcribed and integrated into a genomic target site. The 3′ sequences of the LINE and the SINE are shown as hatched boxes to emphasize their similarity to each other. The empty box in the LINE represents a LINE-specific sequence. The sequence of unknown origin in the SINE and the tRNA-related region are shown as a solid box and a gray box, respectively. Courtesy of Professor Norihiro Okada.

3′ end of both sequences are remarkably similar, while no discernible similarity is observed upstream of the 3′ region. Since this region is known to be recognized by the LINE-encoded reverse transcriptase, Ohshima et al. (1996) proposed that each SINE is propagated within the genome by using the enzyme from a corresponding LINE (Figure 7.16). In some cases, two or more SINEs were found to share the same 3′ tail as a single LINE. Because each reverse transcriptase of a LINE is specific for the LINE on which the gene resides, there are no universal similarities in either length or sequence composition among the 3′ recognition regions beyond the similarity between the conspecific pairs. Many such **LINE/SINE couples** have been identified in such diverse taxa as ruminants, reptiles, salmonid fishes, sharks, silkworm, tobacco plants, and fungi (Okada et al. 1997).

The above model makes three interesting predictions. First, if there is a tRNA-derived SINE, a LINE partner must be present in the genome of the same organism. LINEs may exist on their own; SINEs cannot. Second, the model predicts that once a LINE partner becomes inactive, the SINE partner will lose its ability to retrotranspose. Finally, a LINE partner should have a longer evolutionary history, and hence a broader phylogenetic distribution, than its SINE partner. All these predictions have been validated for the *LINE2/MIR* couple in humans and the *CiLINE2/AFC* couple in cichlid fishes (Terai et al. 1998).

TABLE 7.3 Degrees of sequence divergence between *Alu* sequences and between η-globin pseudogenes

Species pair	Percent divergence	
	Alu sequence[a]	η pseudogene
Human vs. chimpanzee	2.2 ± 1.4	1.7
Human vs. orangutan	3.7 ± 1.9	3.1

Data from Koop et al. (1986a) and Li et al. (1987a).

[a]Seven orthologous sequences were used to compute the means and standard deviations.

DNA-mediated transposable elements and transposable fossils

In addition to the retrotransposed SINEs that have been discussed previously, the mammalian genome contains many interspersed repeats that have been transposed without the aid of reverse transcriptase. The human genome was found to contain at least 14 families and more than 100,000 copies of simple symmetrical transposons, 180–2,500 bp in length (Smit and Riggs 1996). Interestingly, many such transposable elements are very ancient, and are often referred to as **transposon fossils**. For example, members of the human *Tigger* family were found to be very similar to *pogo*, a DNA transposon in *Drosophila*, and to a lesser extent to five transposons from fungi and nematodes, a finding that attests to the antiquity of these elements. Intriguingly, the transposase encoded by the members of the *Tigger* family was found to be similar to the major mammalian centromere protein B, which may indicate that this important component of meiosis and mitosis may have originally been derived from a transposable element.

Rate of SINE evolution

Because of their sheer number, SINEs were initially thought to have a biological function. Whether or not their mere presence in the genome serves a function will be discussed in the next section. However, in terms of sequence evolution, they seem to evolve about as rapidly as other well-characterized functionless pseudogenes. One such example is shown in Table 7.3.

GENETIC AND EVOLUTIONARY EFFECTS OF TRANSPOSITION

Transposition and retroposition can have profound effects on the size and structure of genomes. In particular, transposable elements have been considered examples of **selfish DNA**, i.e., sequences which do not perform a function nor confer an advantage on the host, but can spread within the genome because they multiply faster than other genomic sequences (Doolittle and Sapienza 1980; Orgel and Crick 1980). For this reason, transposition can greatly increase the genome size, an effect we will deal with in Chapter 8.

Here we concern ourselves with the effects of transposable elements on gene evolution and expression.

First, as mentioned above, transposons in bacteria often carry genes that confer antibiotic or other forms of resistance on their carriers. Plasmids can carry such transposons from cell to cell, either within a species or between different bacterial species, so that resistance can quickly spread throughout a population or an entire bacterial ecosystem. Thus, transposons may help species to survive adverse environments.

Second, the expression of a gene may be altered by the presence of a transposable element either within the gene or adjacent to it. In the simplest case, the insertion of a transposable element into the coding region of a protein-coding gene will most probably alter (or obliterate) the reading frame, and that in turn may have drastic phenotypic effects (e.g., Kazazian et al. 1988). Similarly, excision of transposable elements may be imprecise, resulting in the addition or deletion of bases. One such example is of particular historical interest because it concerns the wrinkled pea seeds that were studied by Gregor Mendel. Bhattacharyya et al. (1990, 1993) discovered that the wrinkled variant arose through a mutation at the *rugosus* locus by the insertion of transposon into the reading frame of a gene encoding a starch-branching enzyme (Figure 7.17). Due to the inactivation of the gene, the total amount of starch and the proportion of amylopectin (branched starch) are greatly reduced in homozygotes, while the amount of sucrose is increased. Increased sucrose in seeds causes a greater uptake of water, thereby increasing the seed size. During seed desiccation, these seeds lose more water than seeds from plants possessing a functional starch-branching enzyme, resulting in a wrinkled pea.

There are, however, unexpected effects of transposition on gene expression. For example, a transposable element may contain regulatory elements, such as promoters, which may affect the mode and rate of transcription of nearby genes. For example, the presence of an insertion sequence in the promoter region of the *gal* operon of *E. coli* results in the operon being expressed constitutively, i.e., the regulation of the operon is disrupted (Shapiro 1983). Another example involves retroviruses. The long terminal repeats of retroviruses often contain strong enhancers, which greatly influence the expression of nearby genes. Similarly, *Ty* (which stands for "transposon yeast") elements in *Saccharomyces cerevisiae* are known to increase the expression of downstream genes. This may be beneficial in some specific circumstances, although in most cases the metabolic imbalance produced by such a change is probably detrimental. Transposable elements that contain splice donor or acceptor sites may affect the processing of the primary RNA transcript even if the element has incorporated itself within a noncoding region of a gene, such as an intron. For example, a *MIR* element was found to determine the site of alternative splicing in the human acetylcholine receptor gene (Murnane and Morales 1995). Obviously, the insertion of a transposable element or a retrosequence into an intron or an intergenic region are selectively neutral in the vast majority of cases. For example, there are 13 *Alu* repeats in the introns of the human

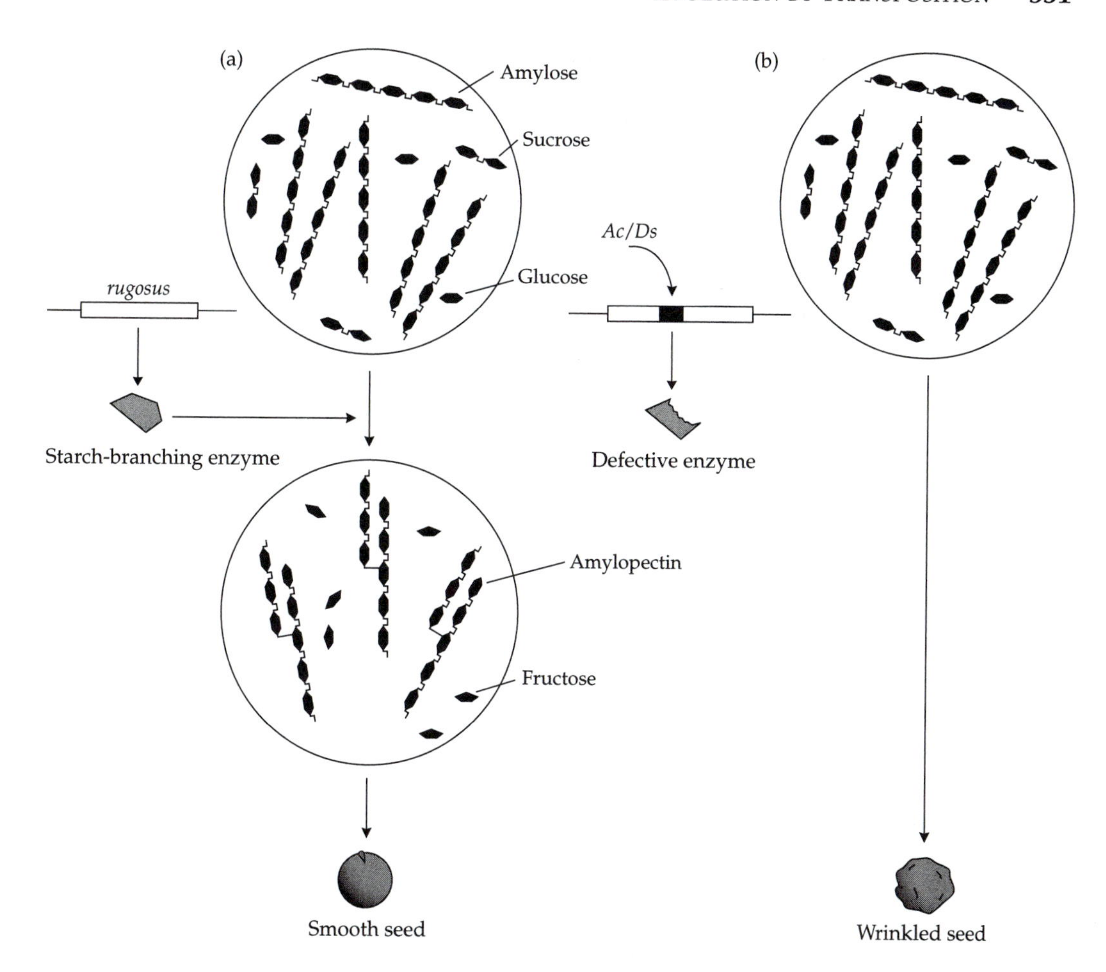

FIGURE 7.17 Mendel's smooth and wrinkled pea seeds. (a) The *rugosus* locus (box) of the pea (*Pisum sativum*) encodes isoform I of the starch-branching enzyme. This enzyme adds branches to starch to create amylopectin. Amylopectin does not absorb much water, so the seeds remain smooth following desiccation. (b) Insertion of an *Ac/Ds* transposable element into *rugosus* disrupts the reading frame and an inactive starch-branching enzyme is produced. The amount of amylopectin is greatly reduced and the amount of sucrose is increased in homozygotes. A greater water uptake increases the size of hydrated seeds, and water loss through desiccation causes the seeds to become wrinkled.

thymidine kinase gene (Fleminton et al. 1987), and yet there is no evidence of deleterious effects.

Third, many transposable elements promote gross genomic rearrangements. Inversions, translocations, duplications, and large deletions and insertions can be mediated by transposable elements. These rearrangements

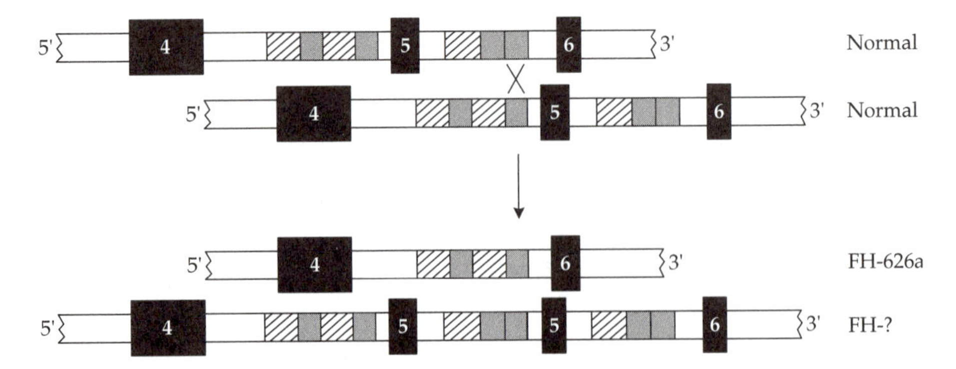

FIGURE 7.18 Unequal crossing over in the low-density lipoprotein receptor gene. Exons are indicated by solid bars and are numbered. *Alu* sequences in the introns are indicated in gray (left arms) and by hatching (right arms); an *Alu* sequence may consist of two arms (dimeric) or one arm (monomeric). The position of the postulated crossover is indicated by the ×. The deleted (observed) and inserted (inferred) products of the recombination are depicted as FH-626a and FH-?, respectively. From Hobbs et al. (1986).

can take place as a direct result of transposition (i.e., by moving a DNA sequence from one genomic location to another), or indirectly if, as a result of transposition, two sequences that previously had little similarity with each other are now sharing a similar transposable element so that unequal crossing over between them is possible. This may occasionally produce a beneficial gene duplication or gene rearrangement. For example, a duplication of the entire growth hormone gene early in human evolution might have occurred via an *Alu-Alu* recombination event (Barsh et al. 1983). Similarly, the duplication that gave rise to the $^{G}\gamma$- and $^{A}\gamma$-globin genes may have resulted from recombination between two *L1* LINEs in an early ancestor of simians (Maeda and Smithies 1986; Fitch et al. 1991). In the majority of cases, however, such an event would constitute a deleterious mutation. Figure 7.18 illustrates how an unequal crossing over event, facilitated by the presence of multiple *Alu* sequences within the introns flanking exon 5 of the gene encoding the low-density lipoprotein receptor, has given rise to a mutant gene lacking this exon (Hobbs et al. 1986). In this gene, the deletion of exon 14 by the same mechanism has also been observed (Lehrman et al. 1986). Patients homozygous for either of these deletions have a high level of cholesterol in the blood (hypercholesterolemia). Recombination between two *Alu* elements has also been shown to be responsible for the deletion of the promoter and the first exon of the adenosine deaminase gene in patients with adenosine deaminase deficiency (Markert et al. 1988). The chromosomal distribution of *Alu* sequences is not uniform (Soriano et al. 1983), and genomic instability has been demonstrated for all regions containing *Alu* repeat sequences (Calabretta et al. 1982).

Fourth, insertion of transposable elements into members of a multigene family can reduce the rate or limit the extent of gene conversion between the members of the family, thereby increasing the rate of divergence between duplicate genes (Hess et al. 1983; Schimenti and Duncan 1984). This may happen because the process of gene conversion involves pairing of homologous DNA (Chapter 6), and nonhomologous regions created by the uneven insertion of transposable elements might reduce the frequency of such pairings and consequently the chance of gene conversion between duplicate genes.

Fifth, there is evidence that some transposable elements may cause an increase in the rate of mutation. For example, strains of *E. coli* that contain the transposable element *Tn10* were found to have elevated rates of insertions (Chao et al. 1983), while the insertion sequence *IS1* increases the probability of deletion in adjacent regions. Under most conditions, this trait will be deleterious to the carrier. However, under severe environmental stress it is possible that an elevated rate of mutation might be advantageous because some mutations may be better suited to the new circumstances and their carriers will have a higher fitness than non-carriers.

Sixth, the presence of a transposable element can turn an otherwise immobile piece of DNA into a mobile one. For example, the presence of *IS1* elements on both sides of a gene may turn a gene into a complex transposon-like entity, and may allow the gene to transpose into plasmids, via which it can be transferred to other cells. Similarly, the spread of SINEs was made possible by reverse transcriptases encoded by transposable elements (see page 347).

Seventh, retrosequences may retain their functionality or even acquire a novel function or a novel form of ontogenic expression (see page 336). Alternatively, they may be inserted in the neighborhood of unrelated functional exons and introns, where they become part of new, chimeric genes. For example, a processed alcohol dehydrogenase pseudogene in *Drosophila* became part of a new four-exon protein-coding gene *jingwei* through its incorporation downstream of three functional exons of unrelated origin* (Long and Langley 1993).

Finally, some sequences in transposable elements may have functions that are unrelated to transposition. For example, a transposable element may contain a sequence motif that is recognized by a certain enzyme. Through replicative transposition, such elements may become "sequence motif donors." Indeed, retrosequences and retroelements were identified as donors of sequence motifs for such biological processes as nucleosome positioning, DNA methylation, transcription enhancement and silencing, polyadenylation signaling, retinoic acid receptors, and RNA splicing, stability, and transport (Brosius and Tiedge 1995; Vansant and Reynolds 1995). The incorporation of such motifs into the genome may even add open-reading frames for novel protein domains in mosaic proteins (Chapter 6). These putative adaptations waiting to be used were called **exaptations** by Gould and Vrba (1982).

*The gene was aptly named *jingwei* after the mythical daughter of the Chinese sun god who, after she drowned, was reincarnated as a bird with a speckled head, a white beak, and red claws.

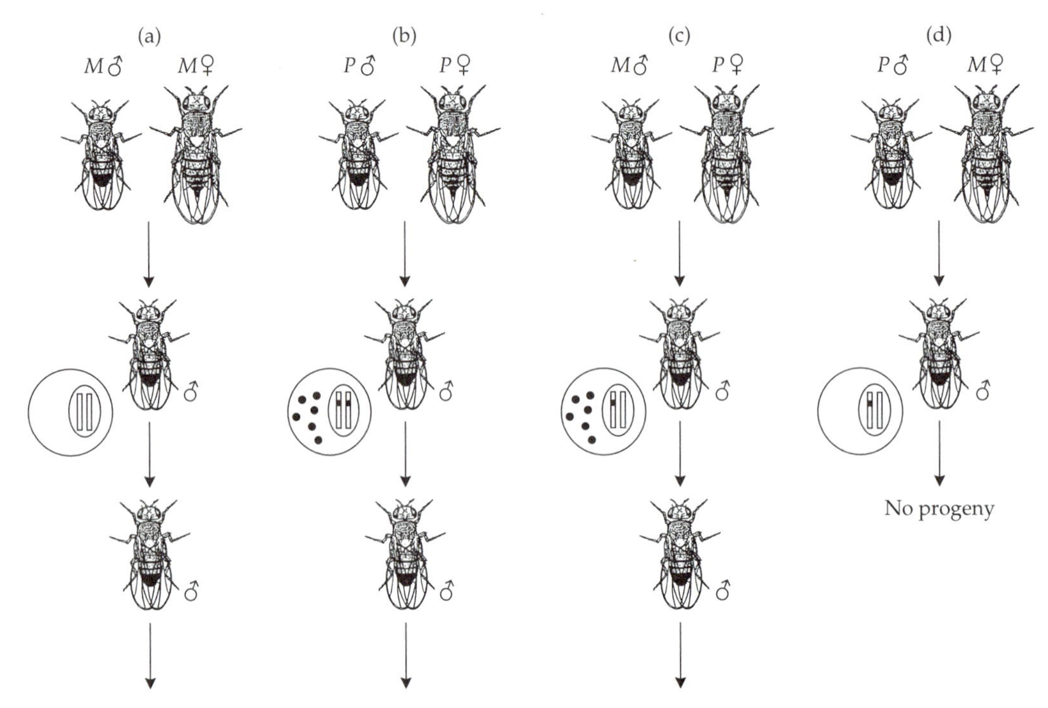

FIGURE 7.19 **Hybrid dysgenesis in *Drosophila*. Matings within *M* strains (a) and matings within *P* strains (b) produce normal progeny. (c) Crosses between *M* males and *P* females also produce normal progeny. (d) Crosses between *P* males and *M* females produce dysgenic offspring, many of which are sterile. The cytotypes are shown schematically on the left-hand side of the F_1 flies. *P*-carrying chromosomes are denoted by a solid box. The maternally inherited cytoplasm may or may not carry *P*-encoded repressors (solid circles). In the absence of such repressors, *P* elements tend to transpose uncontrollably, resulting in dysgenic traits. Such a situation can only occur in *P* male × *M* female matings.**

Hybrid dysgenesis

Hybrid dysgenesis in *Drosophila* is a syndrome of correlated abnormal genetic traits that is spontaneously induced in one type of hybrid between certain mutually interactive strains, but usually not in the reciprocal hybrid (Sved 1976; Kidwell and Kidwell 1976). Hybrid dysgenesis has attracted much attention from molecular and evolutionary biologists because its main manifestation is the creation of a barrier against hybridization between strains or populations, which has been speculated to be a cause of speciation.

There are several dysgenic systems in different *Drosophila* species, some caused by a single transposable element, and others, like the *Helena-Paris-Penelope-Ulysses-Telemac* system in *D. virilis*, caused by "coalitions" of transposable elements (Vieira et al. 1998). In the following we shall deal with just one system, the *P-M* system in *D. melanogaster*.

The asymmetry of hybrid dysgenesis is shown in Figure 7.19. When a male from a *P* strain mates with a female from an *M* strain, the offspring are dysgenic; in the reciprocal mating, the offspring are normal. The dysgenic traits of the *P-M* system include (1) failure of the gonads to develop if the first instar is exposed to temperatures above 27°C, (2) recombination in males (an unnatural occurrence in *Drosophila*, in which recombination is usually restricted to females), (3) chromosomal breakages and rearrangements, (4) distortion of Mendelian transmission proportions (including sex ratios), and (5) high frequencies of lethal and non-lethal mutations, mostly due to chromosomal nondisjunction or the insertion and excision of the mobile elements.

The cause of the *P-M* dysgenesis is a family of transposable elements called *P* elements (Figure 7.20). In *P*-carrying strains, there are 30–50 *P* elements in the genome, although many of them contain deletions and are therefore either non-autonomous or completely inactive. They are distributed throughout all chromosomes, although there may be some subtle chromosomal preferences for *P* element insertion (Kelly et al. 1987; Engels 1989; Ronsseray et al. 1989; Berg and Spradling 1991; O'Hare et al. 1992). *M* strains do not carry *P* elements. The asymmetry of the hybrid dysgenesis system is thought to result from the maternal inheritance of a *P* element-encoded repressor in the cytoplasm of the F_1 progeny of *P* female × *M* male matings, and the absence of such a repressor in the F_1 progeny of the reciprocal *P* male × *M* female matings (Figure 7.19). Dysgenic traits are thought to be due to the uncontrolled transposition of *P* elements in germline cells whose maternally derived cytoplasms lack repressors. In the context of hybrid dysgenesis, the presence or absence of repressors in the cytoplasm defines the type of reaction following the formation of the zygote, and is referred to as the **cytotype**. In the presence of repressors, *P* element transposition may be wholly or partially inhibited.

The transposase-coding region in the *P* element consists of four exons separated from one another by three short introns (each less than 100 bp in length). In germline-cell mRNA, these three introns are excised and a func-

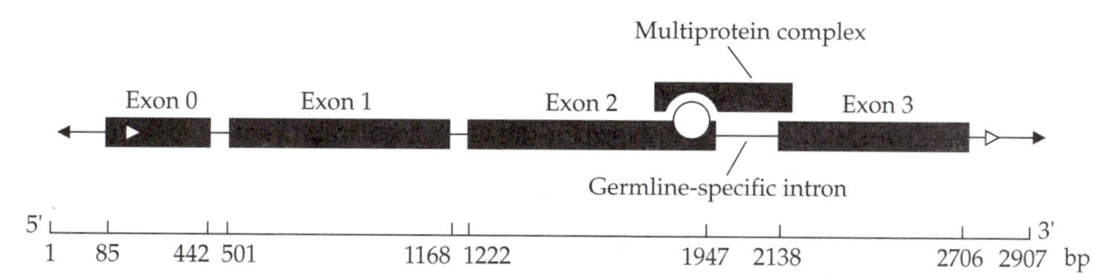

FIGURE 7.20 Schematic structure of a complete 2,907-bp *P* element in *Drosophila melanogaster*. The element is flanked by 31-bp terminal inverted repeats (solid triangles), and by 11-bp subterminal inverted repeats (white triangles). The element contains a single gene encoding a transposase (766 amino acids). The coding region contains four exons (black boxes) interrupted by three spliceosomal introns (lines). Production of the functional protein depends on the splicing of the third intron from the pre-mRNA. The splicing of this intron is prevented in somatic cells by the binding of a multiprotein complex to a site located in exon 2 (white circle). One component of this complex has a very low abundance in germline cells, effectively limiting the production of functional transposase to these cells only (Adams et al. 1997). In somatic cells, a shorter and inactive polypeptide is encoded by the mRNA that retains the third intron (Rio et al. 1986).

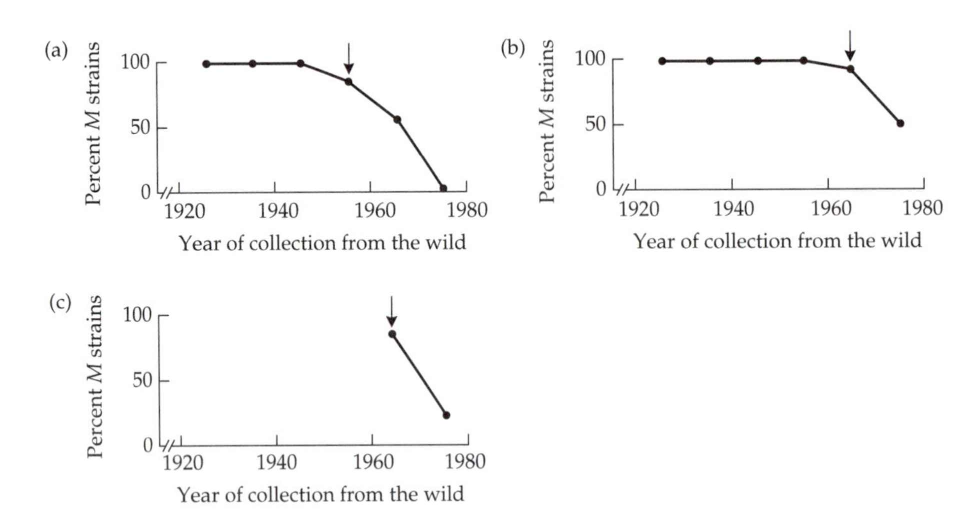

FIGURE 7.21 Temporal changes in the frequencies of *M* strains (i.e., strains devoid of *P* elements) in natural populations in (a) North and South America; (b) Europe, Africa, and the Middle East; and (c) Australia and the Far East. Arrows denote the first appearance of *P* strains. Note that the proportion of *M* strains decreased first in the American continents and only later in the other continents. Modified from Kidwell (1983).

tional transposase is made. In somatic-cell mRNA, the third intron is retained, and as a consequence a truncated protein is produced. This truncated protein acts as a repressor, and hence transposition of *P* elements does not take place in germline cells (Roche et al. 1995).

An interesting observation has been made by Kidwell (1983) concerning the distribution of *P*-carrying strains. *P* characteristics were not found in any *D. melanogaster* strains collected before 1950, and collections made subsequently showed increasing frequencies of *P* with decreasing age (Figure 7.21). A similar observation was made with another dysgenic system, *I-R*. Two hypotheses were suggested to explain this distribution of *P* elements. Engels (1981b) proposed that most strains in nature are of the *P* type but that they tend to lose the transposable elements in the laboratory. The second hypothesis postulated a recent introduction of *P* transposable elements into *D. melanogaster* populations followed by their rapid spread in formerly *M* populations (Kidwell 1979). There are several reasons why the latter hypothesis is favored over the first one. First, *P*-carrying laboratory strains that have been monitored for close to 15 years were never observed to lose their *P* characteristics. Second, there seems to be a geographical cline in the distribution of *P* strains, with American populations showing earlier signs of carrying *P* elements than European, African, and Middle Eastern strains, which in turn seem to have acquired *P* elements before Australasian populations. Lastly, there is now evidence that *P* elements have been recently acquired from a distantly related *Drosophila* species (see page 363).

Transposition and speciation

Speciation or **cladogenesis** (i.e., the creation of two or more species from a parental one) is one of the most important evolutionary processes. Unfortunately, at the molecular level, it is also one of the least understood processes. We know little of the means by which new species arise from old ones. What we do know is that the process of speciation requires the creation of a **reproductive barrier** between two populations belonging to the same species, so that they can no longer interbreed. Hybrid dysgenesis has been thought for a while to represent an early stage in the process of speciation, by acting as a postmating reproductive isolation mechanism between different populations belonging to the same species (e.g., Periquet et al. 1989). Indeed, the sterility of hybrids produced by crosses between the sibling species *D. melanogaster* and *D. simulans* is very similar to that due to dysgenesis (e.g., rudimentary gonads, segregation distortion).

There are problems with this view, however. First, although hybrids exhibit reduced fitness and are therefore partially isolated in terms of reproduction, the transposition of *P* elements in the germline virtually ensures that most of the chromosomes transmitted to the hybrids will bear *P* elements, and the cytotype will eventually change to the *P* type as well. Thus, provided the reduction in fitness in hybrids is not too great, the *P* element will quickly spread through the entire population. Indeed, theoretical considerations indicate that the hybrid must be almost completely sterile for an effective reproduction isolation to last. Second, *P* elements have the ability to transpose themselves horizontally (see page 363) as an infectious agent from individual to individual. Thus, an entire population may be rapidly taken over by *P*, so that hybrid dysgenesis is likely to last for very short periods of time in nature. Indeed, many species belonging to the genus *Drosophila* are known in which all individuals and all populations carry *P* or *P*-like elements, and consequently hybrid dysgenesis does not occur in any of these species. Finally, as far as we know now, hybrid dysgenesis is restricted to *Drosophila* and may not represent a universal phenomenon in nature. Moreover, even in the genus *Drosophila* no evidence has been found for the involvement of mobile elements as barriers to gene flow between sibling species.

Since the discovery of transposable elements, numerous other mechanisms for speciation by transposition have been proposed in the literature. For example, it has been suggested that mass replicative transposition of elements containing regulatory sequences in one population may cause a so-called **genetic resetting** of the genome, whereby many genes will be subject to a novel mode of gene regulation. Such a population would obviously become reproductively isolated from a population that retained the old form of gene regulation. The discovery that some *Alu* repeats contain functional retinoic acid response elements (Vansant and Reynolds 1995), which are known to function as transcription factors, indicates that it is possible to alter the expression of numerous genes through the genomic dispersion of transposable elements.

Another suggestion invoked a mechanism of **mechanical incompatibility**, also caused by mass replicative transposition. In this case, it is assumed that in one population the transposable elements may expand their numbers to such an extent as to cause a significant increase in the size of the chromosomes. A hybrid organism that inherits a set of large chromosomes from one parent and a set of small chromosomes from the other would experience difficulties in chromosome pairing during meiosis, and would most probably be sterile.

Unfortunately, with the exception of some very innovative speculations (e.g., Cohen and Shapiro 1980; Bingham et al. 1982; Ginzburg et al. 1984; Ratner and Vasil'eva 1992; McFadden and Knowles 1997), none of the above speciation models has been substantiated by empirical data.

Evolutionary dynamics of transposable element copy number

There are two opposing views regarding the widespread existence of transposable elements in natural populations. One view maintains that transposable elements are numerous because they confer a selective advantage on their hosts (e.g., Cohen 1976; Never and Saedler 1977; Syvanen 1994). Obvious examples include resistance to antibiotics in bacteria. The other view maintains that transposable elements are intragenomic parasites, or "selfish DNA," and as such they are more likely to be deleterious than beneficial. The spread of transposable elements is therefore explained by their ability to replicate faster than the host genome. Indeed, Hickey (1982) has shown that a high rate of transposition can overcome the effects of even the most stringent purifying selection regime against the spread of transposable elements. In the following, we shall consider the second model quantitatively.

The number of transposable elements within a genome is determined by three factors: (1) u, the probability that a transposable element produces a new genomic copy by replicative transposition, (2) v, the probability that the element is excised, and (3) the intensity of selection against increased numbers of transposable elements within the genome. The values of u and v have been determined experimentally for several transposable elements in populations of *D. melanogaster*. Transposition rates were found to vary among the transposable elements, but on the average were on the order of 10^{-4} per element per generation (Charlesworth et al. 1992). Excision rates were about one order of magnitude lower (Charlesworth and Langley 1989). Consequently, in the absence of selection against the transposable element, the number of copies in the genome is expected to increase indefinitely.

If the number of transposable elements is maintained at an equilibrium—an assumption that may not hold in nature—then selection must operate against an increase in copy number. Indeed, there are indications that viability decreases slightly with copy number (Simmons and Crow 1977). In the simplest deterministic model, we assume that the fitness of an individual, w, decreases with copy number, n. The justification for this assumption is that insertions of transposable elements frequently have deleterious effects, for instance, by altering the expression of adjacent genes. With increasing numbers of transposable elements, the probability of a deleterious alteration of gene expression increases. It can be shown that as long as w decreases with n in a monotonic fashion, then regardless of the exact relationship between n and w, the mean fitness of a population at equilibrium relative to an individual lacking transposable elements is

$$\overline{W} = e^{-n(u-v)} \quad \textbf{(7.1)}$$

(Charlesworth 1985).

In the case of *D. melanogaster*, there are approximately 50 families of transposons, each appearing in the genome on the average 10 times (Finnegan and Fawcett 1986). Thus, $n \approx 500$. Since v is smaller than u by at least one order of magnitude, $u - v \approx u = 10^{-4}$. Solving Equation 7.1, we obtain $\overline{W} = 0.95$. The reduction in fitness is $s = 1 - 0.95 = 0.05$. The stability of the equilibrium given by Equation 7.1 requires the logarithm of the fitness to decline more steeply than linearly with increasing n (Charlesworth 1985). For simplicity, however, if we assume linearity, then the reduction in fitness with each additional transposon is approximately $0.05/500 = 10^{-4}$. Such a small selection coefficient indicates that (1) the selection coefficient needed to control copy number need not be large, and (2) the number of copies of transposable elements within an organism is strongly influenced by random genetic drift. If an organism contains larger numbers of transposable elements, and the possibility exists that even in *Drosophila* the number of transposable elements greatly exceeds 500 (Rubin 1983; Hey 1989), then the effect of a single transposon on fitness might be even smaller than 10^{-4}. Since selection is weak, the effect of random genetic drift is important.

An alternative to selection against increase in copy number would be a mechanism of self-regulated transposition, i.e., a rate of transposition that decreases with copy number or a rate of excision that increases with copy number (Charlesworth 1988; Charlesworth and Langley 1989).

HORIZONTAL GENE TRANSFER

Horizontal gene transfer is defined as the transfer of genetic information from one genome to another, specifically between two species. This term has been coined to distinguish this type of transfer from the usual "vertical" transfer, in which the parental generation passes genetic information on to the progeny. (When genes are transferred from one individual to another within the same species, the process is sometimes referred to as **lateral gene transfer**.)

Horizontal gene transfer among bacteria can occur through one of three possible mechanisms. The first mechanism, **transformation**, involves the uptake of free DNA from the environment. This mechanism does not require a molecular vehicle to transport the genetic information from one organism to another. The other two mechanisms, conjugation and transduction, require a vector or an infectious agent. **Conjugation**, i.e., the direct transfer from one organism to another, requires the help of a plasmid. The cross-cellular transport for **transduction** is provided by a bacteriophage, or bacterial virus. The bacteriophage particle may encapsulate a DNA sequence from the host, and when the particle attaches to another cell, the DNA may be injected into it, eventually becoming part of the host genome. All three mechanisms are DNA-mediated; there are no reports of RNA-mediated cross-species gene transfer in bacteria.

Horizontal gene transfer from a eukaryote to a bacterium may occur by transformation. The same process in the opposite direction probably requires conjugation or transduction. It is not clear how horizontal gene transfer between eukaryotes can be accomplished. Transformation and conjugation are most certainly out of the question, because it is difficult to imagine how these processes could transfer genetic information into the germline. Transduction is imaginable, although the evidence for such a process in eukaryotes is nonexistent. Most probably, horizontal gene transfer between eukaryotes is accomplished through retroviruses, which are capable of both incorporating foreign genetic information into their genomes and crossing species boundaries (Benveniste and Todaro 1976; Bishop 1981).

An organism into which genetic information from a different organism has been incorporated as a stable part of its genome is called a **transgenic organism**. Sequence homology due to horizontal gene transfer is called **xenology**, as opposed to paralogy and orthology, which are due to gene duplication and speciation, respectively.

One may be able to detect a horizontal gene transfer event through the discovery of an outstanding discontinuity in the phylogenetic distribution of a certain gene. One notable example of this so-called **patchy distribution** is the *GlnRS* gene in *E. coli* and related enterobacterial species. An important step in protein synthesis is the aminoacylation of tRNAs by aminoacyl-tRNA synthetases, a process catalyzed by 20 enzymes, one for each amino acid. Interestingly, the formation of Gln-$tRNA^{Gln}$ can be accomplished by two different pathways: aminoacylation of $tRNA^{Gln}$ with Gln by glutaminyl-tRNA synthetase (encoded by the *GlnRS* gene), or transamidation of Glu from Glu-$tRNA^{Gln}$ that has been mischarged by glutamyl-tRNA synthetase (encoded by *GluRS*). The former pathway is common in eukaryotes, while the latter is widespread among bacteria and organelles, which correspondingly lack *GlnRS*. However, a few closely related enterobacterial species, including *E. coli*, do possess *GlnRS*. Such a conspicuous taxonomic discontinuity can only be explained by a horizontal gene transfer from eukaryotes to the ancestor of these enterobacteria, a conclusion strongly supported by several phylogenetic analyses (Lamour et al. 1994; Gagnon et al. 1996).

Horizontal gene transfer may also be suspected when a notable discrepancy between gene phylogeny and species phylogeny is discovered, in particular when sequence similarity seems to reflect geographical proximity rather than phylogenetic affinity. Consider, for example, the phylogenetic tree in Figure 7.22a. Let us assume that a piece of DNA was transferred from species B to species C after the divergence of A from B. On the basis of sequence comparisons involving any gene other than the horizontally transferred one, we expect to be able to reconstruct the correct phylogenetic relationships among

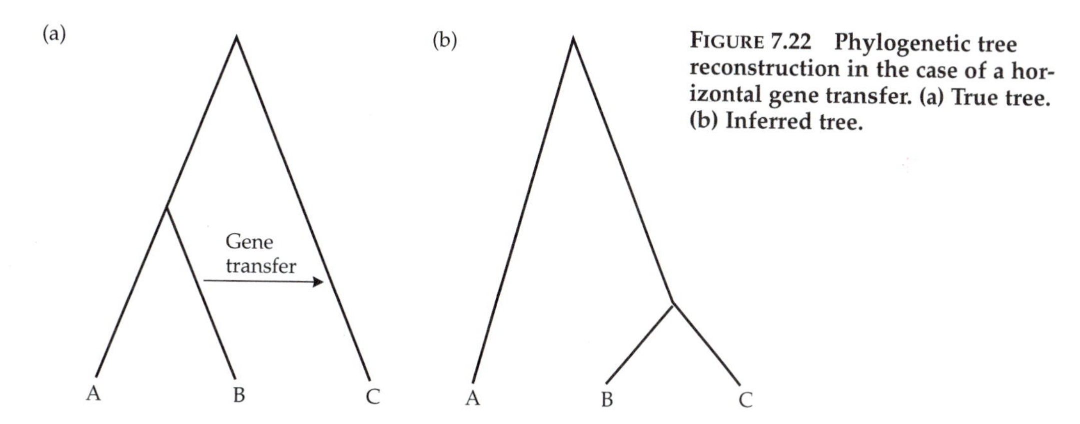

FIGURE 7.22 Phylogenetic tree reconstruction in the case of a horizontal gene transfer. (a) True tree. (b) Inferred tree.

the species. In contrast, if we use the horizontally transferred piece of DNA, we obtain the erroneous tree in Figure 7.22b. On the basis of such considerations, Moens et al. (1996) suggested that two horizontal gene transfers occurred during globin evolution in nonvertebrate species, once from the common ancestor of the ciliates (e.g., *Paramecium* and *Tetrahymena*) and green algae (e.g. *Chlamydomonas*) to the ancestor of cyanobacteria, and once from the ancestor of budding yeasts (e.g., *Saccharomyces* and *Candida*) to the ancestor of the β subdivision (e.g. *Vitreoscilla* and *Alcaligenes*) and the γ subdivision (e.g., *Escherichia* and *Klebsiella*) of proteobacteria. We note, however, that many factors other than horizontal gene transfer may conspire to yield a discrepancy between the species tree and the gene tree (Chapter 5), and therefore additional evidence is required for inferring a horizontal transfer event.

Two types of sequences can be transferred horizontally: (1) sequences derived from transposable elements; and (2) genomic sequences. Horizontal gene transfer of functional genomic sequences is much rarer than that of transposable elements.

Horizontal transfer of virogenes from baboons to cats

The vertebrate genome contains many sequences that are homologous to retroviruses. Such sequences, which are normal constituents of the nuclear DNA of eukaryotes, are called **endogeneous retroviral sequences** or **virogenes**. There are several examples of endogeneous retroviral sequences being transferred between vertebrate species (reviewed in Benveniste 1985). One such example involves a type-C virogene from baboons (Figure 7.23).

Sequences homologous to the baboon virogene have been detected in the cellular DNA of all Old World monkeys. The sequence similarity between them is closely correlated with the phylogenetic relationship among the species. Thus, the virogene has existed for at least 30 million years in the genomes of primates. Interestingly, six cat species closely related to the domestic cat (*Felis catus*) also contain this sequence, although it is present in nei-

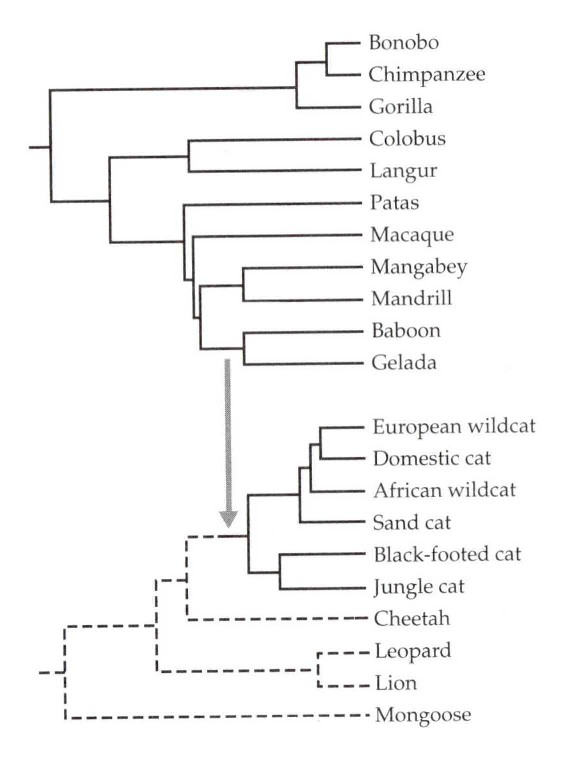

FIGURE 7.23 Phylogenetic trees for Old World monkeys (Catarrhini) and cats (Felidae). The branches leading to taxa that contain the type-C virogene are shown as solid lines. The type-C virogenes from the baboon and the gelada exhibit the highest similarity to the feline type-C virogenes. Therefore, a horizontal gene transfer probably occurred about 10 million years ago, from the ancestor of the baboon and the gelada to the ancestor of the domestic cat group after the divergence of lion, leopard, and cheetah lineages. Species and genera: bonobo, *Pan paniscus*; chimpanzee, *Pan troglodytes*; gorilla, *Gorilla gorilla*; colobus, *Colobus*; langur, *Presbytis*; patas, *Cercopithecus patas*; macaque, *Macaca*; mangabey, *Cercocebus*; mandrill, *Mandrillus*; baboon, *Papio*; gelada, *Theropithecus gelada*; European wildcat, *Felis sylvestris*; domestic cat, *F. catus*; African wildcat, *F. libyca*; sand cat, *F. margarita*; black-footed cat, *F. nigripes*; jungle cat, *F. chaus*; cheetah, *Acinoyx jubatus*; leopard, *Panthera pardus*; lion, *P. leo*; mongoose, *Galidia elegans*. Data from Benveniste (1985), Disotell et al. (1992), and Johnson and O'Brien (1997).

ther more distantly related Felidae, such as lions, leopards, and cheetahs, nor in any other carnivores. It is thus highly probable that this sequence was horizontally transferred between species some time in the past.

The date and direction of the horizontal transmission can be deduced from two types of data: sequence similarity and paleogeographical information. All cat species that contain the baboon virogene are from the Mediterranean area, while Felidae from Southeast Asia, the New World, and Africa lack the sequence. Therefore, the transfer must have occurred after the major radiation of the Felidae and was limited to one zoogeographical area. This conclusion points to a date of about 5–10 million years ago for the horizontal gene transfer event. The direction of transmission can be deduced by considering the distribution of the sequence among primates, on the one hand, and cats, on the other. Since all Old World monkeys possess the virogene, while only a few felid species possess it, it is reasonable to assume that the cats acquired the sequence from the baboons and not vice versa. This conclusion is strengthened by considering that the virogene in cats is more similar to that in three species of baboons (*Papio cynocephalus*, *P. papio*, and *P. hamadryas*) and the closely related gelada (*Theropithecus gelada*) than to that of any other primate sequence. Therefore, the sequence must have been transferred to cats from the ancestor of the baboons and the gelada shortly after their divergence from the mandrill–mangabey clade (Figure 7.23). The date derived from the Old World monkey species tree agrees well with the date derived from the cats.

Horizontal transfer of P elements between Drosophila species

Another example of horizontal gene transfer involves the *P* elements in *D. melanogaster*. As mentioned previously, *P* elements have spread rapidly through natural populations of *D. melanogaster* within the last five decades (see page 356). With the possible exception of the little-studied *Drosophila tsacasi*, *P* elements could not be detected in any of the hundreds of closely related *melanogaster* group species, such as *D. mauritania*, *D. séchellia*, *D. simulans*, and *D. yakuba* (Daniels et al. 1990). Where, then, did these elements come from?

Interestingly, all species of the distantly related *willistoni* and *saltans* groups were found to contain *P* and *P*-like elements. Moreover, drosophilid genera such as *Scaptomyza*, and even nondrosophilid genera such as *Lucilia*, were also found to contain *P*-like elements. In particular, the *P* element from *D. willistoni* was found to be identical to the one in *D. melanogaster* with the exception of a single nucleotide substitution, indicating that *D. willistoni* may have indeed served as the donor species in the horizontal gene transfer of *P* elements to *D. melanogaster*.

There are several reasons to suspect that this horizontal gene transfer occurred quite recently. First, the near-identity between the *P* sequences from *D. melanogaster* and *D. willistoni* suggests a very short time of divergence. Second, the near-absence of genetic variability in the *P* sequences from *D. melanogaster* from even very distant geographic locations indicates that the time since the introduction of *P* elements into *D. melanogaster* is too short for genetic variability to have accumulated. Finally, the geographical pattern of appearance of *P* elements in *D. melanogaster*, with populations in the American continents acquiring it first (Figure 7.21), suggests a very recent invasion, probably within the last 50 years (Kidwell 1983; Anxolabéhère et al. 1988; Gamo et al. 1990).

Clark et al. (1994) and Clark and Kidwell (1997) conducted a phylogenetic analysis of *P* elements from *Drosophila* and other dipteran species. The phylogenetic trees obtained from these analyses were found to be incongruent with the species tree that has been reliably inferred from molecular, morphological,

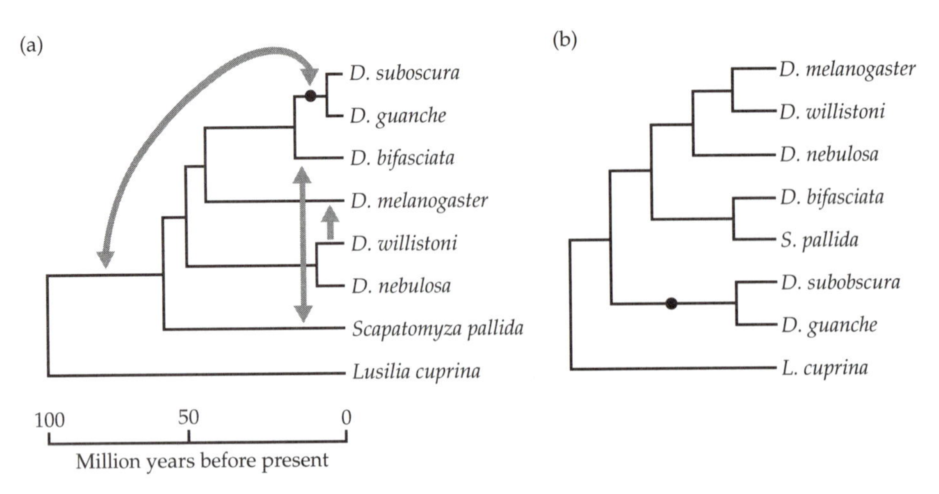

FIGURE 7.24 Schematic comparison of a species tree for *Drosophila* and related taxa (a) with a gene tree derived from *P* element sequences (b). With the exception of the *subobscura–guanche* clade (black dot), the two trees are completely incongruent. To resolve the incongruence, several horizontal gene transfers had to be assumed. Only a few of these transfers could be localized on the branches of the species tree. The only horizontal transfer whose direction could be unambiguously inferred, i.e., from *D. willistoni* to *D. melanogaster*, is marked with a single-headed arrow. The direction of the horizontal gene transfer event involving *D. bifasciata* and *Scaptomyza pallida* could not be determined (double-headed arrow). Some putative transfers proposed by Clark et al. (1994) defy temporal constraints, i.e., the transfers are assumed to have occurred between two biological entities that could not have coexisted at the same time. One such example involving the ancestor of *D. subobscura* and *D. guanche*, on the one hand, and the ancestor of *Drosophila* and *Scaptomyza*, on the other, is shown (curved double-headed arrow). Data from Clark et al. (1994).

and paleontological data (Figure 7.24). Numerous instances of horizontal gene transfer had to be postulated to restore congruence between the two trees. However, in many cases, the horizontal transfers must have been quite ancient, such that they could not be exactly localized to particular internal branches in the species tree. Furthermore, in some cases "impossible" horizontal gene transfer events had to be assumed. For instance, in one of the scenarios presented by Clark et al. (1994), a transfer is assumed to have occurred between the ancestor of all drosophilds to the ancestor of *Drosophila subobscura* and *D. guanche*, which under no circumstances could have coexisted at the same time (Figure 7.24).

Houck et al. (1991) identified the semiparasitic mite *Proctolaelaps regalis* as a potential suspect in the transfer of *P* elements from *willistoni* to *melanogaster*. Although *P. regalis* is not the most common mite co-occurring with *Drosophila*, several of its anatomical, behavioral, and ecological attributes (such as its peripatetic mode of fluid-feeding on the immature stages of flies) and its geographic overlap with North American *Drosophila* species make it a very suitable putative vector in the transfer. Indeed, *P. regalis* possesses the capacity to acquire *P* elements from *P*-carrying strains of *Drosophila*.

Houck et al. (1991) have also defined a minimum number of conditions that have to be satisfied in order for this horizontal gene transfer to take place. These are: (1) Two *Drosophila* females from the donor and the recipient species must lay their eggs in proximity to one another; (2) The recipient egg must be less than 3 hours old, i.e., before the 512-cell stage; (3) The germline of the recipient embryo must incorporate a complete copy of a *P* element DNA before it degrades in the cytoplasm; and (4) The receiving embryo must survive the injuries inflicted by the mite. If, as seems likely, each of these events has a low independent probability in nature, then the combined multiplicative probability must be extremely low, and the interspecific horizontal gene transfer of *P* elements probably occurs only very rarely.

Promiscuous DNA

Of all organisms, plants have the most complex cells. Each plant cell has three autonomous genomic systems: nuclear, chloroplastic, and mitochondrial. The chloroplasts and the mitochondria derive from free-living cells that linked up with the ancestor of the nuclear genome to form an endosymbiotic relationship (Chapter 5), in which the nuclear part became the dominant genetic component. Organelle genomes were suspected for a long time to have lost many of their original genes or to have had them transferred to the nuclear DNA during the long evolutionary time since the beginning of the endosymbiotic relationship. However, until the 1980s, evidence for such transfer was lacking, and the three genetic systems were thought to be largely discrete. Stern and Lonsdale (1982) identified a 12-Kb region in maize mitochondria, specifying 16S rRNA, $tRNA^{Ile}$, and $tRNA^{Val}$, which was similar in sequence to the inverted repeat region in the chloroplast genome. Subsequently, maize mitochondria were also found to contain a protein-coding gene of chloroplastic origin, a ribulose-1,5-biphosphate carboxylase-coding gene (Lonsdale et al. 1983). This type of "disrespect" for genomic barriers has been dubbed **promiscuous DNA** by two prominent science commentators (Lewin 1982; Ellis 1982).

Promiscuous DNA provides evidence for gene flow between organelles, and between organelles and the nucleus. To date, examples have been found for five out of the six possible types of gene transfer among genomes: chloroplast to mitochondria, mitochondria to chloroplast, mitochondria to nucleus, nucleus to mitochondria, and chloroplast to nucleus (Thorsness and Weber 1996).

Gene transfer from the organelles to the nucleus most probably occurs by replicative retroposition (Schuster and Brennicke 1987). Direct evidence for this statement is provided by the *rpl22* gene that encodes the chloroplast ribosomal protein L22. An *rpl22* gene is located in the chloroplast genome of all plants with the exception of legumes (Fabaceae), in which the functional *rpl22* gene is in the nuclear genome. Phylogenetic analysis indicated that the *rpl22* gene was transferred to the nucleus in the common ancestor of all flowering plants at least 100 million years before its loss from the chloroplast in the legume lineage (Gantt et al. 1991).

FURTHER READINGS

Berg, D. E. and M. M. Howe (eds.). 1989. *Mobile DNA*. American Society for Microbiology, Washington, DC.

Capy, P., T. Langin, and D. Anxolabéhère. 1998. *Dynamics and Evolution of Transposable Elements*. Chapman & Hall, New York.

Charlesworth, B. and C. H. Langley. 1989. The population genetics of *Drosophila* transposable elements. Annu. Rev. Genet. 23: 251–287.

Galun, E. and A. Breiman. 1997. *Transgenic Plants*. Imperial College Press, London.

Kazazian, H. H. and J. V. Moran. 1998. The impact of *L1* retrotransposons on the human genome. Nat. Genet. 19: 19–24.

Kempken, F. and U. Kuck. Transposons in filamentous fungi—facts and perspectives. BioEssays 20: 652–659.

Kidwell, M. G. 1993. Lateral transfer in natural populations of eukaryotes. Annu. Rev. Genet. 27: 235–256.

Leib-Mosch, C. and W. Seifarth. 1995. Evolution and biological significance of human retroelements. Virus Genes 11: 133–145.

Mel, S. F. and J. J. Mekalanos. 1996. Modulation of horizontal gene transfer in pathogenic bacteria by *in vivo* signals. Cell 87: 795–798.

Okada, N., M. Hamada, I. Ogiwara, and K. Ohshima. 1997. SINEs and LINEs share common 3′ sequences: A review. Gene 205: 229–243.

Rice, S. A. and B. C. Lampson. 1995. Bacterial reverse transcriptase and msDNA. Virus Genes 11: 95–104.

Smit, A. F. A. 1996. The origin of interspersed repeats in the human genome. Curr. Opin. Genet. Develop. 6: 743–748.

Syvanen, M. and C. I. Kado (eds.). 1998. *Horizontal Gene Transfer*. Chapman & Hall, London.

Chapter 8

Genome Evolution

Until quite recently, genomes could only be studied indirectly, by using partial and sometimes unrepresentative genomic sequences. This situation is changing rapidly as complete genomic sequences are becoming available. Organelle genomes were the first to be sequenced; the first complete mitochondrial sequence (~17,000 bp) was published in 1981, and the first chloroplast genome (~156,000 bp) in 1986. The first complete genome sequence of a free-living organism, the eubacterium *Haemophilus influenzae* (~1,830,000 bp), was completed in 1995, followed in quick succession by the complete sequences of an archaeon, *Methanococcus jannaschii* (~1,660,000 bp), and all 16 chromosomes of a unicellular yeast, *Saccharomyces cerevisiae* (~12,000,000 bp). The first complete genome of a multicellular organism, the nematode *Caenorhabditis elegans* (~97,000,000 bp), was reported in 1998, and the genome projects for *Drosophila melanogaster*, human, mouse, rice, and maize are expected to be completed in the very near future.* We are privileged, therefore, to be able to study genome evolution in a straightforward manner by using genomic sequences.

Our discussion of genome evolution includes five different topics. The first is genome size, which varies enormously among organisms. How did this variation come into existence, and what mechanisms can increase or decrease genome size to produce such variation? The second topic is the genetic information included within genomes. Do genomes contain mostly genic DNA, or is the genome made of mostly nongenic sequences? Does the nongenic fraction have a function, or is it merely "junk"? Are there many repetitive sequences in the genome, and if so, what is their function and pattern of chromosomal distribution? The third topic concerns gene order and the dynamics

*Updated compilations of genomic sequences may be found at www.ncbi.nlm.nih.gov/entrez/genome/org.html

of evolutionary change in gene order. How are genes distributed along and among the chromosomes? What mechanisms are responsible for the reshaping of gene order during evolution? The fourth topic concerns the nucleotide composition of the genome. Is there heterogeneity in composition among different regions of the genome? What mechanisms can give rise to localized differences in nucleotide composition? Finally, we deal with the evolution of the genetic code. How can the rules of translation change without deleterious effects, and under what conditions?

Because prokaryotes and eukaryotes exhibit such distinct genomic structures, we shall discuss them separately.

C VALUES

In haploid organisms such as bacteria, the **genome size** refers to the total amount of DNA in the genome. In diploid or polyploid organisms, the genome size is defined as the amount of DNA in the unreplicated haploid genome, such as that in the sperm nucleus. The genome size is also called the **C value**, where C stands for "constant" or "characteristic" to denote the fact that the size of the haploid genome shows little intraspecific variability, i.e., it is fairly constant within any one species. In contrast, C values vary widely from species to species among both prokaryotes and eukaryotes.

The sizes of nuclear genomes in eukaryotes are usually measured in picograms (pg) of DNA (1 pg = 10^{-12} g). The smaller prokaryotic genomes are more commonly measured in daltons, the unit of relative atomic or molecular mass. The sizes of the still smaller genomes, such as those of organelles and viruses, as well as the sizes of specific stretches of DNA, are most often expressed in base pairs (bp) or kilobase pairs (Kb) of double-stranded DNA or RNA (1 Kb = 1,000 bp). Completely sequenced genomes are customarily measured in megabase pairs (1 Mb = 1,000 Kb). To avoid confusion, we shall only use bp and Kb. The conversion factors are provided in Table 8.1.

THE EVOLUTION OF GENOME SIZE IN PROKARYOTES

Bacterial genome sizes vary over a 20- to 30-fold range, from a little under 6×10^5 bp in some obligatory intracellular parasites, to more than 10^7 bp in several cyanobacteria species (Table 8.2). Mollicutes, which lack a cell wall and are the smallest free-living prokaryotes capable of self-reproduction, generally have very small genomes. (Class Mollicutes consists of six genera, among which *Mycoplasma* is the most famous. In fact, the term *Mycoplasma* has frequently been used to denote all mollicute species.)

The smallest known genome is that of the urogenital pathogen *Mycoplasma genitalium*, which contains about 470 protein-coding genes, 3 rRNA-specifying genes, and 33 tRNA-specifying genes. The genetic information contained within the genome of *M. genitalium* is believed to be slightly larger than the

TABLE 8.1 Conversion of units commonly used to measure genome sizes

	Conversion factor		
Unit	**Picograms**	**Daltons**	**Base pairs**
Picogram	1	6.02×10^{11}	0.98×10^{9}
Dalton	1.66×10^{-12}	1	1.62×10^{-3}
Base pair	1.02×10^{-9}	618	1

minimum required to support autonomous life (see page 371). The numbers of genes in other bacteria range very roughly from about 500 to 8,000 (approximately a 20-fold range). In other words, the variation in the number of genes is about the same as the variation in C values (Herdman 1985).

Since the average size of protein-coding genes in bacteria is about 1 Kb, the size of the genic fraction of the genome is estimated to vary from 500 Kb to about 10^4 Kb. We can therefore conclude that prokaryotes do not contain large quantities of nongenic DNA. Indeed, in the vast majority of completely sequenced bacterial species, protein-coding regions take up 87–94% of the genome, so that the nongenic fraction seems to be rather small. The only exception to date is the genome of the intracellular parasite *Rickettsia prowazekii*, which contains 24% noncoding DNA (Andersson et al. 1998). For the eubacteria that have been completely sequenced, it is possible to compute the correlation between genome size and gene number (Figure 8.1). The almost perfect correlation indicates that the variation in genome size in bacteria can be wholly explained by gene number. The same correlation is seen in Archaea, but the data are currently too limited to draw definite conclusions.

The genomes of bacteria can be divided into three fractions: (1) chromosomal DNA, (2) DNA that originated in plasmids, and (3) transposable elements (Hartl et al. 1986). The chromosomal fraction contains protein-coding genes required for growth and metabolic functions (90–95%), spacers and various signals (~5%), RNA-specifying genes (~1%), and a number of repeated se-

TABLE 8.2 Range of C values in prokaryotes

Taxon	Genome size range (Kb)	Ratio (highest/lowest)
Bacteria	580–13,200	23
Mollicutes	580–2,200	4
Gram negatives[a]	650–9,500	15
Gram positives (Firmicutes)	1,600–11,600	7
Cyanobacteria	3,100–13,200	4
Archaea	1,600–4,100	3

Data from Cavalier-Smith (1985), Römling et al. (1992), Carle et al. (1995), and other sources.
[a]Most probably a paraphyletic group.

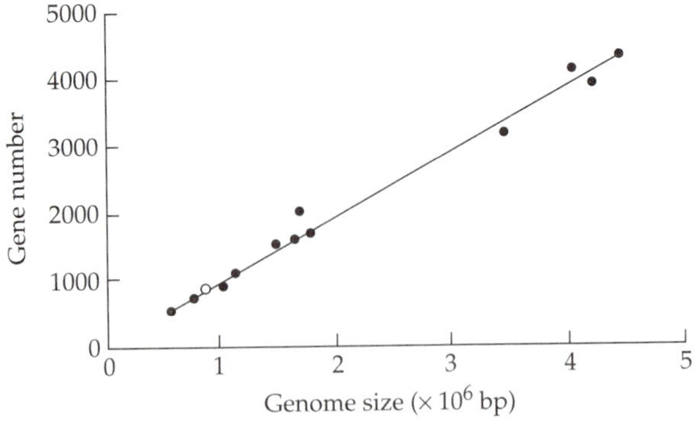

FIGURE 8.1 Relationship between gene number and genome size in twelve completely sequenced eubacterial species with circular genomes (solid circles) and one with a linear genome (empty circle).

quences, usually on the order of a few dozen base pairs in length. Some bacteria may carry plasmids as extrachromosomal genetic elements. In some instances, however, genes derived from plasmids are found integrated in the bacterial chromosome (Davey and Reanney 1980). Transposable elements are common components of the bacterial genome. For example, wild strains of *Escherichia coli* contain 1–10 copies of at least six different types of insertion sequences (Chapter 7). The nongenic fraction of the genome (including insertion sequences, as well as plasmid and bacteriophage-derived genes) seems to be one order of magnitude smaller than the chromosomal fraction. Interestingly, in all bacterial species whose complete genome has been sequenced, we also find evidence for functional genes acquired through horizontal gene transfer (Chapter 7). In many cases, horizontal gene transfer has been inferred through regional peculiarities in GC content and codon usage (Groisman et al. 1992; Lan and Reeves 1996).

The distribution of genome sizes in bacteria is discontinuous, showing major peaks with modal values of about 0.8×10^6, 1.6×10^6, and 4.0×10^6 bp, and several minor peaks at 7.2×10^6 and 8.0×10^6 bp (Herdman 1985). This distribution led Riley and her colleagues to suggest that the larger genomes of such organisms as *E. coli* could have evolved from smaller genomes by successive cycles of genome duplication (Zipkas and Riley 1975). However, as more data on genome sizes accumulated, the peaks in the distribution tended to disappear as the gaps in the distribution were filled in. Indeed, in a series of more recent studies, Labedan and Riley (1995a,b) and Riley and Labedan (1997) found no evidence for genome duplication in the evolutionary history of *E. coli*. At present, only the Gram-negative bacteria exhibit a discontinuous distribution (Trevors 1996), but that may be an artifact of the paraphyly of this group.

Since there seems to be no notable relationship between genome size and bacterial phylogeny, it has been suggested that increases in genome size have occurred frequently in the evolution of bacterial lineages (Wallace and Morowitz 1973). Using a tentative phylogeny of bacteria based on comparisons of rRNA sequences, Herdman (1985) was able to relate changes in genome size to phylogenetic history. His results indicated that increases in genome size occurred independently in different bacterial lineages. Interestingly, many of the major increases in genome size seem to have occurred coincidentally in several bacterial lineages at a rather specific time in the evolutionary history of the planet—to wit, soon after the appearance of appreciable quantities of oxygen in the atmosphere, approximately 1.8 billion years ago (see Appendix I).

The distribution of genome sizes in bacteria can be explained by a combination of several processes: (1) many independent gene and operon duplications, (2) small-scale deletions and insertions, (3) duplicative transposition, (4) horizontal transfer of genes derived mainly from plasmids and bacteriophages, but also from other species, and (5) loss of massive chunks of DNA in many parasitic lines (see page 375).

THE MINIMAL GENOME

The search for the genome of the "smallest autonomous self-replicating entity" was begun in the late 1950s by Morowitz and coworkers (see review in Morowitz 1984). This led to studies on the Mollicutes, which were found to be the cellular organisms with the smallest genomes and the smallest number of genes in nature. There is no evidence, however, that the 468 protein-coding genes in *M. genitalium* actually represent the minimal requirement for sustaining life. It is possible that a certain degree of genetic redundancy exists even in the most streamlined genome. In the following we shall describe two approaches for inferring the minimal gene set for cellular life.

The analytical approach

The rationale behind the analytical method of Koonin and Mushegian (1996) and Mushegian and Koonin (1996a) is quite straightforward. The initial estimate of the minimal gene complement is made by identifying the set of all orthologous genes that are common to a group of organisms. One such example, concerning the comparison of the proteomes of *E. coli*, *H. influenzae*, and *M. genitalium*, is shown in Figure 8.2. From this comparison, the approximate minimal set was inferred to include 239 genes.

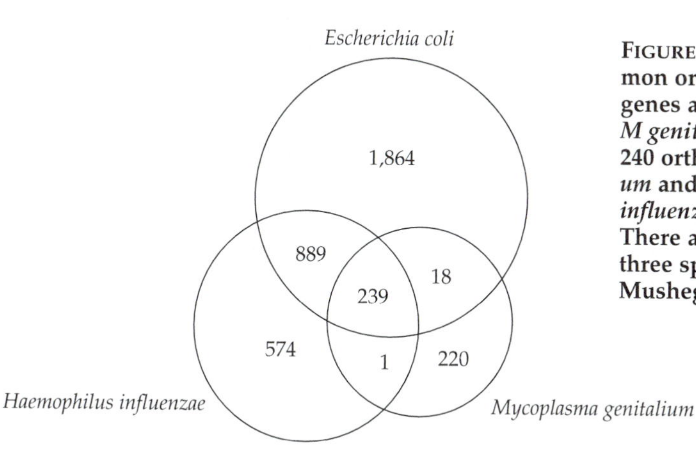

FIGURE 8.2 Venn diagram of common orthologous protein-coding genes among three bacterial species. *M genitalium* and *H. influenzae* have 240 orthologs in common, *M. genitalium* and *E. coli* have 257, and *H. influenzae* and *E. coli* have 1,128. There are 239 orthologs common to all three species. Data from Koonin and Mushegian (1996).

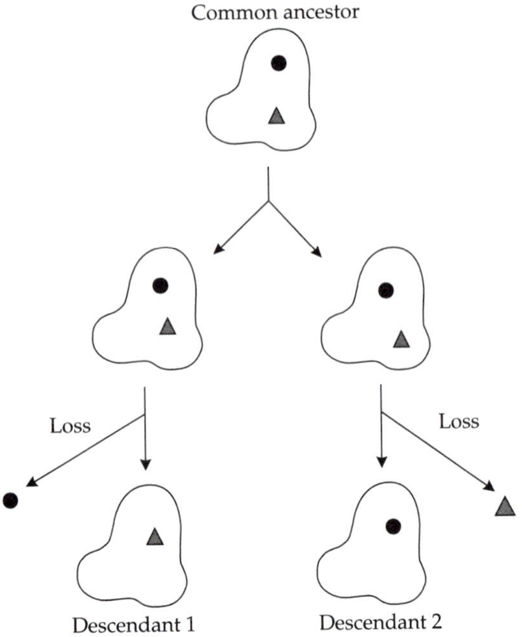

FIGURE 8.3 A scenario of differential gene loss for nonorthologous gene displacement. The common ancestor had two proteins (circle and triangle) performing similar functions. The gene encoding one of them was lost in descendant 1, while the other was lost in descendant 2. The result is functional convergence. Modified from Koonin and Mushegian (1996).

However, in addition to these protein-coding genes, other vital genes must be included in the minimal set. These genes cannot be identified in the first step of the analysis because of the phenomenon of **nonorthologous gene displacement**, which is a form of functional convergence brought about by the use of unrelated proteins for performing the same vital function (Figure 8.3). For example, the function of the glycolytic enzyme phosphoglycerate mutase is performed in different bacteria by two proteins that are unrelated to each other. One is encoded by the *gpm* gene and is 2,3-biphosphoglycerate-dependent, the other is encoded by *yibO* and is 2,3-biphosphoglycerate-independent. In *M. genitalium* the phosphoglycerate mutase function is performed by the *yibO* gene product, whereas in *H. influenzae* the same function is performed by the protein encoded by the *gpm* gene. Because the two phosphoglycerate mutases are unrelated in sequence to each other, the intersection of the two proteome sets would contain neither of them, although their common catalytic function is probably indispensable for life. About two dozen genes involved in such nonorthologous gene displacement were discovered, and these were added to the initial minimal set.

Finally, genes that appear to be specific for parasitic bacteria or to represent functional redundancy were removed, resulting in a bacterial version of a minimal gene set of 256 genes.

From this approach, the minimal gene set was found to include: (1) a nearly complete system of translation; (2) a nearly complete DNA replication machinery; (3) a rudimentary set of genes for recombination and DNA repair; (4) a transcription apparatus consisting of four RNA polymerase units; (5) a large set of chaperone-like proteins; (6) a few protein-coding genes involved in anaerobic metabolism; (7) several genes encoding enzymes for lipid and cofactor biosynthesis; (8) several transmembrane transport proteins; and (9) a set of 18 proteins of unknown function. This minimal set is notable in that it does not contain the necessary machinery for the biosynthesis of amino acids and nucleotides, which presumably must have been procured "ready-made" from the environment.

We note that the smallest known genome in nature (*M. genitalium*) is almost twice the size of the estimate for the minimal genome (Maniloff 1996).

The experimental approach

An elegant experimental approach to the minimal-genome problem was taken by Itaya (1995). Seventy-nine randomly selected protein-coding loci in the Gram-positive bacterium *Bacillus subtilis* were knocked out by mutagenesis (Figure 8.4). Mutations at only six of these loci rendered *B. subtilis* unable to grow and form colonies, while mutants at the rest of the 73 loci retained their ability to multiply. Only three out of the six protein-encoding loci were identified unambiguously in terms of their function. These were *dnaA* and *dnaB*, which are involved in the initiation of DNA replication, and *rpoD*, whose product takes part in RNA synthesis.

To make sure that knocked-out genes that did not affect growth are not redundant members of multigene families, Itaya (1995) also constructed bacteria with multiple mutations. Interestingly, even when 33 loci were incapacitated simultaneously, the bacterium and its progeny retained their ability to form colonies. Thus, 73 out of 79 genes were inferred to be truly dispensable, while only about 7.5% of the genome was deemed indispensable. Given that the length of the genome of *B. subtilis* is 4.2×10^6 bp, and assuming that the ge-

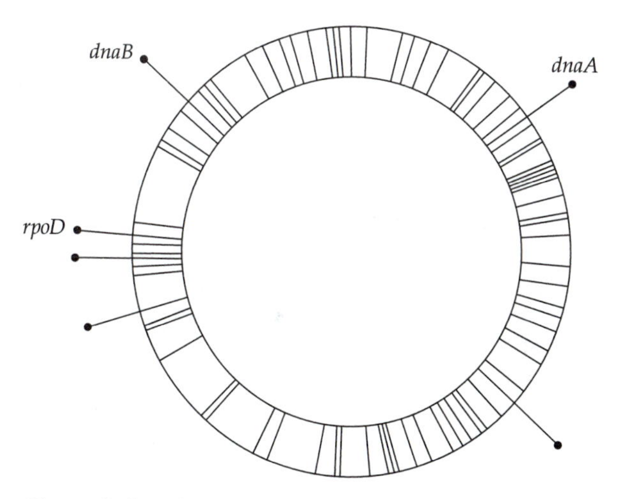

FIGURE 8.4 Genomic locations of the 79 randomly chosen loci (lines) in *Bacillus subtilis* that have been knocked out by mutagenesis. The six solid circles indicate indispensable loci, of which three are identified. Data from Itaya (1995).

nomic ratio of indispensable to dispensable genes is the same as that in the sample, the length of the indispensable genome was estimated to be $4.2 \times 10^6 \times 0.075 = 3.2 \times 10^5$ bp. Using 1.25 Kb as the average size of a protein-coding gene, we obtain an estimate of the minimal gene set of 320,000/1,250 = 254 genes.

Given that the analytical and the experimental approaches used unrelated methodologies and data, the agreement between the two results is astounding.

The quest for the fewest genes necessary for life raises the very real possibility of creating life in the laboratory. A popular account on the advances in the field (Hayden 1999) reads: "One day a scientist will drop gene number 297 into a test tube, then number 298, then 299 … and presto: what was not alive a moment ago will be alive now. The creature will be as simple as life can be. But it will still be life. And humans will have made it, in an ordinary glass tube, from off-the-shelf chemicals."

GENOME MINIATURIZATION

The question of "use and disuse" in evolution is as old as the discipline itself. Few general conclusions have been reached on the subject as far as morphological evolution is concerned. In comparison, at least one unambiguous rule can be deduced concerning the effects of disuse at the molecular level: A drastic reduction in genome size (**genome miniaturization**) is invariably associated with loss of function. In particular, parasitic or endosymbiotic modes of life were found to affect genome size profoundly and, as we have seen previously, the smallest bacterial genome belongs to an endocellular parasite.

Genome miniaturization may occur through two processes: gene transfer or gene loss. In the following we shall discuss genome size reduction due to endosymbiosis and parasitisim separately.

Genome size reduction following endosymbiosis

Wholesale miniaturization of genomes occurred following the endosymbiotic events that gave rise to the mitochondria and the chloroplasts (Chapter 5). Many organelle genes were probably redundant and were lost without replacement through deletions; others have been transferred *en masse* to the nuclear genome (Chapter 7). For example, the yeast nuclear genome contains about 300 protein-coding genes that function exclusively in the mitochondria. Its mitochondrial genome, however, contains only eight protein-coding genes. Presumably, some of the nuclear genes whose products function in mitochondria were once part of the mitochondrial genome, whose current coding capacity is quite limited. Even the mitochondrial genome with the largest coding capacity, that of the heterotrophic flagellate *Reclinomonas americana*, contains only 62 protein-coding genes (Lange et al. 1997), far less than the number of genes required for independent existence (see page 371).

In addition to mitochondria and chloroplasts, many other eukaryotic organelles are thought to have been derived from endosymbiotic events between independent organisms. Margulis et al. (1979) proposed that flagellae, cilia, and other organelles of cell motility were derived from free-living spirochetes that became associated symbiotically with the eukaryote ancestor. If this proposal turns out to be true, then these organelles must have undergone maximal genome miniaturization—i.e., they have lost their entire genome.

An interesting example of genome reduction following endosymbiosis concerns the Chlorarachniophyta, a group of amoeboflagellates that have acquired photosynthetic capacity by engulfing and retaining a flagellate green alga (class Ulvophyceae). The algal endosymbiont has retained its chloroplast, nucleus, cytoplasm, and plasma membrane. Its vestigial nucleus, called the nucleomorph, contains three small linear chromosomes with a total haploid genome size of about 380,000 bp, the smallest known "eukaryotic" genome. The nucleomorph genome is the quintessence of compactness: the average space between adjacent genes is a mere 65 bp, some genes overlap and others are co-transcribed, and the genes are disrupted by the tiniest spliceosomal introns (18–20 bp) ever found (Gilson and McFadden 1996, 1997; Ishida et al. 1997; Gilson et al. 1997). As expected, the majority of the proteins in these endosymbionts are imported from the host (Schwartzbach et al. 1998).

Genome size reduction in parasites

Parasitism involves an intimate association between two organisms: a host that provides many metabolic and physiological requirements for the other, the parasite. Parasitism invariably entails loss of genetic functions in the parasite and a consequent reduction in genome size. For example, the beechdrop *Epiphagus virginiana*, a nonphotosynthetic parasitic relative of lavender, basil, and catnip, has a very small chloroplast genome (~70,000 bp) that contains only 42 genes. Understandably, all genes for photosynthesis and chlororespiration are absent. It is not clear, however, why all chloroplast-encoded RNA polymerase genes, as well as many ribosomal protein-coding genes and tRNA-specifying genes have also been lost (Wolfe et al. 1992a,b).

As we have seen previously, the cellular parasitism of *Mycoplasma genitalium* is accompanied by genome miniaturization due to gene loss. There is, however, a genomic price in the opposite direction that must be paid in order to maintain parasitism: **gene addition**. That is, a significant number of unique genes in *Mycoplasma* are devoted to coding adhesins (adhesive proteins), attachment organelles, and variable membrane-surface antigens directed toward evasion of the immune system (Razin 1997).

GENOME SIZE IN EUKARYOTES AND THE C VALUE PARADOX

C values in eukaryotes are usually much larger than in prokaryotes, but there are exceptions. For instance, the yeast *S. cerevisiae* has a genome that is similar in size to many Gram-positive bacteria, such as *Streptomyces coelicolor* and *S. rimosus*, and smaller than that of most cyanobacterial species, especially those belonging to the genus *Calothrix*. However, since the eukaryotic nuclear

TABLE 8.3 Range of C values in various eukaryotic groups of organisms

Taxon	Genome size range (Kb)	Ratio (highest/lowest)
All eukaryotes	8,800–686,000,000	77,955
Alveolata	23,500–201,000,000	8,553
Apicomplexians	9,400–201,000,000	21,383
Ciliates	23,500–8,620,000	367
Dinoflagellates	1,370,000–98,000,000	72
Diatoms	35,300–24,500,000	694
Amoebae	35,300–686,000,000	19,433
Euglenozoa	98,000–2,350,000	24
Fungi	8,800–1,470,000	167
Animals	49,000–139,000,000	2,837
Sponges	49,000–53,900	1
Cnidarians	323,000–715,000	2
Aschelminthes	80,000–2,450,000	31
Annelids	882,000–5,190,000	6
Mollusks	421,000–5,290,000	13
Crustaceans	686,000–22,100,000	32
Insects	98,000–7,350,000	75
Echinoderms	529,000–3,230,000	6
Non-vertebrate chordates	157,000–1,470,000	9
Agnathes	637,000–2,790,000	4
Elasmobranchs	1,470,000–15,800,000	11
Bony fishes	340,000–139,000,000	409
Amphibians	931,000–84,300,000	91
Reptiles	1,230,000–5,340,000	4
Birds	1,670,000–2,250,000	1
Mammals	1,700,000–6,700,000	4
Monotremes	3,470,000–3,700,000	1
Marsupials	3,470,000–4,560,000	1
Placentals	1,700,000–6,700,000	4
Plants	50,000–307,000,000	6,140
Algae	80,000–30,000,000	375
Pteridophytes	98,000–307,000,000	3,133
Gymnosperms	4,120,000–76,900,000	17
Angiosperms	50,000–125,000,000	2,500

Data from Sparrow et al. (1972), Cavalier-Smith (1985), and many other sources.

genome has multiple origins of replication while most prokaryotes seem to have only one, the eukaryotes are able to replicate much larger amounts of DNA per unit time than prokaryotes.

The variation in C values in eukaryotes is much larger than that in bacteria, from 8.8×10^6 bp to 6.9×10^{11} bp, approximately an 80,000-fold range (Table 8.3). Unicellular protists, particularly sarcodine amoebae, show the greatest variation in C values, close to a 20,000-fold range. In comparison, the range of C values in the entire animal kingdom, from sponges to humans, is only about 3,000-fold. The three amniote classes (mammals, birds and reptiles) are exceptional among eukaryotes in their small variation in genome size (up to only fourfold). Other classes, for which a substantial body of C value data exist, show variation of at least 100-fold.

Interestingly, the huge interspecific variation in genome sizes among eukaryotes seems to bear no relationship to either organismic complexity or the likely number of genes encoded by the organisms. For example, several unicellular protozoans possess much more DNA than mammals, which are presumably more complex (Table 8.4). Moreover, organisms that are similar in morphological and anatomical complexity (e.g., flies and locusts, onion and lily, *Paramecium aurelia* and *P. caudatum*) exhibit vastly different C values (Table 8.4). This lack of correspondence between C values and the presumed amount of genetic information contained within the genome has become known in the literature as the **C value paradox**. The C value paradox is also evident in comparisons of sibling species (i.e., species that are so similar to each other morphologically as to be indistinguishable phenotypically). In protists, bony fishes, amphibians, and flowering plants, many sibling species differ greatly in their C values, even though by definition no difference in organismic complexity exists. Since we cannot assume that an organism possesses less DNA than the amount required for its vital functions, we have to explain why so many species contain seemingly vast excesses of DNA.

The first question to be clarified is whether a correlation exists between genome size and gene number. In other words, are the interspecific differences in genome size attributable to genic DNA or to nongenic DNA? If the variation in C values is attributed to genes, it can be due to interspecific differences in (1) the number of protein-coding genes, (2) the size of proteins, (3) the size of protein-coding genes, and (4) the number and sizes of genes other than protein-coding ones.

Of course, we must realize that in the absence of completely determined genomic sequences, ascertaining the number of genes in a species is a very difficult task. For protein-coding genes, this task is frequently accomplished by using two-dimensional gel electrophoresis, in which the proteins are separated by weight in one dimension and by isoelectric point (the pH at which the protein is uncharged) in the second dimension. The result is a collection of blots of different sizes dispersed all over the gel. In principle, counting the blots will allow us to estimate the number of proteins in a cell. In practice, the resolution is often quite poor, as many of the blots are faint or obscure one another. Thus, the numbers of genes inferred by this method are usually underestimated to a

TABLE 8.4 C values from eukaryotic organisms ranked by genome size

Species	C value (Kb)
Saccharomyces cerevisiae (baker's yeast)	12,000
Neurospora crassa (fungus)	17,000
Navicula pelliculosa (pennate diatom)	35,000
Dysidea crawshagi (sponge)	54,000
Caenorhabditis elegans (nematode)	80,000
Chlorella ellipsoide (green alga)	80,000
Ascidia atra (sea squirt)	160,000
Drosophila melanogaster (fruitfly)	180,000
Paramecium aurelia (ciliate)	190,000
Oryza sativa (rice)	590,000
Strongylocentrotus purpuratus (sea urchin)	870,000
Scomber scombrus (mackerel)	950,000
Gallus domesticus (chicken)	1,200,000
Erysiphe cichoracearum (powdery mildew)	1,500,000
Cyprinus carpio (common carp)	1,700,000
Lampetra planeri (brook lamprey)	1,900,000
Boa constrictor (snake)	2,100,000
Parascaris equorum (roundworm)	2,500,000
Carcharias obscurus (sand-tiger shark)	2,700,000
Canis familiaris (dog)	2,900,000
Rattus norvegicus (rat)	2,900,000
Xenopus laevis (African clawed frog)	3,100,000
***Homo sapiens* (human)**	**3,600,000**
Nicotiana tabacum (tobacco plant)	3,800,000
Locusta migratoria (migratory locust)	6,600,000
Spirogyra setiformis (desmid alga)	7,000,000
Paramecium caudatum (ciliate)	8,600,000
Schistocerca gregaria (desert locust)	9,300,000
Allium cepa (onion)	15,000,000
Triturus cristatus (warty newt)	19,000,000
Thuja occidentalis (western giant cedar)	19,000,000
Coscinodiscus asteromphalus (centric diatom)	25,000,000
Lilium formosanum (lily)	36,000,000
Amphiuma means (two-toed salamander)	84,000,000
Pinus resinosa (Canadian red pine)	68,000,000
Lepidosiren paradoxa (South American lungfish)	120,000,000
Protopterus aethiopicus (marbled lungfish)	140,000,000
Ophioglossum petiolatum (adder's tongue fern)	160,000,000
Amoeba proteus (amoeba)	290,000,000
Amoeba dubia (amoeba)[a]	690,000,000

Data from Sparrow et al. (1972), Cavalier-Smith (1985), and many other sources.

[a]The ploidy of the sarcodine amoeba *Chaos chaos* is not known, but it is highly probable that its C value is even higher than that of *Amoeba dubia* (Sparrow et al. 1972).

considerable extent. For instance, the number of protein-coding genes in *S. cerevisiae* was estimated by two-dimensional electrophoresis to be about 3,000. The number of protein-coding genes actually identified in the genomic sequence is more than double (about 6,200 putative genes). Nevertheless, if we consistently use estimates derived from the same method for comparative purposes, we may use these numbers as relative indices of the true number of genes.

The number of protein-coding genes in eukaryotes is thought to vary over a 50-fold range (Cavalier-Smith 1985a). This variation is obviously insufficient to explain the 80,000-fold variation in nuclear DNA content. Moreover, gene number is positively correlated with structural complexity, whereas genome size is not. (We note that "complexity" is a variable that is quite difficult to define, let alone quantify.) Nor can the interspecific variation in the lengths of mRNA molecules explain the C value paradox. While there are small differences in the mean length of both coding and noncoding regions among different organisms, no correlation exists between mean gene length and the size of the genome. For instance, mRNAs are only slightly longer in multicellular organisms than in protists (1,400–2,200 bp versus 1,200–1,500 bp). Moreover, organisms with larger genomes do not always produce larger proteins. Similarly, differences in gene sizes (i.e., the lengths of the introns and other noncoding regions) cannot account for the variation in genome size. While the genes of animals are indeed 3–7 times longer on the average than those of protists, and the genes of vertebrates are 2–4 times larger than those of invertebrates, no correlation was ever found between genome size and average gene length.

As to the other types of genic DNA, a positive correlation between the degree of repetition of several RNA-specifying genes and genome size has indeed been found (Chapter 6); similarly, a correlation exists between genome size and the number of copies of some untranscribed genes involved in chromosome replication, segregation, and recombination during meiosis and mitosis. However, all these genes constitute only a negligible fraction of the genome, such that the variation in the number of RNA-specifying genes and untranscribed genes cannot explain the variation in genome size.

Another way to compare the gene numbers between two genomes is to compare **polysomal polyadenylated RNA complexity**, i.e., the total length of different mRNA molecules produced by a certain tissue type. These comparisons also reveal no correlation between gene number and genome size. For example, the polysomal RNA complexity in chicken liver is 2×10^6 nucleotides, whereas the polysomal RNA complexity in mouse liver is half this amount, despite the fact that the size of the mouse genome is more than double that of chicken (John and Miklos 1988).

In summary, we are left with the nongenic DNA fraction as the sole culprit for the C value paradox. In other words, a substantial portion of the eukaryotic genome consists of DNA that does not contain genetic information. It has been estimated that the amount of nongenic DNA per genome varies in eukaryotes from about 3.0×10^3 Kb to over 10^8 Kb (a 300,000-fold range), and constitutes anything from less than 30% to about 99.998% of the genome (Cavalier-Smith 1985a).

MECHANISMS FOR GLOBAL INCREASES IN GENOME SIZE

In attempting to explain the existence of the vast amounts of nongenic DNA in the genome of eukaryotes, we must first deal with the processes that may bring about an increase in the size of genomes. We distinguish between two types of genome increase: (1) **global increases**, in which the entire genome or a major part of it, such as a chromosome, is duplicated; and (2) **regional increases**, in which a particular sequence is multiplied to generate repetitive DNA. In this section we shall only concern ourselves with global increases of the genome; regional increases will be dealt with in the context of the repetitive structure of the eukaryotic genome (see page 389).

Polyploidization

Since the genomes of eukaryotes are significantly larger than those of bacteria, the evolution of eukaryotes from prokaryote-like ancestors must have involved an increase in the size of the genome. There are several molecular mechanisms by which an increase in genome size can be brought about. One such mechanism is **polyploidization**: the addition of one or more complete sets of chromosomes to the original set. An organism whose cells contain four copies of each autosome is a tetraploid; one with six copies is a hexaploid, and so on. The gametes of polyploid organisms are not haploid, and organisms with an odd number of autosomes, e.g., the triploid domestic banana plant (*Musa acuminata*) cannot undergo meiosis and reproduce sexually.

There are two main types of polyploidy: **allopolyploidy**, the condition that arises from the combination of genetically distinct chromosome sets; and **autopolyploidy**, the multiplication of one basic set of chromosomes. Allopolyploidy is quite common in plants. For instance, common wheat (*Triticum aestivum)* is an allohexaploid containing three distinct sets of chromosomes derived from three different diploid species of goat-grass (*Aegilops*). In this section we shall mainly deal with **autotetraploidy** (or simply **tetraploidy**), also called **genome duplication** or **genome doubling**. Genome duplication occurs as a consequence of a lack of disjunction between all the daughter chromosomes following DNA replication.

Tetraploidy is a common mutational occurrence in nature. Indeed, somatic tetraploidy is found in almost all organisms, including protists, algae, plants, mollusks, insects, and mammals (Nagl 1990). However, during evolutionary history tetraploids seem to have survived only rarely. The reason is that, in many cases, tetraploidy is deleterious and will be strongly selected against. Deleterious effects include (1) prolongation of cell division time, (2) increase in the volume of the nucleus, (3) increase in the number of chromosome disjunctions during meiosis, (4) genetic imbalances, and (5) interference with sexual differentiation when the sex of the organisms is determined by either the ratio between the number of sex chromosomes and the number of autosomes (as in *Drosophila*), or by degree of ploidy (as in Hymenoptera).

In some cases, however, tetraploidy (or higher degrees of ploidy) seem to have no effect on the phenotype. For example, diploid and polyploid *Chrysanthemum* species vary in chromosome number from 18 to 198, yet they are almost indistinguishable from one another. Similar situations are known in roses (*Rosa*), leptodactylid toads (*Odontophrynus*), and goldfish (*Carasius*). Surprisingly, in some cases, tetraploidization may even be beneficial. In plants, for instance, polyploidy reduces hybrid infertility (Cavalier-Smith 1985a) and, in many cases, results in the loss of self-incompatibility, so that individual plants at the edges of a habitat can reproduce by self-pollination (Stebbins 1974).

In a recently formed tetraploid, one cannot speak of an increase in the C value, since this value refers to the size of the haploid genome and does not depend on degree of ploidy. However, as the two genomes undergo mutations, translocations, chromosomal rearrangements, and changes in chromosome number, they will eventually become a single new genome, a situation that has been dubbed **cryptopolyploidy**. In other words, an ancient polyploid will no longer be distinguishable from a diploid (Cavalier-Smith 1985a). Cryptopolyploidy may explain much of the genome size variation in plants, amphibians, and bony fishes (Table 8.3).

A polymodal distribution of genome sizes has been registered in many groups of eukaryotes (Rees and Jones 1972; Sparrow and Nauman 1976; Grime and Mowforth 1982). This is particularly evident in monocotyledons, where genome sizes exhibit a polymodal distribution with peaks at 0.60×10^6, 1.18×10^6, 4.51×10^6, and 8.53×10^6 Kb (Figure 8.5). Similar distributions have been observed in echinoderms, insects, and fungi, and to a lesser extent in amphibians and bony fishes. Thus, genome duplication seems to be a major mechanism in the evolution of genome size in eukaryotes. Interestingly, each round of genome duplication appears to have involved small losses of DNA, such that the amount of DNA after each round increases by a factor slightly smaller than two (Sparrow and Nauman 1976).

Given that the mammalian genome is about 1,000 times larger than the genome of bacteria, and assuming that genome duplication is solely responsible for genome enlargement, we can deduce that only about ten rounds of genome

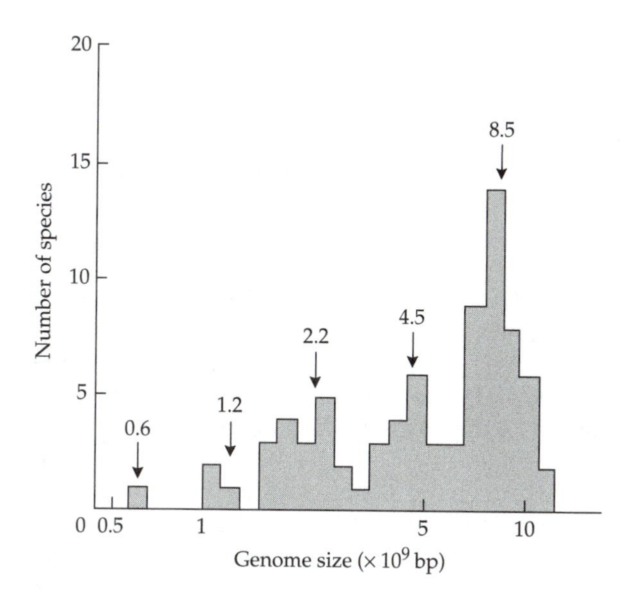

FIGURE 8.5 Frequency distribution of genome sizes in 80 grass species (family Poaceae). Peaks in the multimodal distribution are marked with arrows. Note that the abscissa is in logarithmic scale. Modified from Sparrow and Nauman (1976).

duplications were required to enlarge the genome from a primordial bacterial size to its present size in mammals (Nei 1969). To put it in another way, genome duplication has occurred an average of once every 300–350 million years. If, on the other hand, DNA content increased in a continuous fashion by the addition of small pieces of DNA, say by means of transposition or unequal crossing over, then the rate of genome growth from bacterial size to mammalian size should have been approximately 6–7 nucleotides per year (Nei 1969). We note, however, that genome doubling and nucleotide addition are not mutually elusive processes.

Following polyploidization, a very rapid process of duplicate-gene loss ensues (Feldman et al. 1997). For example, the common wheat *Triticum aestivum* is an allohexaploid that originated about 10,000 years ago. In this very short time, many of the triplicated loci have been silenced. Aragoncillo et al. (1978) estimated that the proportion of enzymes produced by triplicate, duplicate, and single loci in wheat is 57%, 25%, and 18%, respectively.

Polyploidy may be an important factor in speciation. In particular, sexually reproducing autotetraploids are automatically isolated from their diploid progenitors because they produce diploid gametes; were these to combine with the haploid gametes of the diploids, they would give rise to triploid progeny. As mentioned previously, organisms with an odd number of autosomes cannot reproduce sexually, so polyploidy represents an effective mechanism of reproductive isolation.

Polysomy

Aneuploidy refers to the condition in which the number of chromosomes in a cell is not an integral multiple of the typical haploid set for the species. (**Euploidy** refers to a chromosome number that is an exact multiple of the haploid chromosome number.) Since we are dealing with mechanisms responsible for increases in genome size, we shall only deal with two types of aneuploidy: the duplication of a complete chromosome (**polysomy**), and the duplication of a major part of a chromosome (**partial polysomy**).

Polysomy is most often deleterious. In mammals, for instance, it is frequently associated with lethality or infertility. In humans, well-known examples of polysomy include such anomalies as Down's syndrome (trisomy 21), and trisomy 18. Similarly, severe deleterious manifestations are often associated with partial polysomy (e.g., cat-eye syndrome). Therefore, chromosomal duplication—either complete or partial—is not expected to contribute significantly to genome size increase.

The yeast genome: Tetraploidy or regional duplications?

Saccharomyces cerevisiae has long been suspected of being a cryptotetraploid (Smith 1987). Wolfe and Shields (1997) systematically searched the complete yeast proteome for duplicated regions (Figure 8.6). The criteria used for defining two regions as duplicates were: (1) a sequence similarity between the two regions associated with a probability of less than 10^{-18} of it being fortuitous; (2) at least three genes in common, with intergenic distances of less than 50 Kb; and (3) conservation of gene order and relative orientation of the genes. According

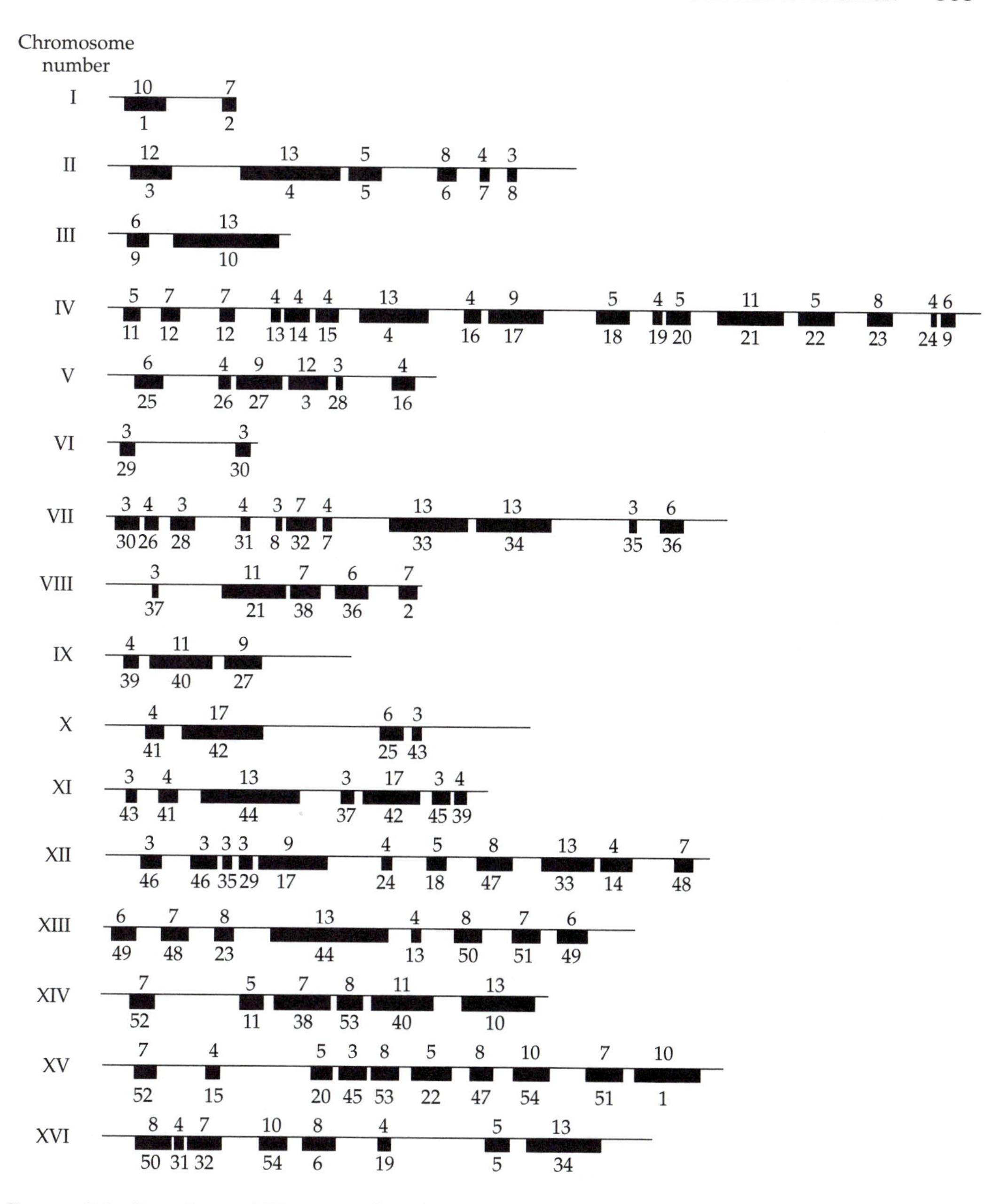

FIGURE 8.6 Locations of 54 nonoverlapping duplicated regions (solid boxes) in the yeast genome. The two copies of each duplicated region are given the same number below their respective boxes. Numbers are listed in order of chromosomal occurrence. The number of homologous genes in each duplicated region is listed above its box. Chromosome numbers are given in roman numerals. Modified from Wolfe and Shields (1997).

to these criteria, Wolfe and Shields (1997) identified 54 nonoverlapping pairs of duplicated regions spanning about 50% of the yeast genome (Figure 8.6). (We note that since the authors' criteria for identifying duplicated regions were quite stringent, the above estimate should be regarded as an underestimate.)

There are two possible explanations for these observations. Either (1) the duplicated regions formed independently by many regional duplications occurring at different times during the evolution of *S. cerevisiae*, or (2) the duplicated regions were produced simultaneously by a single tetraploidization event, followed by massive rearrangements of the genome and loss of many redundant duplicate genes. There are two reasons to favor the latter model. First, 50 of the duplicated regions have maintained the same orientation with respect to the centromere. Second, based on a Poisson distribution (Appendix II), 54 independent regional duplications are expected to result in about seven triplicated regions (i.e., duplicates of duplicates), but none was observed.

Wolfe and Shields (1997) proposed that *S. cerevisiae* is an ancient tetraploid, formed through the fusion of two ancestral diploid yeast genomes, each containing about 5,000 genes. They estimated the tetraploidization event to have occurred approximately 100 million years ago in the ancestor of four *Saccharomyces* species after the divergence of *S. kluyveri*. The new species then became a cryptotetraploid, and about 92% of the duplicate gene copies were lost through sequence decay or deletion. Some 70–100 subsequent map disruptions (i.e., regional translocations) were inferred to have been required to explain the current chromosomal distribution of the duplicate genes (Seoighe and Wolfe 1998). A schematic scenario of gene number and gene order evolution following genome duplication is shown in Figure 8.7.

Polyploidy of the vertebrate genome

It has been known for quite some time that vertebrates possess more genes than invertebrates. Indeed, an extensive survey of gene families from aldolases to zinc-finger transcription factors revealed that a single invertebrate gene usually corresponds to up to four vertebrate genes on different chromosomes. Moreover, it seems that the sequences of many of the quadruple copies are equidistant from one another. This pattern was first observed for the *Hox* gene clusters, but, according to Spring (1997), this phenomenon is a general one. He put forward a hypothesis, according to which the emergence of vertebrates was made possible by two rounds of tetraploidization, resulting in genome quadruplication. Thus, vertebrates (including the readers of this book) may in fact be cryptooctoploids!

MAINTENANCE OF NONGENIC DNA

Accounting for the vast quantities of seemingly superfluous nongenic DNA requires that we address the question of what function this DNA might have, if any. Numerous attempts have been made to solve the C value paradox, and in the following, we present four such hypotheses and pertinent empirical evidence.

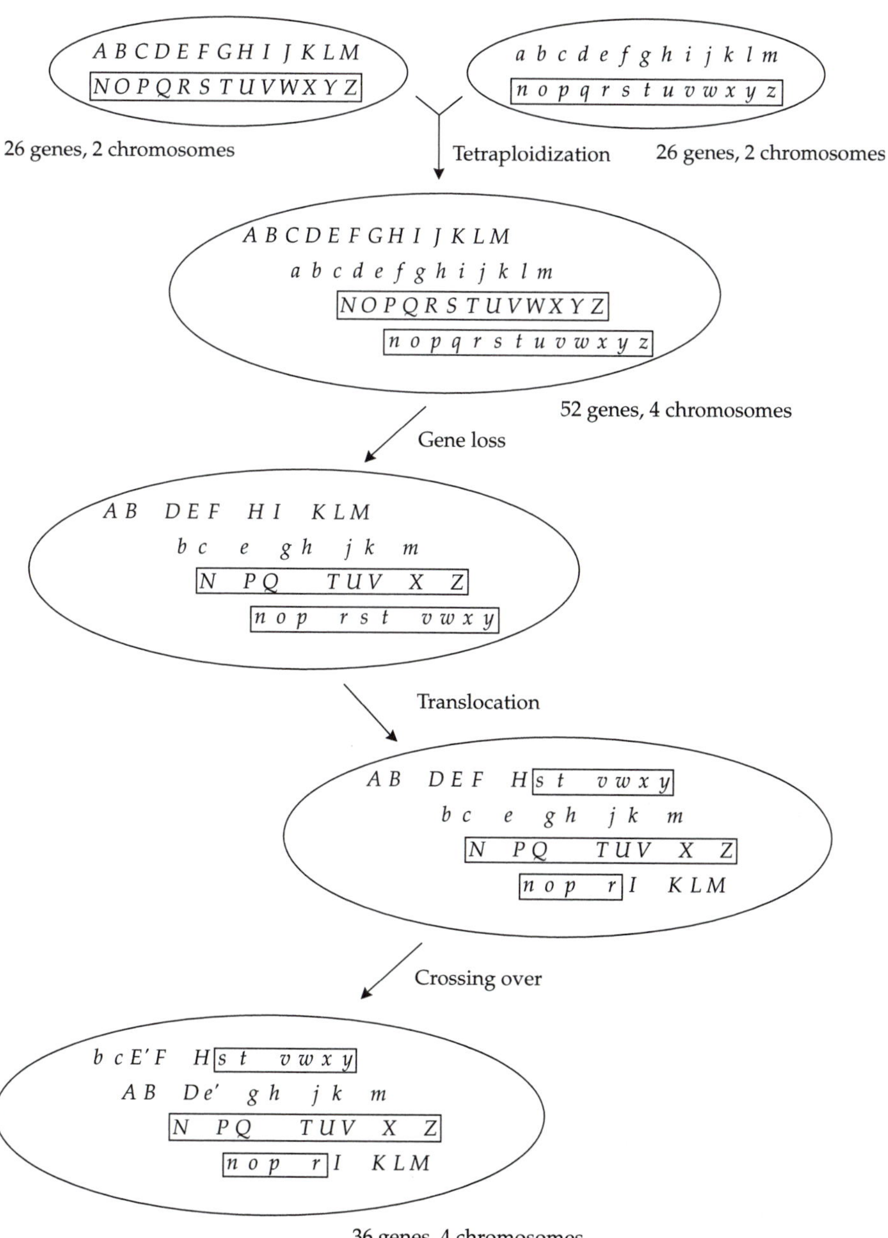

FIGURE 8.7 Schematic scenario of gene number and gene order evolution in a duplicated genome such as yeast. A schematic genome is shown with two chromosomes (one boxed) and 26 genes (*A* to *Z*). Upper- and lowercase letters are used to distinguish between the two original sets of chromosomes. In the last stage, the effect of a recombination event within two paralogous genes is shown. The event produces two new hybrid genes (designated *E′* and *e′*) and a new gene order. Modified from Keogh et al. (1998).

The hypotheses

The **selectionist hypothesis** asserts that the so-called nongenic DNA performs essential functions, such as the global regulation of gene expression (Zuckerkandl 1976a). According to this hypothesis, the excess of DNA is only apparent, and the DNA is wholly functional. Consequently, deletion of such DNA will have a deleterious effect on fitness.

The **neutralist hypothesis** proposes that the nongenic DNA fraction in the eukaryotic genome is genetically and physiologically inert (Darlington 1937). Ohno (1972) deliberately chose the provocative term **junk DNA** for the nongenic fraction of the genome to emphasize its uselessness. Junk DNA is carried passively by the chromosomes merely because of its physical linkage to functional genes (Rees and Jones 1972). According to this view, the excess DNA is an incidental result of evolutionary processes and, as long as it does not affect the fitness of the organism, it will be carried on from generation to generation indefinitely.

The **intragenomic selectionist hypothesis** regards nongenic DNA as a "functional parasite" (Östergren 1945), or "genetic symbiont" (Cavalier-Smith 1983) that accumulates in the genome and is actively maintained by intragenomic selection due to its elevated rate of reproduction in comparison to that of the genomic fraction (Cavalier-Smith 1980). In the literature, it is common to find the term **selfish DNA** applied to the nongenic fraction (Orgel and Crick 1980; Doolittle and Sapienza 1980). Selfish DNA has two distinct properties: (1) it arises when a DNA sequence spreads by forming additional copies of itself within the genome, and (2) it either makes no specific contribution to the fitness of the host organism, or is actually detrimental. The major mechanism for amplifying selfish DNA is duplicative transposition (Chapter 7), and the most abundant type of selfish DNA are transposable and retrotransposable elements. A crucial distinction between selfish DNA and junk DNA is that the former is capable of promoting its own amplification, whereas the latter is carried passively in the genome. Thus, junk DNA is maintained in the population by random genetic drift, whereas selfish DNA is maintained by a type of insertion–deletion quasi-equilibrium, whereby the process of elimination by selection of selfish DNA is too slow to offset the rate of its accumulation. Selfish DNA has a tendency to increase in the genome. However, it cannot increase indefinitely, because an organism with excessive amounts of nongenic DNA would be at a metabolic, and hence selective, disadvantage relative to one with lesser amounts.

The **nucleotypic hypothesis** (Bennett 1971) attributes a structural function to nongenic DNA, that is, a function unrelated to the task of carrying genetic information. One such nucleotypic scheme has been proposed by Cavalier-Smith (1978, 1985a), who argued that there must be a "major evolutionary force" maintaining large genomes. His hypothesis was that the DNA acts as a "nucleoskeleton" that maintains the volume of the nucleus at a size proportional to the volume of the cytoplasm. Since larger cells require larger nuclei, selection for a particular cell volume will secondarily result in selection for a particular genome size. According to this scheme, excess DNA is maintained by selection, but its nucleotide composition may change at random. Many additional nucleotypic functions have been attributed to the nongenic fraction, but all nucleotypic hypotheses have one thing in common: they all regard the genome as a structural unit of nuclear architecture—a building block made of nucleic acid, rather than a mere carrier of genetic information.

The evidence

There is very little evidence for the selectionist hypothesis. In fact, most indications are that the bulk of what is now considered nongenic DNA is indeed devoid of genetic information, and can be deleted without discernible phenotypic effects. It therefore seems that excess DNA in eukaryotes does not tax the metabolic system to a significant extent, and that the cost (e.g., in energy and nutrients) of maintaining and replicating large amounts of nongenic DNA is not excessive. However, there may be some drawbacks in maintaining large amounts of nongenic DNA. First, large genomes have been found to exhibit greater sensitivity to mutagens than small genomes (Heddle and Athanasiou 1975). Second, maintaining and replicating large amounts of nongenic DNA may impose a certain burden on the organism, especially when the vast majority of the genome is nongenic. It is therefore accepted that nongenic DNA can only accumulate until the cost to the organism of replicating it becomes significant.

It is very difficult to distinguish between the intragenomic selectionist hypothesis and the neutralist hypothesis at the conceptual level, let alone to test them against real data. Selfish DNA may indeed be a major contributor of nongenic DNA, although there are other important mechanisms of generating such DNA (see page 395). However, it is equally true that most of the nongenic fraction of the genome originating as selfish DNA is no longer selfish. Much of it is currently in a degenerate state—dead and moribund transposable elements that are no longer capable of transposition.

Distinguishing experimentally between the junk DNA and the nucleoskeletal explanations has been quite difficult. Pagel and Johnstone (1992) proposed two expectations derived from each of the two theories. According to these authors, a major cost of junk DNA is the time required to replicate it. Organisms that develop at a slower pace may therefore be able to "tolerate" greater amounts of junk DNA, and thus a negative correlation across species between genome size and developmental rate is predicted. In contrast, the prediction of the nucleoskeletal hypothesis is for a positive correlation between genome size and cell size. Unfortunately, organisms with large cells also tend to develop slowly, whereas faster-growing organisms typically have smaller cells. Thus, according to the skeletal DNA hypothesis, a negative correlation between developmental rate and the C value is also expected. However, according to the nucleotypic hypothesis, the relation between developmental rate and genome size occurs only secondarily, as a result of the relationship between developmental rate and cell size.

Pagel and Johnstone (1992) studied 24 salamander species. The size of the nuclear genome was found to be negatively correlated with developmental rate, even after the effects of nuclear and cytoplasmic volume have been removed. However, the correlations between genome size, on the one hand, and nuclear and cytoplasmic volumes, on the other, become statistically insignificant once the effects of developmental rates have been removed. These results support the junk DNA theory. Whether Pagel and Johnstone's results represent a general phenomenon or one restricted to *Salamandra* is not known at present (Martin and Gordon 1995; Jockusch 1997).

No single explanation is likely to solve the C value paradox. All the above mechanisms, and many additional ones—working alone or in synergy (Xia 1995)—may contribute to the maintenance of excess genomic size, and our task in the future will be to determine the relative contribution of each.

Why do similar species have different genome sizes?

We are essentially left with one aspect of the C value paradox that we have not yet properly addressed. At issue is the difference in genome sizes between closely related organisms, in which the C value paradox cannot be explained away by invoking nucleotypic functions, because no nucleotypic differences exist. We are left with two mechanistic possibilities: either there is a difference in the rate of accumulation of junk DNA, or there is a difference in the rate with which different organisms get rid of junk DNA.

For quite a long time it has been known that the genomes of *Drosophila* species contain very few pseudogenes (Vanin 1985; Weiner et al. 1986; Wilde 1986). More recently, Petrov et al. (1996) and Petrov and Hartl (1998) found that dead *Helena* retroposons lose DNA at unusually high rates during evolution. They put two and two together, and suggested that DNA regions that are not subject to selective constraints are deleted at "rampant" rates, and they further extrapolated that different deletion rates, rather than accumulation rates, may contribute to the divergence in genome size among taxa. Their assumption was that high rates of deletion are not confined to *Helena* elements alone, but that the phenomenon is of general application to all selectionally unconstrained regions.

To test this assumption, they compared intron sizes between two *Drosophila* species. *D. virilis* has a genome twice as large as that of *D. melanogaster* (Moriyama et al. 1998). Admittedly, part of the difference could be attributed to heterochromatin (see page 390), but even if this factor is taken into account, the genome of *D. virilis* is still about 36% larger than that of *D. melanogaster*. In their comparison of 115 complete introns collected from 42 orthologous genes, they found that the difference in intron length between the two *Drosophila* species was statistically significant. The differences in the mean length of the introns between *D. virilis* and *D. melanogaster* (394 and 283 bp, respectively) was 39%, which was surprisingly close to the size difference in the nonrepetitive fraction between the genomes. Thus, it seems that some organisms are simply more efficient at "throwing out the trash" than others (Petrov and Hartl 1997).

THE REPETITIVE STRUCTURE OF THE EUKARYOTIC GENOME

The eukaryotic genome is characterized by two major features: the repetition of sequences, and compositional compartmentalization into distinct fragments characterized by specific nucleotide compositions.

Repetitive DNA consists of nucleotide sequences of various lengths and compositions that occur several times in the genome, either in tandem or in a dispersed fashion. Segments of DNA that do not repeat themselves are referred to as **single-copy** or **unique DNA**. The proportion of the genome taken up by repetitive sequences varies widely among taxa. In yeast, this proportion amounts to about 20% of the genome. In animals, the proportion ranges from about 5% in the nonbiting midge *Chironomus tetans* to close to 90% in the newt *Necturus masculosus*. In mammals, up to 60% of the DNA is repetitive. In plants, the proportion can exceed 80%, and much higher values have also been registered (Flavell 1986).

Classic studies of the kinetics of DNA reassociation by Britten and Kohne (1968) showed that the genome of higher eukaryotes can be divided roughly into four fractions (Figure 8.8). The first fraction is called **foldback DNA**, and it consists of palindromic sequences that can form hairpin double-stranded structures as soon as the denatured DNA is allowed to renature. The foldback DNA fraction is usually very small, although in some organisms it may reach values in excess of 10%.

Some DNA only reanneals at high C_0t (pronounced "cot") values. This fraction is comprised of single-copy sequences, and because of its light stain-

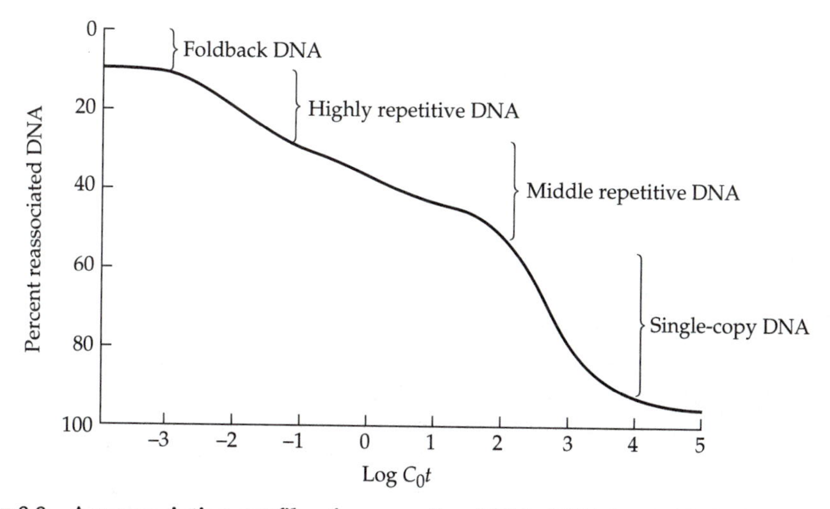

FIGURE 8.8 A reassociation profile of mammalian DNA. DNA is purified, sheared, thermally melted into single strands, and then allowed to reassociate through gradual cooling. The percentage of reassociated double-stranded DNA on the vertical axis is shown as a function of the product of DNA concentration and time (C_0t) on the horizontal axis. Modified from Schmid and Deiniger (1975).

ing properties in karyological preparations, it is sometimes referred to as **euchromatin**. In between these two more or less well-defined genomic components, there are DNA sequences that reanneal at intermediate C_0t values. It is customary to divide these sequences into **highly repetitive DNA** and **middle repetitive DNA**. The highly repetitive fraction is made up of short sequences, from a few to hundreds of nucleotides long, which are repeated thousands and even millions of times. In karyological preparations, the highly repetitive fraction appears dark and heavily stained and is called **heterochromatin**. The middle repetitive fraction consists of much longer sequences, hundreds or thousands of base pairs on average, which appear in the genome up to hundreds of times. We note that as far as repetitive sequences are concerned, there is a continuum of both repeat sizes and numbers of repeats in the genome. Hence, the terms highly repetitive DNA and middle repetitive DNA are terms of convenience; they do not represent truly distinct DNA classes.

On the basis of the pattern of dispersion of repeats, the repetitive fraction was found to consist of two types of repeated families: **localized repeated sequences** and **dispersed repeated sequences**.

Localized repeated sequences

Most eukaryotic genomes contain tandemly arrayed, highly repetitive DNA sequences. In some species, these localized repetitive DNA sequences can account for the majority of the DNA in the genome. For example, in the kangaroo rat, *Dipodomys ordii*, more than 50% of the genome consists of three repeated sequences: AAG, 2.4×10^9 times; TTAGGG, 2.2×10^9 times; and ACACAGCGGG, 1.2×10^9 times (Salser et al. 1976). Of course, these families are not completely homogeneous but contain many variants that differ from the consensus sequence in one or two nucleotides. For example, some sequences in the "TTAGGG" family are actually TTAGAG.

Even much smaller genomes may contain a huge proportion of highly repetitive sequences. For example, 40% of the *Drosophila virilis* genome consists of three highly repeated sequences: ACAAACT, 1.1×10^7 times; ATAAACT, 3.6×10^6 times; and ACAAATT, 3.6×10^6 times (Lohe and Roberts 1988). Surprisingly, 35% of the genome of the unicellular pin mould, *Absidia glauca*, which is only nine times larger than that of *E. coli*, is made of repetitive DNA (Wostemeyer and Burmester 1986).

Many of the localized highly repeated sequences have such a uniform nucleotide composition that, upon fractionalization of the genomic DNA and separation by density gradient, they form one or more thick bands that are clearly distinguishable from the smear created by the other DNA fragments of more heterogeneous composition. These bands, which are either much heavier or much lighter than the other genomic sequences, are called **satellite DNA**. Some satellite DNA may be extremely G+C-rich or extremely A+T-rich; GC content in satellites ranges from as low as 1% in the crabs *Cancer gracilis* and *C. antenarius*, to as high as 73% in the trypanosomal pathogen *Leishmania infantum* and the midge *Chironomus plumosus*. Mammalian genomes typically consist of 5–30% satellite DNA. The amount of satellite DNA in plants may reach 40% of the total genome.

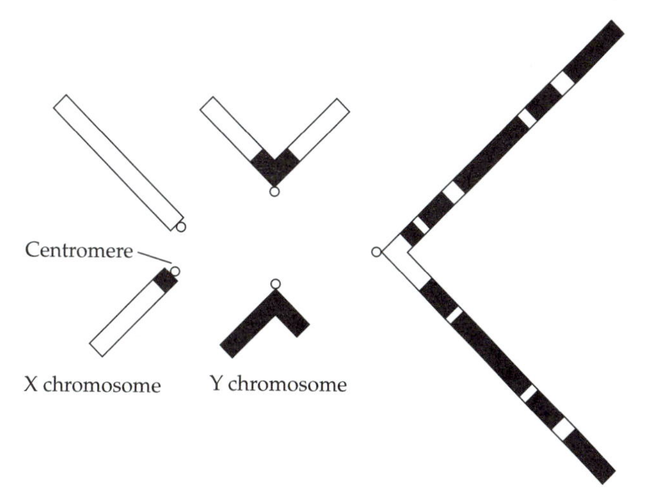

FIGURE 8.9 Highly repetitive DNA sequences (black areas) in *Drosophila nasutoides* are mostly localized on the largest of the three autosomes and on chromosome Y. Data from Wheeler and Altenburg (1977) and Miklos (1985).

In some species, tandemly arrayed highly repetitive sequences are found on all chromosomes, while in others they are restricted to a particular chromosomal location. For example, more than 60% of the genome of *Drosophila nasutoides* consists of satellite DNA, and the vast majority of it is localized on one of the four autosomes and the Y chromosome (Figure 8.9), which seem to contain little else (Miklos 1985). Not all localized highly repeated DNA consists of short repeats. For example, the killer whale, *Orcinus orca*, contains about half a million copies of a sequence 1,579 bp long, accounting for approximately 15% of its genome (Widegren et al. 1985).

Based on the evidence available at the present time, it is highly probable that localized highly repeated sequences are devoid of function. Moreover, it is possible that the amount of localized repeated sequences neither lowers nor increases the fitness of the individual. Consequently, the evolution of such sequences is not affected by natural selection. The number and composition of these repeats vary in time due to the mutational input of such processes as gene conversion and unequal crossing over (Chapters 1 and 6), and fixation in the population occurs via random genetic drift (Chapter 2). Gene conversion and unequal crossing over will result in two outcomes for these sequences: (1) sequence homogeneity, and (2) wide fluctuations in numbers over evolutionary time (Charlesworth et al. 1986). It has also been suggested that the rate of turnover of localized repeated sequences is quite high; that is, existing arrays may be removed by unequal crossing over, while new arrays may be continuously created by processes of DNA duplication (Walsh 1987).

The suggestion that most tandemly repeated sequences are merely junk DNA essentially implies that they have no phenotypic effects. Moreover, it is assumed that their presence or absence in variable numbers does not affect the fitness of the carriers. While this may be true in the majority of cases, there is evidence pertaining to a specific array of highly repeated sequences indicates that this is not always the case. The *Responder* (*Rsp*) locus in natural populations of *Drosophila melanogaster* consists of 20–2,500 copies of an AT-rich, 120-bp-long sequence (Wu et al. 1988). In a competition experiment involving a mixed population consisting of flies with 700 copies of the repeat and flies with only 20 copies, it was observed that the frequency of the flies with 20 repeats decreased over time (Wu et al. 1989). Therefore, it was concluded that flies with 700 copies have a higher fitness than flies with only 20 copies. Except for its role in the segregation distortion system, the function of the *Rsp* locus is not currently known, but it is clearly not junk DNA, since its absence affects the fitness of the organism. However, we are not aware of any other cases in which tandemly repeated sequences were shown to affect fitness.

Dispersed repeated sequences

The second class of highly repetitive DNA consists of sequences that are dispersed throughout the genome. Copies of dispersed highly repetitive sequences are found in introns, flanking regions of genes, intergenic regions, and nongenic DNA. In the following we shall mainly discuss the human genome.

There are two major categories of dispersed repeated sequences: **simple tandem repetitive sequences** and **interspersed repeats**. Table 8.5 shows a classification of simple tandem repetitive sequences according to the size of the repeat unit, the number of repeat units per array, and the genomic location of the tandem arrays. Note that satellites and **minisatellites** are mostly localized repeated sequences, although a small fraction of minisatellites is dispersed. It has been estimated that there are 300,000 trinucleotide and tetranucleotide **short tandem repeats** in the human genome, or about one array every 10 Kb of genomic DNA (Beckmann and Weber 1992). The most common human **microsatellite** consists of CA dinucleotide repeats. There are about 50,000 copies of this microsatellite in the human genome, i.e., one array every 30 Kb (Hudson et al. 1992).

The human genome also contains four major classes of interspersed repeats: (1) SINEs, (2) LINEs, (3) retrovirus-like and retrotransposon-like elements, and (4) DNA-mediated transposable fossils (Chapter 7). The relative abundance and genomic distributions of these interspersed classes of repeats are shown in Figure 8.10.

TABLE 8.5 Four classes of simple tandem repetitive sequences

Class	Repeat size (bp)	Array size (number of units)	Genomic distribution
Satellites	2–2,000	≥1,000	Centromeric, heterochromatic
Minisatellites	9–100	10–100	Subtelomeric, dispersed
Short tandem repeats	3–5	10–100	Dispersed
Microsatellites	1–2	10–100	Dispersed

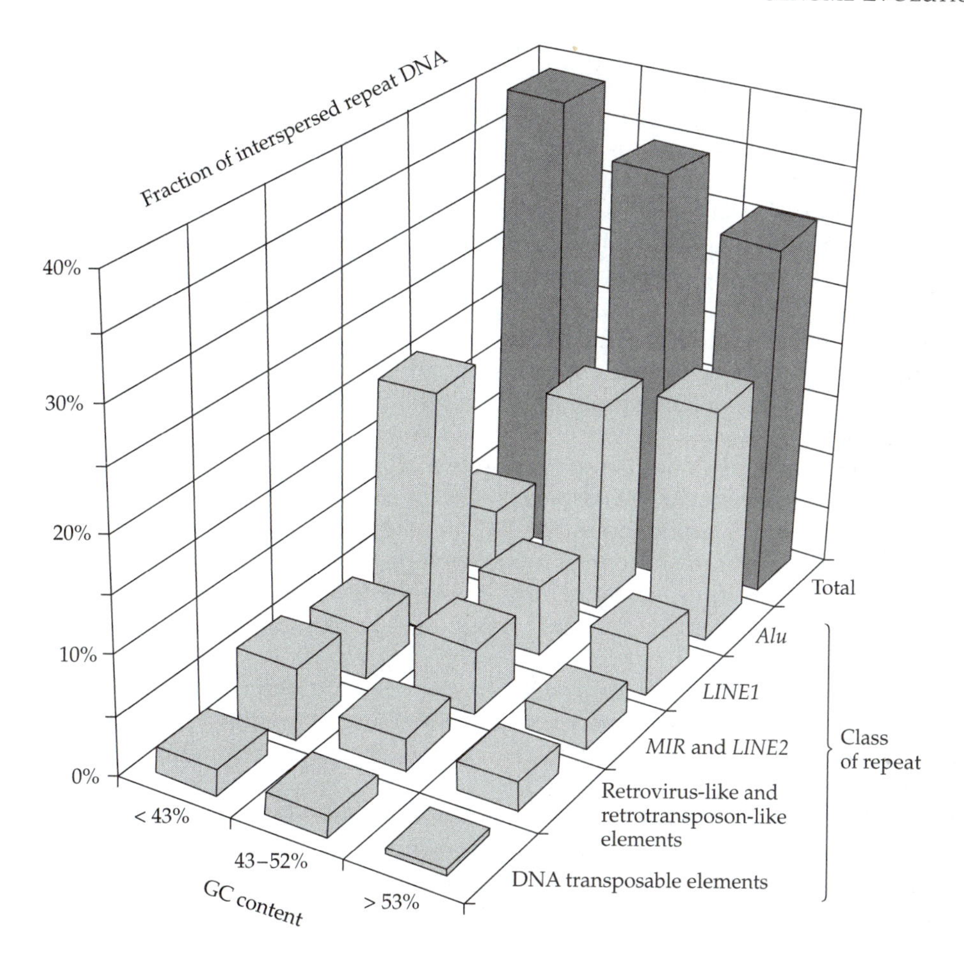

FIGURE 8.10 Relative abundance and genomic distribution by regional GC content of human interspersed repeat classes. Note the almost complementary distribution of *Alu* and *LINE1* repeats. Modified from Smit (1996).

The human genome contains two LINE families, *LINE1* (*L1*) and *LINE2* (*L2*). There are about 600,000 *L1* repeats in the human genome, or about 15% of the genome. The *L1* family has been active in mammalian genomes since before the divergence between marsupials and placentals. The origin of the much smaller *L2* family (~271,000 repeats) may be very ancient, most probably pre-dating the divergence of amphibians from amniote vertebrates. About 95% of all *L1* sequences are truncated at their 5′ end and are neither transcribed nor retrotransposed. The degree of *L1* sequence divergence among species is much greater than the degree of divergence among conspecific *L1* copies. For example, *L1* sequences from mice and humans differ from each other by about 30% on average, in comparison to a sequence divergence of about 4% within mice (Hutchison et al. 1989).

Defective *L1* elements were found to evolve much more rapidly than intact elements. Moreover, evolutionary lineages of defective *L1* sequences were found to contain no branches, indicating that these elements are incapable of replicative transposition. They are thus pseudogenes of retroposons, on which functional constraints no longer operate, and as such are subject to compositional assimilation and length abridgment (Chapter 7) until they are no longer recognizable as LINEs. The fact that most *L1* sequences are defective implies that the propagation of *L1* elements within the genome depends on only a small number of source elements. As a consequence, *L1* elements within the genome are highly homogeneous and the rate of sequence turnover is very high. Indeed, in rodents it has been estimated that more than half of the *L1* elements are only 3 million years old or younger (Hardies et al. 1986).

The human genome also contains two SINE families, the 7SL-derived *Alu* family, with about 1,100,000 copies or 10% of the genome, and the tRNA-derived *MIR* family, with about 400,000 copies. To complete the list of interspersed repeats in the human genome, we must also mention retrovirus-like and retrotransposon-like elements (~5% of the genome), remnants of DNA transposable elements (~2%), and about 60,000 copies of unclassified interspersed repeats (~1%).

In conclusion, over a third of the human genome is derived from mobile elements belonging to a handful of families. The vast majority of these interspersed repetitive sequences, however, no longer possess the ability to move.

Repetitive sequences: A cause of variation in genome size

As mentioned previously, a major component of the C value paradox is the fact that organisms that are morphologically and anatomically similar exhibit vastly different C values. Nowhere is this fact more evident than in comparisons among species belonging to the same genus. It seems, however, that a general cause has been found at the root of the paradox. In all closely related organisms so far studied, the differences in genome size could be explained by differences in the repetitive fractions. From rodents such as *Ctenomys* (tuco-tucos) to plants such as *Avena* (oats), and from *Hylobates* (gibbons) to *Drosophila*, whenever congeneric species differ from one another in C value, the difference can be wholly explained by the repetitive nongenic fraction of the genome, frequently by differences in the quantity of simple tandem repeats. Moreover, whenever a taxon is found in which the genome size is much smaller than that in related taxa, we invariably find that the difference is entirely due to repetitive sequences. For example, some bats possess genomes that are approximately 50% the size of other eutherian mammals. The difference was attributed to a paucity of AT and GC microsatellites, which in other mammals are quite common (van den Bussche et al. 1995). Similarly, the relative lack of genome size variation in birds (Table 8.3) may be due to a scarcity of microsatellites in the avian genome (Primmer et al. 1997). In Figure 8.11, we

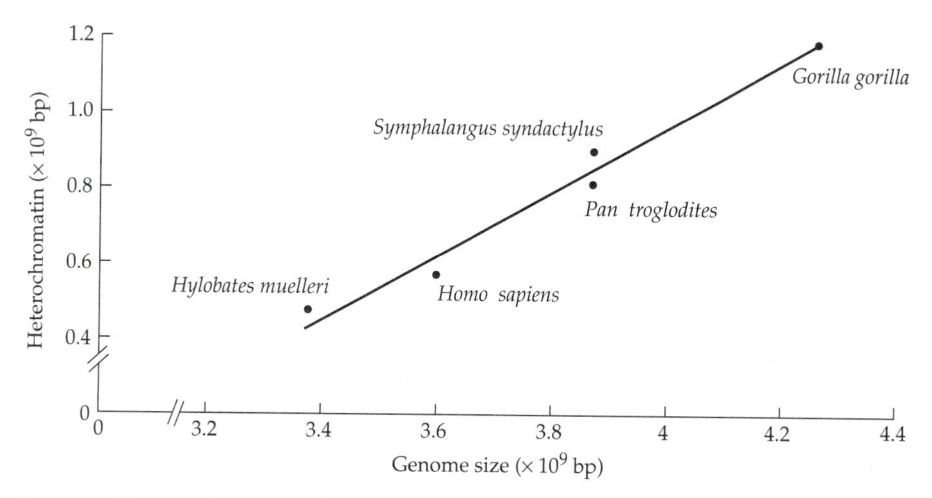

FIGURE 8.11 An almost perfect correlation exists between genome size and amount of heterochromatin in Hominoidea. Data from Manfredi-Romanini et al. (1994).

see that about 98% of the variation in genome sizes in apes is explained by the variation in tandemly repetitive sequences (heterochromatin).

MECHANISMS FOR REGIONAL INCREASES IN GENOME SIZE

Regional increases in genome size can be brought about by several mechanisms. Duplicative transposition (Chapter 7), however, is the only known mechanism that can produce dispersed repetitive sequences; all other mechanisms result in localized repetitive sequences. It has been suggested that the entire middle repetitive DNA fraction in eukaryotes originated in transposable elements. Most of the elements, however, are no longer mobile, their ability to transpose being destroyed by mutations or the insertion of other elements. A large part of the heterochromatin in *Drosophila*, for instance, may in fact be a graveyard for dead members of about 30 major families of transposable elements. Inactivated mobile elements can, however, multiply locally through such processes as unequal crossing over. Comparison of dispersed repeats between closely related species suggests that transposition and retroposition can quickly increase the fraction of dispersed repeats in the genome. For example, the sibling species *D. melanogaster* and *D. simulans* diverged from each other less than 2.5 million years ago. During this relatively short period of time, their genomes have become very different in terms of the fraction of dispersed repeats (21% in *D. melanogaster* and 3% in *D. simulans*). This difference is most likely explained by an increase in copy number in D. melanogaster rather than a decrease in *D. simulans*, because loss of genomic transposable elements is known to be a very slow process. We recall, however, that length abridgment is faster in *D. simulans* than in *D. melanogaster* (see page 388), so we must conclude that the contribution of copy number to genome size is greater than that of repeat length.

Unequal crossing over is most probably the primary mechanism responsible for increases and decreases in copy number of satellites and minisatellites. This fact notwithstanding, unequal crossing over events usually create sequences consisting of relatively long repeats. In contrast, many localized repeated sequences, such as microsatellites and short tandem repeats, consist of very short, simple, repeated motifs. In addition, the pattern of microsatellite variability is similar in recombining regions of the genome and in the nonrecombining portion of the mammalian Y chromosome (Nachman 1998). It seems, therefore, that replication slippage (Chapter 1) is the primary force in the evolution of these sequences.

There is evidence that the copy number at minisatellite loci may increase rapidly. For example, in humans, the *MS32* locus may contain more than 600 repeats. In Old World monkeys, in contrast, the homologous locus contains only 3–4 repeats. The latter character state presumably represents the ancestral state, whereas the high number of repeats in humans most probably represents a recently derived state.

It has been noted that replication slippage and unequal crossing over tend to remove tandem arrays more often than to increase their size and copy number, so these processes cannot explain the existence of all the localized repeated DNA (Walsh 1987). To explain the existence of such sequences, DNA amplification has been suggested. DNA amplification refers to any event that increases the number of copies of a gene or a DNA sequence far above the level characteristic for an organism. In particular, DNA amplification refers to events that occur within the lifespan of an organism and cause a sudden increase in the copy number of a DNA sequence. We distinguish between vertical amplification and horizontal amplification. **Vertical amplification** refers to processes through which a certain sequence is multiplied outside the chromosome. **Horizontal amplification** refers to a process of creation of multiple copies of a certain DNA sequence and their incorporation within the heritable genome of the organism.

One of the most powerful methods of amplification is the **rolling circle** mode of DNA replication (Figure 8.12). This type of replication (Bostock 1986) is used in the amplification of rRNA genes in amphibian oocytes. In this case, amplification involves the formation of an extrachromosomal circular copy of a DNA sequence, which can then produce many additional extrachromosomal units containing tandem repeats of the original sequence. If such units become integrated back into the chromosome, there will be an addition to the genome consisting of identical repeated sequences.

Replication slippage, or slipped-strand mispairing, is a process in which the DNA polymerase turns back and uses the same template again to produce a repeat (see Figure 1.18). Existing tandem repetitive sequences are particularly prone to replication slippage, and the process can therefore produce very

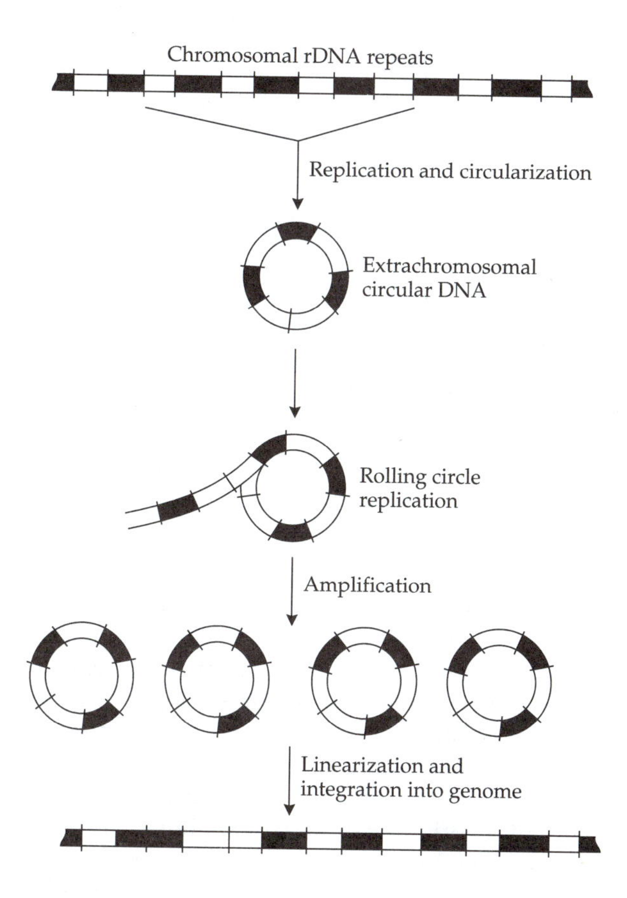

FIGURE 8.12 The rolling circle model of gene amplification in amphibian oocytes. The chromosomal rRNA is arranged in a tandem array containing transcribed (black) and nontranscribed (white) parts. Amplification involves the formation of an extrachromosomal circular copy containing a variable number of repeats, which is then amplified by multiple rounds of rolling circle replication. Note that the periodicity may change following rolling circle amplification. Modified from Bostock (1986).

long tandem arrays of short repeats. Both rolling circle replication and replication slippage can provide mechanisms for the rapid proliferation of tandemly repeated sequences within the genome. However, the empirical evidence for these processes is limited.

GENE DISTRIBUTION

So far, we have only dealt with portions of the DNA that may or may not have a function, but if they do, this function is certainly not a protein-coding one. It is therefore time to ask, "Where are the protein-coding genes?" In this section we shall deal with five interconnected issues: (1) the number of genes, (2) their genomic location, (3) gene density, (4) chromosome number variability, and (5) evolutionary processes affecting gene order.

How many genes are there, where are they, and do we need them?

The eukaryotic organisms for which we have the most information to answer these three questions are baker's yeast, *Saccharomyces cerevisiae*, and the nema-

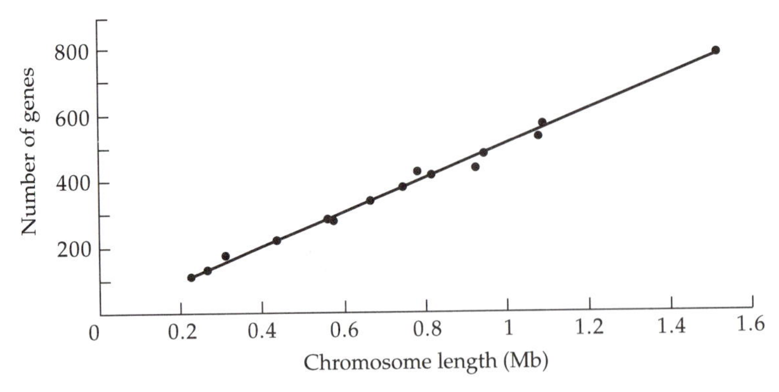

FIGURE 8.13 Relationship between gene number and chromosome length in *Saccharomyces cerevisiae*. The negligibly small variation around the regression line indicates that genes are distributed evenly among the 16 chromosomes.

tode *Caenorhabditis elegans*, whose entire genomes have been sequenced. (We note, however, that these organisms are not representative of the eukaryotic domain, since their genomes were chosen for sequencing because of their exceptional small sizes.) *S. cerevisiae* has just over 6,000 protein-coding genes distributed about evenly among 16 chromosomes, i.e., the number of genes on each chromosome is proportional to its length (Figure 8.13). On the other hand, the distribution of genes along the chromosomes is not even. There are regions with high gene density and regions with low gene density (Figure 8.14). In *C. elegans*, there are over 19,000 genes distributed among 6 chromosomes with a total of about 97 Mb in length. The chromosomal distribution is less uniform than that in the yeast, with the X chromosome having a lower gene density than that of the other chromosomes, but the departure from uniformity is not very large.

Our knowledge of the genomes of multicellular organisms, including our own, is much more limited. What we do know for certain, however, is that most of the genome does not contain protein-coding information. If we subtract from the length of the genome all the repetitive sequences, all the pseudogenes, all the introns, and all the intergenic regions, very little is left. In humans, RNA–DNA hybridization experiments long ago showed that there are almost no protein-coding genes within the repeated fractions of the genome, and even within the unique DNA fraction only about 3% of the DNA is transcribed (see Lewin 1997). By using transcription-mapping data, Gardiner (1997) estimated that less than 10% of the human genome is genic. These experiments constitute further support for the view that the vast majority of the eukaryotic genome is devoid of genetic information.

The distribution of protein-coding genes among human chromosomes is extremely uneven. Some chromosomes, such as chromosomes 1, 19, and 20, are predicted to be very gene-rich; others, such as chromosomes 4 and 18, may contain much sparser genetic information. For example, human chromo-

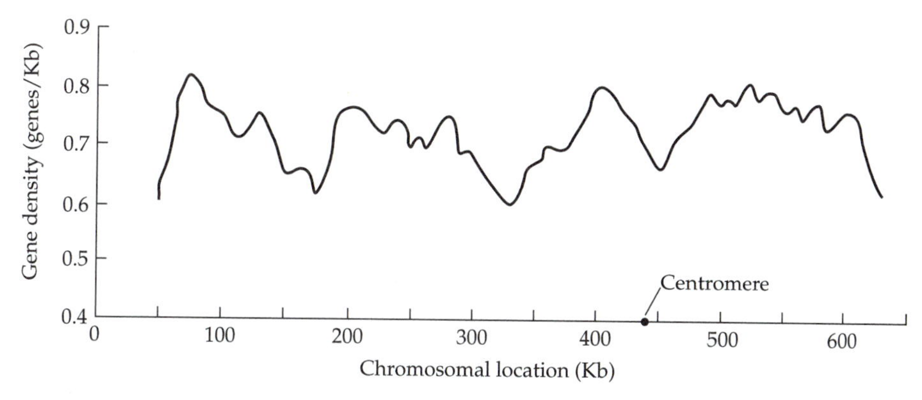

FIGURE 8.14 Periodicity in gene density along chromosome XI of *Saccharomyces cerevisiae*. Modified from Sharp and Matassi (1994).

some 19 is the most gene-rich chromosome, with an estimated 2,000 genes contained within a euchromatic region of about 60 million bp (Mohrnweiser et al. 1996). Its **gene density** is therefore 0.03 genes/Kb. We note that this value is an overestimate even for chromosome 19, let alone for other chromosomes. There are three main reasons for this statement: (1) only the euchromatic region has been taken into account, (2) some of the genes may in fact be pseudogenes, and (3) as mentioned previously, chromosome 19 is the chromosome with the highest gene density.

Gene density, and by extension the genic fraction, seems to be negatively correlated with genome size (Figure 8.15). For example, gene density in *Mycoplasma genitalium* is 0.8 genes/Kb. The density drops to 0.6 genes/Kb in *Escherichia coli*, which has a genome that is 8 times larger. In eukaryotes, the density is approximately 0.5 genes/Kb in the yeast and 0.2 genes/Kb in *Caenorhabditis*, which has a genome that is about 8 times larger. Our estimates of gene density in other organisms are less certain, but the same trend is evident. For example, gene density in *Arabidopsis thaliana* is 0.2 genes/Kb in a gene-rich region on chromosome 1, but only 0.03 genes/Kb in the euchromatin of the most gene-rich human chromosome. The last value compares quite unfavorably with the estimate for *Alu* density in the same chromosome (1.1 elements/Kb).

In large plant genomes, such as rice, maize, and barley, most protein-coding genes are clustered in long DNA segments (collectively called the **gene space**) that represent a small fraction (12–24%) of the nuclear genome, separated by vast expanses of **gene-empty regions** (Barakat et al. 1998). Interestingly, the realization that genes are so sparse and so unevenly distributed within the genomes of multicellular organisms has led to a call to abandon the "factory approach" to the sequencing of the human genome, and to adopt instead a "boutique approach," whereby only gene-rich regions will be sequenced.

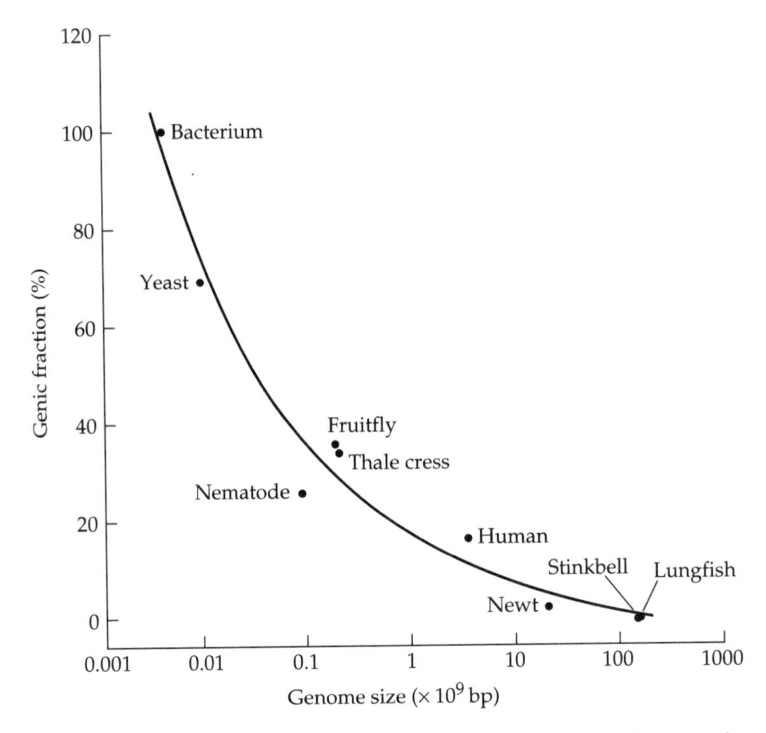

8.15 Relationship between genic fraction and genome size. The organisms included in the figure are: *Escherichia coli* (bacterium), *Saccharomyces cerevisiae* (baker's yeast), *Caenorhabditis elegans* (nematode), *Triturus cristatus* (warty newt), *Drosophila melanogaster* (fruitfly), *Homo sapiens*, *Protopterus aethiopicus* (marbled lungfish), *Arabidopsis thaliana* (thale cress), and *Fritillaria agrestis* (stinkbell). Note the logarithmic scale of the abscissa. Data from Szathmáry and Maynard Smith (1995).

The last question to be addressed is, What proportion of genes is essential? In experimental and analytical approaches quite similar to those employed to estimate the size of the minimal genome, Miklos and Rubin (1996) used frequencies of loci known to experience lethal mutations in several model organisms to estimate the proportion of genes that are indispensable. Their conclusion was that only about one in three genes is essential for viability. Interestingly, the proportion did not vary much between organisms, and remained around 25–35% in organisms with a large number of genes (e.g., humans, fish), organisms with an intermediate number of genes (e.g., nematodes, *Drosophila*), and organisms with a low gene number (e.g., yeast).

Gene number evolution

As mentioned previously, there is no generally accepted measure of biological complexity. Two possible candidates are the number of protein-coding genes, and "the richness and variety of morphology and behavior" (Szathmáry and Maynard Smith 1995). The latter measure is beyond the scope of this book, so we shall concentrate on the former.

There are no theoretical *a priori* reasons to expect gene number to increase with evolutionary time. However, the empirical evidence indicates that, in some lineages, genic complexity has increased enormously. It has been suggested that gene number has not increased continuously during evolution, but has risen in discrete steps (Bird and Tweedie 1995). Szathmáry and Maynard Smith (1995) suggested that the biggest steps occurred at the transition from prokaryotes to eukaryotes and at the transition from invertebrates to vertebrates. The first step is presumed to have been facilitated by the invention of nucleosomes, whereas the second by the spread of gene methylation as a mechanism controlling gene expression throughout the genome.

In recent years, reliable estimates of gene numbers based on large-sequence sampling are accumulating. Interestingly, these data indicate that gene number increases indeed occurred in "quantum" steps (Simmen et al. 1998). In fact, for one such event it is possible to pinpoint quite accurately the time at which it occurred. In animals, it seems that the last "great leap forward" in gene number occurred sometime in the Silurian, before the divergence of vertebrates, but after the divergence of the invertebrate chordates from the vertebrate ancestor (Appendix I). In Figure 8.16, we see that all vertebrates possess 50,000–100,000 genes, whereas all invertebrates, from nematodes through fruitflies to sea squirts, have considerably fewer than 25,000 genes.

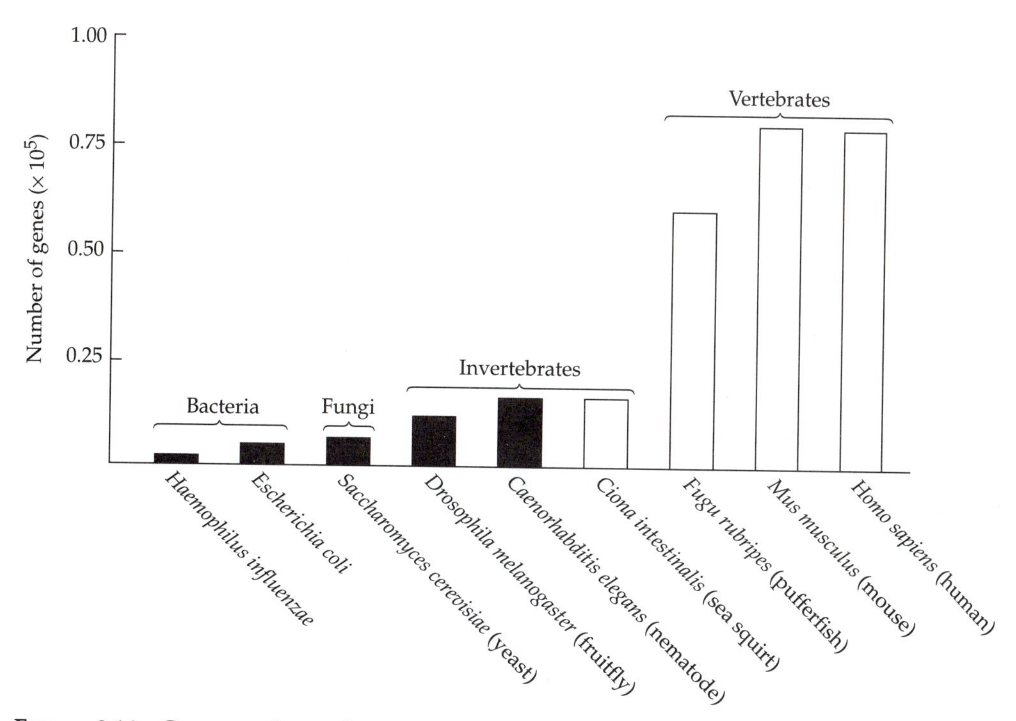

FIGURE 8.16 Gene number estimates in bacterial, fungal, invertebrate, and vertebrate species. Solid bars indicate results based on completed genomic sequences. White bars are estimates based on sample sequencing of genomic DNA. Data from Elgar (1996), Miklos and Rubin (1996), and Simmen et al. (1998).

CHROMOSOMAL EVOLUTION

Despite the fact that cytogenetics is a much older scientific discipline than molecular biology, we know very little about chromosomal evolution beyond descriptive phenomenology. However, with the advent of the genomics era, we are beginning to glean some insight into such issues as the evolution of chromosome number and the dynamics of gene order rearrangements.

Chromosomes, plasmids, and episomes

Organisms and organelles contain two types of genetic material: chromosomes and extrachromosomal elements. **Chromosomal DNA** contains genes, at least some of which are unconditionally essential. **Extrachromosomal elements**, on the other hand, contain genetic information that, although it may have important phenotypic effects, is not necessary under all conditions. Among the best-known phenotypic effects of extrachromosomal elements are: (1) antibiotic, heavy-metal, and heat resistance; (2) virulence and pathogenicity; (3) autotrophy; and (4) antigenic plasticity.

The main classes of extrachromosomal elements are plasmids and episomes. **Plasmids** are autonomously replicating extrachromosomal molecules distinct from the chromosomal genome. They exist solely in an autonomous state and are inherited independently of chromosomes. Their replication rate may be considerably higher than that of the chromosomal DNA, although in many instances there seems to be a 1:1 relationship between plasmid and chromosomal replication rates.

The plasmid genome can be circular or linear, and it may range in size from about 1,000 nucleotides (**cryptic plasmids**) to 400 Kb (**giant plasmids**). For unknown reasons, cryptic plasmids are always circular, and giant plasmids always linear. So far, plasmids have been discovered in all three domains (Bacteria, Eucarya, and Archaea), as well as in mitochondria. In eukaryotes, however, they have only been found in fungi, slime molds, red algae, and plants (although, depending on definition, a mammalian virus such as SV40 may be considered a plasmid). An organism may carry many types of plasmids simultaneously.

Episomes, too, contain only nonessential genetic information, but they are capable of alternating between two states: independently replicating within a cell, or integrated into the chromosome. An example of episomes are viral prophages. An episome that loses its ability to become attached to the chromosome becomes a plasmid; one that loses its ability to detach itself from the chromosome becomes part of the chromosome.

Evolution of chromosome number in prokaryotes

The vast majority of bacteria contain a single chromosome. There are, however, exceptions. Within the genus *Brucella*, a Gram-negative bacterial group pathogenic to animals and humans, we find species with either a single chromosome or with two. For example, *B. melitensis*, a pathogen of sheep and goats, has two circular chromosomes, 2,100 Kb and 1,150 Kb in size. In other *Brucella* species with two chromosomes, the sizes of the chromosomes may be different, e.g., 1,850 Kb and 1,350 Kb in the porcine pathogen *B. suis*. Interestingly, the size of the chromosome in single-chromosome strains is about the same as the total chromosome size for two-chromosome strains. Jumas-Bilak et al. (1998) have shown unambiguously that these are *bona fide* chromosomes rather than extrachromosomal elements, because the researchers could map all known genes of the two chromosomes on the one chromosome of single-chromosome *Brucella* species. Moreover, they could explain all the chromosomal variation in *Brucella* by paralogous recombination among three rRNA-specifying loci.

A similar situation was found in another member of the α subdivision of proteobacteria. *Rhodobacter sphaeroides*, a facultative photosynthetic bacterium, was found to possess two true circular chromosomes (3,000 and 900 Kb), each containing genes that are essential for metabolic function (Mouncey et al. 1997).

Interestingly, the genome of *Methanococcus jannaschii*, the first completely sequenced archaeon, was found to consist of three physically distinct elements: (1) a large circular chromosome of about 1,700 Kb, containing about 1,700 protein-coding genes; (2) a large element of close to 60 Kb, containing 43 predicted protein-coding regions; and (3) a small element of about 17 Kb with a coding capacity for 12 proteins. However, we do not know as yet whether the last two elements are chromosomal or extrachromosomal.

Using an elaborate laboratory protocol, Itaya and Tanaka (1997) succeeded in dividing a bacterial chromosome into two independently replicating **subgenomes**. The resulting two-chromosome bacteria were viable. This finding indicates that the evolution of chromosome number in bacteria may be restricted by mutational input rather than by selection against multichromosomality. That is, the generation of two viable chromosomes out of one chromosome requires many low-probability steps in a particular order. We note, however, that given the long evolutionary history and the diversity of lineages in prokaryotes, even a low rate of mutational input should have led to a significant increase in chromosome number. The rarity of multichromosomal prokaryotic species indicates that selection against mutlichromosomality must be quite powerful.

Chromosome number variation in eukaryotes

In eukaryotes, chromosome numbers vary as extensively as genome sizes, and they also do not exhibit any relationship with biological complexity. Even within closely related taxa, the haploid number of chromosomes (n) may vary enormously. In insects, n varies from 1 in the Australian ant *Myrmecia pilosula* to almost 250 in the butterfly *Lysandra atlantica*. In plants, within a single family (Asteraceae) we find that n varies from 2 in *Haplopappus gracilis* to circa 90 in *Senecio roberti-friesii*. Even in mammals, which have quite a narrow range of genome sizes (Table 8.3), n varies from 5 in the hystricognath rodent *Ctenomys steinbachi* to 102 in another hystricognath, *Tympanoctomys barrerae*.

Surprisingly, however, chromosome number does not correlate at all with DNA content. In jawless fish, for instance, the lamprey *Eptatretus stoutii* (Pacific hagfish) with 24 chromosomes in the haploid genome has twice the amount of DNA of *Lampetra planeri* (brook lamprey), which has 73 chromosomes in the haploid nucleus. Similarly, the yeast *Saccharomyces cerevisiae* has 16 chromosomes and about 1.2×10^4 Kb, while the lily *Lilium longiflorum* has 10,000 times as much DNA and only 12 chromosomes. The lack of correlation is nicely illustrated by pea species belonging to the genus *Lathyrus*. Despite the fact that nuclear DNA content was shown to vary widely among species, all species have the same number of chromosomes (Narayan and Rees 1976).

Mechanisms for Changes in Gene Order and Gene Distribution Among Chromosomes

Depending on the position of the centromere, eukaryotic chromosomes can be divided into three types: **telocentric**, **acrocentric**, and **metacentric** (Figure 8.17). Many processes can bring about changes in gene order (Figure 8.18). **Chromosomal inversions** involve rotating a segment 180°, with the result that the gene order for the segment is reversed with respect to its original order. There are two types of inversions: **pericentric** and **paracentric**. In the former, the inverted segment includes the centromere. **Chromosomal deletions** may be **terminal** or **interstitial**. Alternatively, parts of a chromosome may be duplicated, a process we have previously referred to as partial polysomy. A chromosome may also be eliminated, a process that has only been documented in somatic cells of arthropods and is, therefore, of little evolutionary interest. Finally, a chromosome may break into two, with each of the resulting chromosomes assuming independent existence. This process, however, can only occur if the chromosome has a diffuse centromere, i.e., if during mitosis and meiosis the spindle traction fibers attach to several sites along the length of the chromosome.

Different chromosomes may exchange genetic information through such processes as **reciprocal** and **nonreciprocal translocation**, and **centric fusion** (Figure 8.19). Depending on the types of chromosomes involved in the process, centric fusion may or may not be accompanied by the loss of genes.

Chromosome number reduction by fusion seems to be a recurrent evolutionary occurrence. As recently as 12 million years ago, for instance, the ances-

FIGURE 8.17 Classification of eukaryotic chromosomes by centromere position (circle).

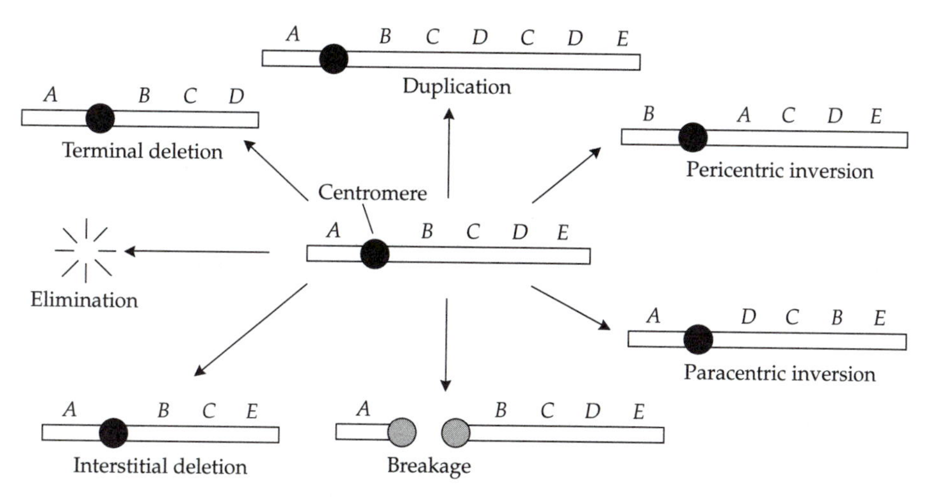

FIGURE 8.18 A chromosome with five genes (center) may undergo several process of gene rearrangement. Genes may be lost through terminal and interstitial deletions, or through outright chromosome elimination. The number of gene may increase through duplication. The order of genes may change following pericentric and paracentric inversions. Gene distribution among chromosomes may change through chromosome breakage. The last process, however, can only occur if the centromere is diffuse (gray circles) and, hence, divisible.

tors of pig (*Sus scrofa*) experienced two centromere–telomere fusions, creating two large chromosomes out of four smaller ones (Thomsen et al. 1996). The ancestral state is still retained in the closely related babirusa (*Babyrousa babyrussa*). The most dramatic reduction in chromosome number in mammals has been registered for the Indian and black muntjacs (*Munitacus muntiacus vaginalis* and *M. crinifrons*), whose males have a haploid chromosome set of 6 and 8, respectively. Yang et al. (1997) proposed that the reduction in chromosome number within the family Cervidae (deer) was accomplished through 12 fusions

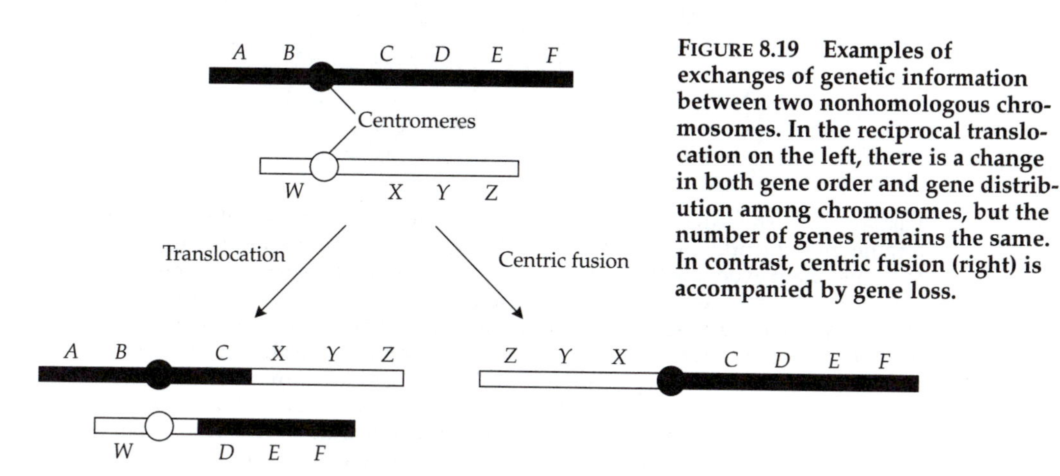

FIGURE 8.19 Examples of exchanges of genetic information between two nonhomologous chromosomes. In the reciprocal translocation on the left, there is a change in both gene order and gene distribution among chromosomes, but the number of genes remains the same. In contrast, centric fusion (right) is accompanied by gene loss.

from an ancestral state of $n = 70$ that is still retained in the brown-brocket deer (*Mazama gouazoubira*) and the Chinese water deer (*Hydropotes inermis*) through an intermediate state of $n = 46$ as in the Chinese muntjac (*M. reevesi*).

Counting gene order rearrangement events

To be able to study the evolution of gene order rearrangements, we must first be able to estimate the number of events (e.g., inversions, transpositions, and deletions) that are necessary to change the gene order of one extant genome into another. This will give us an estimate of the number of gene order rearrangement events that have occurred since the divergence of two genomes from each other. A simple method, called the **alignment reduction method**, has been suggested by Sankoff et al. (1992). In this method we compute a so-called **evolutionary edit distance** (E) between two genomes, say A and B. E has two components: the **deletion distance** (D), which is the minimal number of deletions or insertions necessary for the A and B genomes to have identical sets of genes, albeit in a different order, and the **rearrangement distance** (R), i.e., the minimal number of inversions and transpositions necessary to convert the gene order of A into the gene order of B.

$$E = D + R \tag{8.1}$$

To estimate E, we employ three simple geometrical procedures: **deletion**, **bundling**, and **inversion** (Figure 8.20). We start by connecting homologous genes by lines. In this stage, we distinguish between homologous pairs that have the same genomic orientation and those that are inverted relative to one another. The easiest procedure is deletion: all genes that are absent in either of the two genomes are removed. Thus, D is equal to the number of segments

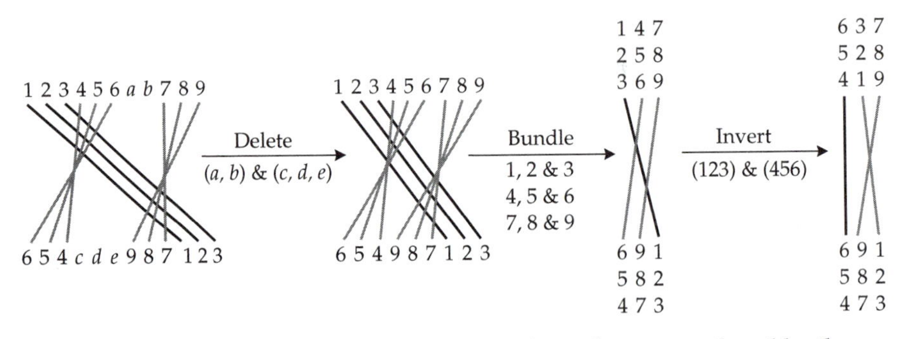

FIGURE 8.20 The three basic geometrical procedures that are employed by the alignment reduction method (Sankoff et al. 1992) to infer the number of gene order rearrangement events between two genomes. Homologous genes are connected by lines. Black lines indicate the same orientation; gray lines indicate reverse orientation. Genes that are absent in either of the two genomes (lowercase letters) are deleted. Five genes but only two segments were removed and, therefore, the deletion distance is 2. Bundling involves the clustering of adjacent genes if the genes have the same relative order and the same orientation in both genomes. Bundling carries no weight in computing the rearrangement distance. The inversion of segments (123) and (456) causes a change in their relative orientation.

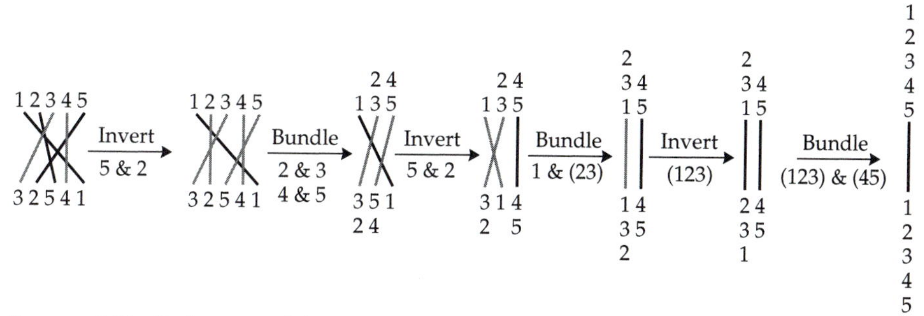

FIGURE 8.21 Inference of the rearrangement distance for two genomes of five genes each by the alignment reduction method. Homologous genes are connected by lines. Black lines indicate the same orientation; gray lines indicate reverse orientation. The total number of inversions is 3. Modified from Sankoff et al. (1992).

that were removed. Note that in our case, we removed 5 genes but only 2 segments, so $D = 2$.

If certain genes are adjacent to one another in both genomes, and if all these genes have the same relative order, whether in the same or in the reverse orientation, we may combine them into one bundle. We note that bundling is just an algorithmic procedure that carries no weight in computing the rearrangement distance. The third possible procedure is inversion, i.e., the conceptual rotation of a segment 180° without changing the genomic location. An inversion will, of course, change the relative orientation of the homologous segments, and we must keep track of such changes. In the alignment reduction method, we invert and bundle until the alignment of the two genomes reduces to a single segment in the same orientation in both genomes. R is the sum of all inversions along the way.

A hypothetical case of inferring the number of events that have occurred since the divergence of two (very small) genomes from each other is shown in Figure 8.21. In this case, we do not bother with D, since its computation is trivial. The minimal solution to the problem in Figure 8.21 turns out to be $R = 3$.

Unfortunately, finding the smallest possible R is what computer scientists call a computationally hard problem—a problem requiring computing time that increases exponentially with the number of genes in the two genomes. Thus, the method is only applicable to small and evolutionarily conserved genomes. Sankoff et al. (1992) have applied the method to infer the number of indels and segmental changes that have occurred in the evolution of animal mitochondrial genomes (Table 8.6). We note that even though animal mitochondrial genomes have only about 30 genes in common, computing R required more than 300 hours of computing time. From the comparison of the deletion distances and the rearrangement distances (Table 8.6), we deduce that gene number has been quite a conserved trait throughout animal evolution, whereas rearrangements occurred very frequently. For example, while mitochondrial gene content has not changed at all from sea urchins to hu-

TABLE 8.6 Evolutionary edit distance between pairs of animal mitochondrial genomes[a]

OTUs[b]	*Hs*	*Gg*	*Sp*	*Ap*	*Po*	*Dy*	*As*
Hs		1	18	16	19	13	25
Gg	0		19	17	17	12	26
Sp	0	0		2	1	26	27
Ap	4	4	4		1	22	24
Po	1	1	1	5		23	24
Dy	0	0	0	4	1		28
As	1	1	1	5	2	1	

Modified from Sankoff et al. (1992).

[a]Deletion distances and rearrangment distances are below and above the diagonal line, respectively.

[b]*Hs, Homo sapiens* (human); *Gg, Gallus gallus* (chicken); *Sp, Strongylocentrotus purpuratus* (sea urchin); *Ap, Asterina pectinifera* (starfish); *Po, Pisaster ochraceus* (starfish); *Drosophila yakuba* (fruitfly); *As, Ascaris suum* (pig roundworm).

mans, a minimum of 16 rearrangements of the 30 genes are inferred to have occurred since they last shared a common ancestor.

Gene order rearrangements in bacteria

Because bacteria are severely restricted in recombination and because many of their genes function as units (operons), Ochman and Wilson (1987) suggested that bacterial evolution is characterized by spatial and temporal stability of gene order. With the completion of the sequencing of the first bacterial genomes, however, it became clear that gene order in bacteria is anything but conserved (Mushegian and Koonin 1996b).

To illustrate this lack of conservation, let us compare the gene order between *Haemophilus influenzae* and *Mycoplasma genitalium*. Watanabe et al. (1997) identified 184 orthologous genes between these two species. The correspondence of orthologous gene location is shown in Figure 8.22. In this representation, informally called a **Tsuzumi graph** because of its resemblance to a traditional Japanese drum, adjoining parallel lines are indicative of gene order conservation. (For circular genomes, Tsuzumi graphs are three-dimensional, but they can be rendered two-dimensional for convenience of presentation without much loss in visual information.) It is very obvious from Figure 8.22 that lines indicative of gene order conservation are rare. In fact, the genes in the two genomes seem to have been reshuffled so frequently as to appear to be randomly ordered. The situation is even worse in a comparison between *H. influenzae* and *E. coli*, despite the fact that these two species are more closely related to each other than *H. influenzae* and *M. genitalium*. Nevertheless, several short regions of conservation were found, including 11 regions composed of 2 genes, 5 regions composed of 3 genes, and 1 region consisting of 4 genes. The only large conserved region was the so-called *S10* region, con-

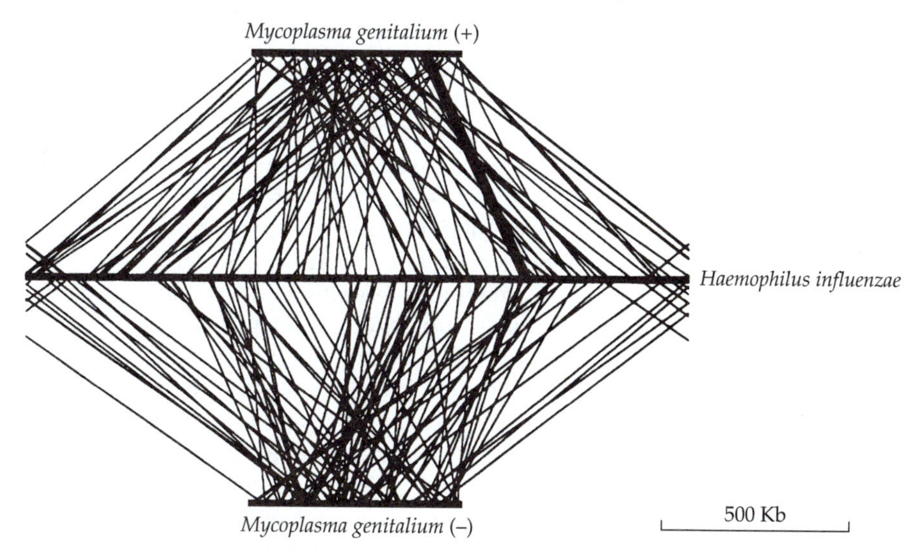

FIGURE 8.22 Comparison of the physical positions of orthologous genes between the genomes of *Mycoplasma genitalium* and *Haemophilus influenzae* by means of a two-dimensional Tsuzumi graph. The plus (+) and minus (–) strands of *M. genitalium* are compared separately to the *H. influenzae* genome. Orthologous genes are connected by lines. The heaviest line represents the conserved *S10* region (see Figure 8.23). A scale bar is provided for reference. From Watanabe et al. (1997).

sisting of 3 operons with 17 genes (Figure 8.23). Surprisingly, this region was also found to be conserved in much more distant comparisons, for example, between the cyanobacterium *Synechocystis* and the archaeon *Methanococcus*. This exceptional conservation indicates not only that a prototype *S10* region

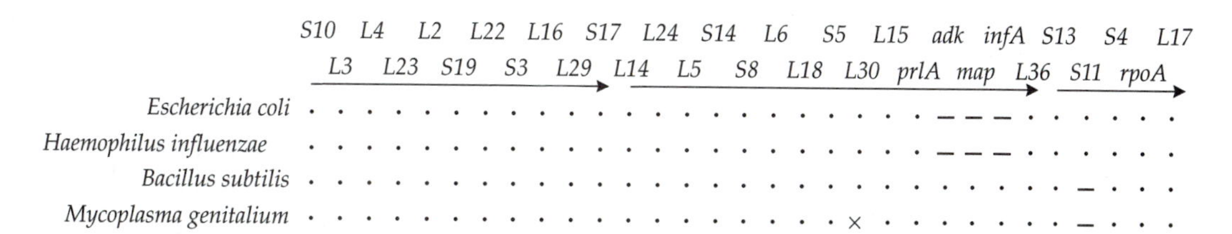

FIGURE 8.23 The extensively conserved *S10* region among *Escherichia coli*, *Haemophilus influenzae*, *Bacillus subtilis*, and *Mycoplasma genitalium*. The three arrows shown below the gene names represent operons in *E. coli*. A dot (•) indicates the existence of a gene at a site; a minus sign (–) indicates that the gene has been translocated elsewhere in the genome; × indicates that the gene was not found in the genome. Gene abbreviations: *L*- and *S*-prefixed genes, large and small ribosomal protein subunits, respectively; *prlA*, preprotein-translocation secY subunit; *adk*, adenylate kinase; *map*, methionine aminopeptidase; *infA*, initiation factor 1; *rpoA*, DNA-directed RNA polymerase α chain. From Watanabe et al. (1997).

may have already existed in the cenancestor, but also that extraordinary selective constraints on gene order are operating in this region.

Gene order rearrangements in eukaryotes

Studying gene order rearrangements in eukaryotes is infinitely more complicated than in bacteria for several reasons. First, the eukaryotic genome contains many repeated genes, and deciding whether two genes from two organisms are orthologous or paralogous is quite complicated. Second, gene order rearrangements involve the movement and exchange of genetic information between chromosomes as well as within chromosomes. Third, gene order seems to be a very unstable character. For example, genes from human chromosome 1 were found on nine different chromosomes in mice. Finally, eukaryotes have larger numbers of genes than prokaryotes, an insurmountable impediment for all computer algorithms designed to deal with the problem.

A simple quantitative approach to the problem of gene order conservation is to count how many gene neighbors in one species are also neighbors in another species (Keogh et al. 1998). By using this measure, Keogh et al. (1998) found that gene order conservation decreases uniformly with evolutionary distance, so much so that not even one common pair of neighbors was found between the baker's yeast *Saccharomyces cerevisiae* and the fission yeast *Schizosaccharomyces pombe*, which diverged from each other approximately 420 million years ago.

In comparisons between multichromosomal organisms, it is convenient to define several terms that may help us tackle the problem of gene movement among chromosomes. The following definitions have been taken from Ehrlich et al. (1997). **Synteny** refers to the occurrence of two or more genes on the same chromosome. **Conserved synteny** refers to two or more homologous genes that are syntenic in two species. **Conserved linkage** pertains to the conservation of both synteny and gene order in homologous genes between species. A **disrupted synteny** refers to cases where two genes are located on the same chromosome in one species but their orthologs are located on different chromosomes in the second species. Finally, a **disrupted linkage** refers to a difference in gene order between two species. Obviously, a disrupted synteny will also be counted as a disrupted linkage (Figure 8.24).

From the comparison of any two multichromosomal genome maps, it is possible to obtain information on the following variables: (1) the number of conserved syntenies; (2) the distribution of number of genes among conserved syntenies; (3) the number of conserved linkages; and (4) the distribution of number of genes among conserved linkages. By assuming a uniform distribution of genes over the genome, and by using a maximum likelihood approach with gene order data from nine well-mapped mammalian genomes (cow, chimpanzee, human, baboon, hamster, mouse, rat, mink, and cat), Ehrlich et al. (1997) estimated the number of breakpoints or synteny disruptions required to explain the differences in karyotypes between any two genomes. Their main findings were: (1) gene order rearrangements occur with amazing rapidity; (2) the rates of synteny disruption vary widely among

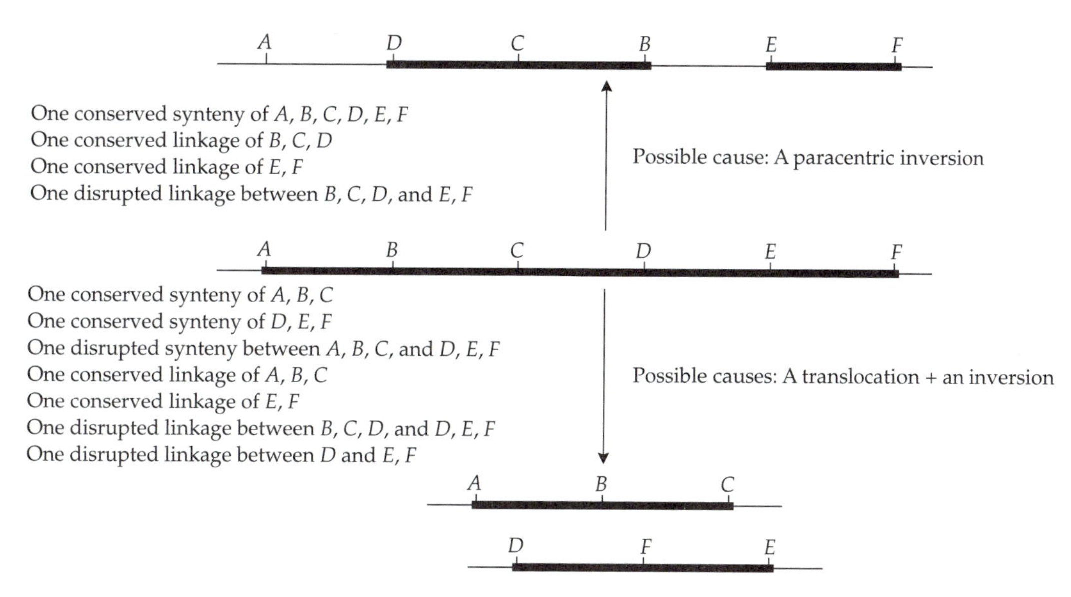

FIGURE 8.24 Examples of conserved and disrupted syntenies and linkages, and possible reasons for disruptions. Genes are indicated by capital letters. Linkages are indicated by solid bars. The reference sequence is shown in the middle. Modified from Ehrlich et al. (1997).

mammalian lineages, with the mouse lineage experiencing a rate of synteny disruptions 25 times higher than that of the cat lineage; and (3) despite theoretical *a priori* considerations implying that interchromosomal rearrangements should be strongly selected against, they were found to occur approximately four times more frequently than intrachromosomal rearrangements.

Gene order as a phylogenetic character

Evolutionary biologists are increasingly drawn to structural features of the genome, such as gene order, as a phylogenetic marker. Unfortunately, the rapid rate of gene order rearrangements and the rate variation among evolutionary lineages essentially spells bad news for phylogenetic reconstructions based on gene order. For example, in a study of mitochondrial gene order in 137 species of birds, Mindell et al. (1998) showed that parallel evolution in gene order is quite common, and that the same gene order may arise independently several times even in closely related taxa. Because of this unsavory phylogenetic property and the difficulty in reconstructing the sequence of events leading to gene order rearrangements, phylogenetic trees based on gene order frequently contain glaring errors. For example, by using synteny conservation and chromosome rearrangements in nine species of mammals, Ehrlich et al. (1997) inferred an absurd phylogenetic tree in which baboons are closer to humans than humans are to chimpanzees, and carnivores are paraphyletic. A more "optimistic" view on the utility of gene order in phylogenetic reconstruction, at least as far as small mitochondrial genomes are concerned, is presented by Boore and Brown (1998).

GC CONTENT IN BACTERIA

Among eubacterial genomes (including organelle genomes), the mean percentage of guanine and cytosine, or **GC content**, varies from approximately 25% to 75%. In many cases, bacterial GC content seems to be related to phylogeny, with closely related bacteria having similar GC contents (Figure 8.25).

There are essentially two types of hypotheses to explain the variation in GC content in bacteria. The **selectionist view** regards the GC content as a form of adaptation to environmental conditions. For example, G:C pairs are more stable than A:T pairs because of an additional hydrogen bond (Chapter 1), and therefore there may be a relationship between GC content and the temperatures to which bacteria are exposed. Indeed, initial studies by Argos et al. (1979), Kagawa et al. (1984), and Kushiro et al. (1987) seemed to indicate that in thermophilic bacteria, which inhabit very hot niches, there may be a preferential usage of amino acids encoded by GC-rich codons (e.g., alanine and arginine) and an avoidance of amino acids encoded by GC-poor codons (e.g., serine and lysine). However, in an extensive study of 764 prokaryotic species, Galtier and Lobry (1997) found no correlation between GC content and optimal growth temperatures.

Another selectionist scenario invokes UV radiation as the selective force. Since T–T dimers are sensitive to radiation, microorganisms in the upper layers of the soil, which are exposed to sunlight, should have a higher GC content than bacteria that are not exposed, e.g., intestinal bacteria (Singer and Ames 1970; Ellison and Childs 1981).

The **mutationist view** invokes biases in the mutation patterns to explain the variation in GC contents (Sueoka 1964; Muto and Osawa 1987). According to this view, the GC content of a given bacterial species is determined by the balance between (1) the rate of substitution from G or C to T or A, denoted as u; and (2) the rate of substitution from A or T to G or C, denoted as v. At equilibrium, the GC content is expected to be

$$P_{GC} = \frac{v}{v+u} \quad \textbf{(8.2)}$$

Therefore,

$$u/v = \frac{1-P_{GC}}{P_{GC}} \quad \textbf{(8.3)}$$

The ratio u/v is also called the **GC mutational pressure**. When u/v is 3.0, the GC content at equilibrium will be 25%. Such is the situation in *Mycoplasma capricolum*. When the ratio is 1, the GC content will be 50%, as in *Escherichia coli*. When it is 0.33, the GC content will be 75%, as in *Micrococcus luteus*. To estimate the GC mutational pressure, it is advisable to use DNA sites at which selective constraints are absent or very weak. For example, the GC content at

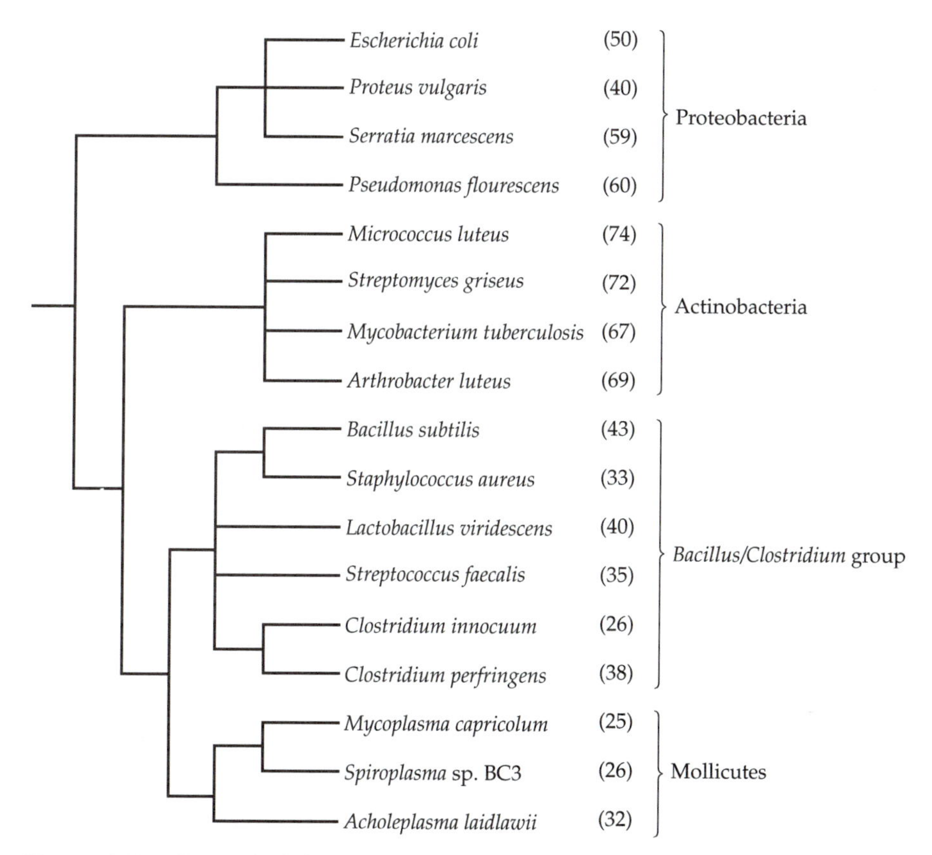

FIGURE 8.25 Genomic GC contents (parentheses) in several eubacteria. The unscaled phylogenetic tree is based on data from Galtier and Gouy (1994) and Ludwig et al. (1998). GC content data from Hori and Osawa (1986) and Muto et al. (1986, 1987).

fourfold degenerate sites in *M. capricolum* protein-coding genes is lower than 10%. Consequently, u/v must be higher than 9. Similarly, in *M. luteus*, $P_{GC} > 0.9$ at fourfold degenerate sites; therefore, $u/v < 0.11$.

In addition to GC mutational pressure, mutational changes are also subject to selective constraint. In other words, the pattern of substitution is determined by the pattern of mutation and the pattern of purifying selection against certain mutations (Chapter 4). The weaker the selective constraint is in a particular region, the stronger the effect the GC mutational pressure will have on the GC composition. Figure 8.26 shows the correlation between the total GC content and the GC content at the three codon positions for 11 bacterial species covering a broad range of GC content values. We see that the correlation at the third position resembles the expectation for the case of no selection. On the other hand, the correlations at the first and second positions, while positive, show a more moderate slope. This is easily explained by the

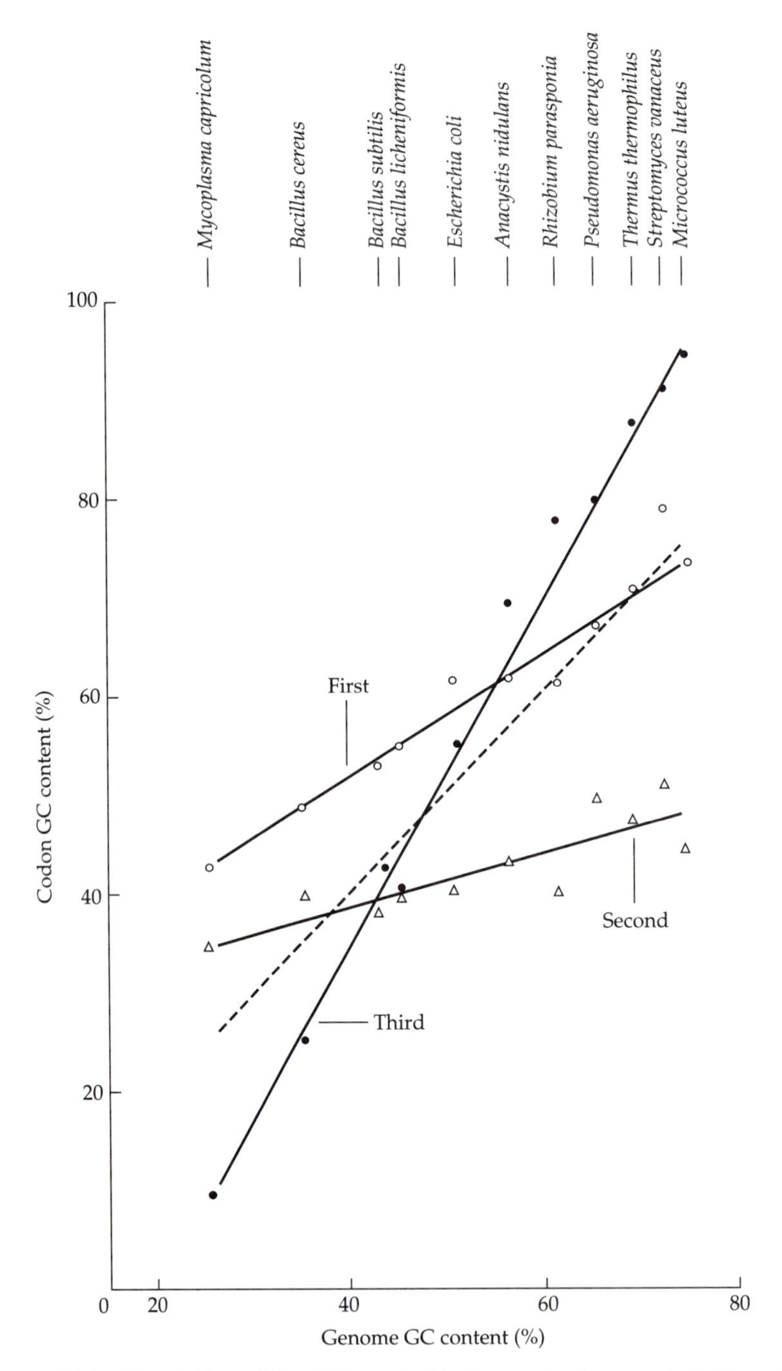

FIGURE 8.26 Correlation of the GC content between total genomic DNA and the first (open circles), second (open triangles), and third (solid circles) codon positions. The dashed line represents the theoretical expectation of a perfect correspondence between the GC content in the genome and that in the codons. Modified from Muto and Osawa (1987).

fact that selection at the mostly degenerate third position of codons is expected to be much less stringent than that at the first and second positions (Chapter 4), so that the GC level at the third position is largely determined by mutation pressure.

CHIROCHORES

The differences in the way the leading and lagging strands of DNA are replicated (Chapter 1) can result in strand-dependent mutation patterns. In the absence of any selection bias between the two strands, differences in the patterns of replication error may lead to differences in the substitution patterns. During evolutionary times, such differences may accumulate to yield equilibrium frequencies within each strand that differ from those expected under no-strand-bias conditions, i.e., $f_A = f_T$ and $f_C = f_G$ (Chapter 4). Deviations from equal mutation rates between the two strands are quantified by using a variable called the **skew**, $S_{X=Y}$, which is a measure of inequality between the frequencies of nucleotides X and Y on a strand. It is calculated as

$$S_{X=Y} = \frac{f_X - f_Y}{f_X + f_Y} \tag{8.4}$$

If there are no violations of the no-strand-bias conditions, $S_{X=Y} = 0$. Skew values are calculated for a sliding window of a predetermined length and are plotted on a **skew diagram** (Figure 8.27).

Lobry (1996c) examined three bacterial genomes and found considerable deviation from $f_C = f_G$. That in itself was not particularly surprising. What was surprising, however, was the spatial distribution of the skews (Figure 8.27). $S_{C=G}$ deviations abruptly switched direction at the origin and the terminus of replication. The abruptness of the switch can be illustrated by means of a vectorial representation suggested by Mizraji and Ninio (1985). In this represen-

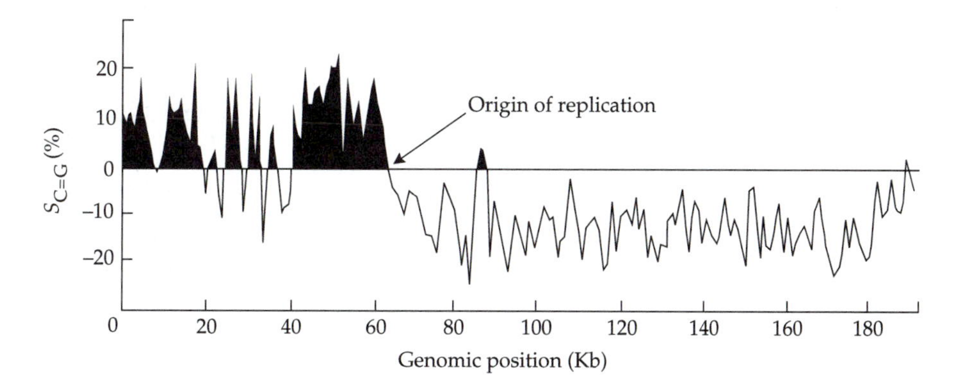

FIGURE 8.27 Skew diagram for C and G in a 200-Kb genomic sequence of *Bacillus subtilis*. Positive deviations from 0 are indicated in black. From Lobry (1996b).

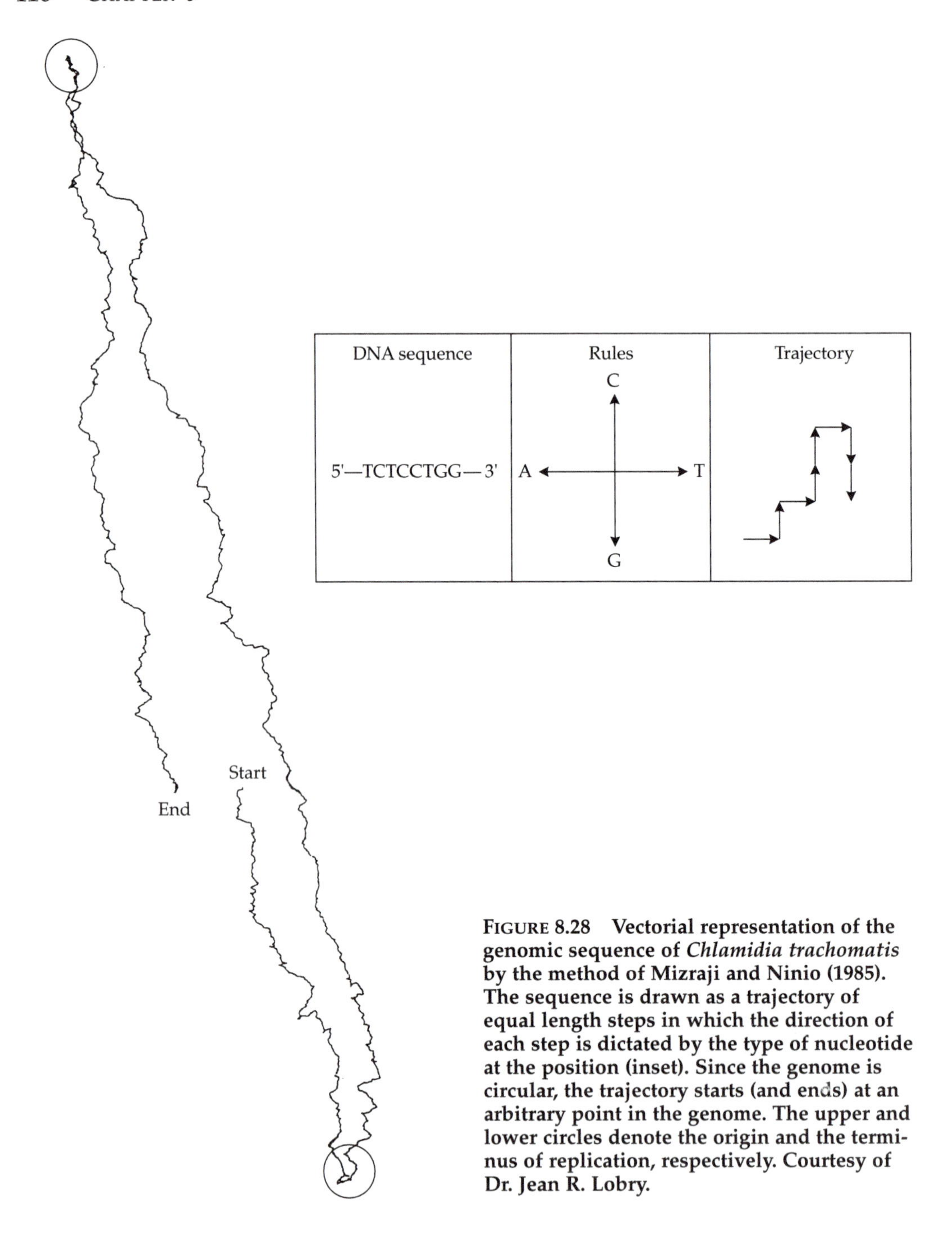

FIGURE 8.28 Vectorial representation of the genomic sequence of *Chlamidia trachomatis* by the method of Mizraji and Ninio (1985). The sequence is drawn as a trajectory of equal length steps in which the direction of each step is dictated by the type of nucleotide at the position (inset). Since the genome is circular, the trajectory starts (and ends) at an arbitrary point in the genome. The upper and lower circles denote the origin and the terminus of replication, respectively. Courtesy of Dr. Jean R. Lobry.

tation the sequence is shown as a trajectory composed of equal-length steps, in which the rules dictate a movement leftward, upward, rightward, and downward for A, C, T, and G, respectively (Figure 8.28; Lobry 1996a).

The switch in $S_{C=G}$ values at the origin of replication proved to be of such a general pertinence in bacteria that they have been used successfully to identify the origin of replication in species in which the lack of consensus sequences prevented their precise identification before (e.g., Lobry 1996b).

Bacterial genomes are thus divided into two segments defined by a symmetry break in the racemic proportions of complementary nucleotides, mainly C and G, between the two strands. In analogy with isochores (discussed in the next section), they were called **chirochores**. Two possible explanations have been put forward to explain the existence of chirochores. The selectionist hypothesis of Forsdyke (1995) claims that there are global selective pressures on genomes in favor of close complementary oligonucleotides that may form secondary structures, such as double-stranded hairpins, for protecting the genome against the ravages of temperature. However, in a study of a wide range of bacterial species with different optimal growth temperatures, Galtier and Lobry (1997) found no evidence for selective pressures that may affect chirochore structure. Again, the most plausible explanation seems to be mutation bias. There are two lines of evidence for this claim. First, the boundaries between chirochores coincide with the origin and terminus of replication, and this immediately suggests a link with processes of replication and DNA repair. Second, the relative bias from $S_{C=G} = 0$ is larger for intergenic regions and third codon positions then for first and second codon positions, as expected if the bias is mutational.

COMPOSITIONAL ORGANIZATION OF THE VERTEBRATE GENOME

Figure 8.29 shows the GC content in different groups of organisms. Interestingly, while the genome sizes of multicellular eukaryotes are generally larger

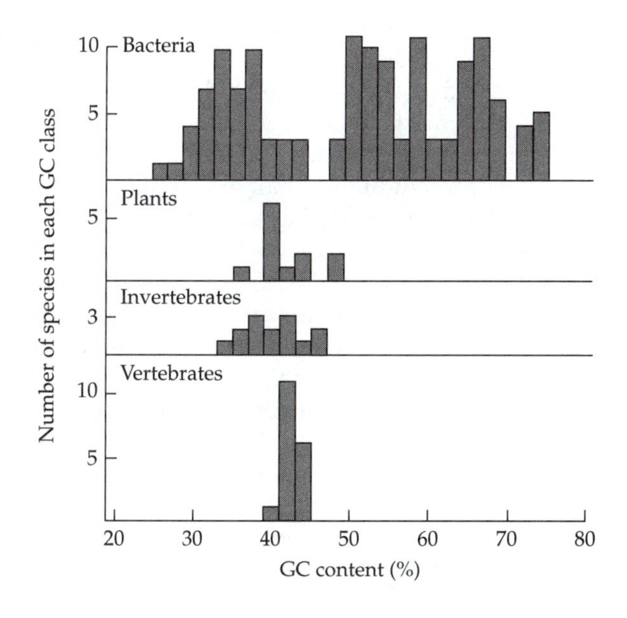

FIGURE 8.29 GC composition of different groups of organisms. From Sueoka (1964).

and more variable in length than those of prokaryotes, GC content exhibits a much smaller variation in eukaryotes. In particular, vertebrate genomes show quite a uniform GC content, ranging from about 40% to 45% (Sueoka 1964). Part of the reason for the small range in GC content in vertebrates might be that vertebrate species, unlike bacteria, have not diverged long enough from one another to allow considerable differences in GC content to accumulate.

The uniformity of the genomic GC content notwithstanding, vertebrate genomes have a much more complex compositional organization than prokaryotic genomes. When vertebrate genomic DNA is randomly sheared into fragments 30–100 Kb in size and the fragments are separated by their base composition, the fragments cluster into a small number of classes, distinguished from each other by their GC content (GC-rich fragments being heavier than AT-rich fragments). Each class is characterized by bands of similar, although not identical, base compositions (Bernardi et al. 1985; Bernardi 1989).

There are conspicuous differences in the compositional organization among vertebrate genomes, which Bernardi et al. (1985, 1988) interpreted as a distinction between warm-blooded and cold-blooded vertebrates. Figure 8.30a shows the relative amounts and buoyant densities of the major DNA components from the carp, *Cyprinus carpio* (left panel), as well as from three warm-blooded vertebrates: chicken, mouse, and human (right panel). In the genomes of chickens, mice, and humans, there are two light components (L_1 and L_2), representing about two-thirds of the genome, and two to four heavy components (H_1, H_2, H_3, and H_4), representing the remaining third. In contrast, the genomic DNA from most fish and amphibians comprises mostly light components.

The compositional distribution of DNA fragments is largely independent of the size of the fragments, indicating a compositional homogeneity over very long DNA stretches. Such homogeneous stretches have been termed **isochores**. Figure 8.30b shows the mosaic organization of nuclear DNA from birds and mammals (i.e., the alternation of light and heavy isochores). When the isochores break during DNA shearing, four major families of molecules with different GC contents are generated. Bernardi et al. (1985) concluded that the GC-rich (heavy) isochores represent about one-third of the genome of birds and mammals but are nearly absent in fish and amphibians. For example, in *Xenopus laevis*, DNA fragments with a GC content higher than 42% represent less than 10% of the genome, as compared with more than 40% in mouse.

In mammals, isochores can be roughly assigned to a small number of families (i.e., L_1, L_2, H_1, H_2, H_3, and H_4) characterized by GC contents ranging from 30–60% (Bernardi 1989). The basic isochore organization in mammals is quite conserved, but several taxa have been shown to possess a different isochore structure than the so-called generalized mammalian one. Thus, mammalian nuclear genomes may be divided into **genomic phenotypes** (Mouchiroud 1995). The most deviant genomic phenotype in mammals is that of mouse and rat (Figure 8.30a). This murid genomic phenotype is characterized by a lower heterogeneity in nucleotide composition, i.e., the GC-rich isochores are not so rich and the GC-poor isochores are not so poor in GC (Mouchiroud and Gautier 1990).

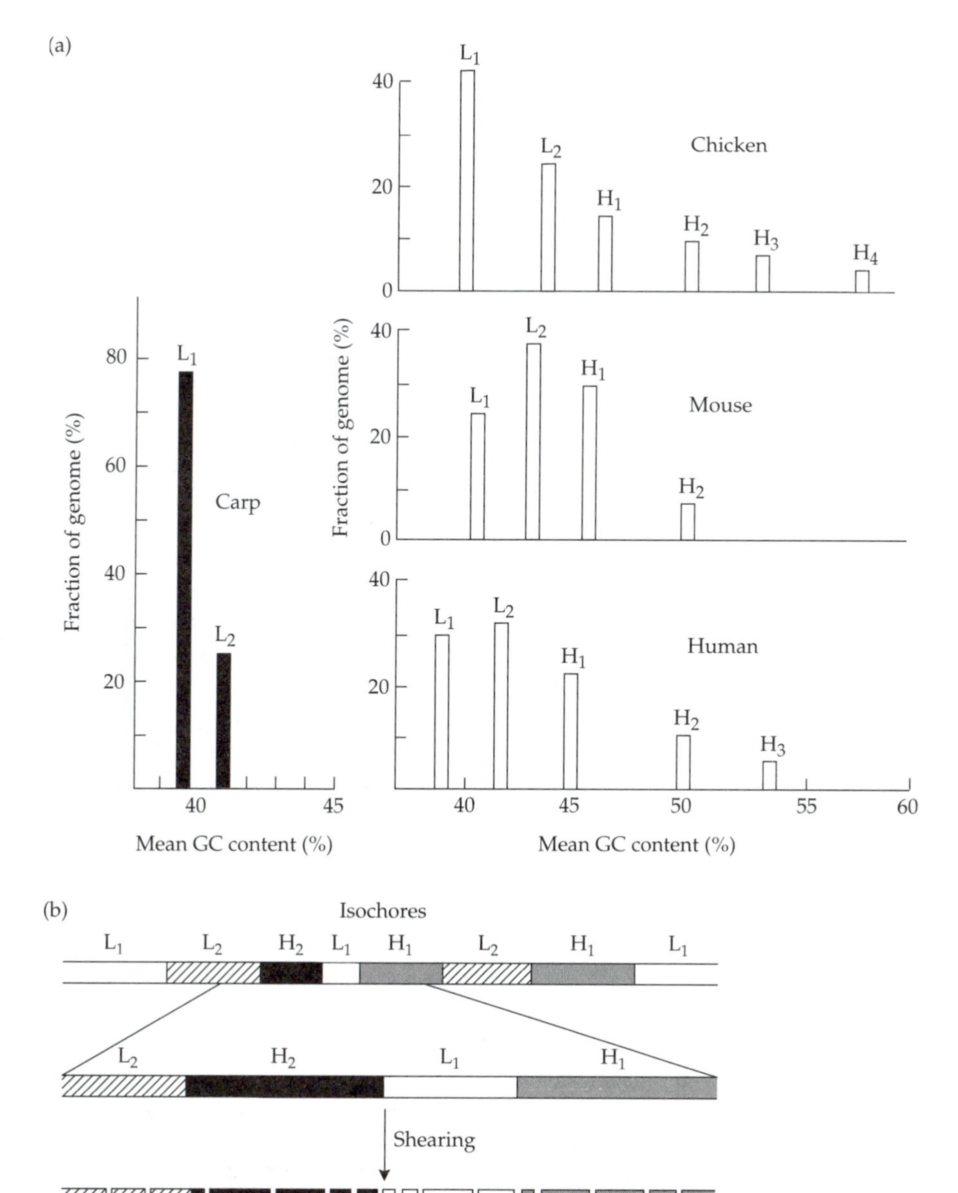

FIGURE 8.30 (a) Histograms showing the relative amounts and mean GC content of the major DNA isochores from a poikilotherm, *Cyprinus carpio* (solid bars in left panel) and three homeotherms: chicken, mouse, and human (white bars in right panel). (b) Scheme depicting the mosaic organization of nuclear DNA from mammals and birds. When the isochores undergo random breakage during DNA preparation, four major families of molecules with different GC contents are generated. Several minor hybrid families (e.g., L_2H_2) are also generated. Modified from Bernardi et al. (1985) and Bernardi (1995).

The increase in GC content in birds and mammals in comparison with that in fish and amphibians is also supported by DNA sequence data (Bernardi et al. 1985, 1988; Mouchiroud et al. 1987). Protein-coding genes from organisms with GC-rich isochores generally have higher GC levels at all codon positions than genes from vertebrates lacking heavy isochores.

The finding that the genome of birds and mammals is mosaic is also consistent with chromosome banding studies. The metaphase chromosomes of these organisms exhibit distinct dark **Giemsa bands** (**G bands**) and light **reverse bands** (**R bands**) when treated with fluorescent dyes, proteolytic digestion, or differential denaturing conditions. The R bands consist of three subsets: **R′ bands**, and two types of **telomeric bands** (**T** and **T′ bands**). In contrast, the metaphase chromosomes of fish and amphibians show either little banding or no banding at all. In human and mouse, GC-poor isochores are distributed among G bands and R bands, GC-rich isochores of the H_1 family correspond roughly to the R′ bands, and the T and T′ bands contain all the H_2 and H_3 isochores (Saccone et al. 1997).

Initial studies on replication time showed that genes localized in GC-rich isochores replicate early in the cell cycle, whereas genes localized in GC-poor isochores replicate late (Goldman et al. 1984; Bernardi et al. 1985; Bernardi 1989), but more recent studies (e.g., Eyre-Walker 1992) failed to unravel such a clear relationship.

The distribution of genes and other genetic elements among isochores

The isochoric positions of many genes from humans and other vertebrates have been determined by various methods. These studies suggest a highly nonrandom distribution of genes throughout the genome. For example, close to 30% of all human genes reside in the heaviest component (H_3), which represents only 3–5% of the genome. Interestingly, long genes are scarce in GC-rich isochores. The difference between genes in GC-rich regions and those in GC-poor regions is almost entirely explained by introns, which are on average three times longer in GC-poor regions (Duret et al. 1995).

In most cases, a gene is embedded in DNA fragments that have a GC content similar to that of the gene itself. This finding provides independent evidence for the existence of isochores, as well as for the large size of the individual isochores. Indeed, since the fragments making up the DNA preparations were produced by random degradation, the narrow compositional range of the gene-carrying fragments indicates that they are very homogeneous in base composition over sizes roughly twice as large as the fragments themselves. These observations are valid not only for isolated genes but also for clustered genes, indicating again that isochores are large in comparison with the gene clusters explored, some of which were 40 Kb or larger. It is estimated that isochores are larger than 300 Kb. Occasionally, a gene is found in fractions covering a wide range of GC levels. This can happen if the probed gene is located close to a border between two isochores, so that random breakage produces gene-carrying fragments of different compositions.

Further support for the existence of isochores in the genome of birds and mammals came from analyses of DNA sequence data, which revealed a positive correlation between the GC level of genes, exons and introns, and the GC level in the large DNA regions in which they are embedded (Bernardi and Bernardi 1985; Bernardi et al. 1985; Ikemura 1985; Aota and Ikemura 1986). Figure 8.31 contrasts the α- and β-globin clusters in humans. The β- and β-like globin genes are low in GC content, and are embedded in a low-GC region. The α- and α-like globin genes, on the other hand, are GC-rich and are embedded in a GC-rich region. The same situation is found in rabbits, goats, and mice. In chickens, both α- and β-globin genes are GC-rich, and both are embedded in GC-rich regions. In contrast, the α- and β-globin genes in *Xenopus* are GC-poor, and both are embedded in a GC-poor region.

In the vast majority of cases, the GC content in coding regions tends to be higher than that in the flanking regions (Figure 8.32). We also see that the GC level at the third codon position is on average higher than that in introns, which in turn is higher than that in the 5′ and 3′ flanking regions. The GC level in the 5′ flanking region tends to be higher than that in the 3′ flanking region, probably because the promoter and its surrounding regions tend to be GC-rich.

As we have noted previously (see Figure 8.10), the distribution of SINEs, LINEs, and other remnants of transposition and retroposition are also not randomly distributed with respect to GC content. Uneven distributions were also noted for the integration sites of active retroviruses (Salinas et al. 1987; Zouback et al. 1994), and patterns of CpG methylation (Caccio et al. 1997).

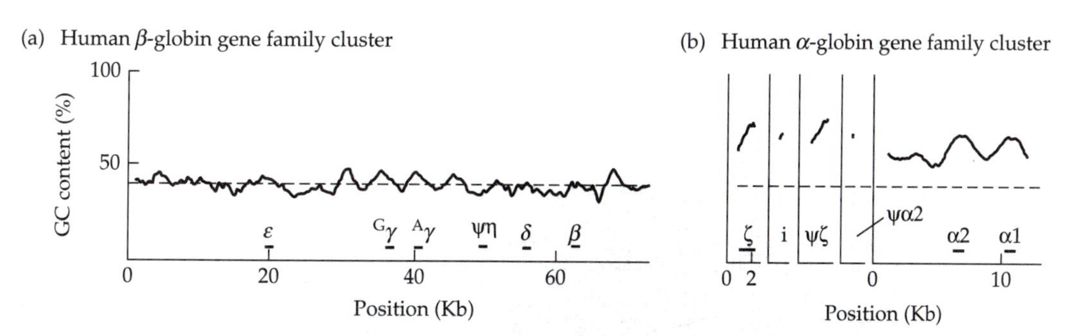

FIGURE 8.31 Distribution of GC content along human globin DNA sequences. (a) The β-globin gene cluster; (b) the α-globin gene cluster (incomplete). The functional genes (bars) and the pseudogenes are arranged in the same order as in Figure 6.11. The gene names are shown at the bottom of the figure; region *i* is the intergenic region between ζ and ψζ. In the β-globin gene family and the region covering the α1- and α2-globin genes, each point represents the average of the GC composition of the 2,001 nucleotides surrounding the point, while in the other regions each point represents the average of 1,401 nucleotides. The horizontal broken line represents the overall GC content of the human genome (40%). Modified from Ikemura and Aota (1988).

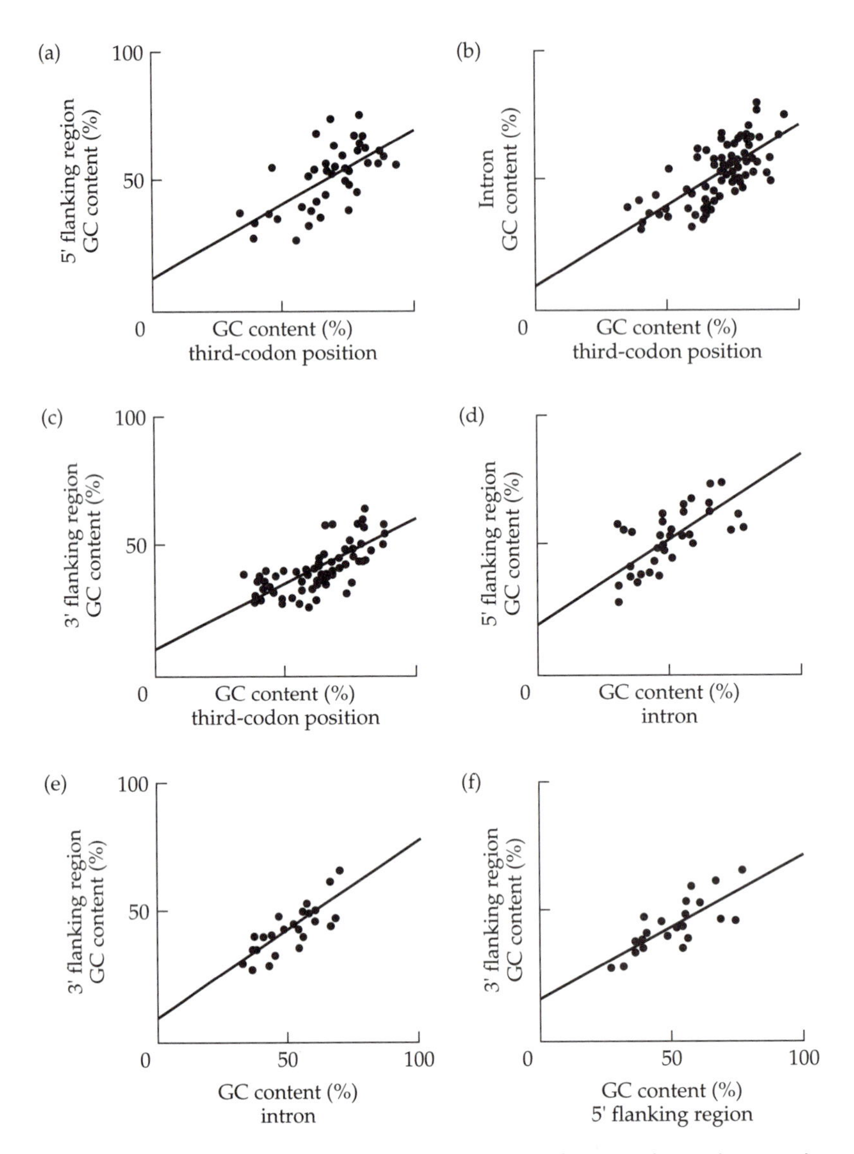

FIGURE 8.32 Relationships among the percentages of GC in the various regions of the gene. (a) The third codon position and the 5′ flanking region. (b) The third codon position and introns. (c) The third codon position and the 3′ flanking region. (d) Introns and the 5′ flanking region. (e) Introns and the 3′ flanking region. (f) 5′ and 3′ flanking regions. From Aota and Ikemura (1986).

Origin of isochores

The origin of GC-rich isochores is as mysterious as it is controversial. Note that what is at issue is the general tendency of long DNA segments (300 Kb or more) to be either GC-rich or GC-poor, not the localized variation in GC content, such as that observed among the various regions of a gene. Bernardi et al. (1985, 1988) proposed that isochores arose because of a functional (i.e., selective) advantage. For example, it has been claimed that in warm-blooded organisms, an increase in GC content can protect DNA, RNA, and proteins from heat damage. We call this view the **selectionist hypothesis**.

Wolfe et al. (1989a) proposed that isochores arose from mutational biases due to compositional changes in the precursor nucleotide pool during the replication of germline DNA. The GC-rich isochores are carried on replicons that replicate early in the germline cell cycle, during which the precursor pool has a high GC content, and thus have a propensity to mutate to GC. The AT-rich isochores, on the other hand, are replicated late in the cell cycle, when the precursor nucleotide pool has a high AT content. These isochores will have a propensity to mutate to AT. We call this the **mutationist hypothesis**. This hypothesis is based on the observation that the composition of the nucleotides precursor pool changes during the cell cycle, and that such changes can in fact lead to altered base ratios in the newly synthesized DNA (Leeds et al. 1985), especially since the replication of the mammalian genome is such a lengthy process, taking eight hours or more (Holmquist 1987).

Let us now examine the pros and cons for each of the theories and their variants. One variant of the selectionist hypothesis maintains that GC increases in the first and second positions of codons may confer thermal stability to proteins because proteins encoded by GC-rich codons contain amino acids that are more resistant to heat degradation than those encoded by AT-rich codons (Argos et al. 1979). High temperatures are also invoked in other variants of the selectionist hypothesis, which claim that GC-rich introns, third-codon positions, and untranslated regions may increase the thermal stability of primary mRNA transcripts, either because G—C bonds are stronger than A—T bonds (Wada and Suyama 1986), or through the stabilization of chromosomal structures by DNA–protein interactions (Bernardi et al. 1988). Indeed, in some thermophilic bacteria, a preferential usage of GC-rich codons has been reported (but see Galtier and Lobry 1997). However, the body temperature of warm-blooded and even cold-blooded vertebrates does not vary nearly as much as that experienced by bacteria, and so temperature may not be a very important factor in the evolution of isochores.

The selectionist theories put a great deal of emphasis on the GC-rich isochores because that is where many of the human genes happen to be. Thus, the selectionist hypothesis postulates that the GC-rich isochores are maintained by natural selection whereas the GC-poor isochores are not. Under this hypothesis, the rate of silent substitution should be higher in GC-poor regions than in GC-rich regions because silent changes are subject to selective constraints in the latter regions but not in the former. This prediction is supported by the study of Ticher and Graur (1989) on human genes.

The selectionist hypothesis is also compatible with the much higher concentration of housekeeping genes in GC-rich regions. However, there may be an anthropocentric bias to this argument. Indeed, many genes reside in GC-rich regions, but that only typifies humans and a handful of mammalian species studied to date, and there are important genes that reside in GC-poor isochores. Moreover, the discovery of a novel genomic phenotype in the guinea pig (Robinson et al. 1997) indicates that homologous genes may reside in GC-rich isochores in one organism but in GC-poor isochores in another, and that the reordering of genes among isochores may take place without changes in the overall isochore structure. Moreover, the selectionist hypothesis cannot explain why some duplicate genes have opposite GC contents. For example, in mammals, the β-globin cluster is low in GC, while the α-globin cluster is GC-rich (Figure 8.31), although the two types of globin genes are expressed in the same cells, at the same time, and serve the same function. Similarly, some immunoglobulin genes are located in GC-rich regions while others are located in AT-rich regions. Bernardi et al.'s (1985, 1988) explanation is that isochores represent the unit of selection and the α cluster in mammals has been translocated to a high-GC isochore, whereas the β cluster remained in a low-GC isochore. According to this argument, the α and β clusters in chicken should have both been translocated to GC-rich isochores. This argument, however, raises the issue of the functional advantage of GC-rich isochores in the first place. If the advantage does not come from the genes in the isochore, where does it come from? Finally, we must emphasize that not only genes exhibit a proclivity for GC-rich regions; so do *Alu* sequences and satellite DNA, which lack any function.

Under the mutationist hypothesis, both GC-rich and AT-rich isochores should have a low chance of misincorporation of nucleotides in their replication because the former are replicated early in the replication cycle of germline DNA, during which the GC content in the precursor pool is high, while the latter are replicated late in the replication cycle of germline DNA, during which the AT content in the precursor pool is high. Thus, this hypothesis predicts a low substitution rate in both GC- and AT-rich regions and a high rate in regions with an intermediate GC content. This was indeed the case in some rodent genes but not in others (Wolfe et al. 1989a). This discrepancy is not surprising, because unless the mutational bias is very strong, synonymous rates are likely to be subject to large stochastic effects.

The mutationist hypothesis is not without difficulties, however. As pointed out by Bernardi et al. (1988), constitutive heterochromatin such as satellite DNA, and facultative heterochromatin such as the inactive X chromosome, which are mostly GC-rich, replicate at the end of the cell cycle, and in these cases no connection between changes in nucleotide pools and DNA composition is observed. Finally, if there is no correspondence between isochore type and replication time, as suggested by Eyre-Walker (1992), the mutationist hypothesis will be severely compromised.

A variant mutationist hypothesis postulates that the GC-rich and GC-poor isochores have arisen because in germline DNA, damage in the relaxed and compact chromatin loops carrying transcriptionally active and inactive genes, respectively, is repaired by different polymerases (Filipski 1987). Indeed, the removal of UV-induced pyrimidine dimers is much more efficient in transcriptionally active genes than in the inactive ones (Bohr et al. 1985).

There are, however, some difficulties with this hypothesis. First, dimer removal occurs only in transcribed regions and only on the transcribed strand (Mellon et al. 1987), and it is highly unlikely that all GC-rich regions are transcribed in the germline. Second, it is difficult to explain the large differences in GC levels between duplicate genes such as the α- and β-globins (discussed previously).

In conclusion, the presently available data seem to be insufficient to distinguish between the two hypotheses, and it is also possible that both mutation pressure and natural selection have played a role in shaping the compositional organization of the vertebrate genome.

EMERGENCE OF NONUNIVERSAL GENETIC CODES

Almost all organisms use the universal genetic code (Chapter 1). Two explanations have been proposed for this phenomenon. The **amino acid–codon interaction theory** (Woese 1969) states that the origin of the specific codon assignments was through direct chemical interactions between nucleic acids and amino acids, and the universal genetic code is therefore a relic of a world in which translation did not require tRNA intermediates.

The **frozen accident theory** (Crick 1968) states that the code is universal not because of any chemical or physical imperative, but because the genetic code happened to evolve to a certain point by chance or by selective optimization (e.g., Figureau 1989), and when the genome grew to such an extent that it specified the production of many proteins, the rules of translation could not be further altered without affecting many proteins at once. Such a drastic change would not stand a chance of being benign, let alone beneficial. Both theories have one thing in common: they preclude the existence of nonuniversal codes. However, nonuniversal genetic codes have been found, and they are invariably associated with genome miniaturization—many of them with mitochondrial genomes—as well as with biased GC-mutational pressures. The question, then, is how a change in the genetic code may evolve.

The best answer so far seems to be the **codon-capture hypothesis** (Jukes 1985), which postulates that a codon may disappear from the genome—say, because of biased GC-mutational pressures or random genetic drift—and may later reappear (as a rare codon) with a different amino acid assignment.

To illustrate this phenomenon, let us take codon AAA, which codes for lysine in the universal genetic code, whereas in the mitochondria of echinoderms (e.g., starfish and sea urchins) it is used for asparagine, and in the mitochondria of hemichordates (e.g., acorn worms) it is unassigned (Castresana et al. 1998). Let us start with the situation in the universal genetic code, where lysine is coded by two codons (AAA and AAG) that are recognized by a $tRNA^{Lys}$ with the anticodon UUU (Figure 8.33a). GC pressure may change many AAA codons into AAG without affecting amino acid sequences. In small genomes, AAA may disappear altogether (Figure 8.33b). Anticodon UUU does not pair strongly with AAG, so a mutation in the $tRNA^{Lys}$-specifying

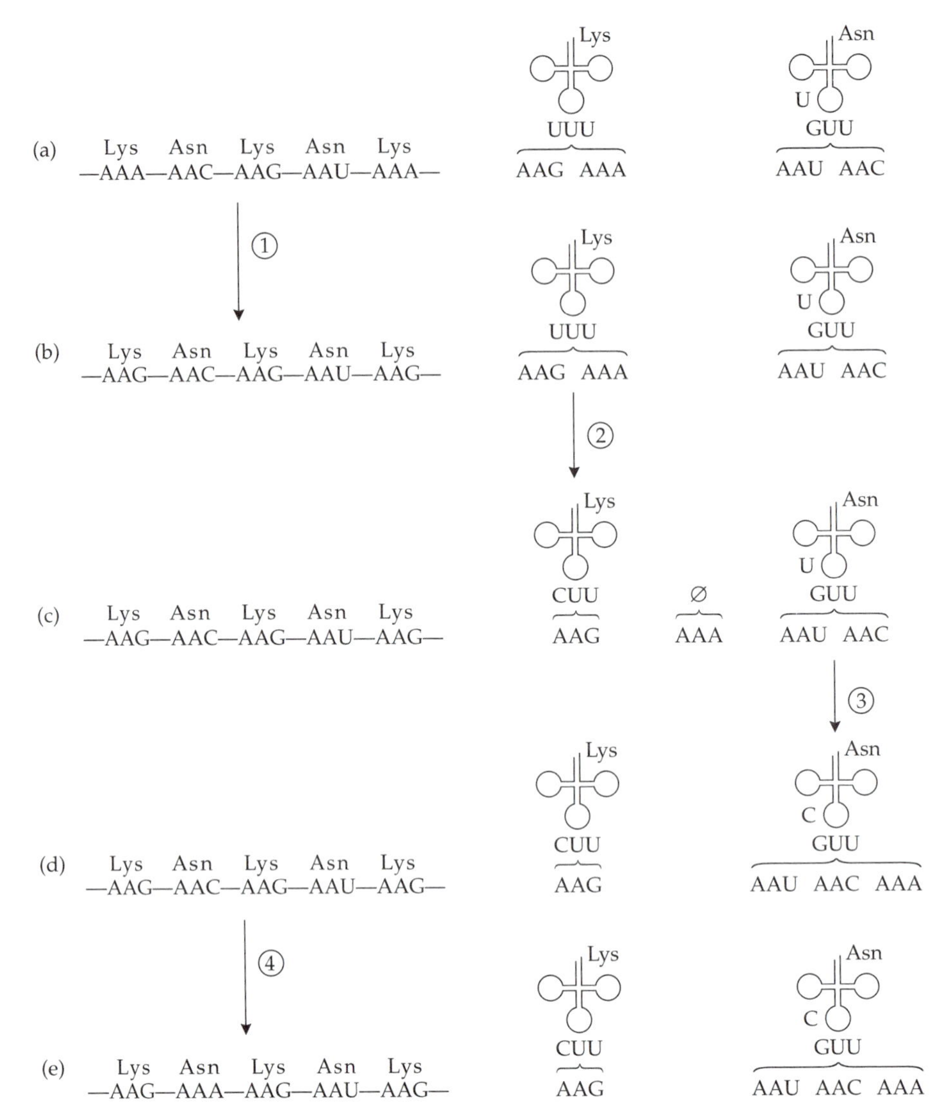

FIGURE 8.33 Possible scenario for the evolutionary steps (numbered arrows) predicted for the change in the assignment of the AAA codon from lysine in the universal genetic code to asparagine in the mitochondrial genetic code of echinoderms by the codon-capture hypothesis. (a) In the universal genetic code, lysine is coded by two codons (AAA and AAG) that are recognized by $tRNA^{Lys}$ with anticodon UUU (cloverleaf). Asparagine is coded by two codons (AAU and AAC) that are recognized by $tRNA^{Asn}$ with anticodon UUU and a uridine (U) at position 33. (b) In step 1, codons AAA disappear from the genome. (c) A mutation changes anticodon UUU into CUU, thereby restricting its recognition to AAG only. AAA becomes an unassigned codon (Ø). (d) A mutation changes position 33 in $tRNA^{Asn}$ from U to C, thereby enabling it to recognize AAA as well. (e) an AAA codon reappears in the genome and is translated into asparagine. The situation in (c) represents the mitochondrial genome of hemichordates. The situation in (e) represents the mitochondrial genome of echinoderms. Note that the amino acid sequence remained unchanged throughout the evolutionary course. Modified from Castresana et al. (1998).

gene changing its anticodon to CUU will be advantageous (Figure 8.33c). Anticodon GUU of $tRNA^{Asn}$ is used in the universal genetic code to pair with codons AAU and AAC (asparagine). According to the wobble rules (Chapter 1), $tRNA^{Asn}$ should have also recognized codon AAA. However, a uridine (U) at position 33 of $tRNA^{Asn}$ limits its recognition to AAU and AAC codons only. In the absence of AAA codons from the genome, however, the selective constraint on position 33 is removed, and it may change by chance to C. This change would allow $tRNA^{Asn}$ to recognize AAA codons, should such codons reappear in the genome (Figure 8.33d). If an AAA codon reappears in the genome, it will be translated into asparagine (Figure 8.33e).

FURTHER READINGS

Bernardi, G. 1995. The human genome: Organization and evolutionary history. Annu. Rev. Genet. 29: 445–476.

Cavalier-Smith, T. (ed.). 1985. *The Evolution of Genome Size*. Wiley, New York.

Cold Spring Harbor Symposia on Quantitative Biology. 1986. *Molecular Biology of Homo sapiens*. Vol. 51. Cold Spring Harbor Laboratory, Cold Spring Harbor, NY.

Cooper, N. (ed.). 1994. *The Human Genome Project: Deciphering the Blueprint of Heredity*. University Science Books, Mill Valley, CA.

Doolittle, R. F. 1986. *Of URFs and ORFs: A Primer on How to Analyze Derived Amino Acid Sequences*. University Science Books, Mill Valley, CA.

Dover, G. A. and R. B. Flavell (eds.). 1982. *Genome Evolution*. Academic Press, New York.

John, B. and G. L. Miklos. 1988. *The Eukaryotic Genome in Development and Evolution*. Allen & Unwin, London.

Karlin, S., J. Mrazek, and A. M. Campbell. 1997. Compositional biases of bacterial genomes and evolutionary implications. J. Bacteriol. 179: 3899–3913.

Osawa, S. 1995. *Evolution of the Genetic Code*. Oxford University Press, Oxford.

Palmer, J. D. 1997. The mitochondrion that time forgot. *Nature* 387: 454–455.

Primrose, S. B. 1998. *Principles of Genome Analysis*. Blackwell Science, Oxford.

Singer, M. F. 1982. Highly repeated sequences in mammalian genomes. Int. Rev. Cytol. 76: 67–112.

Singer, M. and P. Berg. 1991. *Genes and Genomes: A Changing Perspective*. University Science Books, Mill Valley, CA.

Smit, A. F. A. 1996. The origin of interspersed repeats in the human genome. Curr. Opin. Genet. Develop. 6: 743–748.

Appendix I

Spatial and Temporal Frameworks of the Evolutionary Process

Understanding the evolutionary process requires that we familiarize ourselves with the time scales and geographical history of our planet. Indeed, these temporal and spatial constraints provide the framework for conducting evolutionary research and for testing evolutionary theories. Suppose, for instance, that based on molecular data we conclude that birds and mammals last shared a common ancestor 320 million years ago. This conclusion can be easily refuted if we find both bird and mammal fossil remains that are older than 320 million years. Or, let us assume a claim is made that the flightless rhea of South America and the African ostrich diverged from a common ancestor 30 million years ago. Given that Africa and South America were last in physical contact about 80 million years ago, and that these birds cannot easily transport themselves across water barriers, we can refute this evolutionary claim as contrary to physical feasibility.

Timetables of Evolution

The age of the universe is estimated at 10–16.5 billion years. Earth's geological history starts approximately 4.5 billion years ago, and is usually divided into four eons: Priscoan (4.5–4 billion years ago), Archean (4–2.5 billion years ago),

TABLE 1 Main subdivisions of geological time and a synopsis of Earth's evolutionary history

Eon/ Era	Period	Epoch	MYA	Major events
Cenozoic	Pleistogene[a]	Holocene	0.01–Present	Spread of modern *Homo sapiens* populations throughout the world; historic times.
		Pleistocene	2–0.01	Emergence of *Homo sapiens*; several glaciations and retreats; extinctions of large mammals.
	Neogene[a]	Pliocene	5–2	North and South America meet at the Panamanian land bridge; emergence of *Australopithecus* and *Homo*.
		Miocene	25–5	Continued radiation of mammals; spread of angiosperms.
	Paleogene[a]	Oligocene	38–25	Emergence of apes, whales, and grazing mammals; rise of Alps and Himalayas.
		Eocene	55–38	Divergence of primates; angiosperm dominance.
		Paleocene	65–55	Major radiation of mammals, birds, hymenopterans, and dipterans; emergence of primates and carnivores; disappearance of shallow continental seas.
Mesozoic	Cretaceous		144–65	Extinction of dinosaurs; separation of continents; emergence of marsupial and placental mammals.
	Jurassic		213–144	Radiation of dinosaurs, marine and flying reptiles, and ammonites; spread of gymnosperms (mainly cycads) and ferns.
	Triassic		248–213	Emergence of dinosaurs, flowering plants, birds, and egg-laying mammals; mass extinction at end of period.
Paleozoic	Permian		286–248	Radiation of reptiles and insects; creation of Pangaea; glaciations, mass extinction at end of period.
	Carboniferous		360–286	Radiation of vascular plants; emergence of reptiles and winged insects.
	Devonian		408–360	Emergence of ammonites, amphibians and insects; Fresnian extinction in late Devonian.
	Silurian		438–408	Emergence of jawed fishes; invasion of land by vascular plants and arthropods.
	Ordovician		505–438	Emergence of chordates; mass extinction at end of period.
	Cambrian		590–505	Appearance of all but one eukaryotic phyla; repeated mass extinctions throughout period.
Proterozoic			2,500–590	Oxygen in the atmosphere; emergence of eukaryotes, photosynthetic, and multicellular organisms.
Archean			4,000–2,500	Origin of life.
Priscoan			4,500–4,000	Formation of Earth and the solar system.

Data from Harland et al. (1982), Calder (1983), and later sources.

[a]In the literature, the Pleistogene period is also referred to as the Quaternary, and the Neogene and Paleogene are referred to as the Tertiary. The terms Primary and Secondary are now obsolete.

Proterozoic or Precambrian (2,500–590 million years ago), and Phanerozoic (590 million years ago to the present). These eons are in turn subdivided into eras, periods, subperiods, and epochs. The major geological and biological events affecting the evolutionary process are shown in Table 1. Note that the geological nomenclature tends to vary with geographical region, and other divisions may be found in the literature.

Although the evidence for or against the presence of life prior to 3.5 billion years ago is tenuous, it is widely accepted that life must have arisen approximately 4 billion years ago. The first paleontologically recognizable living forms resemble simple prokaryotes—they were unicellular organisms devoid of a nucleus. The first eukaryotes (organisms possessing a nuclear envelope) probably appeared in the Proterozoic, around 1.8 billion years ago. This chronology is intriguing, since it means that life evolved out of inanimate matter in a "mere" half a billion years, but because that it took four times longer for a membrane-enclosed nucleus to emerge.

In the following, we list a somewhat anthropocentric chronology of some notable events in evolutionary history. We note that paleontological dates are frequently and hotly disputed, as befits a vibrant scientific discipline.

The first multicellular organisms, macroscopic algae, appeared 670–900 million years ago. Invertebrates emerged at the beginning of the Phanerozoic, and all animal phyla except the chordates are already represented in abundance at the beginning of the Cambrian.

The first chordates, the agnathes, appeared about 500 million years ago, followed by the appearance of sharks and bony fishes in the Silurian. The Silurian is also marked by the appearance of the first terrestrial organisms, the vascular plants. The first amphibians and the first insects appeared in the late stages of the Devonian, approximately 360 million years ago.

Reptiles appeared in the Carboniferous, about 340 million years ago—a period otherwise known for the profusion and diversity of its vascular plants. Winged insects appeared in the early Carboniferous, and in the Permian, both reptiles and insects diversified greatly. Reptiles continued to diversify in the Triassic, followed by an impressive radiation of species throughout the Mesozoic. The first flowering plants appeared in the early Jurassic, and the age of dinosaurs began. The "reign" of the dinosaurs ended abruptly about 65 million years ago.

Egg-laying mammals, the monotremes, appeared 240 million years ago. Birds may have appeared at about the same time, although the early avian fossil record is ambiguous. The first placental mammals emerged 100 million years later. The Paleogene and Neogene epochs saw the start of the principal radiations of mammals and birds. The paleontological record of primates starts in the late Eocene, about 40 million years ago, although they may have existed earlier. The earliest apes date back 27 million years, and our australopithecine ancestor appeared approximately 5 million years ago. The first member of the genus *Homo*, presumably an early form of *Homo habilis*, appeared 2.5 million years ago, and our own species, *Homo sapiens*, made its appearance about 400,000 years ago. The Neandertals emerged 100,000 years ago, at about the same time as modern *Homo sapiens*.

GEOLOGICAL HISTORY

Earth's crust is dynamic. Continents continuously split, drift laterally, and occasionally merge. At the beginning of the Cambrian there were probably four major paleocontinents and several minor ones, but in the Triassic, 250–200 million years ago, they merged into a single continuous land mass, Pangaea (Figure 1). Pangaea straddled the equator and was surrounded by the ocean Panthalassa. Soon afterwards, Pangaea broke into two "supercontinents," Laurasia

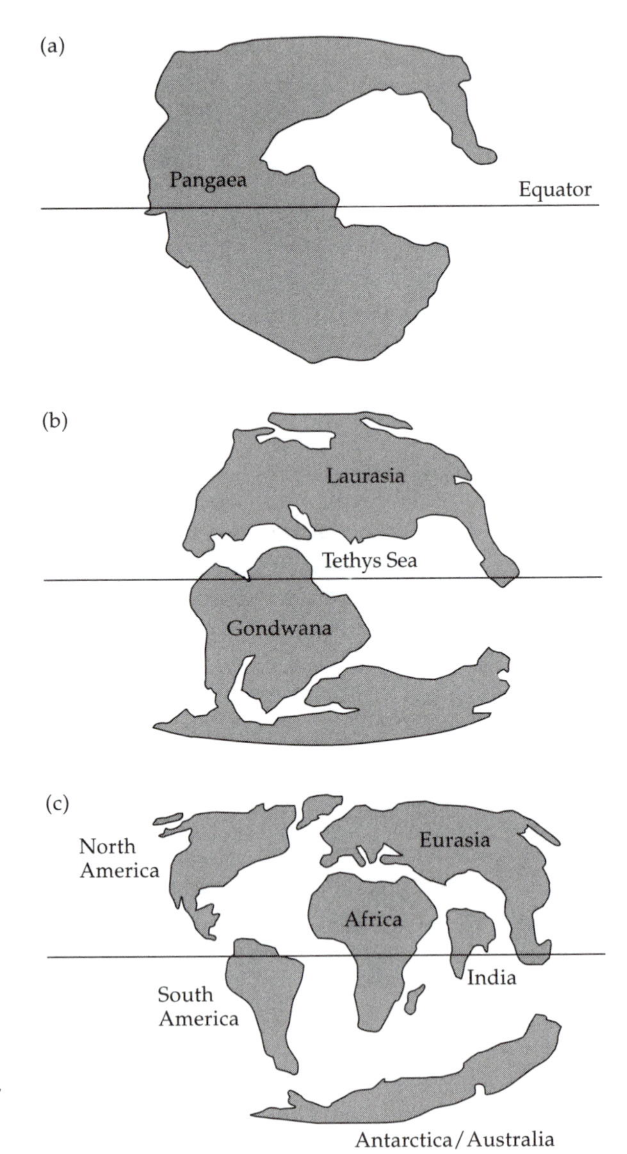

FIGURE 1 Distribution of the world's land masses. (a) 180 million years ago. Drifting continental fragments have assembled into one supercontinent, Pangaea. (b) 125 million years ago. Laurasia and Gondwana are separated by the Tethys Sea. (c) 55 million years ago. Six major land masses and numerous smaller islands drift apart; some eventually merge.

and Gondwana (or Gondwanaland), separated by the Tethys Sea. These supercontinents, in turn, split into smaller land masses, and became effectively separated in the Cretaceous. During the last 40 million years Africa touched Eurasia, India collided with Asia, and North and South America met at the Panamanian land bridge. The last event occurred only 3 million years ago.

Geological processes have a profound impact on evolutionary history. The separation of land and water masses creates barriers to reproduction, effectively isolating populations from one another. However, the opposite process—the merger of lands or seas—creates opportunities for migrations, irrevocably changing the floristic and faunistic makeup of the areas affected. For instance, continental breakup isolates terrestrial organisms from each other by marine barriers, thus increasing endemism on land. The Triassic reptile faunas were cosmopolitan because migration was free throughout Pangaea. In contrast, Cenozoic mammals exhibit significant endemism because migration was restricted to the continuous land masses. In the past several hundred years, humans have often accelerated these processes by wittingly or unwittingly importing species such as mice and flies to far-away islands, and bringing into contact species that would otherwise have remained separated. The Lessepsian migrations of marine organisms between the Mediterranean Sea and the Indian Ocean following the opening of the Suez Canal in 1869 constitute an example of a traumatic influence on evolutionary biology brought about by human interference.

It has been estimated that the half-life of a biological species is on the average 1–10 million years. Thus, virtually all plant and animal species that have ever lived are now extinct. Extinctions are classified into different types according to their intensity. In addition to "background extinctions," which are regarded as the spontaneous rate of replacement of one species by another, several extremely large extinctions stand out consistently, and they are conventionally labeled "mass extinctions" or "episodic extinctions" (Figure 2). Mass extinctions are defined according to their ferocity, their short duration on a geological time scale, and their global geographical extent.

The largest mass extinction occurred at the transition between the Permian and the Triassic, in which it is estimated that as many as 96% out of the 45,000–240,000 marine species existing at the time died out. Of all the mass extinctions, the one at the Cretaceous–Tertiary boundary, during which all dinosaurs died out, is by far the most thoroughly documented. This event affected terrestrial and marine organisms equally, with up to 75% of all marine species becoming extinct. The duration of the Cretaceous–Tertiary cataclysm is estimated to have lasted no more than 10,000–450,000 years.

The possibility that extinctions of plant and animal species and abrupt faunal turnovers have been mostly episodic rather than gradual has prompted researchers to suggest that extinctions are driven by extraterrestrial impacts. The first such suggestions were made in the late 1960s, but it was not until the findings by Alvarez et al. (1980) that the impact theory received serious consideration. According to the prevailing view, the mass extinction at the end of the Cretaceous was triggered by the impact in the Mexican Yucatan of an extrater-

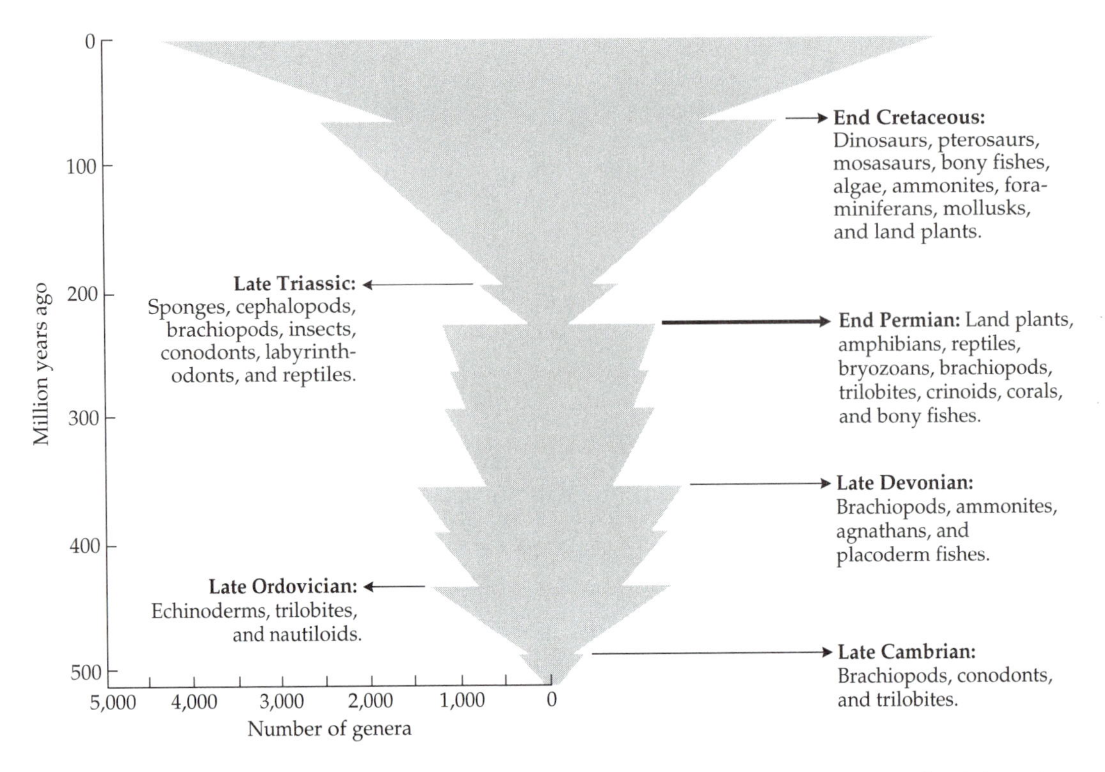

Figure 2. Effects of mass extinctions on taxonomic diversity during the Paleozoic, Mesozoic, and Cenozoic eras. For the six largest mass extinction events, the taxa that suffered the heaviest losses are indicated. Note that the single largest extinction event occurred at the transition between the Permian and the Triassic (heavier arrow).

restrial body about 10 Km in diameter that deposited a layer of iridium and other meteoritic elements across the entire planet. The effects of this impact included a global fire, subsequent darkness and cooling, atmospheric pollution, destruction of stratospheric ozone, lowering of seawater pH, tsunami waves, acid rain, suppression of primary production on land and sea, and weakening of the Earth's mantle that resulted in increased volcanic activity.

Extraterrestrial causes have been implicated in other mass extinctions, such as the late Devonian and the Permian–Triassic, although the evidence for these is less overwhelming. According to a number of analyses, major extinction events in the last 250 million years are regularly spaced throughout time, with a period of 26–32 million years (Raup and Sepkoski 1984). The last such event occurred 13 million years ago. If a periodicity in the record of extinctions can indeed be established, a common trigger most probably caused all such events. Several astrophysical explanations have been proposed, most of which implicate astronomical perturbations as the reason for an increased probability of an extraterrestrial object colliding with Earth. Periodicity, however, is still controversial (e.g., Newman 1996; Sole et al. 1997), and the Cretaceous–Tertiary event remains the only mass extinction whose causes cannot be attributed to earthly factors.

SPECIES DIVERSITY

One of the most striking features of the biological world is the immense diversity of its life forms. Over a million species belonging to the animal kingdom have been described in the literature. Plants constitute the second-largest kingdom with just over 300,000 described species, approximately 80% of which are flowering plants, or angiosperms. In total, the number of known species is approximately 1.5 million (Table 2).

Taxonomic diversity in nature is characterized by a marked unevenness in the distribution of species among taxa. For example, the great majority of all organisms are insects. Moreover, the unevenness extends to the class Insecta, with one order out of 26, Coleoptera, comprising over 350,000 described species (45%). Wilson (1971) noted that there are more social hymenopteran species in the world than all the known species of birds (8,600) and mammals (3,700) combined. The fact that molecular evolution, as all other biological disciplines, deals mainly with mammals reflects more on our anthropocentric proclivities than on anything else. It is sobering to realize that most biological knowledge is based

TABLE 2 Approximate number of extant species described in the literature

Group	Number of species
Animals	1,099,000
Vertebrates	54,000
Non-vertebrate chordates	1,000
Arthropods	881,000
Insects	793,000
Crustaceans	25,000
Arachnids and myriapods	63,000
Mollusks	107,000
Other invertebrates	56,000
Plants	328,000
Flowering plants	275,000
Other plants	53,000
Fungi	44,000
Protists	30,000
Prokaryotes	3,000
Total	1,504,000

Updated from Grant (1985).

on research done on this quite minor taxonomic division. In fact, mammals may not even be representative of the vertebrates; bony fishes, with more than 30,000 species, would have been a more representative choice.

The task of taxonomic description, while fairly advanced in some groups, is still far from completion in many others. It has been suggested, for instance, that the total number of described insects (approximately 750,000) represents less than one-fifth of the insect species actually in existence, and that the total number of extant species is 4.5–10 million. Erwin (1997) suggested that a dramatic revision upwards of this number is necessary. According to his estimates, there are over 30 million species of terrestrial arthropods in existence, and thus the number of all living organisms must be in excess of 50 million.

About 250,000 extinct plant and animal species have been described in the paleontological literature (Raup 1987), and the total number of species, both extinct and extant, has been estimated to be at least 4 billion (Grant 1985). However, estimates pertaining to extinct organisms must be treated with extreme caution. First, 90% of Earth's history, from its formation to the Cambrian, is virtually undocumented, and those periods for which we do have a fossil record contain many gaps. In general, the fossil record is biased in favor of overrepresentation of the more recent taxa, a phenomenon that has been called "the pull of the recent." Second, the fossil record is incomplete not only because of gaps in time—periods during which a particular lineage is not represented—but also because each fossil is at best a small fragment of an organism's anatomy. This fragmentary representation leads in many cases to either ambiguous or outright erroneous assignments of taxonomic affinity.

Appendix II

Basics of Probability

Because of the stochastic component in heredity and evolution, throughout this book we have relied heavily on the theory of probability. In the following, we present a short introduction to the subject.

TERMINOLOGY

The possible outcomes of a stochastic process are called **events**. (A deterministic process has only one possible outcome.) A stochastic process may have a finite or an infinite number of outcomes. The **probability** of a particular event is the fraction of outcomes in which the event occurs. The probability of event A is denoted by $P(A)$. Probability values are between 0 (the event never occurs) and 1 (the event always occurs). Events may or may not be **mutually exclusive**. Events that are not mutually exclusive are called **independent events**. For example, the birth of a son or a daughter are mutually exclusive events. In contrast, the birth of a daughter and the birth of carrier of the sickle-cell anemia allele are not mutually exclusive. The sum of probabilities of all mutually exclusive events in a process is 1. For example, if there are n possible mutually exclusive outcomes, then

$$\sum_{i=1}^{n} P(i) = 1 \qquad \textbf{(1)}$$

Simple and Conditional Probabilities

If A and B are mutually exclusive events, then the probability of either A or B occurring is

$$P(A \cup B) = P(A) + P(B) \quad \textbf{(2)}$$

where $\cup$ symbolizes "or" (**union**).

If A and B are independent events, then the probability that both A and B will occur is

$$P(A \cap B) = P(A) \times P(B) \quad \textbf{(3)}$$

where $\cap$ symbolizes "and" (**intersection**).

For independent events, we are sometimes interested in the **conditional probability**, i.e., what is the probability that A will occur, given that event B did occur. The conditional probability of A given B is

$$P(B|A) = \frac{P(A \cap B)}{P(A)} \quad \textbf{(4)}$$

Permutations and Combinations

For a series of outcomes, we may sometimes be interested in the order in which the particular events occurred. In such a case we deal with **permutations**. The number of possible permutations, N_p, is

$$N_p = \frac{n!}{(n-r)!} \quad \textbf{(5)}$$

where r is the number of events in the series, n is the number of possible events, and $n!$ denotes the factorial of n (the product of all the positive integers from 1 to n).

When the order in which the events occurred is of no interest, we are dealing with **combinations**. The number of possible combinations, N_c, is

$$N_c = \binom{n}{r} = \frac{n!}{r!(n-r)!} \quad \textbf{(6)}$$

Probability Distributions

The **probability distribution** refers to the frequency with which all possible outcomes occur. The mathematical and statistical literature is replete with numerous types of probability distributions. In the following we shall only describe two distributions that have been used in the book, the **binomial distribution** and the **Poisson distribution**.

The binomial distribution

A process that has only two possible outcomes is called a **binomial process**. In statistics, the two outcomes are frequently denoted as **success** and **failure**. The probabilities of a success or a failure are denoted by p and q, respectively. Note that $p + q = 1$. The binomial distribution gives the probability of exactly k successes in n trials as

$$P(k) = \binom{n}{r} p^k (1-p)^{n-k} \tag{7}$$

The mean and variance of the number of successes in n trials are given by

$$\mu = np \tag{8}$$

and

$$V = npq \tag{9}$$

The Poisson distribution

When the probability of success, p, is very small and n is large, then p^k and $(1-p)^{n-k}$ may become too tedious to calculate exactly by the binomial distribution. In such cases, the Poisson distribution becomes useful. Let λ be the expected number of successes in a process consisting of n trials, i.e., $\lambda = np$. The probability of observing k successes is given by the Poisson distribution as

$$P(k) = \frac{\lambda^k e^{-\lambda}}{k!} \tag{10}$$

The mean and variance of a Poisson variable are given by

$$\mu = \lambda \tag{11}$$

and

$$V = \lambda \tag{12}$$

Literature Cited

The numbers in brackets following each reference denote the chapter in which the reference is cited.

Adams, M. D., R. S. Tarng, and D. C. Rio. 1997. The alternative splicing factor PSI regulates *P*-element third intron splicing *in vivo*. Genes Develop. 11: 129–138. **[7]**

Akashi, H. 1997. Codon bias evolution in *Drosophila*. Population genetics of mutation–selection drift. Gene 205: 269–278. **[4]**

Alber, B. E. and J. G. Ferry. 1994. A carbonic anhydrase from the archaeon *Methanosarcina thermophila*. Proc. Natl. Acad. Sci. USA 91: 6909–6913. **[6]**

Alberts B., D. Bray, J. Lewis, M. Raff, K. Roberts, and J. D. Watson. 1995. *Molecular Biology of the Cell*, 3rd Ed. Garland, New York. **[1]**

Alvarez, L. W., W. Alvarez, F. Asaro, and H. V. Michel. 1980. Extraterrestrial cause for the Createaceous–Tertiary extinction. Science 208: 1095–1108. **[App. I]**

Amy, C. M., B. Williams-Ahlf, J. Naggert, and S. Smith. 1992. Intron–exon organization of the gene for the multifunctional animal fatty acid synthase. Proc. Natl. Acad. Sci. USA 89: 1105–1108. **[6]**

Andersen, R. D., B. W. Birren, S. J. Taplitz, and H. R. Herschman. 1986. Rat metallothionein-I structural gene and three pseudogenes, one of which contains 5′ regulatory sequences. Mol. Cell Biol. 6: 302–314. **[7]**

Anderson, S. and 13 others. 1981. Sequence and organization of the human mitochondrial genome. Nature 290: 457–464. **[6]**

Andersson, S. G. E. and 9 others. 1998. The genome sequence of *Rickettsia prowazekii* and the origin of mitochondria. Nature 396: 133–140. **[5, 8]**

Andrews, P. 1987. Aspects of hominoid phylogeny. pp. 23–53. In: C. Patterson (ed.), *Molecules and Morphology in Evolution: Conflict or Compromise?* Cambridge University Press, Cambridge. **[5]**

Antoine, M., C. Erbil, E. Munch, S. Schnell, and J. Niessing. 1987. Genomic organization and primary structure of five homologous pairs of intron-less genes encoding secretory globins from the insect *Chironomus thummi thummi*. Gene 56: 41–51. **[7]**

Anxolabéhère, D., M. G. Kidwell, and G. Periquet. 1988. Molecular characteristics of diverse populations are consistent with the hypothesis of a recent invasion of *Drosophila melanogaster* by mobile *P* elements. Mol. Biol. Evol. 5: 252–269. **[7]**

Aota, S.-I. and T. Ikemura. 1986. Diversity in G+C content at the third position of codons in vertebrate genes and its cause. Nuc. Acids Res. 14: 6345–6355. **[8]**

Aragoncillo, C., M. A. Rodriguez-Loperena, G. Salcedo, P. Carbonero, and F. Garcia-Olmedo. 1978. Influence of homologous chromosomes on gene-dosage effects in allohexaploid wheat (*Triticum aestivum* L.). Proc. Natl. Acad. Sci. USA 75: 1446–1450. **[8]**

Argos, P., M. G. Rossmann, U. M. Grau, A. Zuber, G. Franck, and J. D. Tratschin. 1979. Thermal stability and protein structure. Biochemistry 18: 5698–5703. **[8]**

Argyle, E. 1980. A similarity ring for amino acids based on their evolutionary substitution rates. Orig. Life 10: 357–360. **[4]**

Árnason, U., A. Gullberg, A. Janke, and X. Xu. 1996. Pattern and timing of evolutionary divergences among hominoids based on analyses of complete mtDNAs. J. Mol. Evol. 43: 650–661. **[5]**

Arnheim, N. 1983. Concerted evolution of multigene families. pp. 38–61. In: M. Nei and R. K. Koehn (eds.), *Evolution of Genes and Proteins*. Sinauer Associates, Sunderland, MA. **[6]**

Austin, J. J., A. B. Smith, and R. H. Thomas. 1997. Palaeontology in a molecular world: The search for authentic ancient DNA. Trends Ecol. Evol. 12: 303–306. **[5]**

Avise, J. C. and W. S. Nelson. 1989. Molecular genetic relationships of the extinct dusky seaside sparrow. Science 243: 646–648. [5]

Avise, J. C., B. W. Bowen, T. Lamb, A. B. Meylan, and E. Bermingham. 1992. Mitochondrial DNA evolution at a turtle's pace: Evidence for low genetic variability and reduced microevolutionary rate in the Testudines. Mol. Biol. Evol. 9: 457–473. **[5]**

Axelrod, D. I. 1952. A theory of angiosperm evolution. Evolution 6: 29–60. **[5]**

Axelrod, D. I. 1970. Mesozoic paleogeography and early angiosperm history. Bot. Rev. 36: 277–319. **[5]**

Ayala, F. J. (ed.). 1976. *Molecular Evolution*. Sinauer Associates, Sunderland, MA. **[5]**

Baba, M. L., M. Goodman, J. Berger-Cohn, J. G. Demaille, and G. Matsuda. 1984. The early adaptive evolution of calmodulin. Mol. Biol. Evol. 1: 442–455. **[6]**

Bailey, W. J., D. H. A. Fitch, D. A. Tagle, J. Czelusniak, J. L. Slightom, and M. Goodman. 1991. Molecular evolution of the ψξ-globin gene locus: Gibbon phylogeny and the hominoid slowdown. Mol. Biol. Evol. 8: 155–184. **[5]**

Bailliet, G., F. Rothhammer, F. R. Carnese, C. M. Bravi, and N. O. Bianchi. 1994. Founder mitochondrial haplotypes in Amerindian populations. Am. J. Hum. Genet. 54: 27–33. **[5]**

Bains, W. 1986. The multiple origins of human *Alu* sequences. J. Mol. Evol. 23: 189–199. **[7]**

Baker, B. S. 1989. Sex in flies: The splice of life. Nature 340: 521–524. **[6]**

Baldauf, S. L., J. D. Palmer, and W. F. Doolittle. 1996. The root of the universal tree and the origin of eukaryotes based on elongation factor phylogeny. Proc. Natl. Acad. Sci. USA 93: 7749–7754. **[5]**

Baltimore, D. 1981. Gene conversion: Some implications for immunoglobulin genes. Cell 24: 592–594. **[6]**

Banyai, L., A. Varadi, and L. Patthy. 1983. Common evolutionary origin of the fibrin-binding structures of fibronectin and tissue-type plasminogen activator. FEBS Lett. 163: 37–41. **[6]**

Barakat, A., G. Matassi, and G. Bernardi. 1998. Distribution of genes in the genome of *Arabidopsis thaliana* and its implication for the genome organization of plants. Proc. Natl. Acad. Sci. USA 95: 10044–10049. **[8]**

Barker, W. C. and M. O. Dayhoff. 1980. Evolutionary and functional relationships of homologous physiological mechanisms. BioScience 30: 593–600. **[6]**

Barker, W. C., L. K. Ketcham, and M. O. Dayhoff. 1978. Duplications in protein sequences. pp. 359–362. In: M. O. Dayhoff (ed.), *Atlas of Protein Sequence and Structure*, Vol. 5, Supplement 3. Natl. Biomed. Res. Found., Silver Spring, MD. **[6]**

Barns, S. M., C. F. Delwiche, J. D. Palmer, and N. R. Pace. 1996. Perspectives in archaeal diversity, thermophyly and monophyly from environmental rRNA sequences. Proc. Natl. Acad. Sci. USA. 93: 9188–9193. **[5]**

Barsh, G. S., P. H. Seeburg, and R. E. Gelinas. 1983. The human growth hormone gene family: Structure and evolution of the chromosomal locus. Nuc. Acids Res. 11: 3939–3958. **[7]**

Barton, G. J. and M. J. E. Sternberg. 1987. A strategy for the rapid multiple alignment of protein sequences. Confidence levels from tertiary sequence comparisons. J. Mol. Biol. 198: 327–337. **[3]**

Beale, D. and H. Lehmann. 1965. Abnormal haemoglobin and the genetic code. Nature 207: 259–261. **[4]**

Beckman, J. S. and J. L. Weber. 1992. Survey of human and rat microsatellites. Genomics 12: 627–631. **[8]**

Begun, D. J. and C. F. Aquadro. 1992. Levels of naturally occurring DNA polymorphism correlate with recombination rates in *D. melanogaster*. Nature 356: 519–520. **[2]**

Bennett, D. K. 1980. Stripes do not a zebra make. I. A cladistic analysis of *Equus*. Syst. Zool. 29: 272–287. **[5]**

Bennett, M. D. 1971. The duration of meiosis. Proc. Roy. Soc. 178B: 259–275. **[8]**

Benton, M. J. 1997. *Vertebrate Paleontology*, 2nd Ed., Chapman & Hall, New York. **[5]**

Benveniste, R. E. 1985. The contributions of retroviruses to the study of mammalian evolution. pp. 359–417. In: R. I. MacIntyre (ed.), *Molecular Evolutionary Genetics*. Plenum, New York. **[7]**

Benveniste, R. E. and G. J. Todaro. 1976. Evolution of type C viral genes: Evidence for an Asian origin of man. Nature 261: 101–108. **[7]**

Berg, D. E. and M. M. Howe (eds.). 1989. *Mobile DNA*. American Society for Microbiology, Washington, DC. **[7]**

Berg, C. A. and A. C. Spradling. 1991. Studies on the rate and site-specificity of *P* element transposition. Genetics 127: 515–524. **[7]**

Berg, D. E., C. M. Berg, and C. Sasakawa. 1984. The bacterial transposon *Tn5*: Evolutionary inferences. Mol. Biol. Evol. 1: 411–422. **[7]**

Bernardi, G. 1989. The isochore organization of the human genome. Annu. Rev. Genet. 23: 637–661. **[8]**

Bernardi, G. 1995. The human genome: Organization and evolutionary history. Annu. Rev. Genet. 29: 445–476. **[8]**

Bernardi, G. and G. Bernardi. 1985. Codon usage and genome composition. J. Mol. Evol. 22: 363–365. **[8]**

Bernardi, G. and 7 others. 1985. The mosaic genome of warm-blooded vertebrates. Science 228: 953–958. **[4, 8]**

Bernardi, G., D. Mouchiroud, C. Gautier, and G. Bernardi. 1988. Compositional patterns in vertebrate genomes: Conservation and change in evolution. J. Mol. Evol. 28: 7–18. **[8]**

Berry, V. and O. Gascuel. 1996. On the interpretation of bootstrap trees: Appropriate threshold of clade selection and induced gain. Mol. Biol. Evol. 13: 999–1011. **[5]**

Bhattacharyya, M., A. Smith, T. H. Ellis, C. Hedley, and C. Martin. 1990. The wrinkled-seed character of pea described by Mendel in caused by a transposon-like insertion in a gene encoding starch-branching enzyme. Cell 60: 115–122. **[7]**

Bhattachayya, M., C. Martin, and A. Smith. 1993. The importance of starch biosynthesis in the wrinkled seed shape character of peas studied by Mendel. Plant Mol. Biol. 22: 525–531. **[7]**

Bingham, P. M., M. G. Kidwell, and G. M. Rubin. 1982. The molecular basis of *P–M* hybrid dysgenesis: The role of the *P* element, a *P*-strain-specific transposon family. Cell 29: 995–1004. **[7]**

Bird, A. and S. Tweedie. 1995. Transcriptional noise and the evolution of gene number. Philos. Trans. Roy. Soc. 349B: 249–253. **[8]**

Birky, C. W. and R. V. Skavaril. 1976. Maintenance of genetic homogeneity in systems with multiple genomes. Genet. Res. 27: 249–265. **[6]**

Bishop, J. M. 1981. Enemies within: The genesis of retrovirus oncogenes. Cell 23: 5–6. **[7]**

Black, J. A. and G. H. Dixon. 1968. Amino acid sequence of α chains of human haptoglobins. Nature 218: 736–741. **[6]**

Blaisdell, B. E. 1985. A method for estimating from two aligned present-day DNA sequences, their ancestral composition and subsequent rates of substitution, possibly different in the two lineages, corrected for multiple and parallel substitutions at the same site. J. Mol. Evol. 22: 69–81. **[3]**

Blake, C. C. F. 1978. Do genes-in-pieces imply proteins in pieces? Nature 273: 267. **[6]**

Bodmer, W. F. and L. L. Cavalli-Sforza. 1976. *Genetics, Evolution, and Man*. W.H. Freeman, San Francisco. **[2]**

Bohr, V. A., C. A. Smith, D. S. Okumoto, and P. C. Hanawalt. 1985. DNA repair in an active gene: Removal of the *DHFR* gene of CHO cells is much more efficient than in the genome overall. Cell 40: 359–369. **[8]**

Boissinot, S. and 7 others. 1998. Origins and antiquity of X-linked triallelic color vision systems in New World monkeys. Proc. Natl. Acad. Sci. USA 95:13749–13754. **[6]**

Boore, J. L. and W. M. Brown. 1998. Big trees from little genomes: Mitochondrial gene order as a phylogenetic tool. Curr. Opin. Genet. Develop. 8: 668–674. **[8]**

Bork, P., C. Sander, and A. Valencia. 1993. Convergent evolution of similar enzymatic function on different protein folds: The hexokinase, ribokinase, and galactokinase families of sugar kinases. Prot. Sci. 2: 31–40. **[6]**

Bork, P., A. K. Downing, B. Kieffer, and I. D. Campbell. 1996. Structure and distribution of modules in extracellular proteins. Q. Rev. Biophys. 29: 119–167. **[6]**

Bostock, C. J. 1986. Mechanisms of DNA sequence amplification and their evolutionary consequences. Phil. Trans. Roy. Soc. 312B: 261–273. **[8]**

Bova, M. P., L. L. Ding, J. Horwitz, and B. K. Fung. 1997. Subunit exchange of αA-crystallin. J. Biol. Chem. 272: 29511–29517. **[4]**

Branciforte, D. and S. L. Martin. 1994. Developmental and cell type specificity of *LINE-1* expression in mouse testis: Implication for transposition. Mol. Cell Biol. 14: 2584–2592. **[7]**

Bränden, C.-I. and J. Tooze. 1991. *Introduction to Protein Structure*. Garland, New York. **[1]**

Braunitzer, G., R. Gehring-Muller, N. Hilschmann, K. Hilse, G. Hobom, V. Rudolf, and B. Wittmann-Liebold. 1961. Die konstitution des normalen adulten hämoglobins. Z. Physiol. Chem. Hoppe-Seyler 325: 283–286. **[6]**

Breukelman, H. J., N. van der Munnik, R. G. Kleineidam, A. Furia, and J. J. Beintema. 1998. Secretory ribonuclease genes and pseudogenes in true ruminants. Gene 212: 259–268. **[6]**

Bridges, C. B. 1936. The *Bar* "gene"—a duplication. Science 83: 210–211. **[6]**

Brinkmann, B., M. Klintschar, F. Neuhuber, J. Huhne, and B. Rolf. 1998. Mutation rate in human microsatellites: Influence of the structure and length of the tandem repeat. Am. J. Hum. Genet. 62: 1408–1415. **[1]**

Britten, R. J. 1986. Rates of DNA sequence evolution differ between taxonomic groups. Science 231: 1393–1398. **[4]**

Britten, R. J. and D. E. Kohne. 1968. Repeated sequences in DNA. Science 161: 529–540. **[8]**

Britten, R. J., D. E. Graham, and B. R. Neufeld. 1974. Analysis of repeating DNA sequences by reassociation. Methods Enzymol. 29: 363–418. **[3]**

Britten, R. J., W. F. Baron, D. B. Stout, and E. H. Davidson. 1988. Sources and evolution of human *Alu* repeated sequences. Proc. Natl. Acad. Sci. USA 85: 4770–4774. **[7]**

Brosius, J. and H. Tiedge. 1995. Reverse transcriptase: Mediator of genomic plasticity. Virus Genes 11: 163–179. **[7]**

Brown, A. J. L. and D. Ish-Horowicz. 1981. Evolution of the *87A* and *87C* heat-shock loci in *Drosophila*. Nature 290: 677–682. **[6]**

Brown, D. D. and K. Sugimoto. 1973. 5S DNAs of *Xenopus laevis* and *Xenopus mulleri*: Evolution of tandem genes. J. Mol. Biol. 63: 57–73. **[6]**

Brown, D. D., P. C. Wensink, and E. Jordan. 1972. A comparison of the ribosomal DNAs of *Xenopus laevis* and *Xenopus mulleri*: Evolution of tandem genes. J. Mol. Biol. 63: 57–73. **[6]**

Brown, J. R. and W. F. Doolittle. 1995. Root of the universal tree of life based on ancient aminoacyl-tRNA synthetase gene duplications. Proc. Natl. Acad. Sci. USA 92: 2441–2445. **[5]**

Brown, T. A. and K. A. Brown. 1994. Using molecular biology to explore the past. BioEssays 16: 719–726. **[5]**

Brown, W. M., M. George, and A. C. Wilson. 1979. Rapid evolution of animal mitochondrial DNA. Proc. Natl. Acad. Sci. USA 76: 1967–1971. **[4]**

Brown, W. M., E. M. Prager, A. Wang, and A. C. Wilson. 1982. Mitochondrial DNA sequences of primates: Tempo and mode of evolution. J. Mol. Evol 18: 225–239. **[1, 4, 5]**

Bull, J. J. and H. A. Wichman. 1998. A revolution in evolution. Science 281: 1959. **[Introduction]**

Buneman, P. 1971. The recovery of trees from measurements of dissimilarity. pp. 387–395. In: E. R. Hodson, D. G. Kendall, and P. Tautu (eds.), *Mathematics in the Archaeological and Historical Sciences*. Edinburgh University Press, Edinburgh. **[5]**

Buonagurio, D. A., S. Nakada, W. M. Fitch, and P. Palese. 1986. Epidemiology of influenza C virus in man: Multiple evolutionary lineages and low rate of change. Virology 153: 12–21. **[4]**

Butticé, G., P. Kaytes, J. D'Armiento, G. Vogeli, and M. Kurkinen. 1990. Evolution of collagen IV genes from a 54-base pair exon: A role for introns in gene evolution. J. Mol. Evol. 30: 479–488. **[6]**

Caballero, A. 1994. Developments in the prediction of effective population size. Heredity 73: 657–679. **[2]**

Caccio, S., K. Jabbari, G. Matassi, F. Guermonprez, J. Desgres, and G. Bernardi. 1997. Methylation patterns in the isochores of vertebrate genomes. Gene 205: 119–124. **[8]**

Caccone, A. and J. R. Powell. 1989. DNA divergence among hominoids. Evolution 43: 925–942. **[5]**

Caccone, A. and J. R. Powell. 1990. Extreme rates and heterogeneity in insect DNA evolution. J. Mol. Evol. 30: 273–280. **[4]**

Cairns-Smith, A. G. 1982. *Genetic Takeover and the Mineral Origins of Life*. Cambridge University Press, Cambridge. **[Introduction]**

Calabretta, B., D. L. Robberson, H. A. Barrera-Saldana, T. P. Lambrou, and G. F. Saunders. 1982. Genome instability in a region of human DNA enriched in *Alu* repeat sequences. Nature 296: 219–225. **[7]**

Callaghan, A., T. Guillemaud, N. Makate, and M. Raymond. 1998. Polymorphisms and fluctuations in copy number of amplified esterase genes in *Culex pipiens* mosquitoes. Insect Mol. Biol. 7: 295–300. **[6]**

Camin J. H. and R. R. Sokal. 1965. A method for deducing branching sequences in phylogeny. Evolution 19: 311–326. **[5]**

Cann, R. L., M. Stoneking, and A. C. Wilson. 1987. Mitochondrial DNA and human evolution. Nature 325: 31–36. **[5]**

Cano, R. J., H. N. Poinar, N. J. Pieniazek, A. Acra, and G. O. Poinar. 1993. Amplification and sequencing of DNA from a 120–135-million-year-old weevil. Nature 363: 536–538. **[5]**

Cao, Y., J. Adachi, A. Janke, S. Pääbo, and M. Hasegawa. 1994. Phylogenetic relationships among eutherian orders estimated from inferred sequences of mitochondrial proteins: Instability of a tree based on a single gene. J. Mol. Evol. 39: 519–527. **[5]**

Cappello, J., S. M. Cohen, and H. F. Lodish. 1984. *Dictyostelium* transposable element *DIRS-1* preferentially inserts into *DIRS-1* sequences. Mol. Cell. Biol. 4: 2207–2213. **[7]**

Capy, P., T. Langin, and D. Anxolabéhère. 1998. *Dynamics and Evolution of Transposable Elements*. Chapman & Hall, New York. **[7]**

Carle, P., F. Laigret, J. G. Tully, and J. M. Bove. 1995. Heterogeneity of genome sizes within the genus *Spiroplasma*. Int. J. Syst. Bacteriol. 45: 178–181. **[8]**

Carlson, N. R. 1991. *Physiology of Behavior*, 4th Ed. Allyn & Bacon, Boston. **[6]**

Castresana, J., G. Feldmaier-Fuchs, and S. Pääbo. 1998. Codon reassignment and amino acid composition in hemichordate mitochondria. Proc. Natl. Acad. Sci. USA 95: 3703–3707. **[8]**

Catzeflis, F. M., F. H. Sheldon, J. E. Ahlquist, and C. G. Sibley. 1987. DNA–DNA hybridization evidence of the rapid rate of muroid rodent DNA evolution. Mol. Biol. Evol. 4: 242–253. **[4]**

Cavalier-Smith, T. 1975. The origin of nuclei and of eukaryotic cells. Nature 256: 463–468 **[5]**

Cavalier-Smith, T. 1978. Nuclear volume control by nucleoskeletal DNA, selection for cell volume and cell growth rate and the solution to the DNA C-value paradox. J. Cell. Sci. 34: 247–278. **[8]**

Cavalier-Smith, T. 1980. How selfish is DNA? Nature 285: 617–618. **[8]**

Cavalier-Smith, T. 1983. Genetic symbionts and the origin of split genes and linear chromosomes. pp. 29–45. In: H. E. A. Schenk and W. Schwemmler (eds.), *Endocytobiology II: Intracellular Space as Oligogenetic Ecosystem*. de Gruyter, Berlin. **[8]**

Cavalier-Smith, T. 1985a. Selfish DNA and the origin of introns. Nature 315: 283–284. **[6, 8]**

Cavalier-Smith, T. (ed.). 1985b. *The Evolution of Genome Size*. Wiley, New York. **[8]**

Cavalier-Smith, T. 1991. Intron phylogeny: A new hypothesis. Trends Genet. 7: 145–148. **[6]**

Cavalli-Sforza, L. L. and A. W. F. Edwards. 1967. Phylogenetic analysis: Models and estimation procedures. Am. J. Hum. Genet. 19: 233–257. **[5]**

Cayley, A. 1857. On the theory of the analytical forms called trees. Philos. Mag. 13: 19–30. **[5]**

Cedergren, R., M. W. Gray, Y. Abel, and D. Sankoff. 1988. The evolutionary relationships among known life forms. J. Mol. Evol. 28: 98–112. **[5]**

Celerin, M., J. M. Ray, N. J. Schisler, A. W. Day, W. G. Stetler-Stevenson, and D. E. Laudenbach. 1996. Fungal fimbrae are composed of collagen. EMBO J. 15: 4445–4453. **[6]**

Chang, B. H.-J., L. C. Shimmin, S.-K. Shuye, D. Hewett-Emmett, and W.-H. Li. 1994. Weak male-driven molecular evolution in rodents. Proc. Natl. Acad. Sci. USA 91: 827–831. **[4]**

Chang, L. Y. and J. L. Slightom. 1984. Isolation and nucleotide sequence analysis of the β-type globin pseudogene from human, gorilla and chimpanzee. J. Mol. Biol. 180: 767–784. **[6]**

Chao, L., C. Vargas, B. B. Spear, and E. C. Cox. 1983. Transposable elements as mutator genes in evolution. Nature 303: 633–635. **[7]**

Chao, S., R. Sederoff, and C. S. Levings. 1984. Nucleotide sequence and evolution of the 18S ribosomal RNA gene in maize mitochondria. Nuc. Acids Res. 12: 6629–6615. **[4]**

Chappey, C., A. Danckaert, P. Dessen, and S. Hazout. 1991. MASH: An interactive program for multiple alignment and consensus sequence construction for biological sequences. Comp. Appl. Biosci. 7: 195–202. **[3]**

Charlesworth, B. 1985. The population genetics of transposable elements. pp. 213–232. In: T. Ohta and K. Aoki (eds.), *Population Genetics and Molecular Evolution*. Springer, Berlin. **[7]**

Charlesworth, B. 1988. The maintenance of transposable elements in natural populations. pp. 189–212. In: O. J. Nelson (ed.), *Plant Transposable Elements*. Plenum, New York. **[7]**

Charlesworth, B. and C. H. Langley. 1989. The population genetics of *Drosophila* transposable elements. Annu. Rev. Genet. 23: 251–287. **[7]**

Charlesworth, B., C. Langley, and W. Stephan. 1986. The evolution of restricted recombination and the accumulation of repeated DNA sequences. Genetics 112: 947–962. **[8]**

Charlesworth, B., A. Lapid, and D. Canada. 1992. The distribution of transposable elements within and between chromosomes in a population of *Drosophila melanogaster*. I. Element frequencies and distribution. Genet. Res. 60: 103–114. **[7]**

Chen, L., A. L. DeVries, and C. H. C. Cheng. 1997. Evolution of antifreeze glycoprotein gene from a trypsinogen gene in Antarctic notothenioid fish. Proc. Natl. Acad. Sci. USA 94: 3811–3816. **[6]**

Chen, S.-H. and 12 others. 1987. Apolipoprotein B-48 is the product of a messenger RNA with an organ-specific in-frame stop codon. Science. 238: 363–366. **[6]**

Chevaillier, P. 1993. PEST sequences in nuclear proteins. Int. J. Biochem. 25: 479–482. **[6]**

Chiarelli, A. B. 1973. *Evolution of the Primates: An Introduction to the Biology of Man*. Academic Press, London. **[5]**

Chirala, S. S., M. A. Kuziora, D. M. Spector, and S. J. Wakil. 1987. Complementation of mutations and nucleotide sequence of *FAS1* gene encoding β subunit of yeast fatty acid synthase. J. Biol. Chem. 262: 4231–4240. **[6]**

Christiansen, F. B. and M. W. Feldman. 1986. *Population Genetics*. Blackwell, Cambridge, MA. **[2]**

Ciochon, R. L. 1985. Hominoid cladistics and the ancestry of modern apes and humans. pp. 345–362. In: R. L. Ciochon and J. C. Fleagle (eds.), *Primate Evolution and Human Origin*. Benjamin/Cummings, Menlo Park, CA. **[5]**

Civetta, A. and R. S. Singh. 1998. Sex-related genes, directional sexual selection, and speciation. Mol. Biol. Evol. 15: 901–909. **[4]**

Clark, A. G. and 10 others. 1998. Haplotype structure and population genetic inferences from nucleotide-sequence variation in human lipoprotein lipase. Am. J. Hum. Genet. 63: 595–612. **[2]**

Clark, J. B. and M. G. Kidwell. 1997. A phylogenetic perspective on *P* transposable element evolution in *Drosophila*. Proc. Natl. Acad. Sci. USA 94: 11428–11433. **[7]**

Clark, J. B., W. P. Maddison, and M. G. Kidwell. 1994. Phylogenetic analysis supports horizontal transfer of *P* transposable elements. Mol. Biol. Evol. 11: 40–50. **[7]**

Cleary, M. L., E. A. Schon, and J. B. Lingrel. 1981. Two related pseudogenes are the result of a gene duplication in the goat β-globin locus. Cell 26: 181–190. **[6]**

Clegg, M. T., B. S. Gaut, G. H. Learn, and B. R. Morton. 1994. Rates and patterns of chloroplast DNA evolution. Proc. Natl. Acad. Sci. USA 91: 6795–6801. **[4]**

Coffin, J. M. 1986. Genetic variation in AIDS viruses. Cell 46:1–4. **[4]**

Cohen, S. N. 1976. Transposable genetic elements and plasmid evolution. Nature 263: 731–735. **[7]**

Cohen, S. N. and J. A. Shapiro. 1980. Transposable genetic elements. Sci. Am. 242(2): 40–49. **[7]**

Cold Spring Harbor Symposia on Quantitative Biology. 1986. *Molecular Biology of Homo sapiens*. Vol. 51. Cold Spring Harbor Laboratory, Cold Spring Harbor, NY. **[8]**

Cold Spring Harbor Symposium on Quantitative Biology. 1987. *Evolution of Catalytic Function*. Vol. 52. Cold Spring Harbor Laboratory, Cold Spring Harbor, NY. **[8]**

Collier, S., M. Tassabehji, P. Sinnott, and T. Strachan. 1993. A *de novo* pathological point mutation at the 21-hydroxylase locus: Implication for gene conversion in the human genome. Nat. Genet. 3: 260–265. **[6]**

Comeron, J. M. 1995. A method for estimating the numbers of synonymous and nonsynonymous substitutions per site. J. Mol. Evol. 41: 1152–1159. **[3]**

Confalone, E., J. J. Beintema, M. P. Sasso, A. Carsana, M. Palmieri, M. T. Vento, and A. Furia. 1995. Molecular evolution of genes encoding ribonucleases in ruminant species. J. Mol. Evol. 41: 850–858. **[6]**

Cooper, N. (ed.). 1994. *The Human Genome Project: Deciphering the Blueprint of Heredity*. University Science Books, Mill Valley, CA. **[8]**

Coulondre, C., J. H. Miller, P. J. Farabaugh, and W. Gilbert. 1978. Molecular basis of base substitution hotspots in *Escherichia coli*. Nature 274: 775–780. **[4]**

Crawford, D. J. 1990. *Plant Molecular Systematics: Macromolecular Approaches*. Wiley, San Francisco. **[5]**

Crick, F. H. C. 1968. The origin of the genetic code. J. Mol. Biol. 38: 367–379. **[8]**

Crow, J. F. and M. Kimura. 1970. *An Introduction to Population Genetics*. Harper & Row, New York. **[2]**

Curtis, D. and W. Bender. 1991. Gene conversion in *Drosophila* and the effects of the meiotic mutants *mei–9* and *mei–218*. Genetics 127: 739–746. **[6]**

Curtis, S. E. and M. T. Clegg. 1984. Molecular evolution of chloroplast DNA sequences. Mol. Biol. Evol. 1: 291–301. **[4]**

Czelusniak, J., M. Goodman, D. Hewett-Emmett, M. L. Weiss, P. J. Venta, and R. E. Tashian. 1982. Phylogenetic origins and adaptive evolution of avian and mammalian haemoglobin genes. Nature 298: 297–300. **[4, 6]**

Danciger E., N. Dafni, Y. Bernstein, Z. Laver-Rudich, A. Neer, and Y. Groner. 1986. Human Cu/Zn superoxide dismutase gene family: Molecular structure and characterization of four Cu/Zn superoxide dismutase-related pseudogenes. Proc. Natl. Acad. Sci. USA 83: 3619–3623. **[7]**

Daniels, S. B., K. R. Peterson, L. D. Strausbaugh, M. G. Kidwell, and A. Chovnick. 1990. Evidence for horizontal transmission of the *P* transposable element between *Drosophila* species. Genetics 124: 339–355. **[7]**

Darlington, C. D. 1937. *Recent Advances in Cytology*, 2nd Ed. Churchill, London. **[8]**

Darnell, J. E. 1978. Implications of RNA–RNA splicing in the evolution of eukaryotic cells. Science 202: 1257–1260. **[6]**

Darwin, C. 1859. *The Origin of Species by Means of Natural Selection*. John Murray, London. (Various reprints available.) **[Introduction, 6]**

Darwin, C. 1871. *The Descent of Man and Selection in Relation to Sex*. John Murray, London. (Various reprints available) **[5]**

Davey, R. B. and D. C. Reanney. 1980. Extrachromosomal genetic elements and the adaptive evolution of bacteria. Evol. Biol. 13: 113–147. **[8]**

Dayhoff, M. O. 1972. *Atlas of Protein Sequence and Structure*, Vol. 5. National Biomedical Research Foundation, Silver Spring, MD. **[4]**

Dayhoff, M. O. 1978. *Atlas of Protein Sequence and Structure*, Vol. 5, Supplement 3. National Biomedical Research Foundation, Silver Spring, MD. **[6]**

Dean, A. M. 1998. The molecular anatomy of an ancient adaptive event. Am. Sci. 86: 26–37. **[6]**

de Souza, S. J., M. Long, L. Schoenbach, S. W. Roy, and W. Gilbert. 1997. The correlation between introns and the three-dimensional structure of protein. Gene. 205: 141–144. **[6]**

de Souza, S. J., M. Long, R. J. Klein, S. Roy, S. Lin, and W. Gilbert. 1998. Toward a resolution of the introns early/late debate: Only phase zero introns are correlated with the structures of ancient proteins. Proc. Natl. Acad. Sci. USA 95: 5094–5099. **[6]**

Devos, R., R. Contreras, J. van Emmela, and W. Fiers. 1979. Identification of the translocatable element *IS1* in a molecular chimera constructed with pBR 322 into which MS2 DNA copy was inserted by poly(dA.dT) linked method. J. Mol. Biol. 128: 621–632. **[7]**

Dickerson, R. E. 1971. The structure of cytochrome *c* and the rates of molecular evolution. J. Mol. Evol. 1: 26–45. **[4]**

Disotell, T. R., R. L. Honeycutt, and M. Ruvolo. 1992. Mitochondrial DNA phylogeny of the old-world monkey tribe Papionini. Mol. Biol. Evol. 9: 1–13. **[7]**

Dixon, B., B. Walker, W. Kimmins, and B. Pohajdak. 1991. Isolation and sequencing of a cDNA for an unusual hemoglobin from the parasitic nematode *Pseudoterranova decipiens*. Proc. Natl. Acad. Sci. USA. 88: 5655–5659. **[6]**

Djian, P. and H. Green. 1989. Vectorial expansion of the involucrin gene and the relatedness of the hominoids. Proc. Natl. Acad. Sci. USA 86: 8447–8451. **[5]**

Dobzhansky, T. 1970. *Genetics and the Evolutionary Process*. Columbia University Press, New York. **[1]**

Doege, K. J., K. Garrison, S. N. Coulter, and Y. Yamada. 1994. The structure of the rat aggrecan gene and preliminary characterization of its promoter. J. Biol. Chem. 269: 29232–29240. **[6]**

Doolittle, R. F. 1985. The genealogy of some recently evolved vertebrate proteins. Trends Biochem. Sci. 10: 233–237. **[6]**

Doolittle, R. F. 1986. *Of URFs and ORFs: A Primer on How to Analyze Derived Amino Acid Sequences*. University Science Books, Mill Valley, CA. **[8]**

Doolittle, R. F. 1987. The evolution of the vertebrate plasma proteins. Biol. Bull. 172: 269–283. **[6]**

Doolittle, R. F. 1988. Lens proteins. More molecular opportunism. Nature 336: 18. **[6]**

Doolittle, R. F. (ed.). 1990. *Molecular Evolution: Computer Analyses of Protein and Nucleic Acid Sequences*. Academic Press, San Diego. **[3]**

Doolittle, R. F., D.-F. Feng, M. S. Johnson, and M. A. McClure. 1986. Relationships of human protein sequences to those of other organisms. Cold Spring Harbor Symp. Quant. Biol. 51: 447–455. **[6]**

Doolittle, R. F., D.-F. Feng, M. S. Johnson, and M. A. McClure. 1989. Origins and evolutionary relationships of retroviruses. Q. Rev. Biol. 64: 1–31. **[7]**

Doolittle, R. F., D.-F. Feng, S. Tsang, G. Cho, and E. Little. 1996. Determining divergence times of the major kingdoms of living organisms with a protein clock. Science 271: 470–477. **[5]**

Doolittle, W. F. and J. R. Brown. 1994. Tempo, mode, the progenote, and the universal root. Proc. Natl. Acad. Sci. USA 91: 6721–6728. **[5]**

Doolittle, W. F. and C. Sapienza. 1980. Selfish genes, the phenotype paradigm and genome evolution. Nature 284: 601–603. **[7, 8]**

Douglas, A. M. 1986. Tigers in Western Australia? New Sci. 110(1505): 44–47. **[5]**

Dover, G. A. 1982. Molecular drive: A cohesive mode of species evolution. Nature 299: 111–117. **[6]**

Dover, G. A. 1993. Evolution of genetic redundancy for advanced players. Curr. Opin. Genet. Develop. 3: 902–910. **[6]**

Dover, G. A. and R. B. Flavell (eds.). 1982. *Genome Evolution*. Academic Press, New York. **[8]**

Doyle, J. A. 1978. Origin of angiosperms. Annu. Rev. Ecol. Syst. 9: 365–392. **[5]**

Drake, J. W., B. Charlesworth, D. Charlesworth, and J. F. Crow. 1998. Rates of spontaneous mutation. Genetics 148: 1667–1686. **[1]**

Duboule, D. and A. S. Wilkins. 1998. The evolution of "bricolage." Trends Genet. 14: 54–59. **[6]**

Dujon, B. and 107 others. 1994. Complete DNA sequence of yeast chromosome XI. Nature 369: 371–378. **[4]**

Durbin, R., S. Eddy, A. Krogh, and G. Mitchison. 1998. *Biological Sequence Analysis: Probabilistic Models of Proteins and Nucleic Acids*. Cambridge University Press, Cambridge. **[5]**

Duret, L., D. Mouchiroud, and C. Gautier. 1995. Statistical analysis of vertebrate sequences reveals that long genes are scarce in GC-rich isochores. J. Mol. Evol. 40: 308–317. **[8]**

Dyson, F. 1985. *Origins of Life*. Cambridge University Press, Cambridge. **[Introduction]**

Dytrych, L., D. L. Sherman, C. S. Gillespie, and P. J. Brophy. 1998. Two PDZ domain proteins encoded by the murine periaxin gene are the result of alternative intron retention and are differentially targeted in Schwann cells. J. Biol. Chem. 273: 5794–5800. **[6]**

Eanes, W. F., M. Kirchner, and J. Yoon. 1993. Evidence for adaptive evolution of the *G6PD* gene in the *Drosophila melanogaster* and *Drosophila simulans* lineages. Proc. Natl. Acad. Sci. USA 90: 7475–7479. **[2]**

Easteal, S., C. C. Collet, and D. J. Betty. 1995. *The Mammalian Molecular Clock*. Springer & Landes, Austin, TX. **[4]**

Eck, R. V. and M. O. Dayhoff. 1966. *Atlas of Protein Sequence and Structure*. National Biomedical Research Foundation, Silver Spring, MD. **[5]**

Edelman, G. M. and J. A. Gally. 1970. Arrangement and evolution of eukaryotic genes. pp. 962–972. In: F. O. Schmitt (ed.), *The Neurosciences: Second Study Program*. Rockefeller University Press, New York. **[6]**

Edlind, T. D., J. Li, G. S. Visvesvara, M. H. Vodkin, G. L. McLaughlin, and S. K. Katiyar. 1996. Phylogenetic analysis of β-tubulin sequences from amitochondrial protozoa. Mol. Phylogenet. Evol. 5: 359–367. **[5]**

Edwards, A. W. F. and L. L. Cavalli-Sforza. 1964. Reconstruction of evolutionary trees. pp. 67–76. In: V. H. Heywood and J. McNeill (eds.), *Phenetic and Phylogenetic Classification*. Systematics Association, London. **[5]**

Efron, B. 1982. *The Jackknife, the Bootstrap, and Other Resampling Plans*. Society for Industrial and Applied Mathematics, Philadelphia. **[5]**

Efstratiadis, A. and 14 others. 1980. The structure and evolution of the human β-globin gene family. Cell 21: 653–668. **[6]**

Ehrlich, J., D. Sankoff, and J. H. Nadeau. 1997. Synteny conservation and chromosome rearrangements during mammalian evolution. Genetics 147: 289–296. **[8]**

Eickbush, T. E. 1994. Origin and evolutionary relationships of retroelements. pp. 121–157. In: S. S. Morse (ed.), *The Evolutionary Biology of Viruses*. Raven, New York. **[7]**

Eigen, M. and R. Winkler-Oswattish. 1996. *Steps towards Life: A Perspective on Evolution*. Oxford University Press, Oxford. **[Introduction]**

Eisenmann, V. 1985. Le couagga: Un zèbre aux origines douteuses. La Recherche 16: 254–256. **[5]**

Elder, J. F. and B. J. Turner. 1995. Concerted evolution of repetitive DNA sequences in eukaryotes. Q. Rev. Biol. 70: 297–320. **[6]**

Eldredge, N. and S. J. Gould. 1972. Punctuated equilibria: An alternative to phyletic gradualism. In: T. J. M. Schopf (ed.), *Models in Paleobiology*. W.H. Freeman, San Francisco. **[4]**

Eldredge, N. and J. Cracraft. 1980. *Phylogenetic Patterns and the Evolutionary Process: Method and Theory in Comparative Biology*. Columbia University Press, New York. **[5]**

Elgar, G. 1996. Quality not quantity: The pufferfish genome. Hum. Mol. Genet. 5S: 1437–1442. **[8]**

Elgin, S. C. R. and H. Weintraub. 1975. Chromosomal proteins and chromatin structure. Annu. Rev. Biochem. 44: 725–774. **[6]**

Ellegren, H. and A. K. Fridolfsson. 1997. Male-driven evolution of DNA sequences in birds. Nat. Genet. 17: 182–184. **[4]**

Ellis, J. 1982. Promiscuous DNA chloroplast genes inside plant mitochondria. Nature 299: 678–679. **[7]**

Ellison, M. J. and J. D. Childs. 1981. Pyrimidine dimers induced in *Escherichia coli* DNA by ultraviolet radiation present in sunlight. Photochem. Photobiol. 34: 465–469. **[8]**

Embley, T. M. and R. P. Hirt. 1998. Early branching eukaryotes? Curr. Opin. Genet. Develop. 8: 624–629. **[5]**

Endo, T., K. Ikeo, and T. Gojobori. 1996. Large-scale search for genes on which positive selection may operate. Mol. Biol. Evol. 13: 685–690. **[4]**

Engels, W. R. 1981a. Estimating genetic divergence and genetic variability with restriction endonucleases. Proc. Natl. Acad. Sci. USA 78: 6329–6333. **[3]**

Engels, W. R. 1981b. Hybrid dysgenesis in *Drosophila* and the stochastic loss hypothesis. Cold Spring Harbor Symp. Quant. Biol. 45: 561–565. **[7]**

Engels, W. R. 1992. The origin of *P* elements in *Drosophila melanogaster*. BioEssays 14: 681–686. **[7]**

Ernst, J. F., J. W. Stewart, and F. Sherman. 1982. Formation of composite iso-cytochrome *c* by recombination between non-allelic genes of yeast. J. Mol. Biol. 161: 373–394. **[6]**

Erwin, T. L. 1997. Biodiversity at its utmost: Tropical forest beetles. pp. 27–40. In: M. L. Reaka-Kudla, D. E. Wilson, and E. O. Wilson (eds.), *Biodiversity II*. Joseph Henry Press, Washington, DC. **[App. I]**

Eyre-Walker, A. 1992. Evidence that both G+C rich and G+C poor isochores are replicated early and late in the cell cycle. Nuc. Acids Res. 20: 1497–1501. **[8]**

Farris, J. S. 1970. Methods for computing Wagner trees. Syst. Zool. 34: 21–34. **[5]**

Farris, J. S. 1977. On the phenetic approach to vertebrate classification. pp. 823–850. In: M. K. Hecht, P. C. Goody, and B. M. Hecht (eds.), *Major Patterns in Vertebrate Evolution*. Plenum, New York. **[5]**

Feagin J. E., I. M. Abraham, and K. Stuart. 1988. Extensive editing of the cytochrome *c* oxidase III transcript in *Trypanosoma brucei*. Cell 53: 413–422. **[1]**

Feldman, M., B. Liu, G. Segal, S. Abbo, A. A. Levy, and J. M. Vega. 1997. Rapid elimination of low-copy DNA sequences in polyploid wheat: A possible mechanism for differentiation of homoeologous chromosomes. Genetics 147: 1381–1387. **[8]**

Felsenstein, J. 1973. Maximum-likelihood and minimum-steps methods for estimating evolutionary trees from data on discrete characters. Syst. Zool. 22: 240–249. **[5]**

Felsenstein, J. 1978a. Cases in which parsimony or compatibility methods will be positively misleading. Syst. Zool. 27: 401–410. **[5]**

Felsenstein, J. 1978b. The number of evolutionary trees. Syst. Zool. 27: 27–33. **[5]**

Felsenstein, J. 1981. Evolutionary trees from DNA sequences: A maximum likelihood approach. J. Mol. Evol. 17: 368–376. **[5]**

Felsenstein, J. 1982. Numerical methods for inferring evolutionary trees. Q. Rev. Biol. 57: 379–404. **[5]**

Felsenstein, J. 1985. Confidence limits on phylogenies: An approach using the bootstrap. Evolution 39: 783–791. **[5]**

Felsenstein, J. 1988. Phylogenies from molecular sequences: Inference and reliability. Annu. Rev. Genet. 22: 521–565. **[4, 5]**

Felsenstein, J. 1996. Inferring phylogenies from protein sequences by parsimony, distance, and likelihood methods. Methods Enzymol. 266: 418–427. **[5]**

Felsenstein, J. and H. Kishino. 1993. Is there something wrong with the bootstrap on phylogenies? A reply to Hillis and Bull. Syst. Biol. 42: 193–200. **[5]**

Feng, D. and R. F. Doolittle. 1987. Progressive sequence alignment as a prerequisite to correct phylogenetic trees. J. Mol. Evol. 25: 351–360. **[3]**

Ferris, S. D., A. C. Wilson, and W. M. Brown. 1981. Evolutionary tree for apes and humans based on cleavage maps of mitochondrial DNA. Proc. Natl. Acad. Sci. USA 78: 2432–2436. **[5]**

Field, L. M. and A. L. Devonshire. 1998. Evidence that *E4* and *FE4* esterase genes responsible for insecticide resistance in the aphid *Myzus persicae* (Sulzer) are part of a gene family. Biochem. J. 330: 169–173. **[6]**

Figueroa, F., E. Günther, and J. Klein. 1988. MHC polymorphisms pre-dating speciation. Nature 335: 265–271. **[6]**

Figureau, A. 1989. Optimization and the genetic code. Orig. Life Evol. Biosphere 19: 57–67. **[8]**

Filipski, J. 1987. Correlation between molecular clock ticking, codon usage, fidelity of DNA repair, chromosome banding and chromatin compactness in germline cells. FEES Lett. 217: 184–186. **[8]**

Finnegan, D. J. and D. H. Fawcett. 1986. Transposable elements in *Drosophila*. Oxford Surv. Eukaryotic Genes 3: 1–62. **[7]**

Fisher, W. K. and E. O. Thompson. 1979. Myoglobin of the shark *Heterodontus portusjacksoni*: Isolation and amino acid sequence. Austral. J. Biol. Sci. 32: 277–294. **[4]**

Fitch, D. H., W. J. Bailey, D. A. Tagle, M. Goodman, L. Sieu, and J. L. Slightom. 1991. Duplication of the γ-globin gene mediated by *L1* long interspersed repetitive elements in an early ancestor of simian primates. Proc. Natl. Acad. Sci. USA 88: 7396–7400. **[7]**

Fitch, W. M. 1967. Evidence suggesting a non-random character to nucleotide replacements in naturally occurring mutations. J. Mol. Biol. 26: 499–507. **[4]**

Fitch, W. M. 1971. Toward defining the course of evolution: Minimum change for a specific tree topology. Syst. Zool. 20: 406–416. **[4]**

Fitch, W. M. 1977. On the problem of discovering the most parsimonious tree. Am. Nat. 111: 223–257. **[4]**

Fitch, W. M. 1981. A non-sequential method for constructing trees and hierarchical classifications. J. Mol. Evol. 18: 3037. **[4]**

Fitch, W. M. and E. Margoliash. 1967. Construction of phylogenetic trees. A method based on mutation distances as estimated from cytochrome *c* sequences is of general applicability. Science 155: 279–284. **[5]**

Fitch, W. M. and K. Upper. 1987. The phylogeny of tRNA sequences provides for ambiguity reduction in the origin of the genetic code. Cold Spring Harbor Symp. Quant. Biol. 52: 759–767. **[5]**

Flavell, R. B. 1986. Repetitive DNA and chromosome evolution in plants. Phil. Trans. Roy. Soc. 312B: 227–242. **[8]**

Fleagle, J. G., T. M. Bown, J. D. Obradovich, and E. L. Simons. 1986. Age of the earliest African anthropoids. Science 234: 1247–1249. **[4]**

Flemington, E., H. D. Bradshaw, V. Traina-Dorge, V. Slagel, and P. L. Deininger. 1987. Sequence, structure and promoter characterization of the human thymidine kinase gene. Gene 52: 267–277. **[7]**

Flint, J., A. M. Taylor, and J. B. Clegg. 1988. Structure and evolution of the horse ζ globin locus. J. Mol. Biol. 199: 427–437. **[6]**

Flower, W. H. 1883. On whales, present and past and their probable origin. Proc. Zool. Soc. London 1883: 466–513. **[5]**

Flower, W. H. and J. G. Garson. 1884. *Catalogue of the Specimens Illustrating the Osteology and Dentition of Vertebrate Animals Recent and Extinct Contained in the Museum of the Royal College of Surgeons of England. II. Mammalia other than Man*. Royal College of Surgeons, London. **[5]**

Fogel, S., R. K. Mortimer, K. Lusnak, and F. Tavares. 1978. Meiotic gene conversion: A signal of the basic recombination event in yeast. Cold Spring Harbor Symp. Quant. Biol. 43: 1325–1341. **[6]**

Forsdyke, D. R. 1995. Relative roles of primary sequence and (G+C)% in determining the hierarchy of frequencies of complementary trinucleotide pairs in DNAs of different species. J. Mol. Evol. 41: 573–581. **[8]**

Fotaki, M. E. and K. Iatrou. 1993. Silk moth chorion pseudogenes: Hallmarks of genomic evolution by sequence duplication and gene conversion. J. Mol. Evol. 37: 211–220. **[6]**

Fox, G. E. and 18 others. 1980. The phylogeny of prokaryotes. Science 209: 457–463. **[5]**

Freund A.-M., M. Bichara, and R. P. P. Fuchs. 1989. Z-DNA-forming sequences are spontaneous deletion hot spots. Proc. Natl. Acad. Sci. USA 86: 7465–7469. **[1]**

Fryxell, K. J. 1996. The coevolution of gene family trees. Trends Genet. 12: 364–369. **[6]**

Gagnon, Y., L. Lacoste, N. Campagne, and J. Lapointe. 1996. Widespread use of the glu-tRNA^{Gln} transamidation pathway among bacteria. A member of the α-purple bacteria lacks glutaminyl-tRNA synthetase. J. Biol. Chem. 271: 14856–14863. **[7]**

Gallo, R. C. 1987. The AIDS virus. Sci. Am. 256(1): 46–56. **[4]**

Galtier, N. and M. Gouy. 1994. Molecular phylogeny of Eubacteria: A new multiple tree analysis method applied to 15 sequence data sets questions the monophyly of Gram-positive bacteria. Res. Microbiol. 145: 531–541. **[8]**

Galtier, N. and M. Gouy 1995. Inferring phylogenies from DNA sequences of unequal base compositions. Proc. Natl. Acad. Sci. USA 92: 11317–11321. **[5]**

Galtier, N. and M. Gouy 1998. Inferring pattern and process: Maximum-likelihood implementation of a nonhomogeneous model of DNA sequence evolution for phylogenetic analysis. Mol. Biol. Evol. 15: 871–879. **[5]**

Galtier, N. and J. R. Lobry. 1997. Relationships between genomic G+C content, RNA secondary structure, and optimal growth temperature in prokaryotes. J. Mol. Evol. 44: 632–636. **[5, 8]**

Galtier, N., N. Tourasse, and M. Gouy. 1999. A nonhyperthermophilic common ancestor to extant life forms. Science 283: 220–221. **[5]**

Galili, U. and K. Swanson. 1991. Gene sequences suggest inactivation of α-1,3-galactosyltransferase in catarrhines after the divergence of apes from monkeys. Proc. Natl. Acad. Sci. USA 88: 7401–7404. **[6]**

Galun, E. and A. Breiman. 1997. *Transgenic Plants*. Imperial College Press, London. **[7]**

Gamo, S., M. Sakajo, K. Ikeda, Y. H. Inoue, Y. Sakoyama, and E. Nakashima-Tanaka. 1990. Temporal distribution of *P* elements in *Drosophila melanogaster* strains from natural populations in Japan. Jpn. J. Genet. 65: 277–285. **[7]**

Gantt, J. S., S. L. Baldauf, J. P. Calie, N. F. Weeden, and J. D. Palmer. 1991. Transfer of *rpl22* to the nucleus greatly preceded its loss from the chloroplast and involved the gain of an intron. EMBO J. 10: 3073–3078. **[7]**

Gardiner, K. 1997. Clonability and gene distribution on human chromosome 21: Reflections of junk DNA content? Gene 205: 39–46. **[8]**

Gaut, B. S., S. V. Muse, W. D. Clark, and M. T. Clegg. 1992. Relative rates of nucleotide substitution at the *rbcL* locus of monocotydelonous plants. J. Mol. Evol. 35: 292–303. **[4]**

Gensel, P. G. and H. N. Andrews. 1984. *Plant Life in the Devonian*. Praeger, New York. **[5]**

Gibbs, A. J. and G. A. McIntyre. 1970. The diagram, a method for comparing sequences. Its use with amino acid and nucleotide sequences. Eur. J. Biochem. 16: 1–11. **[3]**

Gilbert, W. 1978. Why genes in pieces? Nature 271: 501. **[6]**

Gilbert, W. 1987. The exon theory of genes. Cold Spring Harbor Symp. Quant. Biol. 52: 901–905. **[6]**

Gilbert, W., M. Marchionni, and G. McKnight. 1986. On the antiquity of introns. Cell 46: 151–153. **[6]**

Gillespie, J. H. 1991. *The Causes of Molecular Evolution*. Oxford University Press, New York. **[2]**

Gillespie, J. H. 1998. *Population Genetics: A Concise Guide*. Johns Hopkins University Press, Baltimore. **[4]**

Gilson, P. R. and G. I. McFadden. 1996. The miniaturized nuclear genome of eukaryotic endosymbiont contains genes that overlap, genes that are cotranscibed, and the smallest known spliceosmal introns. Proc. Natl. Acad. Sci. USA 93: 7737–7742. **[8]**

Gilson, P. R. and McFadden G. I. 1997. Good things in small packages: The tiny genomes of chlorarachniophyte endosymbionts. BioEssays 19: 167–173. **[8]**

Gilson, P. R., U. G. Maier and G. I. McFadden. 1997. Size isn't everything: Lessons in genetic miniaturisation from nucleomorphs. Curr. Opin. Genet. Develop. 7: 800–806 **[8]**

Gingerich, P. D. 1984. Primate evolution: Evidence from the fossil record, comparative morphology, and molecular biology. Yearbook Phys. Anthropol. 27: 57–72. **[4]**

Gingerich, P. D., B. H. Smith, and E. L. Simons. 1990. Hind limbs of Eocene *Basilosaurus*: Evidence of feet in whales. Science 249: 154–157. **[5]**

Gingerich, P. D., S. M. Raza, M. Arif, M. Anwar, and X. Zhou. 1994. New whale from the Eocene of Pakistan and the origin of cetacean swimming. Nature 368: 844–847. **[5]**

Ginzburg, L. R., P. M. Bingham, and S. Yoo. 1984. On the theory of speciation induced by transposable elements. Genetics 107: 331–341. **[7]**

Giovannoni, S. J., S. Turner, G. J. Olsen, S. Barns, D. J. Lane, and N. R. Pace. 1988. Evolutionary relationships among cyanobacteria and green chloroplasts. J. Bacteriol. 170: 3584–3592. **[5]**

Gō, M. 1981. Correlation of DNA exonic regions with protein structural units in haemoglobin. Nature 291: 90–92. **[6]**

Gō, M. and M. Nosaka. 1987. Protein architecture and the origin of introns. Cold Spring Harbor Symp. Quant. Biol. 52: 915–924. **[6]**

Gogarten, J. P. and 12 others. 1989. Evolution of the vacuolar H^+-ATPase: Implications for the origin of eukaryotes. Proc. Natl. Acad. Sci. USA 86: 6661–6665. **[5]**

Gojobori, T. and M. Nei. 1984. Concerted evolution of the immunoglobulin V_H gene family. Mol. Biol. Evol. 1: 195–211. **[6]**

Gojobori, T. and S. Yokoyama. 1985. Rates of evolution of the retroviral oncogene of Moloney murine sarcoma virus and of its cellular homologues. Proc. Natl. Acad. Sci. USA 82: 4198–4201. **[1, 4]**

Gojobori, T., W.-H. Li, and D. Graur. 1982. Patterns of nucleotide substitution in pseudogenes and functional genes. J. Mol. Evol. 18: 360–369. **[4]**

Golding, G. B. and R. S. Gupta. 1995. Protein-based phylogenies support a chimeric origin for the eukaryotic genome. Mol. Biol. Evol. 12: 1–6. **[5]**

Golding, G. B. and C. Strobeck. 1983. Increased number of alleles found in hybrid populations due to intragenic recombination. Evolution 17: 17–19. **[1]**

Goldman, D., P. R. Giri, S. J. O'Brien. 1987. A molecular phylogeny of the hominoid primates as indicated by two-dimensional protein electrophoresis. Proc. Natl. Acad. Sci. USA 84: 3307–3311. **[5]**

Goldman, M. A., G. P. Holmquist, M. C. Gray, L. A. Caston, and A. Nag. 1984. Replication timing of genes and middle repetitive sequences. Science 224: 686–692. **[8]**

Goldman, N. 1993. Statistical tests of models of DNA substitution. J. Mol. Evol. 36: 182–198. **[4]**

Goldsmith, M. E., R. K. Humphries, T. Ley, A. Cline, J. A. Kantor, and A. W. Nienhuis. 1983. "Silent" nucleotide substitution in a β^+-thalassemia globin gene activates splice site in coding sequence RNA. Proc. Natl. Acad. Sci. USA 80: 2318–2322. **[1, 6]**

Golenberg, E. M., D. E. Giannasi, M. T. Clegg, C. J. Smiley, M. Durbin, D. Henderson, and G. Zurawsky. 1990. Chloroplast DNA sequence from a Miocene *Magnolia* species. Nature 344: 656–658. **[5]**

Golic, K. G. 1994. Local transposition of *P* elements in *Drosophila melanogaster* and recombination between duplicated elements using a site-specific recombinase. Genetics 137: 551–563. **[7]**

Goodman, M. 1961. The role of immunochemical differences in the phyletic development of human behavior. Hum. Biol. 33: 131–162. **[4]**

Goodman, M. 1962. Immunochemistry of the primates and primate evolution. Ann. N. Y. Acad. Sci. 102: 219–234. **[5]**

Goodman, M. 1963. Serological analysis of the systematics of recent hominoids. Hum. Biol. 35: 377–424. **[5]**

Goodman, M. 1981a. Decoding the pattern of protein evolution. Prog. Biophys. Mol. Biol. 38: 105–164. **[4]**

Goodman, M. 1981b. Globin evolution was apparently very rapid in early vertebrates: A reasonable case against the rate-constancy hypothesis. J. Mol. Evol. 17: 114–120. **[4]**

Goodman, M. 1999. The genomic record of humankind's evolutionary roots. Am. J. Hum. Genet. 64: 31–39. **[5]**

Goodman, M., J. Barnabas, G. Matsuda, and G. W. Moor. 1971. Molecular evolution in the descent of man. Nature 233: 604–613. **[4]**

Goodman, M., G. W. Moore, and J. Barnabas. 1974. The phylogeny of human globin genes investigated by the maximum parsimony method. J. Mol. Evol. 3: 1–48. **[4]**

Goodman, M., G. W. Moore, and G. Matsuda. 1975. Darwinian evolution in the genealogy of hemoglobin. Nature 253: 603–608. **[4]**

Goodman, M., A. E. Romero-Herrera, H. Dene, J. Czelusniak, and R. E. Tashian. 1982. Amino acid sequence evidence on the phylogeny of primates and other eutherians. pp. 115–191. In: M. Goodman (ed.), *Macromolecular Sequences in Systematic and Evolutionary Biology*. Plenum, New York. **[5]**

Goodman, M., B. F. Koop, J. Czelusniak, M. L. Weiss, and J. L. Slightom. 1984. The η-globin gene: Its long evolutionary history in the β-globin gene family of mammals. J. Mol. Biol. 180: 803–823. **[6]**

Goodman, M. and 8 others. 1988. An evolutionary tree for invertebrate globin sequences. J. Mol. Evol. 27: 236–249. **[6]**

Goremykin, V. V., S. Hansmann, and W. F. Martin. 1997. Evolutionary analysis of 68 proteins encoded in six completely sequenced chloroplast genomes: Revised molecular estimates of two seed plant divergent times. Plant Syst. Evol. 206: 337–351. **[5]**

Gould, S. J. and E. S. Vrba. 1982. Exaptation—a missing term in the science of form. Paleobiology 8: 4–15. **[7]**

Grant, V. 1985. *The Evolutionary Process: A Critical Review of Evolutionary Theory*. Columbia University Press, New York. **[App. I]**

Grantham, R. 1974. Amino acid difference formula to help explain protein evolution. Science 185: 862–864. **[4]**

Grantham, R., C. Gautier, M. Gouy, R. Mercier, and A. Pavé. 1980. Codon catalog usage and the genome hypothesis. Nuc. Acids Res. 8: r49–r62. **[4]**

Graur, D. 1985. Amino acid composition and the evolutionary rates of protein-coding genes. J. Mol. Evol. 22: 53–63. **[4]**

Graur, D. and D. G. Higgins. 1994. Molecular evidence for the inclusion of cetaceans within the order Artiodactyla. Mol. Biol. Evol. 11: 357–364. **[5]**

Graur, D., M. Bogher, and A. Breiman. 1989a. Restriction endonuclease profiles of mitochondrial DNA and the origin of the B genome of bread wheat, *Triticum aestivum*. Heredity 62: 335–342. **[7]**

Graur, D., Y. Shuali, and W.-H. Li. 1989b. Deletions in processed pseudogenes accumulate faster in rodents than in humans. J. Mol. Evol. 28: 279–285. **[7]**

Graw, J. 1997. The crystallins: Genes, proteins and diseases. Biol. Chem. 378: 1331–1348. **[6]**

Gray, M. W., R. Cedergren, Y. Abel, and D. Sankoff. 1989. On the evolutionary origin of the plant mitochondrion and its genome. Proc. Natl. Acad. Sci. USA 86: 2267–2271. **[5]**

Gregory, W. K. 1910. The orders of mammals. Bull. Am. Mus. Nat. Hist. 27: 1–524. **[5]**

Gribskov, M. and J. Devereux (eds.). 1991. *Sequence Analysis Primer*. Stockton Press, New York. **[3]**

Griffiths, A. J. F., J. H. Miller, and D. T. Suzuki. 1996. *An Introduction to Genetic Analysis*, 6th Ed. W.H. Freeman, New York. **[1]**

Grime, J. P. and M. A. Mowforth. 1982. Variation in genome size and ecological interpretation. Nature 299: 151–153. **[8]**

Groisman, E. A., M. H. Saier, and H. Ochman. 1992. Horizontal transfer of a phosphatase gene as evidence for mosaic structure of the *Salmonella* genome. EMBO J. 11: 1309–1316. **[8]**

Gruskin, K. D., T. F. Smith, and M. Goodman. 1987. Possible origin of a calmodulin gene that lacks intervening sequences. Proc. Natl. Acad. Sci. USA 84: 1605–1608. **[7]**

Gu, X. and W.-H. Li. 1992. Higher rates of amino acid substitution in rodents than in humans. Mol. Phylogenet. Evol. 1: 211–214. **[4]**

Gu, X. and W.-H. Li. 1995. The size distribution of insertions and deletions in human and rodent pseudogenes suggests the logarithmic gap penalty for sequence alignment. J. Mol. Evol. 40: 464–473. **[4]**

Gueiros-Filho, F. J. and S. M. Beverley. 1997. Trans-kingdom transportation of the *Drosophila* element *mariner* within the protozoan *Leishmania*. Science 276: 1716–1719. **[7]**

Gupta, R. S. and G. B. Golding. 1993. Evolution of *HSP70* gene and its implications regarding relationships between archaebacteria, eubacteria, and eukaryotes. J. Mol. Evol. 37: 573–582. **[5]**

Gutiérrez, G. and A. Marín. 1998. The most ancient DNA recovered from an amber-preserved specimen may not be as ancient as it seems. Mol. Biol. Evol. 15: 926–929. **[5]**

Hahn, B. H. and 9 others. 1986. Genetic variation in HTLV-III/LAV over time in patients with AIDS or at risk for AIDS. Science 232: 1548–1553. **[4]**

Haldane, J. B. S. 1932. *The Causes of Evolution*. Longmans & Green, London. **[6]**

Haldane, J. B. S. 1947. The mutation rate of the gene for hemophilia, and its segregation ratios in males and females. Ann. Eugen. 13: 262–271. **[4]**

Hall, B. G. 1990. Directed evolution of a bacterial operon. BioEssays 12: 551–557. **[1]**

Hammer, M. F. 1995. A recent common ancestry for human Y chromosomes. Nature 378: 376–378. **[2]**

Hänni, C., V. Laudet, M. Sakka, A. Begue, and D. Stehelin. 1990. Amplification de fragments d'ADN mitochondrial a partir de dents de d'os humains anciens. Compt. Rend. Acad. Sci. Paris, Ser. III 310: 109–125. **[5]**

Hardies, S. C., S. L. Martin, C. F. Voliva, C. A. Hutchison, and M. H. Edgell. 1986. An analysis of replacement and synonymous changes in the rodent *L1* repeat family. Mol. Biol. Evol. 3: 109–125. **[8]**

Harding, R. M. and 7 others. 1997. Archaic African and Asian lineages in the genetic ancestry of modern humans. Am. J. Hum. Genet. 60: 772–789. **[2]**

Hardison, R. C. and J. B. Margot. 1984. Rabbit globin pseudogene ψβ2 is a hybrid of δ- and β-globin gene sequences. Mol. Biol. Evol. 1: 302–316. **[6]**

Harland, W. B., A. V. Cox, P. G. Llewellyn, C. A. G. Pickton, A. G. Smith, and R. Walters. 1982. *A Geologic Time Scale*. Cambridge University Press, Cambridge. **[App. I]**

Harley, E. H. 1988. The retrieval of the quagga. S. Afr. J. Sci. 84: 158–159. **[5]**

Harlow, P., S. Litwin, and M. Nemer. 1988. Synonymous nucleotide substitution rates of β-tubulin and histone genes conform to high overall genomic rates in rodents but not in sea urchins. J. Mol. Evol. 27: 56–64. **[4]**

Harris, H. 1979. Multilocus enzymes in man. Ciba Foundation Symp. 27/29: 187–204. **[6]**

Harris, H. 1980/1981. Multilocus enzyme systems and the evolution of gene expression: The alkaline phosphatases as a model example. Harvey Lect. 76: 95–124. **[6]**

Harris, S., P. A. Barrie, M. L. Weiss, and A. J. Jeffreys. 1984. The primate *ψβ1* gene. An ancient β-globin pseudogene. J. Mol. Biol. 180: 785–801. **[6]**

Hartl, D. L. and A. G. Clark. 1997. *Principles of Population Genetics*, 3rd Ed. Sinauer Associates, Sunderland, MA. **[2]**

Hartl, D. L., M. Medhora, L. Green, and D. E. Dykhuizen. 1986. The evolution of DNA sequences in *Escherichia coli*. Phil. Trans. Roy. Soc. 312B: 191–204. **[8]**

Harvey, P. H. and M. D. Pagel. 1991. *The Comparative Method in Evolutionary Biology*. Oxford University Press, Oxford. **[3]**

Harvey, P. H., A. J. Leigh Brown, J. Maynard Smith, and S. Nee (eds.). 1996. *New Uses for New Phylogenies*. Oxford University Press, Oxford. **[5]**

Hasegawa, M., H. Kishino, and T. Yano. 1987. Man's place in Hominoidea as inferred from molecular clocks of DNA. J. Mol. Evol. 26: 132–147. **[5]**

Hayasaka, K., D. H. Fitch, J. L. Slightom, and M. Goodman. 1992. Fetal recruitment of anthropoid γ-globin genes. Findings from phylogenetic analyses involving the 5'-flanking sequences of the ψγ1 globin gene of spider monkey, *Ateles geoffroyi*. J. Mol. Biol. 224: 875–881. **[6]**

Hayashida, H., H. Toh, R. Kikuno, and T. Miyata. 1985. Evolution of influenza virus genes. Mol. Biol. Evol. 2: 289–303. **[4]**

Hayden, T. 1999. How low can you go? Seeking the fewest genes necessary for life. Newsweek 133(8): 52. **[8]**

Heddle, J. A. and K. Athanasiou. 1975. Mutation rate, genome size and their relation to the *rec*. concept. Nature 258: 359–361. **[8]**

Hedges, S. B. and L. L. Poling. 1999. A molecular phylogeny of reptiles. Science 283: 998–1001. **[5]**

Hedges, S. B., S. Kumar, K. Tamura, and M. Stoneking. 1992. Human origins and analysis of mitochondrial DNA sequences. Science 255: 737–739. **[5]**

Hedrick, P. W. 1983. *Genetics of Populations*. Science Books International, Portola Valley, CA. **[2]**

Hegyi, H. and P. Bork. 1997. On the classification and evolution of protein modules. J. Prot. Chem. 16: 545–551. **[6]**

Hein, J. 1989. A new method that simultaneously aligns and reconstructs ancestral sequences for any number of homologous sequences when the phylogeny is given. Mol. Biol. Evol. 6: 649–68. **[3]**

Hendriks, W., J. Leunissen, E. Nevo, H. Bloemendal, and W. W. de Jong. 1987. The lens protein αA-crystallin of the blind mole rat, *Spalax ehrenbergi*: Evolutionary change and functional constraints. Proc. Natl. Acad. Sci. USA 84: 5320–5324. **[4]**

Hendriks, W., J. W. M. Mulders, M. A. Bibby, C. Slingsby, H. Bloemendal, and W. W. de Jong. 1988. Duck lens ε-crystallin and lactate dehydrogenase B_4 are identical: A single-copy gene product with two distinct functions. Proc. Natl. Acad. Sci. USA 85: 7114–7118. **[6]**

Hendy, M. D. and D. Penny. 1982. Branch and bound algorithms to determine minimal evolutionary trees. Math. Biosci. 59: 277–290. **[5]**

Henikoff, S., M. A. Keene, K. Fechtel, and J. W. Fristrom. 1986. Gene within a gene: Nested *Drosophila* genes encode unrelated proteins on opposite strands. Cell 44: 33–42. **[4]**

Henikoff, S., E. A. Greene, S. Pietrokovski, P. Bork, T. K. Attwood, and L. Hood. 1997. Gene families: The taxonomy of protein paralogs and chimeras. Science 278: 609–614. **[6]**

Hensche, P. E. 1975. Gene duplication as a mechanism of genetic adaptation in *Saccharomyces cerevisiae*. Genetics 79: 661–674. **[6]**

Hentzen, D., A. Chevallier, and J. P. Garel. 1981. Differential usage of iso-accepting $tRNA^{Ser}$ species in silk glands of *Bombyx mori*. Nature 290: 267–269. **[4]**

Herdman, M. 1985. The evolution of bacterial genomes. pp. 37–68. In: T. Cavalier-Smith (ed.), *The Evolution of Genome Size*. Wiley, New York. **[8]**

Hess, J. F., M. Fox, C. Schmid, and C.-K. J. Shen. 1983. Molecular evolution of the human adult α-globin-like gene region: Insertion and deletion of *Alu* family repeats and non-*Alu* DNA sequences. Proc. Natl. Acad. Sci. USA 80: 5970–5974. **[7]**

Hewett-Emmett, D. and R. E. Tashian. 1996. Functional diversity, conservation, and convergence in the evolution of the α-, β-, and γ-carbonic anhydrase gene families. Mol. Phylogenet. Evol. 5: 50–77. **[6]**

Hey, J. 1989. The transposable portion of the genome of *Drosophila algonquin* is very different from that in *D. melanogaster*. Mol. Biol. Evol. 6: 66–79. **[7]**

Hey, J. 1999. The neutralist, the fly, and the selectionist. Trends Ecol. Evol. 14: 35–38. **[2]**

Hickey, D. A. 1982. Selfish DNA: A sexually-transmitted nuclear parasite. Genetics 101: 519–531. **[7]**

Hickey, L. J. and J. A. Doyle. 1977. Early Cretaceous fossil evidence for angiosperm evolution. Bot. Rev. 43: 2–104. **[5]**

Higgins, D. G. and P. M. Sharp. 1988. CLUSTAL: A package for performing multiple sequence alignment on a microcomputer. Gene 73: 237–44. **[3]**

Higgins, D. G. and P. M. Sharp. 1989. Fast and sensitive multiple sequence alignments on a microcomputer. Comput. Appl. Biosci. 5: 151–3. **[3]**

Higuchi, R. G., L. A. Wrischnik, E. Oakes, M. George, B. Tong, and A. C. Wilson. 1987. Mitochondrial DNA of the extinct quagga: Relatedness and extent of post-mortem changes. J. Mol. Evol. 25: 283–287. **[5]**

Hillis, D. M. and J. J. Bull. 1993. An empirical test of bootstrapping as a method for assessing confidence on phylogenetic analysis. Syst. Biol. 42: 182–192. **[5]**

Hillis, D. M., C. Moritz, and B. K. Mable (eds.). 1996. *Molecular Systematics*, 2nd ed. Sinauer Associates, Sunderland, MA. **[5]**

Hiraoka, B. Y., F. S. Sharief, Y.-W. Yang, W.-H. Li, and S. S.-L. Li. 1990. The cDNA and protein sequences of mouse lactate dehydrogenase B. Molecular evolution of vertebrate lactate dehydrogenase A (muscle), B (heart) and C (testis). Eur. J. Biochem. 189: 215–220. **[6]**

Hixon, J. E. and W. M. Brown. 1986. A comparison of the small ribosomal RNA genes from the mitochondrial DNA of the great apes and humans: Sequence, structure, evolution, and phylogenetic implications. Mol. Biol. Evol. 3: 1–18. **[5]**

Hobbs, H. H., M. S. Brown, J. L. Goldstein, and D. W. Russell. 1986. Deletion of exon encoding cysteine rich repeat of low density lipoprotein receptor alters its binding specificity in a subject with familial hypercholesterolemia J. Biol. Chem. 261: 13114–13120. **[7]**

Holland, J., K. Spindler, E. Horodyski, E. Grabau, S. Nichol, and S. VandePol. 1982. Rapid evolution of RNA genomes. Science 215: 1577–1585. **[4]**

Holmquist, G. P. 1987. Role of replication time in the control of tissue-specific gene expression. Am. J. Hum. Genet. 40: 151–173. **[8]**

Holmquist, R. 1972. Theoretical foundations for a quantitative approach to paleogenetics. I: DNA. J. Mol. Evol. 1: 115–133. **[3]**

Holmquist, R. and D. Pearl. 1980. Theoretical foundations for quantitative paleogenetics III. The molecular divergence of nucleic acids and proteins for the case of genetic events of unequal probability. J. Mol. Evol. 16: 211–267. **[3]**

Holum, J. R. 1978. *Organic and Biological Chemistry*. John Wiley, New York. **[1]**

Hood, L., J. H. Campbell, and S. C. R. Elgin. 1975. The organization, expression and evolution of antibody genes and other multigene families. Annu. Rev. Genet. 9: 305–353. **[6]**

Horai, S. and 7 others. 1992. Man's place in Hominoidea revealed by mitochondrial DNA genealogy. J. Mol. Evol. 35: 32–43. **[5]**

Horai, S., R. Kondo, Y. Nakagawa-Hattorri, S. Hayashi, S. Sonoda, and K. Tajima. 1993. Peopling of the Americas founded by four major lineages of mitochondrial DNA. Mol. Biol. Evol. 10: 23–47. **[5]**

Hori, H. and S. Osawa. 1986. Evolutionary change in 5S rRNA secondary structure and a phylogenetic tree of 352 rRNA species. BioSystems 19: 163–172. **[8]**

Houck, M. A., J. B. Clark, K. R. Peterson, and M. G. Kidwell. 1991. Possible horizontal transfer of *Drosophila* genes by the mite *Proctolaelaps regalis*. Science 253: 1125–1129. **[7]**

Hudson, R. R., M. Kreitman, and M. Aguadé. 1987. A test of neutral evolution based on nucleotide data. Genetics 116: 153–159. **[2]**

Hudson, T. J. and 7 others. 1992. Isolation and chromosomal assignment of 100 highly informative human simple sequence repeat polymorphisms. Genomics 13: 622–629. **[8]**

Huelsenbeck, J. P. and K. A. Crandall. 1997. Phylogeny estimation and hypothesis testing using maximum likelihood. Annu. Rev. Ecol. Syst. 28: 437–466. **[5]**

Hughes, A. L. 1988. The quagga case: Molecular evolution of an extinct species. Trends Ecol. Evol. 3: 95–96. **[5]**

Hughes, A. L. 1994. The evolution of functionally novel proteins after gene duplication. Proc. Roy. Soc. 256B: 119–124. **[6]**

Hughes, M. K. and A. L. Hughes. 1993. Evolution of duplicate genes in a tetraploid animal, *Xenopus laevis*. Mol. Biol. Evol. 10: 1360–1369. **[6]**

Hughes, A. L. and M. Nei. 1989. Nucleotide substitution at major histocompatibility complex class II loci: Evidence for overdominant selection. Proc. Natl. Acad. Sci. USA 86: 958–962. **[2, 4, 6]**

Hurst, D. L. and G. T. McVean. 1996. A difficult phase for introns-early. Molecular evolution. Curr. Biol. 6: 533–536. **[6]**

Hutchison, C. A., S. C. Hardies, D. D. Loeb, W. R. Shehee, and M. H. Edgell. 1989. LINEs and related retroposons: Long interspersed repeated sequences in the eukaryotic genome. pp. 593–617. In: D. E. Berg and M. M. Howe (eds.), *Mobile DNA*. American Society for Microbiology, Washington, DC. **[7, 8]**

Ikemura, T. 1981. Correlation between the abundance of *Escherichia coli* transfer RNAs and the occurrence of the respective codons in its protein genes: A proposal for a synonymous codon choice that is optimal for the *E. coli* translational system. J. Mol. Biol. 151: 389–409. **[4]**

Ikemura, T. 1982. Correlation between the abundance of yeast transfer RNAs and the occurrence of the respective codons in protein genes: Differences in synonymous codon choice patterns of yeast and *Escherichia coli* with reference to the abundance of isoaccepting transfer RNAs. J. Mol. Biol. 158: 573–697. **[4]**

Ikemura, T. 1985. Codon usage and tRNA content in unicellular and multicellular organisms. Mol. Biol. Evol. 2: 13–34. **[8]**

Ikemura, T. and S.-I. Aota. 1988. Global variation in G+C content along vertebrate genome DNA: Possible correlation with chromosome band structures. J. Mol. Biol. 203: 1–13. **[8]**

Ikemura T. and H. Ozeki. 1983. Codon usage and transfer RNA contents: Organism-specific codon-choice patterns in reference to the isoacceptor contents. Cold Spring Harbor Symp. Quant. Biol. 47: 1987–1097. **[4]**

Ina, Y. 1995. New methods for estimating the numbers of synonymous and nonsynonymous substitutions. J. Mol. Evol. 40: 190–226. **[3]**

Inouye, M. and S. Inouye. 1991. msDNA and bacterial reverse transcriptase. Annu. Rev. Microbiol. 45: 163–186. **[7]**

Inouye, S., M.-Y. Hsu, S. Eagle, and M. Inouye. 1989. Reverse transcriptase associated with the biosynthesis of the branched RNA-linked msDNA in *Myxococcus xanthus*. Cell 56: 709–717. **[7]**

Irwin, D. M., T. D. Kocher, and A. C. Wilson. 1991. Evolution of the cytochrome *b* gene of mammals. J. Mol. Evol. 32: 128–144. **[5]**

Ishida, K., Y. Cao, M. Hasegawa, N. Okada, and Y. Hara. 1997. The origin of chlorarachniophyte plastids, as inferred from phylogenetic comparisons of amino acid sequences of *EF-Tu*. J. Mol. Evol. 45: 682–687. **[8]**

Itano, H. A. 1957. The human hemoglobins: Their properties and genetic control. Adv. Prot. Chem. 12: 216–268. **[6]**

Itaya, M. 1995. An estimation of minimal genome size required for life. FEBS Lett. 362: 257–260. **[8]**

Itaya, M. and T. Tanaka. 1997. Experimental surgery to create subgenomes of *Bacillus subtilis* 168. Proc. Natl. Acad. Sci. USA 94: 5378–5382. **[8]**

Iwabe, N., K. Kuma, M. Hasegawa, S. Osawa, and T. Miyata. 1989. Evolutionary relationship of archaebacteria, eubacteria, and eukaryotes inferred from phylogenetic trees of duplicated genes. Proc. Natl. Acad. Sci. USA 86: 9355–9359. **[5]**

Jackson, J. A. and G. R. Fink. 1985. Meiotic recombination between duplicated genetic elements in *Saccharomyces cerevisiae*. Genetics 109: 303–332. **[6]**

Jacob, F. 1977. Evolution and tinkering. Science 196: 1161–1166. **[6]**

Jacob, F. 1983. Molecular tinkering in evolution. pp. 131–144. In: D. S. Bendall (ed.), *Evolution from Molecules to Man*. Cambridge University Press, Cambridge. **[6]**

Jacobs G. H., J. Neitz, and M. Neitz. 1993. Genetic basis of polymorphism in the color vision of platyrrhine monkeys. Vision Res. 33: 269–74. **[6]**

Jacobs, G. H., M. Neitz, J. F. Deegan, and J. Neitz. 1996. Trichromatic colour vision in new world monkeys. Nature 382: 156–158. **[6]**

Janke, A. and U. Árnason. 1997. The complete mitochondrial genome of *Alligator mississippiensis* and the separation between recent Archosauria (birds and crocodiles). Mol. Biol. Evol. 14: 1266–1272. **[4]**

Jeffreys, A. 1979. DNA sequence variants in $^{G}\gamma$-, $^{A}\gamma$-, δ- and β-globin genes of man. Cell 18: 1–10. **[6]**

Jin, L. and Nei, M. 1990. Limitations of the evolutionary parsimony method of phylogenetic analysis. Mol. Biol. Evol. 7: 82–102. **[3]**

Jockusch, E. L. 1997. An evolutionary correlate of genome size changes in plethodontid salamanders. Proc. R. Soc. London 264B: 597–604. **[8]**

John, B. and G. L. Miklos. 1988. *The Eukaryote Genome in Development and Evolution*. Allen & Unwin, London. **[8]**

Johnson, W. E. and S. J. O'Brien. 1997. Phylogenetic reconstruction of the Felidae using 16S rRNA and NADH–5 mitochondrial genes. J. Mol. Evol. 44: S98-S116. **[7]**

Jukes, T. H. 1985. A change in the genetic code in *Mycoplasma capricolum*. J. Mol. Evol. 22: 361–362. **[8]**

Jukes, T. H. and C. R. Cantor. 1969. Evolution of protein molecules. pp. 21–132. In: H. N. Munro (ed.), *Mammalian Protein Metabolism*. Academic Press, New York. **[3]**

Jukes, T. H. and M. Kimura. 1984. Evolutionary constraints and the neutral theory. J. Mol. Evol. 21: 90–92. **[4]**

Jumas-Bilak, E., S. Michaux-Charachon, G. Bourg, D. O'Callaghan, and M. Ramuz. 1998. Differences in chromosome number and genome rearrangements in the genus *Brucella*. Mol. Microbiol. 27: 99–106. **[8]**

Kagawa, Y. and 7 others. 1984. High guanine plus cytosine content in the third letter of codons of an extreme thermophile. J. Biol. Chem. 259: 2956–2960. **[8]**

Kano, A., T. Ohama, R. Abe, and Osawa, S. 1993. Unassigned or nonsense codons in *Micrococcus luteus*. J. Mol. Biol. 230: 51–6. **[1]**

Kaplan, N. 1983. Statistical analysis of restriction enzyme map data and nucleotide sequence data. pp. 75–106. In: B. S. Weir (ed.), *Statistical Analysis of DNA Sequence Data*. Marcel Dekker, New York. **[3]**

Kaplan, N. and K. Risko. 1982. A method for estimating rates of nucleotide substitution using DNA sequence data. Theor. Pop. Biol. 21: 318–328. **[3]**

Karin, M. and R. I. Richards. 1984. The human metallothionein gene family: Structure and expression. Environ. Health Persp. 54: 111–315. **[7]**

Karlin, S., J. Mrazek, and A. M. Campbell. 1997. Compositional biases of bacterial genomes and evolutionary implications. J. Bacteriol. 179: 3899–3913. **[8]**

Kazazian, H. H. and J. V. Moran. 1998. The impact of *L1* retrotransposons on the human genome. Nat. Genet. 19: 19–24. **[7]**

Kazazian, H. H., C. Wong, H. Youssoufian, A. F. Scott, D. G. Phillips, and S. E. Antonarakis. 1988. Haemophilia A resulting from *de novo* insertion of *L1* sequences represents a novel mechanism of mutation in man. Nature 332: 164–166 **[7]**

Kelly, M. R., S. Kidd, R. L. Berg, and M. W. Young. 1987. Restriction of *P* element insertions at the *Notch* locus of *Drosophila melanogaster*. Mol. Cell Biol. 7: 1545–1548. **[7]**

Kempken, F. and U. Kuck. Transposons in filamentous fungi—facts and perspectives. BioEssays 20: 652– 659. **[7]**

Kenrick, P. and P. R. Crane. 1997. The origin and early evolution of plants on land. Nature 389: 33–39. **[5]**

Keogh, R. S., C. Seoighe, and K. H. Wolfe. 1998. Evolution of gene order and chromosome number in *Saccharomyces*, *Kluyveromyces*, and related fungi. Yeast 14: 443–457. **[8]**

Kidwell, M. G. 1979. Hybrid dysgenesis in *Drosophila melanogaster*: The relationship between the *P-M* and *I-R* interaction systems. Genet. Res. 33: 205–217. **[7]**

Kidwell, M. G. 1983. Evolution of hybrid dysgenesis determinants in *Drosophila melanogaster*. Proc. Natl. Acad. Sci. USA 80: 1655–1659. **[7]**

Kidwell, M. G. 1993. Lateral transfer in natural populations of eukaryotes. Annu. Rev. Genet. 27: 235–256. **[7]**

Kidwell, M. and J. F. Kidwell. 1976. Selection for male recombination in *Drosophila melanogaster*. Genetics 84: 333–351. **[7]**

Kimura, M. 1955. Solution of a process of random genetic drift with a continuous model. Proc. Natl. Acad. Sci. USA 41: 144–155. **[2]**

Kimura, M. 1962. On the probability of fixation of mutant genes in populations. Genetics 47: 713–719. **[2]**

Kimura, M. 1968a. Evolutionary rate at the molecular level. Nature 217: 624–626. **[2]**

Kimura, M. 1968b. Genetic variability maintained in a finite population due to mutational production of neutral and nearly neutral isoalleles. Genet. Res. 11: 247–269. **[2, 4]**

Kimura, M. 1969. The rate of molecular evolution considered from the standpoint of population genetics. Proc. Natl. Acad. Sci. USA 63: 1181–1188. **[4]**

Kimura, M. 1977. Preponderance of synonymous changes as evidence for the neutral theory of molecular evolution. Nature 267: 275–276. **[4]**

Kimura, M. 1980. A simple method for estimating evolutionary rate of base substitution through comparative studies of nucleotide sequences. J. Mol. Evol. 16: 111–120. **[3]**

Kimura, M. 1981. Estimation of evolutionary distances between homologous nucleotide sequences. Proc. Natl. Acad. Sci. USA 78: 454–458. **[3]**

Kimura M. 1983. *The Neutral Theory of Molecular Evolution*. Cambridge University Press, Cambridge. **[2, 4, 6]**

Kimura, M. 1989. The neutral theory of molecular evolution and the worldview of the neutralists. Genome 31: 24–31. **[4]**

Kimura, M. and T. Ohta. 1969. The average number of generations until fixation of a mutant gene in a finite population. Genetics 61: 763–771. **[2]**

Kimura, M. and T. Ohta. 1971. Protein polymorphism as a phase of molecular evolution. Nature 229: 467–469. **[2]**

Kimura, M. and T. Ohta. 1972. On the stochastic model for estimation of mutational distance between homologous proteins. J. Mol. Evol. 2: 87–90. **[2]**

King, J. L. and T. H. Jukes. 1969. Non-Darwinian evolution. Science 164: 788–798. **[2, 4]**

King, L. M. 1998. The role of gene conversion in determining sequence variation and divergence in the *Est-5* gene family in *Drosophila pseudoobscura*. Genetics 148: 305–315. **[2]**

Kishino, H. and M. Hasegawa. 1989. Evaluation of the maximum likelihood estimate of the evolutionary tree topologies from DNA sequence data, and the branching order in Hominoidea. J. Mol. Evol. 29: 170–179. **[5]**

Kishino, H., T. Miyata, and M. Hasegawa. 1990. Maximum likelihood inference of protein phylogeny and the origin of chloroplasts. J. Mol. Evol. 31: 151–160. **[5]**

Klaer, R., S. Kühn, E. Tillmann, H. J. Fritz, and P. Starlinger. 1981. The sequence of *IS4*. Mol. Gen. Genet. 181: 169–175. **[7]**

Klein, H. L. and T. D. Petes. 1981. Intrachromosomal gene conversion in yeast. Nature 289: 144–148. **[6]**

Kloek, A. P., D. R. Sherman, and D. E. Goldberg. 1993. Novel gene structure and evolutionary context of *Caenorhabditis elegans* globin. Gene 129: 215–221. **[6]**

Klotz, L. C., N. Komar, R. L. Blanken, and R. M. Mitchell. 1979. Calculation of evolutionary trees from sequence data. Proc. Natl. Acad. Sci. USA 76: 4516–4520. **[5]**

Kocher, T. D., W. K. Thomas, A. Meyer, S. V. Edwards, S. Pääbo, F. X. Villablanca, and A. C. Wilson. 1989. Dynamics of mitochondrial DNA evolution in animals: Amplification and sequencing with conserved primers. Proc. Natl. Acad. Sci. USA 86: 6196–6200. **[5]**

Kohne, D. E. 1970. Evolution of higher organism DNA. Q. Rev. Biophys. 33: 327–375. **[4]**

Kohne, D. E., J. A. Chiscon, and B. H. Hoyer. 1972. Evolution of primate DNA sequences. J. Hum. Evol. 1: 627–644. **[4]**

Kondrashov, A. S. and J. F. Crow. 1993. A molecular approach to estimating human deleterious mutation rate. Hum. Mut. 2: 229–234. **[1]**

Koonin, E. V. and A. R. Mushegian. 1996. Complete genome sequences of cellular life forms: Glimpses of theoretical evolutionary genomics. Curr. Opin. Genet. Develop. 6: 757–762. **[8]**

Koonin, E. V., A. R. Mushegian, M. Y. Galperin, and D. R. Walker. 1997. Comparison of archaeal and bacterial genomes: Computer analysis of protein sequences predicts novel function and suggests a chimeric origin for the Archaea. Mol. Microbiol. 25: 619–637. **[5]**

Koop, B. F., M. Goodman, P. Xu, K. Chan, and J. L. Slightom. 1986a. Primate η-globin DNA sequences and man's place among the great apes. Nature 319: 234–238. **[5]**

Koop, B. F., M. M. Miyamoto, J. E. Embury, M. Goodman, J. Czelusniak, and J. L. Slightom. 1986b. Nucleotide sequence and evolution of the orangutan ε-globin gene region and surrounding *Alu* repeats. J. Mol. Evol. 24: 94–102. **[5]**

Kornegay, J. R., J. W. Schilling, and A. C. Wilson. 1994. Molecular adaptation of a leaf-eating bird: Stomach lysozyme of the hoatzin. Mol. Biol. Evol. 11: 921–928. **[4]**

Koshizaka, T., M. Nishikimi, T. Ozawa, and K. Yagi. 1988. Isolation and sequence analysis of cDNA encoding rat liver L-gulono-γ-lactone oxidase, a key enzyme for L-ascorbic acid biosynthesis. J. Biol. Chem. 263: 1619–1621. **[6]**

Krajewski, C., A. C. Driskell, P. R. Baverstok, and M. J. Braun. 1992. Phylogenetic relationships of the thylacine (Mammalia: Thylacinidae) among dasyroid marsupials: Evidence from cytochrome *b* DNA sequences. Proc. Roy. Soc. 250B: 19–27. **[5]**

Krajewski, C., L. Buckley, and M. Westerman. 1997. DNA phylogeny of the marsupial wolf resolved. Proc. Roy. Soc. 264B: 911–917. **[5]**

Kreitman, M. 1983. Nucleotide polymorphism at the alcohol dehydrogenase locus of *Drosophila melanogaster*. Nature 304: 412–417. **[2]**

Kreitman, M. and M. Aguadé. 1986. Excess polymorphism at the *adh* locus in *Drosophila melanogaster*. Genetics 114: 93–110. **[2]**

Krings, M., A. Stone, R. W. Schmitz, H. Krainitzki, M. Stoneking, and S. Pääbo. 1997. Neandertal DNA sequences and the origin of modern humans. Cell 90: 19–30. **[5]**

Kuhner, M. K. and J. Felsenstein. 1994. A simulation comparison of phylogeny algorithms under equal and unequal evolutionary rates. Mol. Biol. Evol. 11: 459–468. **[5]**

Kuriyan, J., T. S. Krishna, L. Wong, B. Guenther, A. Pahler, C. H. Williams, and P. Model. 1991. Convergent evolution of similar function in two structurally divergent enzymes. Nature 352: 172–174. **[6]**

Kushiro, A., M. Shimizu, and K.-I. Tomita. 1987. Molecular cloning and sequence determination of the *tuf* gene coding for the elongation factor *Tu* of *Thermus thermophilus* HB8. Eur. J. Biochem. 170: 93–98. **[8]**

Labedan, B. and M. Riley. 1995a. Gene products of *Escherichia coli*: Sequence comparisons and common ancestries. Mol. Biol. Evol. 12: 980–987. **[8]**

Labedan, B. and M. Riley. 1995b. Widespread protein sequence similarities: Origin of *Escherichia coli* genes. J. Bacteriol. 177: 1585–1588. **[8]**

Laird, C. D., B. L. McConaughy, and B. J. McCarthy. 1969. Rate of fixation of nucleotide substitution in evolution. Nature 224: 149–154. **[4]**

Lake, J. A. 1994. Reconstructing evolutionary trees from DNA and protein sequences: Paralinear distances. Proc. Natl. Acad. Sci. USA 91: 1455–1459. **[3, 5]**

Lamb, B. C. and S. Helmi. 1982. The extent to which gene conversion can change allele frequencies in populations. Genet. Res. 39: 199–217. **[6]**

Lamour, V., S. Quevillon, S. Diriong, V. C. N'guyen, M. Lipinski, and M. Mirande. 1994. Evolution of the Glx-tRNA synthetase family: The glutaminyl enzyme as a case of horizontal gene transfer. Proc. Natl. Acad. Sci. USA 91: 8670–8674. **[7]**

Lampson, B. C., J. Sun, M.-Y. Hsu, J. Vallejo-Ramirez, S. Inouye, and M. Inouye. 1989. Reverse transcriptase in a clinical strain of *Escherichia coli*: Production of branched RNA-linked msDNA. Science 243: 1033–1038. **[7]**

Lan, R. and P. R. Reeves. 1996. Gene transfer is a major factor in bacterial evolution. Mol. Biol. Evol. 13: 47–55. **[8]**

Lanave, C., G. Preparata, C. Saccone, and G. Serio. 1984. A new method for calculating evolutionary substitution rates. J. Mol. Evol. 20: 86–93. **[3]**

Landsman, J., E. Dennis, T. J. V. Higgins, C. A. Appleby, A. A. Kortt, and W. J. Peacock. 1986. Common evolutionary origin of legume and non-legume plant haemoglobins. Nature 324:166–168. **[6]**

Lanfear, J. and P. W. H. Holland. 1991. The molecular evolution of *ZFY*-related genes in birds and mammals. J. Mol. Evol. 32: 310–315. **[4]**

Lang, B. F. and 8 others. 1997. An ancestral mitochondrial DNA resemble a eubacterial genome in miniature. Nature 387: 493–497. **[8]**

Langley, C. H. and W. M. Fitch. 1974. An examination of the constancy of the rate of molecular evolution. J. Mol. Evol. 3: 161–177. **[4]**

Larhammar, D. and C. Risinger. 1994. Molecular genetics aspects of tetraploidy in the common carp, *Cyprinus carpio*. Mol. Phylogenet. Evol. 3: 59–68. **[6]**

Laroche, J., P. Li, and J. Bousquet. 1995. Mitochondrial DNA and monocot–dicot divergence time. Mol. Biol. Evol. 12: 1151–1156. **[5]**

Lawlor, D. A., C. D. Dickel, W. W. Hauswirth, and P. Parham. 1991. Ancient HLA genes from 7,500-year-old archaeological remains. Nature 349: 785–788 **[5]**

Lawson, F. S., R. L. Charlebois, and J. A. Dillon. 1996. Phylogenetic analysis of carbamoylphosphate synthetase genes: Complex evolutionary history includes an internal duplication within a gene which can root the tree of life. Mol. Biol. Evol. 13: 970–977. **[5, 6]**

Leblanc, C., O. Richard, B. Kloareg, S. Viehmann, K. Zetsche, and C. Boyen. 1997. Origin and evolution of mitochondria: What have we learnt from red algae? Curr. Genet. 31: 193–207. **[5]**

Leder, P. 1982. The genetics of antibody diversity. Sci. Am. 246(5): 102–115. **[6]**

Lee, D. C., P. Gonzalez, P. V. Rao, J. S. Zigler, and G. J. Wistow. 1993. Carbonyl-metabolizing enzymes and their relatives recruited as structural proteins in the eye lens. Adv. Exp. Med. Biol. 328: 159–168. **[6]**

Leeds, J. M., M. B. Slabourgh, and C. K. Mathews. 1985. DNA precursor pools and ribonucleotide reductase activity: Distribution between the nucleus and cytoplasm of mammalian cells. Mol. Cell. Biol. 5: 3443–3450. **[8]**

Lehmann, T., W. A. Hawley, H. Grebert, and F. H. Collins. 1998. The effective population size of *Anopheles gambiae* in Kenya: Implications for population structure. Mol. Biol. Evol. 15: 264–276. **[2]**

Lehrman, M. A., D. W. Russell, J. L. Goldsmith, and M. S. Brown. 1986. Exon-*Alu* recombination deletes 5 kilobases from the low density lipoprotein receptor gene, producing a null phenotype in familial hypercholesterolemia. Proc. Natl. Acad. Sci. USA 83: 3679–3683. **[7]**

Leib-Mosch, C. and W. Seifarth. 1995. Evolution and biological significance of human retroelements. Virus Genes 11: 133–145. **[7]**

Leider, J. M., P. Palese, and F. I. Smith. 1988. Determination of the mutation rate of a retrovirus. J. Virol. 62: 3084–3091. **[1]**

Lenski, R. E. and J. E. Mittler. 1993. The directed mutation controversy and neo-Darwinism. Science 259: 188–194. **[1]**

Leunissen, J. A., H. W. van den Hooven, and W. W. de Jong. 1990. Extreme differences in charge changes during protein evolution. J. Mol. Evol. 31: 33–39. **[4]**

Levinson, G. and G. A. Gutman. 1987. Slipped-strand mispairing: A major mechanism for DNA sequence evolution. Mol. Biol. Evol. 4: 203–221. **[1]**

Lewin, B. 1997. *Genes VI*. Oxford University Press, New York. **[1, 8]**

Lewin, R. 1981. Evolutionary history written in globin genes. Science 214: 426–429. **[7]**

Lewin, R. 1982. Promiscuous DNA leaps all barriers. Science 292: 478–479. **[7]**

Lewontin, R. C. *The Genetic Basis of Evolutionary Change*. Columbia University Press, New York. **[2]**

Li, W.-H. 1981. Simple method for constructing phylogenetic trees from distance matrices. Proc. Natl. Acad. Sci. USA 78: 1085–1089. **[5, 7]**

Li, W.-H. 1983. Evolution of duplicate genes and pseudogenes. pp. 14–37. In: M. Nei and R. K. Koehn (eds.), *Evolution of Genes and Proteins*. Sinauer Associates, Sunderland, MA. **[6]**

Li, W.-H. 1993. Unbiased estimation of the rates of synonymous and nonsynonymous substitution. J. Mol. Evol. 36: 96–99. **[3, 4]**

Li, W.-H. 1997. *Molecular Evolution*. Sinauer Associates, Sunderland, MA. **[1–5]**

Li, W.-H. and L. A. Sadler. 1991. Low nucleotide diversity in man. Genetics 129: 513–523. **[2]**

Li, W.-H. and M. Tanimura. 1987a. The molecular clock runs more slowly in man than in apes and monkeys. Nature 326: 93–96. **[4]**

Li, W.-H. and M. Tanimura. 1987b. The molecular clock runs much faster in rodents than in primates and artiodactyls. J. Mol. Evol. 25: 330–342. **[4,5]**

Li, W.-H., T. Gojobori, and M. Nei. 1981. Pseudogenes as a paradigm of neutral evolution. Nature 292: 237–239. **[6]**

Li, W.-H., C.-C. Luo, and C.-I. Wu. 1985a. Evolution of DNA sequences, pp. 1–94. In: R. J. MacIntyre (ed.), *Molecular Evolutionary Genetics*. Plenum, New York. **[1, 6]**

Li, W.-H., C.-I. Wu, and C.-C. Luo. 1985b. A new method for estimating synonymous and nonsynonymous rates of nucleotide substitution considering the relative likelihood of nucleotide and codon changes. Mol. Biol. Evol. 2: 150–174. **[3, 4]**

Li, W.-H., M. Tanimura, and P. M. Sharp. 1987a. An evaluation of the molecular clock hypothesis using mammalian DNA sequences. J. Mol. Evol. 25: 330–342. **[4]**

Li, W.-H., K. H. Wolfe, J. Sourdis, and P. M. Sharp. 1987b. Reconstruction of phylogenetic trees and estimation of divergence times under nonconstant rates of evolution. Cold Spring Harbor Symp. Quant. Biol. 52: 847–856. **[5]**

Li, W.-H., M. Tanimura, and P. M. Sharp. 1988. Rates and dates of divergence between AIDS virus nucleotide sequences. Mol. Biol. Evol. 5: 313–330. **[4]**

Li, W.-H., D. L. Ellsworth, J. Krushkal, B. H.-J. Chang, and D. Hewett-Emmett 1996. Rates of nucleotide substitution in primates and rodents and the generation-time effect hypothesis. Mol. Phylogenet. Evol. 5: 182–187. **[4]**

Lidgard, S. and P. R. Crane. 1988. Quantitative analyses of the early angiosperm radiation. Nature 331: 344–346. **[5]**

Liebhaber, S. A., M. Goosens, and Y. W. Kan. 1981. Homology and concerted evolution at the α1 and α2 loci of human α-globin. Nature 290: 157–184. **[6]**

Lin, Z. and 7 others. 1995. Sex determination by polymerase chain reaction on mummies discovered at

Taklaman Desert in 1912. Forensic Sci. Int. 75: 197–205. [5]

Linial, M. 1987. Creation of a processed pseudogene by retroviral infection. Cell 49: 93–102. [7]

Lobry, J. R. 1996a. A simple vectorial representation of DNA sequences for the detection of replication origin in bacteria. Biochemie 78: 323–326. [8]

Lobry, J. R. 1996b. Asymmetric substitution patterns in the two DNA strands of bacteria. Mol. Biol. Evol. 13: 660–665. [8]

Lobry, J. R. 1996c. Origin of replication of *Mycoplasma genitalium*. Science 272: 745–746. [8]

Lockhart, P. J., M. A. Steel, M. D. Hendy, and D. Penny. 1994. Recovering evolutionary trees under a more realistic model of sequence evolution. Mol. Biol. Evol. 11: 605–612. [3,5]

Lodish, H., D. Baltimore, A. Berk., S. L. Zipursky, P. Matsudaira, and J. Darnell. 1995. *Molecular Cell Biology*, 3rd Ed. W.H. Freeman, New York. [1]

Logsdon, J. M. and R. F. Doolittle. 1997. Origin of antifreeze protein genes: A cool tale in molecular evolution. Proc. Natl. Acad. Sci. USA 94: 3485–3487. [6]

Lohe, A. and P. Roberts. 1988. Evolution of satellite DNA sequences in *Drosophila*. pp. 148–186. In: R. Verma (ed.), *Heterochromatin*. Cambridge University Press, Cambridge. [8]

Lomedico, P., N. Rosenthal, A. Efstratiadis, W. Gilbert, R. Colodner, and R. Tizard. 1979. The structure and evolution of the two nonallelic rat preproinsulin genes. Cell 18: 545–558. [7]

Long, E. O. and I. B. Dawid. 1980. Repeated genes in eukaryotes. Annu. Rev. Biochem. 49: 727–764. [6]

Long, M. and C. H. Langley. 1993. Natural selection and the origin of *jingwei*, a chimeric processed functional gene in *Drosophila*. Science 260: 91–95. [7]

Lonsdale, D. M., T. P. Hodge, C. J. Howe, and D. B. Stern. 1983. Maize mitochondrial DNA contains a sequence homologous to the ribulose–1,5-bisphosphate carboxylase large subunit gene of chloroplast DNA. Cell 34: 1007–1014. [7]

Loomis, W. F. 1988. *Four Billion Years: An Essay on the Evolution of Genes and Organisms*. Sinauer Associates, Sunderland, MA. **[Introduction]**

Lopez, L. C., W.-H. Li, M. L. Frazier, C.-C. Luo, and G. F. Saunders. 1984. Evolution of glucagon genes. Mol. Biol. Evol. 1: 335–344. [5]

Lowenstein, J. M. and O. A. Ryder. 1985. Immunological systematics of the extinct quagga (Equidae). Experientia 41: 1192–1193. [5]

Lowenstein, J. M., V. M. Sarich, and B. J. Richardson. 1981. Albumin systematics and the extinct mammoth and Tasmanian wolf. Nature 291: 409–411. [5]

Ludwig, W. and 7 others. 1998. Bacterial phylogeny based on comparative sequence analysis. Electrophoresis 19: 554–568. [8]

Luo, C.-C., W.-H. Li, and L. Chan. 1989. Structure and expression of dog apolipoprotein A-I, E, and C-I mRNAs: Implications for the evolution and functional constraints of apolipoprotein structure. J. Lipid Res. 30: 1735–1746. [4]

Lynch, M. 1997. Mutation accumulation in nuclear, organelle, and prokaryotic transfer RNA genes. Mol. Biol. Evol. 14: 914–925. [4]

Maeda, N., J. B. Bliska, and O. Smithies. 1983. Recombination and balanced chromosome polymorphism suggested by DNA sequences 5′ to the human δ-globin gene. Proc. Natl. Acad. Sci. USA 80: 5012–5016. [5]

Maeda, N. and O. Smithies. 1986. The evolution of multigene families: Human haptoglobin genes. Annu. Rev. Genet. 20: 81–108. [7]

Maeda, N., C.-I. Wu, J. Bliska, and J. Reneke. 1988. Molecular evolution of intergenic DNA in higher primates: Pattern of DNA changes, molecular clock and evolution of repetitive sequences. Mol. Biol. Evol. 5: 1–20. [5]

Manfredi-Romanini, M. G., D. Formenti, C. Pellicciari, E. Ronchetti, and Y. Rumpler. 1994. Possible meaning of C-heterochromatic DNA (ChDNA) in primates. pp. 371–378. In: B. Thierry, J. R. Anderson, J. J. Roeder, and N. Herrenschmidt (eds.), *Current Primatology*, Vol. 1. *Ecology and Evolution*. Univ. Louis Pasteur, Strasbourg. [8]

Maniloff, J. 1996. The minimal cell genome: "On being the right size." Proc. Natl. Acad. Sci USA 93: 10004–10006. [8]

Margoliash, E. 1963. Primary structure and evolution of cytochrome *c*. Proc. Natl. Acad. Sci. USA 50: 672–679. [4]

Margulis, L. 1981. *Symbiosis in Cell Evolution: Life and its Environment in the Early Earth*. W.H. Freeman, San Francisco. [5]

Margulis, L. and K. V. Schwartz. 1988. *Five Kingdoms*. W.H. Freeman, New York. [5]

Margulis, L., D. Chase, and L. P. To. 1979. Possible evolutionary significance of spirochaetes. Proc. Roy. Soc. 204B: 189–198. [8]

Markert, C. L. and H. Ursprung. 1971. *Developmental Genetics*. Prentice-Hall, Englewood Cliffs, NJ. [6]

Markert, M. L., J. J. Hutton, D. A. Wiginton, J. C. States, and R. E. Kaufman. 1988. Adenosine deaminase (*ADA*) deficiency due to deletion of the *ADA* gene promoter and first exon by homologous recombination between two *Alu* elements. J. Clin. Invest. 81: 1323–1327. [7]

Maroni, G., J. Wise, J. E. Young, and E. Otto. 1987. Metallothionein gene duplication and metal tolerance in natural populations of *Drosophila melanogaster*. Genetics 117: 739–744. [6]

Martin, A. P. and S. R. Palumbi. 1993. Body size, metabolic rate, generation time, and the molecular clock. Proc. Natl. Acad. Sci. USA 90: 4087–4091. [4]

Martin, C. C. and R. Gordon. 1995. Differentiation trees, a junk DNA molecular clock, and the evolution of neoteny in salamanders. J. Evol. Biol. 8: 339–354. [8]

Martin, R. D. 1990. *Primate Origins and Evolution: A Phylogenetic Reconstruction*. Princeton University Press, Princeton. [5]

Martin, W. and M. Müller. 1998. The hydrogen hypothesis for the first eukaryote. Nature 392: 37–41. [5]

Martin, W., A. Gierl, and H. Saedler. 1989. Molecular evidence for pre-Cretaceous angiosperm origins. Nature 339: 46–48. [5]

Martin, W., D. Lydiate, H. Brinkmann, G. Forkmann, H. Saedler, and R. Cerff. 1993. Molecular phylogenies in angiosperm evolution. Mol. Biol. Evol. 10: 140–162. [5]

Maruyama, T. and M. Kimura. 1974. A note on the speed of gene frequency changes in reverse directions in a finite population. Evolution 28: 162–163. [2]

Mathews, C. K. and K. E. van Holde. 1990. *Biochemistry*. Benjamin/Cummings, Menlo Park, CA. [1]

May, E. W. and N. L. Craig. 1996. Switching from cut-and-paste to replicative *Tn7* transposition. Science 272: 401–404. [7]

McCarrey J. R. and K. Thomas. 1987. Human testis-specific *PGK* gene lacks introns and possesses characteristics of a processed gene. Nature 326: 501–505. **[7]**

McClure, M. A., T. K. Vasi, and W. M. Fitch. 1994. Comparative analysis of multiple-sequence alignment methods. Mol. Biol. Evol. 11: 571–592. **[3]**

McDonald, J. and M. Kreitman. 1991. Adaptive protein evolution at the *adh* locus in *Drosophila*. Nature 351: 652–654. **[2]**

McFadden, J. and G. Knowles. 1997. Escape from evolutionary stasis by transposon-mediated deleterious mutations. J. Theor. Biol. 186: 441–447. **[7]**

McKusick, V. A. 1998. *Mendelian Inheritance in Man. Catalogs of Human Genes and Genetic Disorders*. 12th Ed. Johns Hopkins University Press, Baltimore. **[6]**

McVean, G. T. and L. D. Hurst. 1997. Evidence for a selectively favourable reduction in the mutation rate of the X chromosome. Nature 386: 388–392. **[4]**

Mel, S. F. and J. J. Mekalanos. 1996. Modulation of horizontal gene transfer in pathogenic bacteria by *in vivo* signals. Cell 87: 795–798. **[7]**

Mellon, I., G. Spivak, and P. C. Hanawalt. 1987. Selective removal of transcription-blocking DNA damage from the transcribed strand of the mammalian *DHFR* gene. Cell 23: 241–249. **[8]**

Mereschkowsky, C. 1905. Über Natur und Ursprung der Chromatophoren im Pflanzenreiche. Biol. Centralblatt. 25: 593–604. **[5]**

Miklos, G. L. 1985. Localized highly repetitive DNA sequences in vertebrate and invertebrate genomes. pp. 241–321. In: R. J. MacIntyre (ed.), *Molecular Evolutionary Genetics*. Plenum, New York. **[8]**

Miklos, G. L. and G. M. Rubin. 1996. The role of the genome project in determining gene function: Insights from model organisms. Cell 86: 521–529. **[8]**

Milinkovitch, M. C., G. Orti, and A. Meyer. 1993. Revised phylogeny of whales suggested by mitochondrial ribosomal DNA. Nature 361: 346–348. **[5]**

Mindell, D. P., M. D. Sorenson, and D. E. Dimcheff 1998. Multiple independent origins of mitochondrial gene order in birds. Proc. Natl. Acad. Sci. USA 95: 10693–10697. **[8]**

Miyamoto, M. M. and T. Cracraft (eds.). 1991. *Phylogenetic Analysis of DNA Sequences*. Oxford University Press, New York. **[5]**

Miyamoto, M. M., J. L. Slightom, and M. Goodman. 1987. Phylogenetic relationships of humans and African apes from DNA sequences in the ψη-globin region. Science 238: 369–373. **[5]**

Miyata, T. and T. Yasunaga. 1978. Evolution of overlapping genes. Nature 272: 532–535. **[6]**

Miyata, T. and T. Yasunaga. 1980. Molecular evolution of mRNA: A method for estimating evolutionary rates of synonymous and amino acid substitutions from homologous nucleotide sequences and its application. J. Mol. Evol. 16: 23–36. **[3]**

Miyata T. and T. Yasunaga. 1981. Rapidly evolving mouse α-globin-related pseudogene and its evolutionary history. Proc. Natl. Acad. Sci. USA 78: 450–453. **[6]**

Miyata, T., S. Miyazawa, and T. Yasunaga. 1979. Two types of amino acid substitutions in protein evolution. J. Mol. Evol. 12: 219–236. **[4]**

Miyata, T., T. Yasunaga, and T. Nishida. 1980. Nucleotide sequence divergence and functional constraint in mRNA evolution. Proc. Natl. Acad. Sci. USA 77: 7328–7332. **[4]**

Miyata, T., H. Hayashida, R. Kikuno, M. Hasegawa, M. Kobayashi, and K. Koike. 1982. Molecular clock of silent substitution: At least six-fold preponderance of silent changes in mitochondrial genes over those in nuclear genes. J. Mol. Evol. 19: 28–35. **[4]**

Miyata, T., H. Toh, H. Hayashida, R. Kikuno, Y. Inokuchi, and K. Saigo. 1985. Sequence homology among reverse transcriptase-containing viruses and transposable genetic elements: Functional and evolutionary implications. pp. 313–331. In: T. Ohta and K. Aoki (eds.), *Population Genetics and Molecular Evolution*. Japan Scientific Societies Press/Springer, Tokyo. **[7]**

Miyata, T., H. Hayashida, K. Kuma, K. Mitsuyasa, and T. Yasunaga. 1987. Male-driven molecular evolution: A model and nucleotide sequence analysis. Cold Spring Harbor Symp. Quant. Biol. 52: 863–967. **[4]**

Miyata, T., K. Kuma, N. Iwabe, H. Hayashida, and T. Yasunaga. 1990. Different rates of evolution of autosome-, X chromosome-, and Y chromosome-linked genes: Hypothesis of male-driven evolution. pp. 342–357. In: N. Takahata and J. F. Crow (eds.), *Population Biology of Genes and Molecules*. Baifukan, Tokyo. **[4]**

Mizraji, E. and J. Ninio. 1985. Graphical coding of nucleic acid sequences. Biochemie 67: 445–448. **[8]**

Moens, L. and 8 others. 1996. Globins in nonvertebrate species: Dispersal by horizontal gene transfer and evolution of the structure-function relationships. Mol. Biol. Evol. 13: 324–333. **[7]**

Mohamed, A. H., S. S. Chirala, N. H. Mody, W. Y. Huang, and S. J. Wakil. 1988. Primary structure of the multifunctional α subunit protein of yeast fatty acid synthase derived from *FAS2* gene sequence. J. Biol. Chem. 263: 12315–12325. **[6]**

Mohrenweiser, H. and 10 others. 1996. Report on abstracts of the Third International Workshop on human chomosome 19 mapping. Cytogenet. Cell Genet. 74: 161–186. **[8]**

Monod, J.-L. 1975. On the molecular theory of evolution. pp. 11–24. In R. Harré (ed.), *Problems of Scientific Revolution: Progress and Obstacles to Progress in the Sciences*. Clarendon Press, Oxford. **[Introduction]**

Montgelard, C., F. M. Catzeflis, and E. Douzery. 1997. Phylogenetic relationships of artiodactyls and cetaceans as deduced from the comparison of cytochrome *b* and 12S rRNA mitochondrial sequences. Mol. Biol. Evol. 14: 550–559. **[5]**

Morgan, G. J. 1998. Emile Zuckerkandl, Linus Pauling, and the molecular evolutionary clock, 1959–1965. J. Hist. Biol. 31: 155–178. **[4]**

Moriyama, E. N. and T. Gojobori. 1989. Evolution of nested genes with special reference to cuticle proteins in *Drosophila melanogaster*. J. Mol. Evol. 28: 391–397. **[4]**

Moriyama, E. N. and J. R. Powell. 1997. Codon usage bias and tRNA abundance in *Drosophila*. J. Mol. Evol. 45: 514–523. **[4]**

Moriyama, E. N. and J. R. Powell. 1998. Gene length and codon usage bias in *Drosophila melanogaster*, *Saccharomyces cerevisiae* and *Escherichia coli*. Nuc. Acids Res. 26: 3188–3193. **[4]**

Moriyama, E. N., D. A. Petrov, and D. L. Hartl. 1998. Genome size and intron size in *Drosophila*. Mol. Biol. Evol. 15: 770–773. **[8]**

Mornet, E., P. Crete, F. Kuttenn, M. C. Raux-Demay, J. Boue, P. C. White, and A. Boue. 1991. Distribution of deletions and seven point mutation on *CYP21B* genes in three clinical forms of steroid 21-hydroxylase deficiency. Am. J. Hum. Genet. 48: 79–88. **[6]**

Mornon, J. P., D. Halaby, M. Malfois, P. Durand, I. Callebaut, and A. Tardieu. 1998. α-Crystallin C-terminal domain: On the track of an Ig fold. Int. J. Biol. Macromol. 22: 219–227. **[4]**

Morowitz, H. J. 1984. The completeness of molecular biology. Isr. J. Med. Sci. 20: 750–753. **[8]**

Mouchès, C. and 7 others. 1986. Amplification of an esterase gene is responsible for insecticide resistance in a California *Culex* mosquito. Science 233: 778–780. **[6]**

Mouchiroud, D. 1995. *Structure en Isochores et Évolution des Génomes de Vertébrés.* Mémoire d'Habilitation à Diriger des Recherches. Université Claude Bernard Lyon 1, Lyon. **[8]**

Mouchiroud, D. and C. Gautier. 1990. Codon usage changes and sequence dissimilarity between human and rat. J. Mol. Evol. 31: 81–91. **[8]**

Mouchiroud, D., G. Fichant, and G. Bernardi. 1987. Compositional compartmentalization and gene composition in the genome of vertebrates. J. Mol. Evol. 26: 198–204. **[8]**

Mouncey, N. J., M. Choudhary, and S. Kaplan. 1997. Characterization of genes encoding dimethyl sulfoxide reductase of *Rhodobacter sphaeroides* 2.4.1T: An essential metabolic gene function encoded on chromosome II. J. Bacteriol. 179: 7617–7624. **[8]**

Mourant, A. E., A. C. Kopec, and K. Domanievska-Sobczak. 1976. *The Distribution of the Human Blood Groups and Other Polymorphisms*. Oxford University Press, Oxford. **[6]**

Mullany, P., M. Pallen, M. Wilks, J. R. Stephen, and S. Tabaqchali. 1996. A group II intron in a conjugative transposon from the Gram-positive bacterium *Clostridium difficile*. Gene 174: 145–150. **[7]**

Muller, H. J. 1935. The origination of chromatin deficiencies as minute deletions subject to insertion elsewhere. Genetics 17: 237–252. **[6]**

Mullis, K. B. 1990. The unusual origin of the polymerase chain reaction. Sci. Am. 262(4): 56–65. **[5]**

Munn, D. H. and 7 others. 1998. Prevention of allogeneic fetal rejection by tryptophan catabolism. Science 281: 1191–1193. **[6]**

Murnane, J. P. and J. F. Morales. 1995. Use of a mammalian interspersed repetitive (*MIR*) element in the coding and processing sequences of mammalian genes. Nuc. Acids Res. 23: 2837–2839. **[7]**

Muse S. V. and Weir B. S. 1992. Testing for equality of evolutionary rates. Genetics 132: 269–276. **[4]**

Mushegian, A. R. and E. V. Koonin. 1996a. A minimal gene set for cellular life derived by comparison of complete bacterial genomes. Proc. Natl. Acad. Sci. USA 93: 10268–10273. **[8]**

Mushegian, A. R. and E. V. Koonin. 1996b. Gene order is not conserved in bacterial evolution. Trends Genet. 12: 289–290. **[8]**

Muto, A. and S. Osawa. 1987. The guanine and cytosine content of genomic DNA and bacterial evolution. Proc. Natl. Acad. Sci. USA 84: 166–169. **[8]**

Muto, A., F. Yamao, H. Hori, and S. Osawa. 1986. Gene organization of *Mycoplasma capricolum*. Adv. Biophys. 21: 49–56. **[8]**

Muto, A., F. Yamao, and S. Osawa. 1987. The genome of *Mycoplasma capricolum*. Prog. Nuc. Acids Res. Mol. Biol. 34: 29–58. **[8]**

Nachman, M. W. 1998. Deleterious mutations in animal mitochondrial DNA. Genetica 102/103: 61–69. **[2, 8]**

Nadeau, J. H. and D. Sankoff. 1997. Comparable rates of gene loss and functional divergence after genome duplications early in vertebrate evolution. Genetics 147: 1259–1266. **[6]**

Nagel, G. M. and R. F. Doolittle. 1995. Phylogenetic analysis of the aminoacyl-tRNA synthetases. J. Mol. Evol. 40: 487–498. **[6]**

Nagl, W. 1990. Polyploidy in differentiation and evolution. Int. J. Cell Cloning 8: 216–223. **[8]**

Nagylaki, T. 1984a. Evolution of multigene families under interchromosomal gene conversion. Proc. Natl. Acad. Sci. USA 81: 3796–3800. **[6]**

Nagylaki, T. 1984b. The evolution of multigene families under intrachromosomal gene conversion. Genetics 106: 529–548. **[6]**

Nagylaki, T. and T. D. Petes. 1982. Intrachromosomal gene conversion and the maintenance of sequence homogeneity among repeated genes. Genetics 100: 315–337. **[6]**

Nakamura, Y. and 10 others. 1987. Variable number of tandem repeat (VNTR) markers for human genetic mapping. Science 235: 1616–1622. **[6]**

Naor, D., D. Fischer, R. L. Jernigan, H. J. Wolfson, and R. Nussinov. 1996. Amino acid pair interchanges at spatially conserved locations. J. Mol. Biol. 256: 924–938. **[4]**

Narayan, R. K. J. and H. Rees. 1976. Nuclear DNA variation in *Lathyrus*. Chromosoma 54: 141–154. **[8]**

Nathans, J., D. Thomas, and D. S. Hogness. 1986. Molecular genetics of human color vision: The genes encoding blue, green, and red pigments. Science 232: 193–202. **[6]**

Needleman, S. B. and C. D. Wunsch. 1970. A general method applicable to the search of similarities in the amino acid sequence of two proteins. J. Mol. Biol. 48: 443–453. **[3]**

Nei, M. 1969. Gene duplication and nucleotide substitution in evolution. Nature 221: 40–42. **[8]**

Nei, M. 1975. *Molecular Population Genetics and Evolution*. North-Holland, Amsterdam. **[4,5]**

Nei, M. 1987. *Molecular Evolutionary Genetics*. Columbia University Press, New York. **[2,5]**

Nei, M. 1996. Phylogenetic analysis in molecular evolutionary genetics. Annu. Rev. Genet. 30: 371–403. **[5]**

Nei, M. and T. Gojobori. 1986. Simple methods for estimating the number of synonymous and nonsynonymous nucleotide substitutions. Mol. Biol. Evol. 3: 418–426. **[3]**

Nei, M. and D. Graur. 1984. Extent of protein polymorphism and the neutral mutation theory. Evol. Biol. 17: 73–118. **[2]**

Nei, M. and Y. Imaizumi. 1966. Genetic structure of human populations. II. Differentiation of blood group gene frequencies among isolated populations. Heredity 21: 183–190. **[2]**

Nei, M. and R. K. Koehn (eds.). 1983. *Evolution of Genes and Proteins*. Sinauer Associates, Sunderland, MA. **[6]**

Nei, M. and W.-H. Li. 1979. Mathematical model for studying genetic variation in terms of restriction endonucleases. Proc. Natl. Acad. Sci. USA 76: 5269–5273. **[2, 3]**

Nei, M. and F. Tajima. 1983. Maximum likelihood estimation of the number of nucleotide substitutions from restriction data. Genetics 105: 207–217. **[3]**

Never, P. and H. Saedler. 1977. Transposable genetic elements as agents of gene instability and chromosome rearrangements. Nature 268: 109–115. **[7]**

Newman, M. E. J. 1996. Self-organized criticality, evolution and the fossil extinction record. Proc. R. Soc. London 263B: 1605–1610. **[App. I]**

Nguyen, N. Y., A. Suzuki, R. A. Boykins, and T. Y. Liu. 1986. The amino acid sequence of *Limulus* C-reactive protein: Evidence of polymorphism. J. Biol. Chem. 261: 10456–10465. **[4]**

Nishikimi, M., T. Kawai, and K. Yagi. 1992. Guinea pigs possess a highly mutated gene for L-gulono-γ-lactone oxidase, the key enzyme for L-ascorbic acid biosynthesis missing in this species. J. Biol. Chem. 267: 21967–21972. **[6]**

Nishikimi, M., R. Fukuyama, S. Minoshima, N. Shimizu, and K. Yagi. 1994. Cloning and chromosomal mapping of the human nonfunctional gene for L-gulono-γ-lactone oxidase, the enzyme for L-ascorbic acid biosynthesis missing in man. J. Biol. Chem. 269: 13685–13688. **[6]**

Novacek, M. J. 1994. Whales leave the beach. Nature 368: 807. **[5]**

Nuttall, G. H. F. 1902. Progress report upon the biological test for blood as applied to over 500 bloods from various sources, together with a preliminary note upon a method for measuring the degree of reaction. Brit. Med. J. 1: 825–827. **[5]**

Nuttall, G. H. F. 1904. *Blood Immunity and Blood Relationship*. Cambridge University Press, Cambridge. **[5]**

Ny, T., F. Eligh, and B. Lund. 1984. The structure of the human tissue-type plasminogen activator gene: Correlation of intron and exon structures to functional and structural domains. Proc. Natl. Acad. Sci. USA 81: 5355–5359. **[6]**

Oba, T., Y. Andachi, A. Muto, and S. Osawa. 1991. CGG: An unassigned codon in *Mycoplasma capricolum*. Proc. Natl. Acad. Sci. USA 88: 921–925. **[1]**

Ochman, H. and A. C. Wilson. 1987. Evolution in bacteria: Evidence for a universal substitution rate in cellular genomes. J. Mol. Evol. 26: 74–86. **[8]**

Oda, K. and 10 others. 1992. Gene organization deduced from the complete sequence of liverwort *Marchantia polymorpha* mitochondrial DNA. A primitive form of plant mitochondrial genome. J. Mol. Biol. 223: 1–7. **[4]**

O'Hare, K., A. Driver, S. McGrath S, and D. M. Johnson-Schiltz. 1992. Distribution and structure of cloned *P* elements from the *Drosophila melanogaster P* strain π2. Genet. Res. 1992 60: 33–41. **[7]**

Ohno, S. 1970. *Evolution by Gene Duplication*. Springer, Berlin. **[6]**

Ohno, S. 1972. So much "junk" DNA in our genome. Brookhaven Symp. Biol. 23: 366–370. **[6, 8]**

Ohshima, K., R. Koishi, M. Matsuo, and N. Okada. 1993. Several short interspersed repetitive elements (SINEs) in distant species may have originated from a common ancestral retrovirus: Characterization of a squid SINE and a possible mechanism for generation of tRNA-derived retroposons. Proc. Natl. Acad. Sci. USA 90: 6260–6264. **[7]**

Ohshima, K., M. Hamada, Y. Teral, and N. Okada. 1996. The 3′ ends of tRNA-derived short interspersed repetitive elements are derived from the 3′ ends of long interspersed repetitive elements. Mol. Cell. Biol. 16: 3764. **[7]**

980. *Evolution and Variation of Multigene Families*. ger, Berlin. **[6]**

Ohta, T. 1983. On the evolution of multigene families. Theor. Pop. Biol. 23: 216–240. **[6]**

Ohta, T. 1984. Some models of gene conversion for treating the evolution of multigene families. Genetics 106: 517–528. **[6]**

Ohta, T. 1990. How gene families evolve. Theor. Pop. Biol. 37: 213–219. **[6]**

Ohta, T. 1995. Synonymous and nonsynonymous substitutions in mammalian genes and the nearly neutral theory. J. Mol. Evol. 40: 56–63. **[4]**

Ohta, T. 1998. On the pattern of polymorphisms at major histocompatibility complex loci. J. Mol. Evol. 46: 633–638. **[6]**

Ohta, T. and G. A. Dover. 1983. Population genetics of multigene families that are dispersed into two or more chromosomes. Proc. Natl. Acad. Sci. USA 80: 4079–4083. **[6]**

Ohta, T. and J. H. Gillespie. 1996. Development of neutral and nearly neutral theories. Theor. Pop. Biol. 49: 128–142. **[4]**

Ohta, T. and M. Kimura. 1971. On the constancy of the evolutionary rate in cistrons. J. Mol. Evol. 1: 18–25. **[4]**

O'hUigin, C. and W.-H. Li. 1992. The molecular clock ticks regularly in muroid rodents and hamsters. J. Mol. Evol. 35: 377–384. **[4]**

Okada, N., M. Hamada, I. Ogiwara, and K. Ohshima. 1997. SINEs and LINEs share common 3′ sequences: A review. Gene 205: 229–243. **[7]**

Olendzenski, L., E. Hilario, and J. P. Gogarten. 1998. Horizontal gene transfer and fusing lines of descent: The archaebacteria—a chimera? pp. 349–362. In: M. Syvanen and C. I. Kado (eds.), *Horizontal Gene Transfer*. Chapman & Hall, London. **[5]**

Oliviero, S., M. DeMarchi, A. O. Carbonara, L. F. Bernini, G. Bensi, and G. Raugei. 1985. Molecular evidence of triplication in the haptoglobin Johnson variant gene. Hum. Genet. 71: 49–52. **[6]**

Olsen, G. L. and C. R. Woese. 1996. Lessons from an Archaeal genome: What are we learning from *Methanococcus jannaschii*? Trends Genet. 12: 377–379. **[5]**

Oparin, A. I. 1957. *The Origin of Life on Earth*. Academic Press, New York. **[Introduction]**

Ophir, R. and D. Graur. 1997. Patterns and rates of indel evolution in processed pseudogenes from humans and murids. Gene 205: 191–202. **[7]**

Ophir, R., T. Itoh, D. Graur, and T. Gojobori. 1999. A simple method for estimating the intensity of purifying selection in protein-coding genes. Mol. Biol. Evol. 16: 49–53. **[4]**

Orgel, L. E. and F. H. C. Crick. 1980. Selfish DNA: The ultimate parasite. Nature 284: 604–607. **[7, 8]**

Osawa, S. 1995. *Evolution of the Genetic Code*. Oxford University Press, Oxford. **[8]**

Östergren, G. 1945. Parasitic nature of extra fragment chromosomes. Bot. Notiser 2: 157–163. **[8]**

Ouenzar, B., B. Agoutin, F. Reinisch, D. Weill, F. Perin, G. Keith, and T. Heyman. 1988. Distribution of isoaccepting tRNAs and codons for proline and glycine in collageneous and noncollageneous chicken tissues. Biochem. Biophys. Res. Commun. 150: 148–155. **[4]**

Pääbo, S. 1989. Ancient DNA: Extraction, characterization, molecular cloning, and enzymatic amplification. Proc. Natl. Acad. Sci. USA 86: 1939–1943. **[5]**

Pagel, M. 1997. Inferring evolutionary processes from phylogenies. Zool. Scripta 26: 331–348. **[5]**

Pagel, M. and R. A. Johnstone. 1992. Variation across species in the size of the nuclear genome supports the junk-DNA explanation for the C-value paradox. Proc. Roy Soc. 249B: 119–124. **[8]**

Palmer, J. D. 1985 Evolution of chloroplast and mitochondrial DNA in plants and algae. pp. 131–240. In: R. J. MacIntyre (ed.), *Molecular Evolutionary Genetics*. Plenum, New York. **[4]**

Palmer, J. D. 1997. The mitochondrion that time forgot. Nature 387: 454–455. **[8]**

Palmer, J. D. and L. A. Hebron. 1987. Unicircular structure of the *Brassica hirta* mitochondrial genome. Curr. Genet. 11: 565–570. **[4]**

Pamilo, P. and N. O. Bianchi. 1993. Evolution of the *Zfx* and *Zfy* genes: Rates and interdependence between the genes. Mol. Biol. Evol. 10: 271–281. **[3, 4]**

Pamilo, P. and M. Nei. 1988. Relationship between gene trees and species trees. Mol. Biol. Evol. 5: 568–583. **[5]**

Pardue, M. L. 1974. Localization of repeated DNA sequences in *Xenopus* chromosomes. Cold Spring Harbor Symp. Quant. Biol. 38: 475–482. **[6]**

Patthy, L. 1985. Evolution of the proteases of blood coagulation and fibrinolysis by assembly from modules. Cell 41: 657–663. **[6]**

Pearson, W. R. and W. Miller. 1992. Dynamic programming algorithms for biological sequence comparison. Methods Enzymol. 210: 575–601. **[3]**

Penny, D. and M. D. Hendy. 1985. Testing methods of evolutionary tree construction. Cladistics 1: 266–272. **[5]**

Periquet, G., S. Ronsseray, and M. H. Hamelin. 1989. Are *Drosophila melanogaster* populations under a stable geographical differentiation due to the presence of *P* elements? Heredity 63: 47–58. **[7]**

Perler, F., A. Efstratiadis, P. Lomedico, W. Gilbert, R. Kolodner, and J. Dodgeson. 1980. The evolution of genes: The chicken preproinsulin gene. Cell 20: 555–566. **[3]**

Perlman, P. S. and R. A. Butow. 1989. Mobile introns and intron-encoded proteins. Science 246: 1106–1109. **[4, 6]**

Perutz, M. F. 1983. Species adaptation in a protein molecule. Mol. Biol. Evol. 1: 1–28. **[4]**

Petes, T. D. and C. W. Hill. 1988. Recombination between repeated genes in microorganisms. Annu. Rev. Genet. 22: 147–168. **[6]**

Petrov, D. A. and D. L. Hartl. 1997. Trash DNA is what gets thrown away: High rate of DNA loss in *Drosophila*. Gene 205: 279–289. **[7, 8]**

Petrov, D. A. and D. L. Hartl. 1998. High rate of DNA loss in the *Drosophila melanogaster* and *Drosophila virilis* species group. Mol. Biol. Evol. 15: 293–302. **[8]**

Petrov, D. A., E. R. Lozovskaya, and D. L. Hartl. 1996. High intrinsic rate of DNA loss in *Drosophila*. Nature 384: 346–349. **[8]**

Piatigorsky, J. 1998a. Gene sharing in lens and cornea: Facts and implications. Prog. Retin. Eye Res. 17: 145–174. **[6]**

Piatigorsky, J. 1998b. Multifunctional lens crystallins and cornea enzymes: More than meets the eye. Ann. N. Y. Acad. Sci. 842: 7–15. **[6]**

Piatigorsky, J. and G. J. Wistow. 1989. Enzyme/crystallins: Gene sharing as an evolutionary strategy. Cell 57: 197–199. **[6]**

Piatigorsky, J. and 7 others. 1988. Gene sharing by δ-crystallin and argininosuccinate lyase. Proc. Natl. Acad. Sci. USA 85: 3479–3483. **[6]**

Pieber, M. and J. Tohá. 1983. Code dependent conservation of the physico-chemical properties in amino acid substitutions. Orig. Life 13: 139–146. **[4]**

Pilbeam, D. 1984. The descent of hominoids and hominids. Sci. Am. 250(3): 60–69. **[4]**

Portin, P. 1993. The concept of the gene: Short history and present status. Q. Rev. Biol. 68: 173–223. **[1]**

Post, L. E., G. D. Strycharz, M. Nomura, H. Lewis, and P. P. Dennis. 1979. Nucleotide sequence of the ribosomal protein gene cluster adjacent to the gene for RNA polymerase subunit β in *Escherichia coli*. Proc. Natl. Acad. Sci. USA 76: 1697–1701. **[4]**

Powell, J. R. and A. Caccone. 1990. The TEACL method of DNA-DNA hybridization: Technical considerations. J. Mol. Evol. 30: 267–272. **[3]**

Powell, L. M., S. C. Wallis, R. J. Pease, Y. H. Edwards, T. J. Knott, and J. Scott. 1987. A novel form of tissue-specific RNA processing produces apolipoprotein B-48 in intestine. Cell 50: 831–840. **[6]**

Primmer, C. R., T. Raudsepp, B. P. Chowdhary, A. P. Møller, and H. Ellegren. 1997. Low frequency of microsatellites in the avian genome. Genome Res. 7: 471–482. **[8]**

Primrose, S. B. 1998. *Principles of Genome Analysis*. Blackwell, Oxford. **[8]**

Quentin, Y. 1988. The *Alu* family developed through successive waves of fixation closely connected with primate lineage history. J. Mol. Evol. 27: 194–202. **[7]**

Rabson, A. B. and M. A. Martin. 1985. Molecular organization of the AIDS retrovirus. Cell 40: 477–480. **[4]**

Ratner, V. A. and L. A. Vasil'eva. 1992. The role of mobile genetic elements (MGE) in microevolution. Genetika 28: 5–17. **[7]**

Rau, R. E. 1974. Revised list of the preserved material of the extinct Cape colony quagga, *Equus quagga quagga* (Gmelin). Ann. S. Afr. Mus. 65: 41–87. **[5]**

Raup, D. M. 1986. Biological extinction in Earth history. Science 231: 1528–1533. **[App. I]**

Raup, D. M. and J. J. Sepkoski. 1984. Periodicity of extinctions in the geologic past. Proc. Natl. Acad. Sci. USA 81: 801–805. **[App. I]**

Razin, A. and A. D. Riggs. 1980. DNA methylation and gene function. Science 210: 604–610. **[4]**

Razin, S. 1997. The minimal cellular genome of *Mycoplasma*. Indian J. Biochem. Biophys. 34: 124–130. **[8]**

Rees, H. and R. N. Jones. 1972. The origin of the wide species variation in nuclear DNA content. Int. Rev. Cytol. 32: 53–92. **[8]**

Reilly, J. G., R. Ogden, and J. J. Rossi. 1982. Isolation of a mouse pseudo tRNA gene encoding CCA—a possible example of reverse flow of genetic information. Nature 300: 287–289. **[7]**

Ren, D. 1998. Flower-associated Brachycera flies as fossil evidence for Jurassic angiosperm origins. Science 280: 85–88. **[5]**

Retana Salazar, A. P. 1996. Parasitological evidence on the phylogeny of hominids and cebids. Rev. Biol. Trop. 44: 391–394. **[5]**

Rhinesmith, H. S., W. A. Schroeder, and N. Martin. 1958. The N-terminal sequence of the β chains of normal adult hemoglobin. J. Am. Chem. Soc. 80: 3358–3361. **[6]**

Ribeiro, S. and G. B. Golding. 1998. The mosaic nature of the eukaryotic nucleus. Mol. Biol. Evol. 15: 779–788. **[6]**

Rice, N. R. 1972. Change in repeated DNA in evolution. pp. 44–79. In: H. H. Smith (ed.), *Evolution of Genetic Systems*. Gordon & Breach, New York. [4]

Rice, S. A. and B. C. Lampson. 1995. Bacterial reverse transcriptase and msDNA. Virus Genes 11: 95–104. [7]

Rice, S. A., J. Bieber, J. Y. Chun, G. Stacey, and B. C. Lampson. 1993. Diversity of retron elements in a population of rhizobia and other Gram-negative bacteria. J. Bacteriol. 175: 4250–4254. [7]

Riley, M. and B. Labedan. 1997. Protein evolution viewed through *Escherichia coli* protein sequences: Introducing the notion of a structural segment of homology, the module. J. Mol. Biol. 268: 857–868. [8]

Rio, D. C., F. A. Laski, and G. M. Rubin. 1986. Identification and immunochemical analysis of biologically active *Drosophila P* element transposase. Cell 44: 21–32. [7]

Ritossa, F. M., K. C. Atwood, D. L. Lindsley, and S. Spiegelman. 1966. On the chromosomal distribution of DNA complementary to ribosomal and soluble RNA. Natl. Cancer Inst. Monogr. 23: 449–472. [6]

Robinson, M., C. Gautier, and D. Mouchiroud. 1997. Evolution of isochores in rodents. Mol. Biol. Evol. 14: 823–828. [8]

Robinson, M., M. Gouy, C. Gautier, and D. Mouchiroud. 1998. Sensitivity of the relative-rate test to taxonomic sampling. Mol. Biol. Evol. 15: 1091–1098. [4]

Roche, S. E., M. Schiff, and D. C. Rio. 1995. *P*-element repressor autoregulation involves germline transcriptional repression and reduction of third intron splicing. Genes Develop. 9: 1278–1288. [7]

Rodin, S. N. and S. Ohno. 1995. Two types of aminoacyl-tRNA synthetases could be originally encoded by complementary strands of the same nucleic acid. Orig. Life Evol. Biosphere 25: 565–589. [6]

Rogers, S., R. Wells, and M. Rechsteiner. 1986. Amino acid sequences common to rapidly degraded proteins: The PEST hypothesis. Science 234: 364–368. [6]

Romans, P. and R. A. Firtel. 1985. Organization of the actin multigene family of *Dictyostelium discoideum* and analysis of variability in the protein coding regions. J. Mol. Biol. 186: 321–335. [7]

Romer, A. S. 1966. *Vertebrate Paleontology*. Chicago University Press, Chicago. [5]

Römling, U., D. Grothues, T. Heuer, and B. Tummler. 1992. Physical genome analysis of bacteria. Electrophoresis 13: 626–631. [8]

Ronsseray, S., M. Lehmann, and D. Anxolabéhère. 1989. Copy number and distribution of *P* and *I* mobile elements in *Drosophila melanogaster* populations. Chromosoma 98: 207–214. [7]

Rosenberg, S. M., S. Longerich, P. Gee, and R. S. Harris. 1994. Adaptive mutation by deletions in small mononucleotide repeats. Science 265: 405–409. [1]

Roughgarden, J. 1996. *Theory of Population Genetics and Evolutionary Ecology: An Introduction*. Prentice-Hall, New York. [2]

Rubin, G. 1983. Dispersed repetitive DNAs in *Drosophila*. pp. 329–361. In: J. A. Shapiro (ed.), *Mobile Genetic Elements*. Academic Press, New York. [7]

Ruvolo, M. 1997. Molecular phylogeny of the hominoids: Inferences from multiple independent DNA sequence data sets. Mol. Biol. Evol. 14: 248–265. [5]

Ruvolo, M., T. R. Disotell, M. W. Allard, W. M. Brown, and R. Honeycutt. 1991. Resolution of the African hominoid trichotomy by use of a mitochondrial gene sequence. Proc. Natl. Acad. Sci. USA 88: 1570–1574. [5]

Rzhetsky, A. and M. Nei. 1992. A simple method for estimating and testing minimum-evolution trees. Mol. Biol. Evol. 9: 945–967. [5]

Rzhetsky, A., F. J. Ayala, L. C. Hsu, C. Chang, and A. Yoshida. 1997. Exon/intron structure of aldehyde dehydrogenase genes supports the "introns-late" theory. Proc. Natl. Acad. Sci. USA 94: 6820–6825. [6]

Saccone, S., S. Caccio, P. Perani, L. Andreozzi, A. Rapisarda, S. Motta, and G. Bernardi. 1997. Compositional mapping of mouse chromosomes and identification of the gene-rich regions. Chromosome Res. 5: 293–300. [8]

Saitou, N. 1988. Property and efficiency of the maximum likelihood method for molecular phylogeny. J. Mol. Evol. 27: 261–273. [5]

Saitou, N. 1989. A theoretical study of the underestimation of branch lengths by the maximum parsimony principle. Syst. Zool. 38: 1–6. [5]

Saitou, N. and M. Nei. 1986. Polymorphism and evolution of influenza A virus genes. Mol. Biol. Evol. 3: 57–74. [4]

Saitou, N. and M. Nei. 1987. The neighbor-joining method: A new method for reconstructing phylogenetic trees. Mol. Biol. Evol. 4: 406–425. [5]

Salinas, J., M. Zerial, J. Filipski, M. Crepin, and G. Bernardi. 1987. Nonrandom distribution of MMTV proviral sequences in the mouse genome. Nuc. Acids Res. 15: 3009–3022. [8]

Salo, W. L., A. C. Aufderheide, J. Buikstra, and T. A. Holcomb. 1994. Identification of *Mycobacterium tuberculosis* DNA in a pre-Columbian Peruvian mummy. Proc. Natl. Acad. Sci. USA 91: 2091–2094. [5]

Salser, W. and 10 others. 1976. Investigation of the organization of mammalian chromosomes at the DNA sequence level. Fed. Proc. 35: 23–35. [8]

Samollow, P. B., L. M. Cherry, S. M. Witte, and J. Rogers. 1996. Interspecific variation at the Y-linked *RPS4Y* locus in hominoids: Implications for phylogeny. Am. J. Phys. Anthropol. 101: 333–343. [5]

Sanderson, M. J. 1997. A nonparametric approach to estimating divergence times in the absence of rate constancy. Mol. Biol. Evol. 14: 1218–1231. [5]

Sankoff, D. and J. B. Kruskal (eds.). 1983. *Time Warps, String Edits, and Macromolecules: The Theory and Practice of Sequence Comparison*. Addison-Wesley, Reading, MA. [3]

Sankoff, D., G. Leduc, N. Antoine, B. Paquin, B. F. Lang, and R. Cedergren. 1992. Gene order comparisons for phylogenetic inference: Evolution of the mitochondrial genome. Proc. Natl. Acad. Sci. USA 89: 6575–6579. [8]

Saparbaev M. and J. Laval. 1994. Excision of hypoxanthine from DNA containing dIMP residues by the *Escherichia coli*, yeast, rat, and human alkylpurine DNA glycosylases. Proc. Natl. Acad. Sci. USA 91: 5873–5877. [4]

Sarich, V. M. 1972. Generation time and albumin evolution. Biochem. Genet. 7: 205–212. [4]

Sarich, V. M. and A. C. Wilson. 1967. Immunological time scale for hominid evolution. Science 158: 1200–1203. [4, 5]

Sarich, V. M. and A. C. Wilson. 1973. Generation time and genomic evolution in primates. Science 179: 1144–1147. [4]

Sattath, S. and A. Tversky. 1977. Additive similarity trees. Psychometrika 42: 319–345. [5]

Sawyer, S. A. and D. L. Hartl. 1992. Population genetics of polymorphism and divergence. Genetics 132: 1161–1176. **[2]**

Sawyer, S. A., D. E. Dykhuizen, R. F. DuBose, L. Green, T. Mutangadura-Mhlanga, D. F. Wolczyk, and D. L. Hartl. 1987. Distribution and abundance of insertion sequences among natural isolates of *Escherichia coli*. Genetics 115: 51–63. **[7]**

Scherer, S. and R. W. Davis. 1980. Recombination of dispersed repeated DNA sequences in yeast. Science 209: 1380–1384. **[6]**

Schimenti, J. C. 1994. Gene conversion and the evolution of gene families in mammals. Soc. Gen. Physiol. Ser. 49: 85–91. **[6]**

Schimenti, J. C. and C. H. Duncan. 1984. Ruminant globin gene structures suggest an evolutionary role for *Alu*-type repeats. Nuc. Acids Res. 12: 1641–1655. **[7]**

Schmid, C. W. and P. L. Deininger. 1975. Sequence organization of the human genome. Cell 6: 345–358. **[8]**

Schuler, G. D., S. F. Altshul, and D. J. Lipman. 1991. A workbench for multiple alignment construction and analysis. Proteins 9: 180–190. **[3]**

Schultz, A. H. 1963. *Classification and Human Evolution*. Aldine, Chicago. **[5]**

Schultz, J., F. Milpetz, P. Bork, and C. P. Ponting. 1998. SMART, a simple modular architecture research tool: Identification of signaling domains. Proc. Natl. Acad. Sci. USA 95: 5657–5864. **[6]**

Schuster, W. and A. Brennicke. 1987. Plastid, nuclear and reverse transcriptase sequences in the mitochondrial genome of *Oenothera*: Is genetic information transferred between organelles via RNA? EMBO J. 6: 2857–2863. **[7]**

Schwartz, J. H. 1984. The evolutionary relationships of man and orang-utans. Nature 308: 501–505. **[5]**

Schwartz, R. M. and M. O. Dayhoff. 1978. Origins of prokaryotes, eukaryotes, mitochondria, and chloroplasts. Science 199: 395–403. **[5]**

Schwartzbach, S. D., T. Osafune, and W. Loffelhardt. 1998. Protein import into cyanelles and complex chloroplasts. Plant Mol. Biol. 38: 247–263. **[8]**

Schwarz, Z. and H. Kössel. 1980. The primary structure of 16S rDNA from *Zea mays* chloroplast is homologous to *E. coli* 16S rRNA. Nature 283: 739–742. **[5]**

Scott, A. F. and 8 others. 1984. The sequence of the gorilla fetal globin genes: Evidence for multiple gene conversions in human evolution. Mol. Biol. Evol. 1: 371–389. **[6]**

Seddon, J. M., P. R. Baverstock, and A. Georges. 1998. The rate of mitochondrial 12S rRNA gene evolution is similar in freshwater turtles and marsupials. J. Mol. Evol. 46: 460–464. **[4]**

Segovia, L., J. Horwitz, R. Gasser, and G. Wistow. 1997. Two roles for μ-crystallin: A lens structural protein in diurnal marsupials and a possible enzyme in mammalian retinas. Mol. Vision 3: 9. **[3]**

Seino, S., G. I. Bell, and W.-H. Li. 1992. Sequences of primate insulin genes support the hypothesis of a slower rate of molecular evolution in humans and apes than in monkeys. Mol. Biol. Evol. 9: 193–203. **[4]**

Sekino, N., Y. Sekine, and E. Ohtsubo. 1995. *IS1*-encoded proteins, *InsA* and the *InsA-B'-InsB* transframe protein (transposase): Function deduced from their DNA-binding ability. Adv. Biophys. 31: 209–222. **[7]**

Sellers, P. H. 1974. On the theory and computation of evolutionary distances. SIAM J. Appl. Math. 26: 787–793. **[3]**

Seoighe, C. and K. H. Wolfe. 1998. Extent of genomic rearrangement after genome duplication in yeast. Proc. Natl. Acad. Sci. USA 95: 4447–4452. **[8]**

Shapiro, J. A. 1983. Variation as a genetic engineering process. pp. 253–270. In: D. S. Bendall (ed.), *Evolution from Molecules to Man*. Cambridge University Press, Cambridge. **[7]**

Sharp, P. M. and W.-H. Li. 1986. An evolutionary perspective on synonymous codon usage in unicellular organisms. J. Mol. Evol. 24: 28–38. **[4]**

Sharp, P. M. and W.-H. Li. 1987a. The rate of synonymous substitution in enterobacterial genes is inversely related to codon usage bias. Mol. Biol. Evol. 4: 222–230. **[4]**

Sharp, P. M. and W.-H. Li. 1987b. The codon adaptation index—a measure of directional synonymous codon usage bias, and its potential applications. Nuc. Acids Res. 15: 1281–1295. **[4]**

Sharp, P. M. and G. Matassi. 1994. Codon usage and genome evolution. Curr. Opinion Genet. Develop. 4: 851–860. **[4]**

Sharp, P. M., T. M. F. Tuohy, and K. R. Mosurski. 1986. Codon usage in yeast: Cluster analysis clearly differentiates highly and lowly expressed genes. Nuc. Acids Res. 14: 5125–5143. **[4]**

Sharp, P. M., E. Cowe, D. G. Higgins, D. Shields, K. H. Wolfe, and F. Wright. 1988. Codon usage patterns in *Escherichia coli*, *Bacillus subtilis*, *Saccharomyces cerevisiae*, *Schizosaccharomyces pombe*, *Drosophila melanogaster*, and *Homo sapiens*: A review of the considerable within-species diversity. Nuc. Acids Res. 16: 8207–8211. **[4]**

Sharp, P. M., M. Averof, A. T. Lloyd, G. Matassi, and J. F. Peden. 1995. DNA sequence evolution: The sounds of silence. Philos. Trans. Roy. Soc. 349B: 241–247. **[4]**

Sherry, S. T., H. C. Harpending, M. A. Batzer, and M. Stoneking. 1997. *Alu* evolution in human populations: Using the coalescent to estimate effective population size. Genetics 147: 1977–1982. **[2]**

Shimamura, M. and 8 others. 1997. Molecular evidence from retroposons that whales form a clade within even-toed ungulates. Nature 388: 666–670. **[5]**

Shimmin, L. C., B. H.-J. Chang, and W.-H. Li. 1993. Male-driven evolution of DNA sequences. Nature 362: 745–747. **[4]**

Shoshani, J., C. P. Groves, E. L. Simons, and G. F. Gunnell. 1996. Primate phylogeny: Morphological vs. molecular results. Mol. Phylogenet. Evol. 5: 102–154. **[5]**

Sibley, C. G. and J. E. Ahlquist. 1984. The phylogeny of the hominoid primates, as indicated by DNA-DNA hybridization. J. Mol. Evol. 20: 2–15. **[4,5]**

Sibley, C. G. and J. E. Ahlquist. 1987. DNA hybridization evidence of hominoid phylogeny: Results from an expanded data set. J. Mol. Evol. 26: 99–121. **[4,5]**

Sibley, C. G., J. A. Comstock, and J. E. Ahlquist. 1990. DNA hybridization evidence of hominoid phylogeny: A reanalysis of the data. J. Mol. Evol. 30: 202–236. **[5]**

Silke, J. 1997. The majority of long non-stop reading frames on the antisense strand can be explained by biased codon usage. Gene 194: 143–155. **[6]**

Simmen, M. W., S. Leitgeb, V. H. Clark, S. J. M. Jones, and A. Bird. 1998. Gene number in an invertebrate chordate, *Ciona intestinalis*. Proc. Natl. Acad. Sci. USA 95: 4437–4440. **[8]**

Simmons, M. J. and J. F. Crow. 1977. Mutations affecting fitness in *Drosophila* populations. Annu. Rev. Genet. 11: 49–78. **[7]**

Simpson, G. G. 1945. The principles of classification and a classification of mammals. Bull. Am. Mus. Nat. Hist. 85: 1–350. **[5]**

Simpson, G. G. 1961. *Principles of Animal Taxonomy*. Columbia University Press, New York. **[5]**

Sinden, R. R., C. E. Pearson, V. N. Potaman, and D. W. Ussery. 1998. DNA: Structure and function. Adv. Genome Biol. 5A: 1–141. **[1]**

Singer, C. E. and B. N. Ames. 1970. Sunlight ultraviolet and bacterial DNA base ratios. Science 170: 822–826. **[8]**

Singer, M. F. 1982. SINEs and LINEs: Highly repeated short and long interspersed sequences in mammalian genomes. Cell 28: 433–434. **[7]**

Singer, M. F. and P. Berg. 1991. *Genes and Genomes: A Changing Perspective*. University Science Books, Mill Valley, CA. **[8]**

Sinha N. K. and M. D. Haimes. 1981. Molecular mechanisms of substitution mutagenesis. An experimental test of the Watson-Crick and Topal-Fresco models of base mispairing. J. Biol. Chem. 256: 10671–10683. **[4]**

Skiena, S. S. and G. Sundaram. 1994. A partial digest approach to restriction site mapping. Bull. Math. Biol. 56: 275–294. **[3]**

Slightom, J., A. E. Blechl, and O. Smithies. 1980. Human fetal $^{G}\gamma$- and $^{A}\gamma$-globin genes: Complete nucleotide sequences suggest that DNA can be exchanged between these duplicated genes. Cell 21: 627–638. **[6]**

Slightom, J. L, L.-Y. Chang, B. F. Koop, and M. Goodman. 1985. Chimpanzee fetal $^{G}\gamma$- and $^{A}\gamma$-globin gene nucleotide sequences provide further evidence of gene conversions in hominine evolution. Mol. Biol. Evol. 2: 370–389. **[6]**

Smit, A. F. A. 1996. The origin of interspersed repeats in the human genome. Curr. Opin. Genet. Develop. 6: 743–748. **[8]**

Smit, A. F. A. and A. D. Riggs. 1996. *Tiggers* and DNA transposon fossils in the human genome. Proc. Natl. Acad. Sci. USA 93: 1443–1448. **[7]**

Smith, C. W. J., J. G. Patton, and B. Nadal-Ginard. 1989. Alternative splicing in the control of gene expression. Annu. Rev. Genet. 23: 527–577. **[6]**

Smith, G. P. 1974. Unequal crossover and the evolution of multigene families. Cold Spring Harbor Symp. Quant. Biol. 38: 507–513. **[6]**

Smith, M. M. 1987. Molecular evolution of the *Saccharomyces cerevisiae* histone gene loci. J. Mol. Evol. 24: 252–259. **[8]**

Smith, T. F., M. S. Waterman, and W. M. Fitch. 1981. Comparative biosequence metrics. J. Mol. Evol. 18: 38–46. **[3]**

Smithies, O., G. E. Connell, and G. H. Dixon. 1962. Chromosomal rearrangements and the evolution of haptoglobin genes. Nature 196: 232–236. **[6]**

Sneath, P. H. A. 1966. Relations between chemical structure and biological activity in peptides. J. Theor. Biol. 12: 157–195. **[4]**

Sneath, P. H. A. and R. R. Sokal. 1973. *Numerical Taxonomy: The Principles and Practice of Numerical Classification*. W.H. Freeman, San Francisco. **[5]**

Sniegowski, P. D. 1995. The origin of adaptive mutants: Random or nonrandom? J. Mol. Evol. 40: 94–101. **[1]**

Soares, M. B. and 7 others. 1985. RNA-mediated gene duplication: The rat preproinsulin I gene is a functional retroposon. Mol. Cell. Biol. 5: 2090–2103. **[7]**

Sogin, M. L. 1997. History assignment: When was the mitochondrion founded? Curr. Opin. Genet. Develop. 7: 792–799. **[5]**

Sogin, M. L., H. J. Elwood, and J. H. Gunderson. 1986. Evolutionary diversity of eukaryotic small-subunit rRNA genes. Proc. Natl. Acad. Sci. USA 83: 1383–1387. **[5]**

Sogin, M. L., J. H. Gunderson, H. J. Elwood, R. A. Alonso, and D. A. Peattie. 1989. Phylogenetic meaning of the kingdom concept: An unusual ribosomal RNA from *Giardia lamblia*. Science 243: 75–77. **[5]**

Sokal, R. R. and C. D. Michener. 1958. A statistical method for evaluating systematic relationships. Univ. Kansas Sci. Bull. 28: 1409–1438. **[5]**

Sole, R. V., S. C. Manrubia, M. Benton, and P. Bak. 1997. Self-similarity of extinction statistics in the fossil record. Nature 388: 764–767. **[App. I]**

Sommer, S. S. 1995. Recent human germ-line mutation: Inferences from patients with hemophilia B. Trends Genet. 11: 141–147. **[1]**

Soriano, P., M. Meunier-Rotival, and G. Bernardi. 1983. The distribution of interspersed repeats is nonuniform and conserved in the mouse and human genomes. Proc. Natl. Acad. Sci. USA 80: 1816–1820. **[7]**

Soto, M. A. and J. Tohá. 1983. Conservation of physicochemical amino acid properties during the evolution of proteins. Orig. Life 13: 147–152. **[4]**

Sparrow, A. H. and A. F. Nauman. 1976. Evolution of genome size by DNA doublings. Science 192: 524–529. **[8]**

Sparrow, A. H., H. J. Price, and A. G. Underbrink. 1972. A survey of DNA content per cell and per chromosome of prokaryotic and eukaryotic organisms: Some evolutionary considerations. Brookhaven Symp. Biol. 23: 451–494. **[8]**

Spoerel, N. A., H. T. Nguyen, T. H. Eickbush, and F.C. Kafatos. 1989. Gene evolution and regulation in the chorion complex of *Bombyx mori*. Hybridization and sequence analysis of multiple developmentally middle A/B chorion gene pairs. J. Mol. Biol. 209: 1–19. **[6]**

Spring, J. 1997. Vertebrate evolution by interspecific hybridisation—are we polyploid? FEBS Lett. 400: 2–8. **[8]**

Stanhope, M. J. and 7 others. 1998. Molecular evidence for multiple origins of Insectivora and for a new order of endemic African insectivore mammals. Proc. Natl. Acad. Sci. USA 95: 9967–9972. **[4]**

Stanier, R. Y. 1970. Some aspects of the biology of cells and their possible evolutionary significance. Symp. Soc. Gen. Microbiol. 20: 1–38. **[Introduction]**

Stebbins, G. L. 1974. *Flowering Plants: Evolution Above the Species Level*. Harvard University Press, Cambridge, MA. **[8]**

Stebbins, G. L. 1981. Coevolution of grasses and herbivores. Ann. Missouri Bot. Garden 68: 75–86. **[4]**

Steel, M. 1994. Recovering a tree from the Markov leaf colouration it generates under a Markov model. Appl. Math. Lett. 7: 19–23. **[5]**

Stein, J. P., J. F. Catterall, P. Kristo, A. R. Means, and B. W. O'Malley. 1980. Ovomucoid intervening sequences specify functional domains and generate protein polymorphism. Cell 21: 681–687. **[6]**

Stern, D. B. and D. M. Lonsdale. 1982. Mitochondrial and chloroplast genomes of maize have a 12-kilobase DNA sequence in common. Nature 299: 698–702. **[7]**

Stewart, C.-B. and A. C. Wilson. 1987. Sequence convergence and functional adaptation of stomach lysozymes from foregut fermenters. Cold Spring Harbor Symp. Quant. Biol. 52: 891–899. **[4]**

Stewart, W. N. 1983. *Paleobotany and the Evolution of Plants*. Cambridge University Press, Cambridge. **[5]**

Stiller, J. W., E. C. Duffield, and B. D. Hall. 1998. Amitochondriate amoebae and the evolution of DNA-dependent RNA polymerase II. Proc. Natl. Acad. Sci. USA 95: 11769–11774. **[5]**

Stoltzfus, A., J. M. Logsdon, J. D. Palmer, and W. F. Doolittle. 1997. Intron "sliding" and the diversity of intron positions. Proc. Natl. Acad. Sci. USA 94: 10739–10744. **[6]**

Stryer, L. 1995. *Biochemistry*, 4th Ed. W.H. Freeman, New York. **[1]**

Sueoka, N. 1964. On the evolution of informational macromolecules. pp. 479–496. In: V. Bryson and H. J. Vogel (eds.), *Evolving Genes and Proteins*. Academic Press, New York. **[8]**

Sun, G., D. L. Dilcher, S. Zheng, and Z. Zhou. In search of the first flower: A Jurassic angiosperm, *Archaefructus*, from northeast China. Science 282: 1692–1695. **[5]**

Suzuki, H., Y. Kawamoto, O. Takenaka, I. Munechika, H. Hori, and S. Sakurai. 1994. Phylogenetic relationships among *Homo sapiens* and related species based on restriction site variations in rDNA spacers. Biochem. Genet. 32: 257–269. **[5]**

Suzuki, T., H. Yuasa, and K. Imai. 1996. Convergent evolution. The gene structure of *Sulculus* 41 kDa myoglobin is homologous with that of human indoleamine dioxygenase. Biochim. Biophys. Acta 1308: 41–48. **[6]**

Sved, J. A. 1976. Hybrid dysgenesis in *Drosophila melanogaster*: A possible explanation in terms of spatial organization of chromosomes. Austral. J. Biol. Sci. 29: 375–388. **[7]**

Swofford, D. L. 1993. *PAUP: Phylogenetic Analysis Using Parsimony*, Version 3.1.1. Distributed by the Illinois Natural History Survey, Champaign, IL. (*PAUP*: Phylogenetic Analysis Using Parsimony and Other Methods*, Version 4.0, is in beta-test edition. Sinauer Associates, Sunderland, MA.) **[5]**

Swofford, D. L. and W. P. Maddison. 1987. Reconstructing ancestral character states under Wagner parsimony. Math. Biosci 87: 199–229. **[5]**

Swofford, D. L. and G. J. Olsen. 1990. Phylogeny reconstruction. pp. 411–501. In: D. M. Hillis and C. Moritz (eds.), *Molecular Systematics*. Sinauer Associates, Sunderland, MA. **[5]**

Swofford, D. L., G. J. Olsen, P. J. Waddell, and D. M. Hillis. 1996. Phylogenetic inference. pp. 407–543. In: D. M. Hillis, C. Moritz, and B. K. Mable (eds.), *Molecular Systematics*, 2nd Ed. Sinauer Associates, Sunderland, MA. **[5]**

Syvanen, M. 1994. Horizontal gene transfer: Evidence and possible consequences. Annu. Rev. Genet. 28: 237–261. **[7]**

Syvanen, M. and C. I. Kado (eds.). 1998. *Horizontal Gene Transfer*. Chapman & Hall, London. **[7]**

Szalay, E. S. 1969. The Hapalodectinae and a phylogeny of the Mesonychidae (Mammalia, Condylarthra). Am. Mus. Nat. Hist. Novitates 2361: 1–26. **[5]**

Szathmáry, E. and J. Maynard Smith. 1995. The major evolutionary transitions. Nature 374: 227–232. **[8]**

Szostak, J. W. and R. Wu. 1980. Unequal crossing over in the ribosomal DNA of *Saccharomyces cerevisiae*. Nature 284: 426–430. **[6]**

Tajima, F. 1993. Simple methods for testing the molecular evolutionary clock hypothesis. Genetics 135: 599–607. **[4]**

Takahata, N. 1993. Allelic genealogy and human evolution. Mol. Biol. Evol. 10: 2–22. **[2]**

Takahata, N. 1996. Neutral theory of molecular evolution. Curr. Opin. Genet. Develop. 6: 767–772. **[4]**

Takahata, N. and Y. Satta. 1997. Evolution of the primate lineage leading to modern humans: Phylogenetic and demographic inferences from DNA sequences. Proc. Natl. Acad. Sci. USA 94: 4811–4815. **[4]**

Takahata, N., Y. Satta, and J. Klein. 1995. Divergence time and population size in the lineage leading to modern humans. Theor. Pop. Biol. 48: 198–221. **[2]**

Takezaki, N., A. Rzhetsky, and M. Nei. 1995. Phylogenetic test of the molecular clock and linearized trees. Mol. Biol. Evol. 12: 823–833. **[4, 5]**

Takhtajan, A. 1969. *Flowering Plants. Origin and Dispersal*. Oliver & Boyd, Edinburgh. **[5]**

Tamura, K. and M. Nei. 1993. Estimation of the number of nucleotide substitutions in the control region of mitochondrial DNA in humans and chimpanzees. Mol. Biol. Evol. 10: 512–526. **[3, 4]**

Tan, H. and J. B. Whitney. 1993. Gene rearrangement in the α-globin gene complex during mammalian evolution. Biochem. Genet. 31: 473–484. **[4, 6]**

Tanaka, T. and M. Nei. 1989. Positive Darwinian selection observed at the variable region genes of immunoglobulins. Mol. Biol. Evol. 6: 447–459. **[4]**

Tateno, Y., N. Takezaki, and M. Nei. 1994. Relative efficiencies of the maximum-likelihood, neighbor-joining, and maximum-parsimony methods when the substitution rate varies with site. Mol. Biol. Evol. 11: 261–277. **[5]**

Taylor, W. R. 1996. Multiple protein sequence alignment: Algorithms and gap insertion. Methods Enzymol. 266: 343–367. **[3]**

Taylor, W. S. 1986. The classification of amino acid conservation. J. Theor. Biol. 119: 205–218. **[1]**

Temin, H. M. 1986. Retroviruses and evolution. Cell Biophys. 9: 9–16. **[7]**

Temin, H. M. 1989. Retrons in bacteria. Nature 339: 254–255. **[7]**

Templeton, A. R. 1992. Human origins and analysis of mitochondrial DNA sequences. Science 255: 737. **[5]**

Tenzen, T. and E. Ohtsubo. 1991. Preferential transposition of an *IS630*-associated composite transposon to TA in the 5′-CTAG-3′ sequence. J. Bacteriol. 173: 6207–6212. **[7]**

Terai, Y., T. Takahashi, and N. Okada. 1998. SINE cousins: The 3′-end tail of the two oldest and distantly related families of SINEs are descended from the 3′ ends of LINEs with the same genealogical origin. Mol. Biol. Evol. 15: 1460–1471. **[7]**

Thewissen, J. G. M., S. T. Hussain, and M. Arif. 1994. Fossil evidence for the origin of aquatic locomotion in archaeocete whales. Science 263: 210–212. **[5]**

Thiranagama, R., A. T. Chamberlain, and B. A. Wood. 1991. Character phylogeny of the primate forelimb superficial venous system. Folia Primatol. 57: 181–190. **[5]**

Thomas, B. A. and R. A. Spicer. 1987. *The Evolution and Palaeobiology of Land Plants*. Croom Helm, London. **[5]**

Thomas, R. H., W. Schaffner, A. C. Wilson, and S. Pääbo. 1989. DNA phylogeny of the extinct marsupial wolf. Nature 340: 465–467. **[5]**

Thomsen, P. D., B. Hoyheim, and K. Christensen. 1996. Recent fusion events during evolution of pig chromosome 3 and 6 identified by comparison with the babirusa karyotype. Cytogenet Cell Genet. 73: 203–208. **[8]**

Thorsness, P. E. and E. R. Weber. 1996. Escape and migration of nucleic acids between chloroplasts, mitochondria, and the nucleus. Int. Rev. Cytol. 165: 207–234. **[7]**

Ticher, A. and D. Graur. 1989. Nucleic acid composition, codon usage, and the rate of synonymous substitution in protein-coding genes. J. Mol. Evol. 28: 286–298. **[4, 8]**

Tittiger, C., S. Whyard, and V. K. Walker. 1993. A novel intron site in the triosephosphate isomerase gene from the mosquito *Culex tarsalis*. Nature 361: 470–472. **[6]**

Toh, H., H. Hayashida, and T. Miyata. 1983. Sequence homology between retroviral reverse transcriptase and putative polymerases of hepatitis B virus and cauliflower mosaic virus. Nature 305: 827–829. **[7]**

Tokugana, F., T. Muta, S. Iwanaga, A. Ichinose, E. W. Davie, K. Kuma, and T. Miyata. 1993. *Limulus* hemocyte transglutaminase: cDNA cloning, amino acid sequence, and tissue localization. J. Biol. Chem. 268: 262–268. **[4]**

Tomarev, S. I. and R. D. Zinovieva. 1988. Squid major lens polypeptides are homologous to glutathione S-transferases subunits. Nature 336: 86–88. **[6]**

TomHon, C., W. Zhu, D. Millinoff, K. Hayasaka, J. L. Slightom, M. Goodman, and D. L. Gumucio. 1997. Evolution of a fetal expression pattern via *cis* changes near the γ globin gene. J. Biol. Chem. 272: 14062–14066. **[6]**

Topal, M. D. and J. R. Fresco. 1976. Complementary base pairing and the origin of substitution mutations. Nature 263: 285–289. **[1]**

Torroni, A. and 8 others. 1993. Asian affinities and continental radiation of the four founding native American mtDNAs. Am. J. Hum. Genet. 53: 563–590. **[5]**

Tower, J. and R. Kurapati. 1994. Preferential transposition of a *Drosophila P* element to the corresponding region of the homologous chromosome. Mol. Gen. Genet. 244: 484–490. **[7]**

Trabesinger-Ruef, N., T. Jermann, T. Zankel, B. Durrant, G. Frank, and S. A. Benner. 1996. Pseudogenes in ribonuclease evolution: A source of new biomacromolecular function. FEBS Lett. 382: 319–322. **[6]**

Trevors, J. T. 1996. Genome size in bacteria. Antonie van Leeuwenhoek 69: 293–303. **[8]**

Tsaur, S. C. and C.-I. Wu. 1997. Positive selection and the molecular evolution of a gene of male reproduction, *Acp26Aa*, of *Drosophila*. Mol. Biol. Evol. 14: 544–549. **[4]**

Tsaur, S. C., C.-T. Ting, and C.-I. Wu. 1998. Positive selection driving the evolution of a gene of male reproduction, *Acp26Aa*, of *Drosophila*: II. Divergence versus polymorphism. Mol. Biol. Evol. 15: 1040–1046. **[2, 4]**

Tuttle, R. H. 1967. Knuckle-walking and the evolution of hominoid hands. Am. J. Phys. Anthropol. 26: 171–206. **[5]**

Ullu, E. and C. Tschudi. 1984. *Alu* sequences are processed 7SL RNA genes. Nature 312: 171–172. **[7]**

Unseld, M., J. R. Marienfeld, P. Brandt, and A. Brennicke. 1997. The mitochondrial genome of *Arabidopsis thaliana* contains 57 genes in 366,924 nucleotides. Nat. Genet. 15: 57–61. **[4]**

Upholt, W. B. 1977. Estimation of DNA sequence divergence from comparison of restriction endonuclease digests. Nuc. Acids Res. 4: 1257–1265. **[3]**

Vacquier, V. D., W. J. Swanson, and Y. H. Lee. 1997. Positive Darwinian selection on two homologous fertilization proteins: What is the selective pressure driving their divergence? J. Mol. Evol. 44: S15-S22. **[4]**

van den Bussche, R. A., J. L. Longmire, and R. J. Baker. 1995. How bats achieve a small C-value: Frequency of repetitive DNA in *Macrotus*. Mamm. Genome 6: 521–525. **[8]**

Vanin, E. F. 1985. Processed pseudogenes: Characteristics and evolution. Annu. Rev. Genet. 19: 253–272. **[8]**

Vansant, G. and W. F. Reynolds. 1995. The consensus sequence of a major *Alu* subfamily contains a functional retinoic acid response element. Proc. Natl. Acad. Sci. USA 92: 8229–8233. **[7]**

Van Valen, L. 1966. Deltatheridia, a new order of mammals. Bull. Am. Mus. Nat. Hist. 132: 1–126. **[5]**

Vawter, L. and W. M. Brown. 1986. Nuclear and mitochondrial DNA comparisons reveal extreme rate variation in the molecular clock. Science 234: 194–196. **[4]**

Vieira, J., C. P. Vieira, D. L. Hartl, and E. R. Lozovskaya. 1998. Factors contributing to the hybrid dysgenesis syndrome in *Drosophila virilis*. Genet. 71: 109–117. **[7]**

Vigilant, L., M. Stoneking, H. Harpending, K. Hawkes, and A. C. Wilson. 1991. African populations and the evolution of human mitochondrial DNA. Science 253: 1503–1507. **[5]**

Vogel, F. and M. Kopun. 1977. Higher frequencies of transitions among point mutations. J. Mol. Evol. 9:159–180. **[4]**

Wada, A. and A. Suyama. 1986. Local stability of DNA and RNA secondary structure and its relation to biological function. Prog. Biophys. Mol. Biol. 47: 113–157. **[8]**

Wainright, P. O., G. Hinkle, M. L. Sogin, and S. K. Stickel. 1993. Monophyletic origins of the metazoa: An evolutionary link with fungi. Science 260: 340–342. **[5]**

Waldan, K. K. and Robertson, H. M. 1997. Ancient DNA from amber fossil bees? Mol. Biol. Evol. 14: 1057–1077. **[5]**

Wallace, D. C. and H. J. Morowitz. 1973. Genome size and evolution. Chromosoma 40: 121–126. **[8]**

Wallis, M. 1994. Variable evolutionary rates in the molecular evolution of mammalian growth hormone. J. Mol. Evol. 38: 619–627. **[4]**

Wallis, M. 1996. The molecular evolution of vertebrate growth hormones: A pattern of near-stasis interrupted by sustained bursts of rapid change. J. Mol. Evol. 43: 93–100. **[4]**

Walsh, J. B. 1985. Interaction of selection and biased gene conversion in a multigene family. Proc. Natl. Acad. Sci. USA 82: 153–157. **[6]**

Walsh, J. B. 1987. Persistence of tandem arrays: Implication for satellite and simple sequence DNAs. Genetics 115: 553–567. **[8]**

Walsh, J. B. 1995. How often do duplicated genes evolve new functions? Genetics 139: 421–428. **[6]**

Walter, P. and G. Blobel. 1982. Signal recognition particle contains a 7S RNA essential for protein translocation across the endoplasmic reticulum. Nature 299: 691–698. **[7]**

Wang, S., I. L. Pirtle, and R. M. Pirtle. 1997. A human 28S ribosomal RNA retropseudogene. Gene 196: 105–111. **[7]**

Watanabe, H., H. Mori. T. Itoh, and T. Gojobori. 1997. Genome plasticity as a paradigm of eubacterial evolution. J. Mol. Evol. 44: s57-s64. **[8]**

Watson, J. D. and F. H. C. Crick. 1953. Genetical implications of the structure of deoxyribonucleic acid. Nature 171: 964–967. **[1]**

Watson, J. D., N. H. Hopkins, J. W. Roberts, J. A. Steitz, and A. M. Weiner. 1987. *Molecular Biology of the Gene*, 4th Ed. Benjamin/Cummings, Menlo Park, CA. **[7]**

Weatherall, D. J. and J. B. Clegg. 1979. Recent developments in the molecular genetics of human hemoglobin. Cell 16: 467–479. **[6]**

Weigel, R. M. and G. Scherba. 1997. Quantitative assessment of genomic similarity from restriction fragment patterns. 32: 95–110. **[3]**

Weiner, A. M., P. L. Deininger, and A. Efstratiadis. 1986. Nonviral retroposons: Genes, pseudogenes and transposable elements generated by the reverse flow of genetic information. Annu. Rev. Biochem. 55: 631–661. **[7, 8]**

Weiss, E. H., A. Mellor, L. Golden, K. Fahrner, E. Simpson, J. Hurst, and R. A. Flavell. 1983. The structure of a mutant *H-2* gene suggests that the generation of polymorphism in *H-2* genes may occur by gene conversion-like events. Nature 301: 671–674. **[6]**

Wheeler, L. L. and L. S. Altenburg. 1977. Hoechst 33258 banding of *Drosophila nasutoides* metaphase chromosomes. Chromosoma 62: 351–360. **[8]**

Widegren, B., U. Árnason, and G. Akuslarvi. 1985. Characteristics of conserved 1,579-bp highly repetitive component in the killer whale, *Orcinus orca*. Mol. Biol. Evol. 2: 411–419. **[8]**

Wilde, C. D. 1986. Pseudogenes. CRC Crit. Rev. Biochem 19: 323–352. **[8]**

Wiley, E. O. 1981. *Phylogenetics: The Theory and Practice of Phylogenetic Systematics*. Wiley, New York. **[5]**

Wilks, H. M. and 9 others. 1988. A specific, highly active malate dehydrogenase by redesign of a lactate dehydrogenase framework. Science. 242: 1541–1544. **[6]**

Willard, C., H. T. Nguyen, and C. W. Schmid. 1987. Existence of at least three distinct *Alu* subfamilies. J. Mol. Evol. 26: 180–186. **[7]**

Williams, S. A. and M. Goodman. 1989. A statistical test that supports a human/chimpanzee clade based on noncoding DNA sequence data. Mol. Biol. Evol. 6: 325–330. **[5]**

Wilson, A. C., S. S. Carlson, and T. J. White. 1977. Biochemical evolution. Annu. Rev. Biochem. 46: 573–639. **[4, 5]**

Wilson, A. C., H. Ochman, and E. M. Prager. 1987. Molecular time scale for evolution. Trends Genet. 3: 241–247. **[4]**

Wilson, E. O. 1971. *The Insect Societies*. Harvard University Press, Cambridge, MA. **[App. I]**

Winnepenninckx, B., G. Steiner, T. Backeljau, and R. de Wachter. 1998. Details of gastropod phylogeny inferred from 18S rRNA sequences. Mol. Phylogenet. Evol. 9: 55–63. **[6]**

Wistow, G. J., J. W. M. Mulders, and W. W. de Jong. 1987. The enzyme lactate dehydrogenase as a structural protein in avian and crocodilian lenses. Nature 326: 622–624. **[6]**

Woese, C. R. 1969. Models for the evolution of codon assignments. J. Mol. Evol. 14: 235–240. **[8]**

Woese, C. R. 1987. Bacterial evolution. Microbiol. Rev. 51: 221–271. **[5]**

Woese, C. R. 1996. Phylogenetic trees: Whither microbiology? Curr. Biol. 6:1060–1063. **[5]**

Woese, C. R. and G. E. Fox. 1977. Phylogenetic structure of the prokaryotic domain: The primary kingdoms. Proc. Natl. Acad. Sci. USA 74: 5088–5090. **[5]**

Woese, C. R., O. Kandler, and M. L. Wheelis. 1990. Towards a natural system of organisms: Proposal for the domains Archaea, Bacteria and Eucaria. Proc. Natl. Acad. Sci. USA 87: 4576–4579. **[5]**

Wolfe, K. H. and D. C. Shields. 1997. Molecular evidence for an ancient duplication of the entire yeast genome. Nature 387: 708–713. **[8]**

Wolfe, K. H., W.-H. Li, and P. M. Sharp. 1987. Rates of nucleotide substitution vary greatly among plant mitochondrial, chloroplast, and nuclear DNAs. Proc. Natl. Acad. Sci. USA 84: 9054–9058. **[4]**

Wolfe, K. H., P. M. Sharp, and W.-H. Li. 1989a. Mutation rates differ among regions of the mammalian genome. Nature 337: 283–285. **[4, 5, 8]**

Wolfe, K. H., P. M. Sharp, and W.-H. Li. 1989b. Rates of synonymous substitution in plant nuclear genes. J. Mol. Evol. 29: 208–211. **[4, 5]**

Wolfe, K. H., C. W. Morden, S. C. Ems, and J. D. Palmer. 1992a. Rapid evolution of the plastid translational apparatus in a nonphotosynthetic plant: Loss or accelerated sequence evolution of tRNA and ribosomal protein genes. J. Mol. Evol. 35: 304–317. **[8]**

Wolfe, K. H., C. W. Morden, and J. D. Palmer. 1992b. Function and evolution of a minimal plasmid genome from a nonphotosynthetic parasitic plant. Proc. Natl. Acad. Sci. USA 89: 10648–10652. **[8]**

Wolff, G., I. Plante, B. F. Lang, U. Kuck, and G. Burger. 1994. Complete sequence of the mitochondrial DNA of the chlorophyte alga *Prototheca wickerhamii*. Gene content and genome organization. J. Mol. Biol. 237: 75–86. **[4]**

Woodward, S. R., N. J. Weyand, and M. Bunnel. 1994. DNA sequence from Cretaceous period bone fragments. Science 266: 1229–1232. **[5]**

Wostemeyer, J. and A. Burmester. 1986. Structural organization of the genome of the zygomycete *Absidia glauca*: Evidence for high repetitive DNA content. Curr. Genet. 10: 903–907. **[8]**

Wright, F. 1990. The "effective number of codons" used in a gene. Gene 87: 23–29. **[4]**

Wright, S. 1931. Evolution in Mendelian populations. Genetics 16: 97–159. **[2]**

Wright, S. 1942. Statistical genetics and evolution. Bull. Am. Math. Soc. 48: 223–246. **[2]**

Wu, C.-I. and W.-H. Li. 1985. Evidence for higher rates of nucleotide substitution in rodents than in man. Proc. Natl. Acad. Sci. USA 82: 1741–1745. **[4]**

Wu, C.-I., T. W. Lyttle, M.-L. Wu, and G.-F. Lin. 1988. Association between a satellite DNA sequence and the *Responder of Segregation Distorter* in *Drosophila melanogaster*. Cell 54: 179–189. **[8]**

Wu, C.-I., J. R. True, and N. Johnson. 1989. Fitness reduction associated with the deletion of a satellite DNA array. Nature 341: 248–251. **[8]**

Wyss, A. 1990. Clues to the origin of whales. Nature 347: 428–429. **[5]**

Xia, X.-H. 1995. Body temperature, rate of biosynthesis, and evolution of genome size. Mol. Biol. Evol. 12: 834–842. **[8]**

Xia, X. and W.-H. Li. 1998. What amino acid properties affect protein evolution? J. Mol. Evol. 47: 557–564. **[4]**

Xiong, Y. and T. H. Eickbush. 1988. Similarity of reverse transcriptase-like sequences of viruses, transposable elements, and mitochondrial introns. Mol. Biol. Evol. 5: 675–690. **[7]**

Xiong, Y. and T. H. Eickbush. 1990. Origin and evolution of retroelements based upon their reverse transcriptatse sequences. EMBO J. 9: 3353–3362. **[7]**

Yang, F., P. C. M. O'Brien, J. Weinberg, H. Niezel, C.-C. Lin, and M. A. Ferguson-Smith. 1997. Chromosomal evolution of the Chinese muntjac (*Muntiacus reevesi*). Chromosoma 106: 37–43. **[8]**

Yang, Z. 1998. Likelihood ratio tests for detecting positive selection and application to primate lysozyme evolution. Mol. Biol. Evol. 15: 568–573. **[4]**

Yao, M.-C., A. R. Kimmel, and M. A. Gorovsky 1974. A small number of cistrons for ribosomal RNA in the germinal nucleus of a eukaryote, *Tetrahymena pyriformis*. Proc. Natl. Acad. Sci. USA 71: 3082–3086. **[6]**

Yokoyama, S. and T. Gojobori. 1987. Molecular evolution and phylogeny of the human AIDS viruses *LAV*, *HTLV-III*, and *ARV*. J. Mol. Evol. 24: 330–336. **[4]**

Yokoyama, S. and R. Yokoyama. 1989. Molecular evolution of human visual pigment genes. Mol. Biol. Evol. 6: 186–197. **[6]**

Yoshitake, S., B. G. Schach, D. C. Foster, E. W. Davie, and K. Kurachi. 1985. Nucleotide sequence of the gene for human factor IX (antihemophilic factor B). Biochemistry 24: 3736–3750. **[1]**

Zhang, J., S. Kumar, and M. Nei. 1997. Small-sample tests of episodic adaptive evolution: A case study of primate lysozymes. Mol. Biol. Evol. 14: 1335–1338. **[4]**

Zhang, J., H. F. Rosenberg, and M. Nei. 1998. Positive Darwinian selection after gene duplication in primate ribonuclease genes. Proc. Natl. Acad. Sci. USA 95: 3708–3713. **[6]**

Zhang, S., G. Zubay, and E. Goldman. 1991. Low-usage codons in *Escherichia coli*, yeast, fruit fly, and primates. Gene 105: 61–72. **[4]**

Zharkikh, A. and W.-H. Li. 1992a. Statistical properties of bootstrap estimation of phylogenetic variability from nucleotide sequences. I. Four taxa with a molecular clock. Mol. Biol. Evol. 9: 1119–1147. **[5]**

Zharkikh, A. and W.-H. Li. 1992b. Statistical properties of bootstrap estimation of phylogenetic variability from nucleotide sequences. II. Four taxa without a molecular clock. J. Mol. Evol. 35: 356–366. **[5]**

Zharkikh, A. and W.-H. Li. 1995. Estimation of confidence in phylogeny: The complete-and-partial bootstrap technique. Mol. Phylogenet. Evol. 4: 44–63. **[5]**

Zilling, W. 1991. Comparative biochemistry of Archaea and Bacteria. Curr. Opin. Genet. Develop. 1: 544–551. **[5]**

Zimmer, E. A., S. L. Martin, S. M. Beverley, Y. W. Kan, and A. C. Wilson. 1980. Rapid duplication and loss of genes coding for the α chains of hemoglobin. Proc. Natl. Acad. Sci. USA 77: 2158–2162. **[6]**

Zipkas, D. and M. Riley. 1975. Proposal concerning mechanism of evolution of the genome of *Escherichia coli*. Proc. Natl. Acad. Sci. USA 72: 1354–1358. **[8]**

Zoubak, S. and 7 others. 1994. Regional specificity of HTLV-I proviral integration in the human genome. Gene 143: 155–163. **[8]**

Zuckerkandl, E. 1976a. Gene control in eukaryotes and the C-value paradox: "Excess" DNA as an impediment to transcription of coding sequences. J. Mol. Evol. 9: 73–104. **[4]**

Zuckerkandl, E. 1976b. Evolutionary processes and evolutionary noise at the molecular level. I. Functional density in proteins. J. Mol. Evol. 7: 167–183. **[4]**

Zuckerkandl, E. and L. Pauling. 1962. Molecular disease, evolution and genic heterogeneity. pp. 189–225. In: M. Kash and B. Pullman (eds.), *Horizons in Biochemistry*. Academic Press, New York. **[4]**

Zuckerkandl, E. and L. Pauling. 1965. Evolutionary divergence and convergence in proteins. pp. 97–166. In: V. Bryson and H. J. Vogel (eds.), *Evolving Genes and Proteins*. Academic Press, New York. **[4]**

Zuckerkandl, E., J. Derancourt, and H. Vogel. 1971. Mutational trends and random processes in the evolution of informational macromolecules. J. Mol. Biol. 59: 473–490. **[4]**

Zurawski, G. and M. T. Clegg. 1987. Evolution of higher plant chloroplast DNA-encoded genes: Implication for structure-function and phylogenetic studies. Annu. Rev. Plant Physiol. 38: 391–418. **[4]**

Subject Index

Entries in **boldface** are defined terms. Page numbers in *italics* refer to material in an illustration or table.

Taxonomic Index

Page numbers in *italics* refer to material in an illustration or table.

About the Book

Editor: Andrew D. Sinauer

Project Editor: Carol J. Wigg

Production Manager: Christopher Small

Book and Cover Design: Janice Holabird

Illustrations: Michele Ruschhaupt

Book Manufacture: Courier Companies, Inc.

Cover Manufacture: Henry N. Sawyer Company, Inc.